Polymeric Materials with Antimicrobial Activity

From Synthesis to Applications

RSC Polymer Chemistry Series

Series Editors:
Professor Ben Zhong Tang (Editor-in-Chief), *The Hong Kong University of Science and Technology, Hong Kong, China*
Professor Alaa S. Abd-El-Aziz, *University of Prince Edward Island, Canada*
Professor Stephen L. Craig, *Duke University, USA*
Professor Jianhua Dong, *National Natural Science Foundation of China, China*
Professor Toshio Masuda, *Fukui University of Technology, Japan*
Professor Christoph Weder, *University of Fribourg, Switzerland*

Titles in the Series:
 1: Renewable Resources for Functional Polymers and Biomaterials
 2: Molecular Design and Applications of Photofunctional Polymers and Materials
 3: Functional Polymers for Nanomedicine
 4: Fundamentals of Controlled/Living Radical Polymerization
 5: Healable Polymer Systems
 6: Thiol-X Chemistries in Polymer and Materials Science
 7: Natural Rubber Materials: Volume 1: Blends and IPNs
 8: Natural Rubber Materials: Volume 2: Composites and Nanocomposites
 9: Conjugated Polymers: A Practical Guide to Synthesis
10: Polymeric Materials with Antimicrobial Activity: From Synthesis to Applications

How to obtain future titles on publication:
A standing order plan is available for this series. A standing order will bring delivery of each new volume immediately on publication.

For further information please contact:
Book Sales Department, Royal Society of Chemistry, Thomas Graham House, Science Park, Milton Road, Cambridge, CB4 0WF, UK
Telephone: +44 (0)1223 420066, Fax: +44 (0)1223 420247
Email: booksales@rsc.org
Visit our website at www.rsc.org/books

Polymeric Materials with Antimicrobial Activity
From Synthesis to Applications

Edited by

Alexandra Muñoz-Bonilla
Instituto de Ciencia y Tecnología de Polímeros (ICTP-CSIC), Madrid, Spain
Email: sbonilla@ictp.csic.es

María L. Cerrada
Instituto de Ciencia y Tecnología de Polímeros (ICTP-CSIC), Madrid, Spain
Email: mlcerrada@ictp.csic.es

and

Marta Fernández-García
Instituto de Ciencia y Tecnología de Polímeros (ICTP-CSIC), Madrid, Spain
Email: martafg@ictp.csic.es

RSC Publishing

RSC Polymer Chemistry Series No. 10

ISBN: 978-1-84973-807-1
ISSN: 2044-0790

A catalogue record for this book is available from the British Library

Published by The Royal Society of Chemistry,
Thomas Graham House, Science Park, Milton Road,
Cambridge CB4 0WF, UK

Registered Charity Number 207890

For further information see our web site at www.rsc.org

To our families

Preface

Today, it is clear that we are surrounded by those materials that people commonly named as plastics and scientists prefer to designate as macromolecules or polymers. From early in the morning just after getting up we start using these materials, either natural or synthetic: a simple cotton towel to dry off after the shower, (a natural macromolecule); the comb or the toothbrush (mainly made from synthetic macromolecules); passing through the body or facial creams for maintaining the youth of our skin or used after shaving, all of these cosmetics containing polymers as ingredients; our clothing (shirt, pants, skirt, *etc.*) are manufactured using natural or synthetic macromolecules. Once we are ready, we should go to work in most cases not on foot (although shoes are made also with polymers) but by transport. This transportation will be mostly built by polymeric-based materials whether this is public (bus, train or subway) or private (car). The truth is that natural and synthetic macromolecules or polymers, plastics or elastomers, are an indispensable part of our daily lives and in many cases totally necessary. Once upon a time, we could think that all these enveloped systems would allow us to have a better quality of life, keeping us safe from possible attacks from outsiders. Who among us has not bought the "perfect" cleaner that kills all type of bacteria, and has painted his house with antifungal plastic paint, or has purchased a freezer, air conditioning or light switches that prevent bacterial proliferation? In fact, one of our colleagues has a footrest in the office that was sold as antimicrobial.

We are looking for systems that keep clean or fight against possible contamination and, in this way, people feel safer. This is always desirable but more especially in health and/or food areas since in these environments most of the tools and devices undergo sterilization procedures to avoid contamination. This objective has been pursued since antiquity and people have attempted to prevent from becoming infected, diseased or deceased. Today, this aim is more

RSC Polymer Chemistry Series No. 10
Polymeric Materials with Antimicrobial Activity: From Synthesis to Applications
Edited by Alexandra Muñoz-Bonilla, María L. Cerrada and Marta Fernández-García
© The Royal Society of Chemistry 2014
Published by the Royal Society of Chemistry, www.rsc.org

than a necessity, not only from the health-care standpoint, chasing the same goals as our ancestors but also to reach greater social and economic well-being.

When the RSC raised the idea for this book on polymeric materials with antimicrobial activity, at first, we had some doubts. It was clear that the subject was fascinating; we knew that because we have worked on antimicrobial polymeric materials, but what we did not know was whether we would be able to compose an excellent book, screening all the aspects that this field covers. We now have no doubt, we have been very fortunate to rely on well-known researchers in this wonderful and enigmatic field, offering to us not only their knowledge but also their generosity. Thus, this book is intended to serve both researchers working in the field of antimicrobial polymeric materials as well as those planning to enter into this area, but also for the many students of microbiology, chemistry, materials, physics, pharmacy, medicine, engineering, or others, who, not knowing what the future holds, think about the possibility to develop their scientific, technological, educational vocation on it or, simply, to acquire a background knowledge.

The book is divided into several chapters: Chapter 1, introduction to the antimicrobial polymeric materials, discusses the world of micro-organisms, how they are classified, quantified and really act. Understanding that, the reader will be submerged in the classification of antimicrobial polymers and the way to determine the activity of these systems. Chapter 2 describes the antimicrobial activity of chitosan and its derivatives focusing on their main applications in food and biomedical technologies but also in other areas, such as agriculture. The antimicrobial polymers with ammonium and phosphonium groups will be collected in Chapter 3. The different applications of these materials will also be presented, focusing more in their antimicrobial applications. Chapter 4 includes guanidine-based polymers, gemini surfactants and polymers, polymeric complexes formed with polymers and antibiotics and polymers containing antibiotics. These systems present the water-soluble antimicrobial polymers for functional cellulose fibres and hygiene paper products. The foreword on polymeric systems that mimic antimicrobial peptides will be presented in Chapter 5. On it, previous works dealing with mimic antimicrobial peptides are discussed, pointing out that there are some design rules for them and trends in their biological data. Antimicrobial textiles and clothing is detailed in Chapter 6, with a focus on the prevention of infection diseases in hospitals. The potential of different antimicrobial systems, including also halamine polymers, which are approved by the Environmental Protection Agency, are also commented. The synthesis and characterization by X-ray diffraction of polymeric and nonpolymeric metal complexes with special attention on silver ones are described in Chapter 7. The mode of action against microbes is discussed as a function of the used metal character to form the complexes. Chapter 8 is focused on polymer nanofibres with antimicrobial activities obtained by an electrospinning technique. As will be disclosed in this chapter the unique properties of the obtained fibres, with small diameters and large surface area to volume ratio, are of great interest in a variety of applications including filtration, tissue engineering scaffolds, drug-release systems, enzyme stabilization, protective clothes, sensors, carbonaceous materials, and controlled-drug delivery platforms. The preparation and the characteristics of polyurethanes are rationalized in Chapter 9, where we will see that these materials constitute a great fraction of biomedical devises. Thereof, the requirement to understand their antifouling and antimicrobial mechanisms of action has primary significance. One of the great concerns, especially in developed countries, is our dentition. In Chapter 10, the antimicrobial polymeric dental materials including all kinds of applied materials involved in the dental area, in particular those that release chlorhexidine antiseptic are discussed. Moreover, essential oils and natural compounds are used mainly in the food industry as antimicrobial and virucidal systems. Chapter 11 collects all the properties of such systems, with particular attention on those with virucidal potential since those are less explored. The carbon-based polymer systems, which are now powerfully emerging in different fields as exceptional candidates to shake up nanoscience and nanotechnology, are also revised in Chapter 12 as materials able to fight against micro-organisms. The use of copper and copper oxide in polymeric matrices is detailed in Chapter 13. The use of copper nanoparticles is nowadays remerging within the "nano" era. The toxicity of these particles is also discussed. The preparation of polymeric (nano)composites with zinc oxide and titanium dioxide, taking into consideration the photocatalytic activity of particles, is described in Chapter 14. The size and dispersion of particles into the matrix is widely discussed in it. The last chapter intends to expose antimicrobial systems different than those collected in the other chapters, e.g. those based on nitric oxide, combination and synergy of different approaches as innovative alternatives to fight against infections. Finally, we will comment on the expectations and trends of these antimicrobial materials.

Alexandra Muñoz-Bonilla, María L. Cerrada and
Marta Fernández-García
Instituto de Ciencia y Tecnología de Polímeros
(ICTP-CSIC), Madrid, Spain

Contents

RSC Polymer Chemistry Series No. 10
Polymeric Materials with Antimicrobial Activity: From Synthesis to Applications
Edited by Alexandra Muñoz-Bonilla, María L. Cerrada and Marta Fernández-García
© The Royal Society of Chemistry 2014
Published by the Royal Society of Chemistry, www.rsc.org

CHAPTER 1

Introduction to Antimicrobial Polymeric Materials

ALEXANDRA MUÑOZ-BONILLA, MARÍA L. CERRADA
AND MARTA FERNÁNDEZ-GARCÍA*

Instituto de Ciencia y Tecnología de Polímeros (ICTP-CSIC), Juan de la
Cierva 3, 28006 Madrid, Spain
*Email: martafg@ictp.csic.es

This Chapter seeks to bring the readers, in a very brief way, the wonders and the threats that microbes suppose and how human beings fight against them using polymeric materials.

1.1 Short Overview of the World of Micro-Organisms

Microbes are everywhere in the world and their presence constantly affects the environment in which they are growing. The effects of micro-organisms can be beneficial or harmful for their surroundings. Some of them can be positive, sometimes essential, in association with higher forms of organisms (*e.g.* bacteria and other microbes in the intestines of animals and insects digest nutrients and produce vitamins and growth factors). Moreover, microbes are also used in the manufacture of fermented foods, such as yeasts employed in the fabrication of beer, wine or breads, lactic acid bacteria used to make yogurt, cheese, and other fermented milk products. In addition, microbes are a source in medicine of antibiotics (substances produced by micro-organisms that kill or inhibit other microbes and, then, are used in the treatment of infectious diseases) and

RSC Polymer Chemistry Series No. 10
Polymeric Materials with Antimicrobial Activity: From Synthesis to Applications
Edited by Alexandra Muñoz-Bonilla, María L. Cerrada and Marta Fernández-García
© The Royal Society of Chemistry 2014
Published by the Royal Society of Chemistry, www.rsc.org

vaccines (substances derived from micro-organisms developed to immunise against diseases) for the treatment and prevention of infectious diseases. These advantages come also to biotechnological processes, playing a primary role in recombinant DNA technology and genetic engineering.

Despite these benefits, some microbes cause diseases in animals and plants (pathogens), and they are agents of spoilage and decomposition of foods, textiles and dwellings since nothing lasts forever, and the microbial decomposition of any organic substance will occur with time. Fungi and bacteria are the major microbial agents of decomposition in aerobic environments, while only bacteria can act in anaerobic media. Focusing the attention on the human population, microbial infections still cause around one quarter of all deaths worldwide, especially in undeveloped countries where there are contaminated water or food, unsanitary disposal of human waste, poor personal hygiene, inferior sanitary conditions and lack of access to medical assistance. The magnitude of these infectious diseases (*e.g.* cholera, dysentery, human immunodeficiency virus infection /acquired immunodeficiency syndrome, malaria, tuberculosis, *etc.*) is in those countries as significant as they become the first cause of mortality. On the contrary, morbidity and mortality is triggered in developed countries by the increasing incidence of antibiotic-resistant pathogens along with an easily migratory mobility that allow new paths for micro-organisms to run into human hosts to be created.[1,2] Approximately 25 000 people die each year in the European Union from antibiotic-resistant bacterial infections. For example, Gram-positive *Staphylococcus aureus* has evolved from penicillin-resistant phenotypes into a methicillin-resistant strain (MRSA), which has become a global epidemic[3,4] and it is responsible for the main surgical site infections.[5,6] Countries with the highest rates of resistant infections, such as Greece, Cyprus, Italy, Hungary and Bulgaria, also tend to be the ones with the highest uses of antibiotics. One of the Global Strategy Recommendations dictated by the World Health Organization (WHO) is to make the control of antimicrobial resistance[7] a priority for National Governments and Health Systems. Therefore, new prevention and control strategies are urgently required.

1.1.1 Classification of Micro-Organisms

There are five major groups of micro-organisms: bacteria, algae, fungi, protozoa, and viruses. They are divided into prokaryotic ("before nucleus") and eukaryotic (true nucleus). The former are organisms whose cells lack a cell nucleus (karyon), or any other membrane-bound organelles (only the bacteria and the archaea); the eukaryotic micro-organisms have internal membrane-bound structures, membrane bound nucleus and membrane-bound organelles such as mitochondria, chloroplasts and the Golgi apparatus (algae, protozoa, fungi).[8]

In a simple way, bacteria are prokaryotic and unicellular with a size 1000 times less than the volume of a typical eukaryotic cell, exhibiting different shapes: bacillus (rod), coccus (spherical), spirillum (spiral), vibrio (curved rod).

They are usually classified into two distinct types, Gram-positive and Gram-negative, that differ in the properties of their bacterial cell walls. Gram-positive bacteria are those that are stained dark blue or violet by Gram staining because of the high amount of peptidoglycan in the cell wall. On the contrary, the peptidoglycan layer is thinner in Gram-negative bacteria and is protected by an outer membrane. Consequently, they cannot retain the crystal violet stain, turning in this case reddish or pink by counter-stain (safranin, fuchsine or other stains). In general, Gram-negative bacteria are more resistant against antibiotics compared with Gram-positive ones, because of their outer membrane.

Algae are eukaryotic and unicellular or multicellular; fungi are eukaryotic and unicellular (yeasts) or multicellular (moulds); protozoa (first animals) are eukaryotic and unicellular and viruses are acellular and, then, they are forced to live as intracellular parasites.[9]

1.1.2 Methods of Measuring Microbial Growth

It is of great importance to know the population of micro-organisms and the rates of their growth to inhibit or prevent microbes proliferation. There are numerous techniques of counting microbial growth,[10] measuring either cell mass or cell number, the following are examples:

(a) Dry/wet weight measurement:

This method is a direct approach to determine the net weight of cells. A known volume of culture sample is centrifuged to sediment micro-organisms to the bottom of a vessel. The sedimented cells (called a cell pellet) are, then, washed and weighted in case of wet measurements. Dry weight is measured after drying the centrifuged cells. Dry weight is usually about 10–20% of the wet weight, and gives more consistent results and normally is taken as the reference method. These techniques are simple but highly time consuming. In addition, they are not very sensitive and also cannot distinguish between live and dead bacteria.

(b) Absorbance/turbidity:

Absorbance is measured by using a spectrophotometer. Light scattering rises with the increase in cell number. When light is passed through bacterial cell suspension, light is scattered by the cells and transmission decays. At a particular wavelength, light absorbance is proportional to the cell concentration of micro-organisms present in the suspension. This is a nondestructive method that is also very simple, rapid and accurate. Both live and dead cells are, however, able to scatter light and, therefore, both are counted.

(c) Total cell count:

Cell growth is also measured by counting the total cell number of the microbes present in the sample. Total cells (both live and dead) are microscopically counted by using special microscope glass slide or counting chamber, such as, Helber or Petroff–Hausser slides. Typically, this chamber consists of a slide with a grid with etched squares of known area. For example, the surface of the platform is etched with a grid system in the case of Petroff-Hausser chamber. This consists of 25 large squares, each of which is divided into 16 smaller squares. Cells are then counted (normally more than 500) with a phase contrast microscope using a sufficient and appropriate numbers of squares. The number of cells per mL of sample is subsequently calculated from the average cell number per square divided by the volume of a single square. The dilution factors should be taken into account. The main disadvantages of this method are that a high concentration of bacteria is required and the complexity to distinguish between living or dead cells. On the other hand, this is a simple method and does not need any incubation time.

(d) Viable count:

A viable cell is defined as a cell that is able to divide and increase cell numbers. The normal way to perform a viable count is to determine the number of cells in the sample that are capable of forming colonies on a suitable medium. It is assumed that each viable cell will form one colony. Therefore, the viable count is often called the plate count or colony count. This method requires an incubation period of *ca.* 24 h or longer and can be done in selective and differential. There are two ways of forming plate count: (i) Spread count method: a volume of culture is spread over the surface of an agar plate by using a sterile glass spreader. The plate is incubated to develop colonies. Then, the number of colonies is counted. (ii) Pour plate method: a known volume of the culture is added into sterile Petri dishes. Subsequently, the melted agar medium is poured and gently mixed and, next, incubated. After that, the colonies growing on the surface of the agar are counted. The major shortcoming of this method is the assumption that each colony is generated from a single bacterial cell, thus, sometimes it entails underestimation of the true population. Besides, this technique is time consuming and labour intensive. Despite its disadvantages, the viable plate count is the most frequently used method for measuring cell number because it is very sensitive and allows counting only living bacteria.

(e) Cell-counting instruments:

Coulter counters and flow cytometers are extensively used to count total cells in dilute solutions. Coulter counters are based on electrical impedance. That is, the cells contained in the liquid cause a variation in the electrical impedance that is proportional to the size of the particle. Consequently, the size and the number of cells within a solution can be determined. Flow cytometry is a powerful technique for cell counting that simultaneously measures and, then, analyses physical properties of the cells, such as the light scattering or fluorescence. In this technique, cells are carried in a fluid stream to a laser intercept, thus, thousands of cells pass through a laser beam and the light that emerges from each cell is collected and examined. Flow cytometry can also be used to count organisms to which fluorescent dyes or tags have been attached.

Moreover, there are other methods among all of these common measurements, such as: determination of the amount of a given element, usually nitrogen; acid titration with a pH indicator to quantify acid production; the carbon dioxide formation by using a molecule that fluoresces when the medium becomes slightly more acidic (this gas can be trapped in an inverted Durham tube in a tube of broth); and also ATP measurement using firefly luciferase catalyses light-emitting reaction. However, these methods are more tedious and not often used.

1.1.3 Mechanisms of Action against Micro-Organisms

There are different modes of action against micro-organisms: 1) by affecting their proteins, *i.e.* denaturation or alteration of their protein structure. This denaturation can be either permanent and, then, the action mechanism is called bactericidal, fungicidal, *etc.*, or temporary if their initial and standard structure can be restored, being then called bacteriostatic, fungistatic, *etc.* The common mechanisms of denaturation include disruption of hydrogen and disulfide bonds. Another mechanism is 2) by affecting their cell membrane proteins or membrane lipids. Concerning proteins, the mode of action consists of denaturalisation whereas lipids are dissolved, for instance by a surfactant, and their cell membrane turns out to be damaged. Others mechanisms are 3) by affecting the cell-wall formation through blocking its synthesis; 4) by preventing replication, transcription and translation of the nucleic acid structure or 5) by disturbing the metabolism.

1.2 Antimicrobial Polymeric Materials

1.2.1 Brief Introduction

World Health Organization (WHO) has elaborated a catalogue collecting the major concerns about health. The list is the following:

- Alcohol and health.
- Avian influenza A (H5N1).
- Child and adolescent health and development.
- Cholera.
- Environmental and health (household air pollution, outdoor air pollution).
- Expenditures on health (investment on health and on research).
- HIV/AIDS.
- Integrated management of childhood illness (IMCI).
- Influenza.
- Malaria.
- Maternal and reproductive health.
- Meningococcal disease.
- Mortality and burden of disease.

- Neglected tropical disease.
- Noncommunicable diseases (cardiovascular diseases, cancer, diabetes and chronic respiratory diseases).
- Pandemic (H1N1) 2009.
- Poliomyelitis.
- Substance use and substance abuse.
- Tobacco.
- Tuberculosis.
- Violence and injuries.
- Water, sanitation and health.

As can be noticed, many of them are caused by micro-organisms. For example, influenza is a virus and the main strategy to prevent this illness is the vaccination treatment. Cholera is an infection in the small intestine caused by the bacterium *Vibrio cholerae*. The main symptoms are profuse, watery diarrhoea and vomiting. Transmission occurs primarily by drinking water or eating food that has been contaminated by the faeces of an infected person, including one with no apparent symptoms. In this sense, inadequate access to safe water and sanitation services as well as poor hygiene practices, kills and sickens thousands of children every day, and leads to impoverishment and diminished opportunities for thousands more.[11] Various factors lead to water deterioration, including population growth, rapid urbanisation, agricultural land uses, industrial discharge of chemicals, *etc.* During recent years, it has been shown that the pharmaceuticals are only partially removed in wastewater treatment plants and have been detected in water bodies.[1,12,13] Hospitals contribute in a large extent to the load of wastewater treatment plants, they should be taking into consideration for point-source measures, such as implementing methods to decrease pharmaceuticals or introducing single-use pocket urinals to remove X-ray contrast agents (90–94% are excreted with the urine). In this sense and very recently, Kovalova *et al.*[14] described a pilot-scale membrane bioreactor operating during one year at a Swiss hospital where 68 target analytes (56 pharmaceuticals, such as antibiotics, antimycotics, antivirals, iodinated X-ray contrast media, anti-inflammatory, cytostatics, *etc.*, 10 metabolites, and 2 corrosion inhibitors) were studied. They make an improvement in the pollutant elimination, almost 90%, when only pharmaceuticals and metabolites are considered without iodinated X-ray contrast media.

Therefore, the cleaning, disinfection and sterilisation of wastewater and/or air is an important target in the antimicrobial policy. However, the contaminants are not only concentrated at outdoor sources; there are also chemical contaminants from indoor supplies. *Legionella* bacteria and *Legionella pneumophila* causes legionellosis, a collection of infections that emerged in the second half of the 20th century, and cause a high level of morbidity and mortality in the people exposed.[15] These micro-organisms can be raised in stationary places due to inadequate ventilation, from stationary water, principally in humidifiers or cooling towels, spa pools, spread from air vent systems and/or from water that has been collected on carpets or wood furniture.

Although most of the infectious diseases, such as malaria, meningitis, poliomyelitis, among others, are most frequently given in undeveloped countries, the sexual diseases, such as chlamydia, gonorrhoea, genital herpes, HIV/AIDS, human papilloma viruses (HPV), hepatitis, syphilis or trichomoniasis, are of great concern even in technologically advanced countries because these infections are present all over the world. Moreover, fungal keratitis is a leading cause of ocular morbidity throughout the world, and it is also a major eye disease that leads to blindness in Asia.[16,17] In relation to common problems of human beings, bad odour is one of the most recurrent struggles and it is produced by different bacteria when they digest sugars. *Bacillus subtilis* and *Staphylococcus epidermidis*, contribute to foot odour and the latter is also responsible along with *Corynebacterium xerosis* for underarm odour.[18,19]

All these facts indicate the importance of investing not only in prevention but also in new formulations to combat the micro-organisms. At this point, polymeric materials come into play within the game of antimicrobial systems. In general, the antibiotics, antifungals or antivirals are based on natural products or low molecular weight components, but they have the problem of residual toxicity, even when suitable amounts of the agents are added.[20,21] The use of antimicrobial polymeric materials offers a guarantee for enhancing the efficacy of some existing antimicrobial agents and for minimising, at the same time, the environmental problems that escort conventional antimicrobial agents by reducing their residual toxicity, increasing their efficiency and selectivity, and prolonging their lifetime. In addition, these antimicrobial polymeric materials are nonvolatile and chemically stable.[22] This fact is essential for developing antimicrobial systems that are not easily susceptible to resistance.[23] It is important to remark that the effectiveness of an antimicrobial agent is affected by time, temperature, pH, and concentration, among other factors.

1.2.2 Classification of Antimicrobial Polymeric Materials

Antimicrobial polymers involved traditionally covalent linkages of groups with antimicrobial activity. The synthesis of antimicrobial polymers and copolymers based on 2-methacryloxytroponones was described for the first time by Cornell and Dunaruma[24] in 1965. Many of these polymers showed a good and broad spectrum of antibacterial activity. The group of Dunaruma[25–29] worked very actively on the synthesis of new polymers with antimicrobial activity, such as sulfonamide-dimethylolurea copolymers, *N*-acylsulfanilamide, sulfonamide or sulfapyridine-formaldehyde copolymers. Ascoli *et al.*[30] also developed a polymeric nitrofuran derivative with extended antibacterial action. Later, Ackart *et al.*[31] described a series of carboxyl-containing ethylene copolymers that exhibited long term antibacterial and antifungal properties. These authors tried to blend these polymers with commodity polymers and made water emulsions to test their applicability as components of protecting coatings. Panarin *et al.*[32] synthesised copolymers with *N*-vinylpyrrolidone with (2-methacryloxyethyl)triethylammonium iodine or bromide and studied the influence of these macromolecules, their size and the amount of quaternary ammonium groups on the antimicrobial activity. Vogl and Tirrell[33] demonstrated that nondegradable polymers with functional groups, *i.e.* polymers and copolymers of 4-vinylsalicylic acid and 5-vinylsalicylic acid derivatives were very active against Gram-positive and/or Gram-negative bacteria. The activity of poly(4-vinylsalicylic acid) and poly(5-vinyl-salicylic acid) was found to be independent of molecular weight. In addition, selective activity was obtained by preparing copolymers of 4-vinylsalicylic acid or 5-vinylsalicylic acid with methacrylic acid, a comonomer whose homopolymer is inactive.

A large number of antimicrobial polymers have appeared since those days. Recently, several reviews have summarised the state-of-the-art.[22,34–40] Different parameters are discussed, such as molecular weight, type and degree of alkylation, distribution of charge, hydrophobic/hydrophilic ratio and their influence on the activity, and the action mode of antimicrobial polymeric materials is evaluated. Although there are several ways to classify these systems, we have arranged them in this book into four categories[38] (see Figure 1.1): a) polymers

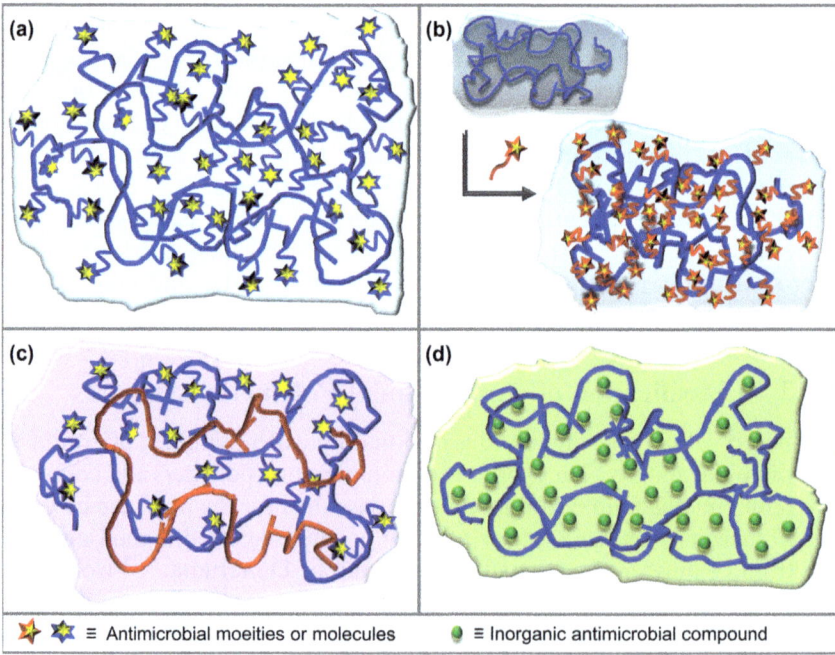

Figure 1.1   Representation of (a) polymers with antimicrobial activity; (b) chemical modification of polymers to achieve antimicrobial properties; (c) incorporation of antimicrobial organic compounds (either of low or high molecular weight) with antimicrobial activity into polymers and; (d) introduction of antimicrobial inorganic materials to polymers to achieve antimicrobial activity.

with antimicrobial activity; b) polymers that undergo chemical modifications to achieve antimicrobial activity; c) polymers containing antimicrobial organic compounds and; d) polymers incorporating antimicrobial inorganic compounds.

(a) Polymers with antimicrobial activity. As their name indicates they display antimicrobial activity by themselves. Usually, their chemical structure is used as the key characteristic to make the categories. Therefore, polymers can be found with quaternary nitrogen atoms (acrylic and methacrylic polymers, cationic conjugated polyelectrolytes, polysiloxanes, polyoxazolines, polyionenes, *etc.*), guanidine-containing polymers, polymers mimicking natural peptides (synthetic peptides, arylamide and phenylene ethynylene backbone polymers, halogen polymers, polynorbornene derivatives), halogen polymers (fluorine or chlorine-containing polymers, polymeric *N*-halamines), polymers containing phospho- and sulfo-derivatives, phenol and benzoic acid derivative polymers, organometallic polymers and others.

(b) Polymers that undergo chemical modifications to achieve antimicrobial activity. There are different approaches to incorporate antimicrobial activity into polymers. In particular, several scenarios can be distinguished if chemical modification is involved: i) a small molecule with antimicrobial activity is covalently attached to the polymer; (ii) antimicrobial peptides are fixed on an inactive polymer and (iii) antimicrobial polymers are grafted to regular polymers. In all these cases, it is desirable that chemical modification does not cause a deterioration of the properties of the final polymeric material.

(c) Polymers containing antimicrobial organic compounds. In this case, the antimicrobial activity is due to (i) noncovalent links between antimicrobial agents, either natural or synthetic, and polymers with the corresponding compound release and (ii) the mixture/blend of antimicrobial polymers and nonactive polymers to confer their biocide characteristics.

(d) Polymers incorporating antimicrobial inorganic compounds. In this category, the antimicrobial activity in the final material is obtained by transfer of the biocidal action of inorganic systems, such as metals, metallic oxides or modified clays, into the polymers.

As we will notice in subsequent chapters the present classification is only performed from a chemical point of view. If we focus our attention on the final applications, we realise that this division shows weak points, since one system depending on its characteristics can be applied for several purposes or functions. Anyway, we would like to give to the reader an extensive picture on how these polymeric materials can be synthesised and their potential applications.

1.2.3   Methods for Determining *In Vitro* Antimicrobial Activity

The increasing resistance to antibiotics and the appearance of new anti-microbial systems make necessary the establishment of methods that can assess the antimicrobial activity of these agents. The killing effect of an antimicrobial agent on a micro-organism can be measured by different approaches. Antimicrobial susceptibility tests (AST) are routinely used in clinical microbiology laboratories to evaluate the microbial pathogens susceptibility or detect resistance to antimicrobial agents. ATS methods commonly include diffusion (disk and E test diffusion) and dilution (agar dilution and broth microdilution) methodologies that measure the inhibitory activity of the tested agent. Other different methods that are not routinely applied to all micro-organisms but very advantageous in some situations, such as time-kill kinetic studies, MBC determination, serum bactericidal test[41] are now described:

(a) Disk diffusion method. This is commonly referred to as the Kirby–Bauer test and gives a quantitative measure of the effect of an antimicrobial agent against bacteria grown in culture. In this test, a 6-mm filter paper disks impregnated with a known amount of antimicrobial agent is placed onto an agar plate and water from the agar is absorbed into the disk. Then, the antimicrobial agent starts to diffuse into the surrounding agar. The plates are incubated overnight, and the inhibition zone of bacterial growth is used as a measure of susceptibility. If the organism is accessible to a specific antimicrobial agent, there will be no growth around the disc containing the antibiotic. Thus, large zones of inhibition indicate that the organism is susceptible, while a small or absent zone of inhibition indicates resistance. This method is simple, economical and there are several commercially available disks. However, not all organisms can be accurately tested, especially slow growing organisms.

(b) Epsilometer test, E test. This is a well-established commercially available method to directly quantify the susceptibility in terms of minimal inhibitory concentration (MIC). The procedure is briefly as follows: bacteria are grown on an agar plate and the E test plastic strip is placed on the top. Subsequently, the antimicrobial agent starts to diffuse into the agar and an exponential gradient of antimicrobial concentration is generated. After the incubation, the MIC can be directly read from the test strip according to the instructions of the manufacturer, where the elliptical zone of inhibition intersects with the MIC scale on the strip.

(c) Broth dilution test. This is one of the earliest antimicrobial susceptibility testing methods. This procedure involves the preparation of serial dilutions of the antimicrobial agents that are inoculated with a standardised number of micro-organisms in tubes and incubated overnight. The tubes are examined for visible bacterial growth by turbidity. The lowest concentration of antimicrobial agent that prevents the appearance of turbidity is considered the MIC. In addition to the determination of the MIC, the minimal bactericidal concentration (MBC) can also be estimated. Its main disadvantage consists in that this method is very tedious and laborious.

(d) Time-kill-kinetic assay. The lethal activity can be expressed as the rate of killing at a given concentration of antimicrobial under controlled conditions. This rate is determined by measuring the number of viable micro-organisms at various time intervals. The resulting graphic representation is known as the time-kill curve. Micro-organism killing rates are, in some way, dependent on the class of antimicrobial system and the concentration of it. The rate of killing rises for certain classes of antimicrobial systems with increasing antimicrobial concentrations up to a point of maximum effect. This is termed *concentration-dependent bactericidal activity*. In contrast, the killing rates of other antimicrobial agents are relatively slow and continue only as long as the concentrations are in excess of the minimal inhibitory concentration, that is, the lowest concentration of antimicrobial agent that inhibits the growth of the micro-organism. This rate of killing is termed the *time-dependent bactericidal activity*. There is also a third category, consisting of antibiotics with both time-dependent and concentration-dependent effects.[42]

(e) MBC determination. The minimal concentration of antimicrobial needed to kill most ($\geq 99.9\%$) of the viable organisms after incubation for a fixed period of time (generally 24 h) under a given set of conditions. It is the most common estimation of bactericidal activity and is known as either the minimal bactericidal concentration (MBC), minimal fungicidal concentration (MFC) or the minimal lethal concentration (MLC). The definition of the MBC (99.9% killing of the final inoculums, which is the growth method or direct suspension equivalent to 0.5 McFarland standard) is somewhat arbitrary and its determination is, moreover, subject to methodological variables. MBC determination is mainly indicated in some cases of endocarditis, osteomyelitis, meningitis, and sepsis in neutropenic patients, since they consist of serious infections in immune-compromised patients requiring antibiotic levels lethal to the infecting organism, or infections located in a site that is difficult to reach with antibiotics.

(f) Serum bactericidal test. The serum of a patient receiving an antimicrobial agent may be tested against the infecting micro-organism. This can be done using time-kill curve methodology (*i.e.* serum bactericidal rate, SBR) or using dilution methodology (*i.e.* serum bactericidal titer, SBT). The principles of these methods and the influence of biological and technical factors are similar.

Tables 1.1–1.4 collect the common micro-organisms tested for analysing the antimicrobial polymer capacity. Micro-organisms can be easily located from these tables, where it can be seen if they are pathogenic or not for humans, the disease that they can cause and where they can be usually found.

Table 1.1 Features of different Gram-positive and Gram-negative bacteria.

Group	Micro-Organisms	Pathogenic Disease	Source
Gram-negative bacteria	Bacteroides forsythus	Periodontal diseases	Oral cavity
	Cellulophaga lytica	Nonpathogenic	Marine environment
	Chlamydia pneumonia	Respiratory illness	Respiratory secretions
	Chlamydia trachomatis	Urethritis, proctitis, trachoma and infertility	Eyes infection, also in genital tract, rectum
	Escherichia coli	Harmless, some serotypes cause food poisoning	Contaminated food
	Haemophilus influenza	Bacterial meningitis, upper respiratory tract infections, pneumonia, bronchitis	Upper respiratory system
	Klebsiella pneumoniae	Pneumonia and other respiratory inflammations	Respiratory infections
	Mycoplasma gallisepticum	Nonpathogenic, chronic respiratory diseases in birds	Chickens, turkeys, birds
	Neisseria gonorrhoeae	Gonorrhoea, ophthalmia neonatorum, septic arthritis	Sexually transmitted infection
	Porphyromonas gingivalis	Periodontal diseases	Oral cavity
	Proteus mirabilis	Urinary tract infections and the formation of stones	Human gastrointestinal tract and also in free living in water and soil
	Proteus vulgaris	Urinary tract infections and wound infections	Intestinal tracts of humans and animals and also in soil, water and faecal matter
	Pseudomonas aeruginosa	Inflammation and sepsis. Bacteraemia, infections of eye, ear skin, urinary, bone, respiratory system, gastrointestinal tract, nervous system. Secondary pneumonia and endocarditis	Soil, water, skin flora, in medical equipment
	Pseudomonas fluorescens	Nonpathogenic	Soil and water
	Pseudomonas putida	Nonpathogenic	Soil
	Pseudoalteromonas haloplanktis	Intoxication by production of tetraodontoxin	Marine environment
	Salmonella enteritidis	Fever, abdominal cramps, and diarrhoea	Food-borne eggs
	Salmonella typhi	Typhoid fever, Dysentery, Colitis	Food-borne faecal contaminated, food or drinking water

Table 1.1 (*Continued*)

Group	Micro-Organisms	Pathogenic Disease	Source
	Salmonella typhimurium	Salmonellosis with gastroenteritis and enterocolitis	Intestinal lumen
	Serratia liquefaciens	Urinary tract infections, bloodstream infections, sepsis, pneumonia, meningocephalitis	Soil, water, plants, and the digestive tracts of rodents, insects, fishes, and humans
	Shigella boydii	Diarrhoea and shigellosis	Intestine and rectum of humans and other primates. Faeces and soil and/or food/water contaminated with faecal matter
	Shigella dysenteriae	Dysentery	Faecal-oral contamination and in contaminated food and water
	Shigella sonnei	Shigellosis in general	Faecal-oral contamination and in contaminated food and water
	Spiroplasma citri	Nonpathogenic	Citrus trees
	Spiroplasma floricola	Nonpathogenic	Flowers of tulip tree
	Spiroplasma melliferum	Nonpathogenic	Honeybees
	Stenotrophomonas maltophilia	Bacteraemia, pneumonia, urinary tract infection	Aqueous environments, soil and plants. Breathing tubes, and indwelling urinary catheters
	Yersinia enterocolitica	Gastric infections, diarrhoea	Food-borne route
	Yersinia pseudotuberculosis	Tuberculosis-like symptoms	Food-borne pathogen, intestinal tract, liver, spleen, and lymph nodes
Gram-positive bacteria	*Acholeplasma laidlawii*	Nonpathogenic	Animals, and some plants and insects
	Actinomyces viscosus	Periodontal disease in dogs and cattle. Human dental calculus and root surface caries	Oral cavity of dogs and humans
	Bacillus coagulans	Nonpathogenic (beneficial bacteria)	Food products
	Bacillus cereus	Severe nausea, vomiting and diarrhoea	Foodborne pathogen (when food is improperly refrigerated)
	Bacillus megaterium	Nonpathogenic	Soil
	Bacillus subtilis	Nonpathogenic for humans	Soil and vegetation

Table 1.1 (*Continued*)

Group	Micro-Organisms	Pathogenic Disease	Source
	Bacillus thuringiensis	Nonpathogenic for humans	Soil
	Bifidobacterium bifidum	Nonpathogenic	Digestive system, mouth, breast milk and the vagina
	Bifidobacterium breve	Nonpathogenic	Lower digestive tract of the human body
	Broehothrix thermosphacta	Nonpathogenic	Chilled raw meats and processed meat products stored aerobically or under modified atmospheres
	Clostridium difficile	Colitis, inflammation and mucosal injury to the colon	Faeces, Foodborne
	Enterococcus faecalis	Lower urinary tract infections (cystitis, prostatitis, and epididymitis) as well as intra-abdominal, pelvic, and soft tissue infections. Less common infections include meningitis, haematogenous, osteomyelitis, septic arthritis, and pneumonia	Nosocomial environments/ long-term care
	Enterococcus faecium	Bacteraemia, surgical wound infection, endocarditis, and urinary tract infections	Nosocomial environments
	Enterococcus hirae	Pathogenic mainly in animals and very rare in human (septicaemia, endocarditis)	Intestinal flora of domestic animals, foods of animal origin and water
	Lactobacillus casei	Nonpathogenic	Fermentation processes (cheese, yogurt, fermented milks, fermented Sicilian green olives, and other products)
	Lactobacillus salivarius	Nonpathogenic	Oral cavities, intestines, and vagina
	Listeria monocytogenes	Listeriosis. Noninvasive: gastroenteritis. Invasive: septicaemia and meningitis	Foodborne pathogen
	Micrococcus luteus	Pruritic eruptions and severe itching. In immuno-compromised patients: septic shock, pneumonia endocarditis or sepsis	Soil, dust, water and air, and as part of the skin flora of mammal

Table 1.1 (*Continued*)

Group	Micro-Organisms	Pathogenic Disease	Source
	Mycobacterium smegmatis	Nonpathogenic	Soil, water, and plants
	Mycobacterium tuberculosis	Tuberculosis	Mammalian respiratory system
	Pediococcus pentosaceus	Nonpathogenic	Plant, ripened cheese, and a variety of processed meats
	Staphylococcus aureus (methicillin-resistant Staphylococcus aureus, MSRA)	Mild skin infections (impetigo, folliculitis, *etc.*), invasive diseases (wound infections, osteomyelitis, bacteremia, *etc.*), toxin mediated diseases (food poisoning, toxic shock syndrome, scaled skin syndrome, *etc.*) and pneumonia.	Contaminated food and nosocomial environments
	Staphylococcus epidermidis	Inflammation and pus secretion, endocarditis, sepsis	Nosocomial infections (sutures, indwelling catheters, and implanted joints)
	Staphylococcus haemolyticus	Septicaemia, peritonitis, infections of urinary tract, wound, bone and joints	Skin flora of humans, in axillae, perineum, and inguinal areas. Nosocomial infections due to insertion of foreign bodies, such as prosthetic valves
	Staphylococcus hominis	Bacteraemia	Human skin especially common in the axillae and the pubic area, where apocrine glands are numerous. Nosocomial infections
	Staphylococcus saprophyticus	Urinary tract infections	Gastrointestinal tracts, meat, cheese and vegetable products
	Streptococcus mutans	Dental caries under frequent and prolonged acidic conditions.	Oral cavity
	Streptococcus pneumoniae	Pneumonia, acute sinusitis, otitis media, meningitis, bacteraemia, sepsis, osteomyelitis, septic arthritis, endocarditis, peritonitis, pericarditis, cellulitis, and brain abscess	Respiratory tracts of mammals

Table 1.1 (*Continued*)

Group	Micro-Organisms	Pathogenic Disease	Source
	Streptococcus pyogenes	Pharyngitis, tonsillitis, sinusitis, otitis, pneumonia, impetigo, erysipelas, cellulitis, joint or bone infections, necrotising fasciitis, myositis, meningitis, endocarditis, scarlet fever, rheumatic fever and glomerulonephritis	Respiratory specimens, skin lesions, blood, sputum and wound

Table 1.2 Features of different fungi, differentiating yeasts.

Group	Micro-Organisms	Pathogenic disease	Source
Fungi	*Alternaria alternate*	Upper respiratory tract infections and asthma	Leaves of trees
	Aspergillus *flavus*	Hepatitis, immunosuppression, hepatocellular carcinoma, and neutropenia	Cereal grains and legumes
	Aspergillus fumigatus	Pulmonary infections	Soil and decaying organic matter
	Aspergillus niger	Lung diseases	Certain fruits and vegetables
	Aspergillus terreus	Rare, only in people with deficient immune systems pulmonary diseases	In tropical and subtropical areas. It is common in stored crops
	Byssochlamys fulva	Nonpathogenic	Acidic canned fruits
	Botrytis cinerea	"Winegrower's lung", a rare form of hypersensitivity pneumonitis	Many plant species, such as wine grapes
	Chaetosphaeridium globosum	Nonpathogenic	Fresh water
	Cladosporium cladosporioides	Very rare, agents of phaeohyphomycosis	Herbaceous and woody plants and dead organic matter
	Eurotium tonophilum (*Aspergillus tonophilus*)	Nonpathogenic	Soybeans, rice
	Fusarium moniliforme	Nonpathogenic	Rice, sugarcane and maize
	Fusarium oxysporum	Rare, some cases of fungal keratitis, Onychomycosis, hyalohyphomycosis	Banana trees and many plants

Table 1.2 (*Continued*)

Group	Micro-Organisms	Pathogenic disease	Source
	Microsporum gypseum	Tinea capitis, tinea corpus, ringworm, and other skin and hair diseases	Warm humid weather
	Mucor circinelloides	Mucormycosis	Plants
	Penicillium citrinum	Nonpathogenic	Cereal plants
	Penicillium digitatum	Nonpathogenic	Citrus fruits
	Penicillium funiculosum	Nonpathogenic	Tropical areas: water, soil, food products, *etc.*
	Penicillium pinophilum	Nonpathogenic	Tomato and cotton plants
	Pyrobaculum islandicum	Nonpathogenic	Submarine hydrothermal systems
	Rhizoctonia bataticola	Rare cases of cutaneous and renal infections	Many plants such as soybean, peanut, and corn
	Rhizopus oryzae	Mucormycosis	Dead organic matter
	Rhizopus stolonifer	Nonpathogenic	Bread, soft fruits such as bananas and grapes, *etc.*
	Sporotrichum pulverulentum	Nonpathogenic	Plants
	Stachybotrys chartarum	Chronic fatigue, fever, irritation to the eyes, nausea, vomiting, pulmonary haemorrhage	Building materials rich in cellulose when they are wet
	Trichoderma lignorum	Nonpathogenic	Soil
	Trichophyton rubrum	Tinea pedis (athlete's foot), tinea cruris, ringworm	Environmental heat and humidity: swimming pools, tight shoes
	Trichophyton mentagrophytes	Cutaneous infections	Soil, floor of swimming pools, animals, footwear
	Trichoderma virens	Nonpathogenic	Soil
	Trichoderma viridis	Nonpathogenic	Soil
Yeasts	*Aureobasidium pullulans*	Dyspnea, cough, fever, chest infiltrates, acute inflammatory reaction and hypersensitivity pneumonitis	In different environments (*e.g.* soil, water, air and limestone)
	Candida albicans	Infections in the skin, mucouses and visceras, causing mild inflammation. Also vaginal thrush	Saprophyte of the human and animal digestive tract
	Candida glabrata	Infections of the urogenital tract and bloodstream	Normal flora of healthy individuals and also nosocomial environments

Table 1.2   (*Continued*)

Group	Micro-Organisms	Pathogenic disease	Source
	Candida parapsilosis	Sepsis and of wound and tissue infections	Domestic animals, insects and soil. Nosocomial infections: hands of health-care workers, indwelling catheters, *etc.*
	Candida tropicalis	Bloodstream infection and less commonly tissue invasive candidiasis. Rarely biofilm infections or oral or vaginal thrush	Tropical and subtropical marine environments. It can also be cultured from various fruits, faeces and soil
	Candida utilis/Pichia jadinii	Urinary tract infections	Cellulose-rich substrates such as wood, leaf litter, and paper pulp
	Cryptococcus neoformans	Lung infection, fungal meningitis and encephalitis	Soil
	Debaryomyces hansenii	Nonpathogenic	All types of cheeses and chilled food
	Hanseniaspora guilliermondii	Nonpathogenic	Dates, grapes, tomatoes, figs and in soil
	Kluyveromyces fragilis or marxianus	Poorly pathogenic pulmonary infection in an immunosuppressed cardiac transplant patient	Grape, milk, cheese and other food products.
	Pichia stipitis	Nonpathogenic	Guts of passalid beetles, forests or areas high in agricultural waste, among others
	Rhodotorula rubra	Fungal peritonitis, fungemia, endocarditis and meningitis in patients on dialysis and chemotherapy	Airborne contaminant of skin, lungs, urine and feces
	Saccharomyces cerevisiae	Nonpathogenic	Skin of grapes

Table 1.3   Features of different algae and protozoa.

Group	Micro-Organism	Pathogenic diseases	Source
Algae	*Amphora coffeaeformis*	Nonpathogenic	Brackish water habitats and can tolerate a very wide spectrum of environmental conditions
	Dunaliella tertiolecta	Nonpathogenic	Marine green flagellate
	Navicula incerta	Nonpathogenic	Phytoplankton

Table 1.4 Features of different viruses.

Group	Micro-Organisms	Pathogenic diseasis	Transmission
Viruses	Herpes simplex	Disorders based on the site of infection	Skin, mucous membrane, body fluid
	Human Immunology Influenza A	Disease of the respiratory tract, flu	People cough and sneeze and by indirect spread from respiratory secretions on hands, tissues
	Simian 40	Tumors, but most often persists as a latent infection	Probably first "from polio vaccine" and then by horizontal infection
	Varicella zoster	Chickenpox (varicella)	Directly touching the blisters, saliva or mucus of an infected person. Also through the air by coughing and sneezing

Acknowledgments

MINECO is fully acknowledged for financial support (MAT2010-17016 and MAT2010-19883). A. Muñoz-Bonilla also thanks MINECO for her Juan de la Cierva postdoctoral contract.

References

1. D. W. Kolpin, E. T. Furlong, M. T. Meyer, E. M. Thurman, S. D. Zaugg, L. B. Barber and H. T. Buxton, *Environ. Sci. Technol.*, 2002, **36**, 1202–1211.
2. E. Topp, R. Chapman, M. Devers-Lamrani, A. Hartmann, R. Marti, F. Martin-Laurent, L. Sabourin, A. Scott and M. Sumarah, *J. Environ. Qual.*, 2013, **42**, 173–178.
3. E. Y. Furuya and F. D. Lowy, *Nature Rev. Microbiol.*, 2006, **4**, 36–45.
4. H. F. Chambers and F. R. DeLeo, *Nature Rev. Microbiol.*, 2009, **7**, 629–641.
5. K. D. Sinclair, T. X. Pham, R. W. Farnsworth, D. L. Williams, C. Loc-Carrillo, L. A. Horne, S. H. Ingebretsen and R. D. Bloebaum, *J. Biomed. Mater. Res. A*, 2012, **100A**, 2732–2738.
6. J. Lora-Tamayo, O. Murillo, J. A. Iribarren, A. Soriano, M. Sánchez-Somolinos, J. M. Baraia-Etxaburu, A. Rico, J. Palomino, D. Rodríguez-Pardo, J. P. Horcajada, N. Benito, A. Bahamonde, A. Granados, M. D. del Toro, J. Cobo, M. Riera, A. Ramos, A. Jover-Sáenz and J. Ariza and on behalf of the REIPI Group for the Study of Prosthetic Infection, *Clin. Infect. Dis.*, 2012.

7. WHO, http://www.who.int/drugresistance/WHO_Global_Strategy_Recommendations/en/index4.html, 2012.
8. R. J. Anderson, P. W. Groundwater, A. Todd, A. J. Worsley, in *Antibacterial Agents: Chemistry, Mode of Action, Mechanisms of Resistance and Clinical Applications*, John Wiley & Sons, Ltd, Chichester, UK, 2012, pp. 1–33.
9. J. Taylor, *Bath Advanced Science - Micro-organisms and Biotechnology 2nd edn*, Nelson Thornes, 2001.
10. C. H. Posten, C. L. Cooney in *Biotechnology*, ed. H. Sahm, Wiley-VCH, Germany, 2nd edn., 2000, vol. 1, pp. 111–120.
11. L. Liu, H. L. Johnson, S. Cousens, J. Perin, S. Scott, J. E. Lawn, I. Rudan, H. Campbell, R. Cibulskis, M. Li, C. Mathers and R. E. Black, *The Lancet*, 2012, **379**, 2151–2161.
12. J. Lienert, M. Koller, J. Konrad, C. S. McArdell and N. Schuwirth, *Environ. Sci. Technol.*, 2011, **45**, 3848–3857.
13. I. Michael, L. Rizzo, C. S. McArdell, C. M. Manaia, C. Merlin, T. Schwartz, C. Dagot and D. Fatta-Kassinos, *Water Res.*, 2013, **47**, 957–995.
14. L. Kovalova, H. Siegrist, H. Singer, A. Wittmer and C. S. McArdell, *Environ. Sci. Technol.*, 2012, **46**, 1536–1545.
15. WHO, http://www.who.int/water_sanitation_health/emerging/legionella_rel/en/, 2007.
16. M. Srinivasan, *Curr. Opin. Opthal.*, 2004, **15**, 321–327.
17. L. Liu, H. Wu, S. N. Riduan, J. Y. Ying and Y. Zhang, *Biomaterials*, 2013, **34**, 1018–1023.
18. K. Ara, M. Hama, S. Akiba, K. Koike, K. Okisaka, T. Hagura, T. Kamiya and F. Tomita, *Can. J. Microbiol.*, 2006, **52**, 357–364.
19. E. R. Dumas, A. E. Michaud, C. Bergeron, J. L. Lafrance, S. Mortillo and S. Gafner, *J. Cosmet. Dermatol.*, 2009, **8**, 197–204.
20. S. Tan, G. Li, J. Shen, Y. Liu and M. Zong, *J. Appl. Polym. Sci.*, 2000, **77**, 1869–1876.
21. G. Li and J. Shen, *J. Appl. Polym. Sci.*, 2000, **78**, 676–684.
22. E.-R. Kenawy, S. D. Worley and R. Broughton, *Biomacromolecules*, 2007, **8**, 1359–1384.
23. A. C. Engler, N. Wiradharma, Z. Y. Ong, D. J. Coady, J. L. Hedrick and Y.-Y. Yang, *Nano Today*, 2012, **7**, 201–222.
24. R. J. Cornell and L. G. Donaruma, *J. Med. Chem.*, 1965, **8**, 388–390.
25. L. G. Donaruma and J. Razzano, *J. Med. Chem.*, 1966, **9**, 258–259.
26. J. R. Dombroski, L. G. Donaruma and J. Razzano, *J. Med. Chem.*, 1967, **10**, 963–964.
27. J. R. Dombroski, L. G. Donaruma and J. Razzano, *J. Med. Chem.*, 1967, **10**, 964–965.
28. L. G. Donaruma and J. R. Dombroski, *J. Med. Chem.*, 1971, **14**, 460–461.
29. L. G. Donaruma and J. Razzano, *J. Med. Chem.*, 1971, **14**, 244–244.
30. F. Ascoli, G. Casini, M. Ferappi and E. Tubaro, *J. Med. Chem.*, 1967, **10**, 97–99.
31. W. B. Ackart, R. L. Camp, W. L. Wheelwright and J. S. Byck, *J. Biomed. Mater. Res.*, 1975, **9**, 55–68.

32. E. F. Panarin, M. V. Solovskii and O. N. Ékzemplyarov, *Pharm. Chem. J.*, 1971, **5**, 406–408.
33. O. Vogl and D. Tirrell, *J. Macromol. Sci., Chem.*, 1979, **13**, 415–439.
34. G. J. Gabriel, A. Som, A. E. Madkour, T. Eren and G. N. Tew, *Mater. Sci. Eng., R*, 2007, **57**, 28–64.
35. F. J. Xu, K. G. Neoh and E. T. Kang, *Prog. Polym. Sci.*, 2009, **34**, 719–761.
36. L. Timofeeva and N. Kleshcheva, *Appl. Microbiol. Biotechnol.*, 2011, **89**, 475–492.
37. F. Siedenbiedel and J. C. Tiller, *Polymers*, 2012, **4**, 46–71.
38. A. Muñoz-Bonilla and M. Fernández-García, *Prog. Polym. Sci.*, 2012, **37**, 281–339.
39. K. Lienkamp, A. Madkour, G. Tew, in *Polymer Composites – Polyolefin Fractionation – Polymeric Peptidomimetics – Collagens*, eds. A. Abe, H.-H. Kausch, M. Möller and H. Pasch, Springer, Berlin Heidelberg, 2013, vol. 251, pp. 141–172.
40. K. Kuroda and G. A. Caputo, *Wiley Interdiscip. Rev. Nanomed. Nanobiotechnol.*, 2013, **5**, 49–66.
41. M. A. Pfaller, D. J. Sheehan and J. H. Rex, *Clin. Microbiol. Rev.*, 2004, **17**, 268–280.
42. M. E. Levison, *Infect. Dis. Clin. North Am.*, 2004, **18**, 451–465.

CHAPTER 2

Antimicrobial Activity of Chitosan in Food, Agriculture and Biomedicine

ALEXANDRA MUÑOZ-BONILLA,* MARÍA L. CERRADA
AND MARTA FERNÁNDEZ-GARCÍA

Instituto de Ciencia y Tecnología de Polímeros (ICTP-CSIC), Juan de la
Cierva 3, 28006 Madrid, Spain
*Email: sbonilla@ictp.csic.es

2.1 Introduction

The earliest known polysaccharide, later named as chitin, was discovered in 1811 by Henry Braconnot in edible fungi. Since that date, there has been a great industrial interest in the use of chitin and its derivatives because of its unique properties that offer a wide range of applications. Moreover, chitin is one of the most abundant biopolymers worldwide. Although chitin can be extracted from some fungi, mollusks, insects and other animals, such as squids, the exoskeleton of crustaceans is currently the principal source for the chitin production. In this context, the large-scale annual production of shell waste from crustacean processing is very high and its accumulation has become an important issue for the global seafood industry. The use of this waste as a renewable resource for the production of chitinous products has attained considerable scientific and industrial interest, while solving the waste accumulation concern at the same time. In addition to its biodegradability, biocompatibility and nontoxicity, associated with the most naturally abounding polymers, such as cellulose or

RSC Polymer Chemistry Series No. 10
Polymeric Materials with Antimicrobial Activity: From Synthesis to Applications
Edited by Alexandra Muñoz-Bonilla, María L. Cerrada and Marta Fernández-García
© The Royal Society of Chemistry 2014
Published by the Royal Society of Chemistry, www.rsc.org

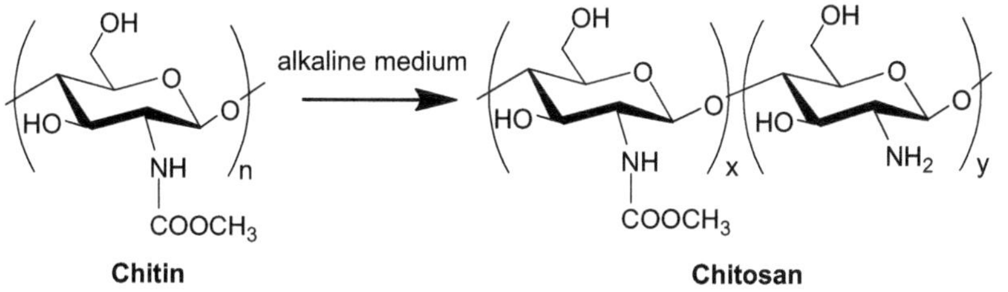

Figure 2.1 Preparation of chitosan from the partial *N*-deacetylation of chitin under alkaline conditions.

dextran, chitin shows high nitrogen content that is especially interesting for its use as chelating agent. Among others, chitosan has become the most commercially employed form of these materials. Chitosan is a polycationic polymer ($pK_a = 6.2$–6.8) obtained by the partial *N*-deacetylation of chitin under alkaline conditions, thus consisting of β-(1,4)-2-acetamido-2-deoxy-D-glucose and β-(1,4)-2-amino-2-deoxy-D-glucose units (Figure 2.1).[1] This cationic character is the basis of most of its applications as antimicrobial compound in many different fields, including the food area, agriculture, biomedicine and textiles. Compared with chitin, chitosan has superior solubility and reactivity as a consequence of the deacetylation, which enhances, in general, its antimicrobial activity. Therefore, the properties and characteristics and, consequently, applicability of chitosan strongly depend on the degree of deacetylation and molecular weight, among others factors.[2]

Chitosan is typically insoluble in water and in most organic solvents, being only soluble in dilute organic acids, such as acetic acid, formic acid or succinic acid, among others. Besides, the viscosity and processability of the solutions are not only affected by the degree of deacetylation and the molecular weight but also by the pH, temperature or ionic strength. In this sense and for certain uses, many scientific efforts have been reported in the literature to increase the solubility in water by chemical modifications. Other studies entail the improvement, for instance, of films and coating formation, particularly for food-packaging applications. In general, chitosan and related products can be easily processed for a desired application into scaffolds,[3] nano- and microparticles,[4] membranes[5] and fibers.[6] And more importantly, many attempts have been also taken up to improve its antimicrobial, antitumoral, antioxidant and hemostatic activities, which are strongly influenced by the physicochemical state of chitosan. In this context, chitosan and its derivatives exhibit a broad antimicrobial spectrum against Gram-negative and Gram-positive bacteria and fungi. This antimicrobial action, as commented, is influenced by the type of chitosan (degree of deacetylation and polymerization, functional groups and kind of modification, *etc.*), environmental conditions, micro-organism species or nutrients. Several possible mechanisms of action have been proposed. The polycationic chitosan generally interacts electrostatically with negatively charged bacterial membranes disrupting the inner and outer membranes. Besides, low molecular weight chitosan derivatives are able to penetrate the cell and bind to

DNA, leading subsequently to the inhibition of RNA and protein synthesis. Other mechanisms imply the chelation of some free metal ions hindering lipid oxidation.[7] The antioxidant activity of chitosan has been associated with its strong hydrogen-donating ability. In spite of the great potential as a anti-microbial component that chitosan and other related compounds have demonstrated, a clear and full understanding of the relationship between the activity as biocide and its molecular structure, chemical functionality or physical aspects is still of paramount importance in order to optimize the biocidal performance of these materials from an applied standpoint.

This chapter is mainly focused on the recent developments and strategies used to enhance the antimicrobial activities of chitosan and derivatives, modes of action and applications in diverse fields, especially in biomedicine, food packaging and agriculture.

2.2   Antimicrobial Activity of Chitosan

The antimicrobial activity of chitin, chitosan and their derivatives against various groups of micro-organisms, such as bacteria, yeasts, viruses and fungi, has received considerable attention in recent years.[8–11] Many investigations have been reported, involving *in vitro* and *in vivo* experiments with chitosan in different formulations, from solutions implemented in agriculture, films used in food sector up to gel structures for some biomedical applications. The physical state of chitosan is a crucial factor that affects its antimicrobial activity. Other intrinsic factors of chitosan are the molecular weight and positive charge density. The environmental conditions, *i.e.* pH, ionic strength and temperature also have an important effect on the antimicrobial activity. Although the broad spectrum of its antimicrobial activity is well known, chitosan exhibits different efficacy depending on the micro-organisms. In general, fungi are more susceptible to the chitosan action than bacteria. Moreover, the mode of action against bacteria differs between Gram-positive and Gram-negative because of cell surface differences. The modes of action and the different parameters affecting the antimicrobial activity of chitosan and its derivatives are described in this section.

2.2.1   Modes of Action

2.2.1.1   *Mechanism of Antibacterial Action*

Investigations describing the potential of chitosan and derivatives consider that they behave as bactericidal (killing the bacteria)[12] and/or bacteriostatic (hindering the growth of bacteria).[13] The exact mechanism is, however, still not fully understood. Several models have been proposed for antibacterial actions of chitosan.

The electrostatic interactions between the positively charged chitosan and the negatively charged bacterium cell membrane[14–16] is the most accepted mechanism. In this model, the electrostatic interactions can promote changes in the wall permeability and, then, internal osmotic imbalances are provoked with the consequent inhibition of the bacterial growth (Figure 2.2).[17] The electrostatic

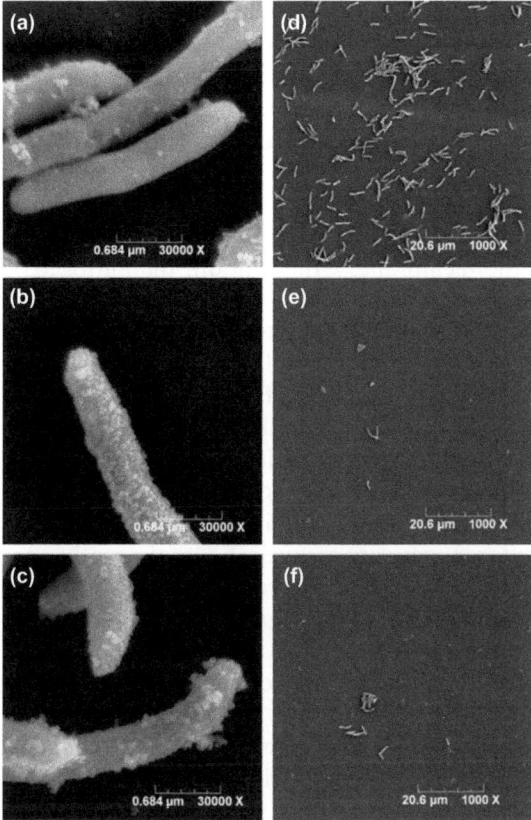

Figure 2.2 SEM of *Pseudomona fluorescens* after incubation with 100 mg/L chitosan-arginine for 3 h. Controls (a and d), cells treated with 6%-substituted chitosan-arginine (b and e), and cells treated with 30%-substituted chitosan-arginine (c and f).[17]
Reprinted with permission from *Acta Biomater.*, **6**, 2562–2571, Copyright 2010, Elsevier.

interactions also may result, for instance, in the hydrolysis of the peptidoglycans in Gram-positive bacteria, leading to the leakage of intracellular compounds, such as proteins, glucose, nucleic acids, or potassium ions.[8] Thus, chitosan modifies the metabolism of the bacteria and prompts their death at the end. Raafat *et al.*[18] visually investigated this model using transmission electron microscopy. They observed that control cells showed an intact plasma membrane and an outer cell wall, which was more or less uniform along the entire cell perimeter. Nevertheless, cells treated with chitosan displayed irregular structures protruding from the cell surface, which might be chitosan deposits still attached to the negatively charged surface. Interestingly, the cell membrane became locally detached from the cell wall, this shrinkage of the membrane suggesting water and ion loss from the cell. Then, cells maintained an intact membrane with impaired membrane function.[18]

The concentration of chitosan and, consequently, the number of positive charges significantly affects the antibacterial efficacy since this model is based on electrostatic interactions. It was reported that at low concentration the antibacterial activity can be attributed to the leakage of intracellular components. However, when the concentration is too high the amount of chitosan bonded with the bacteria is reduced because the larger number of positive charges may tend to form a coating over the bacteria, imparting a net positive charge and maintaining the bacteria in suspension[19] and also blocking the mass transfer across the membrane.[20] In addition, the antibacterial mechanism is different depending on the characteristics of the cell surface, differing between Gram-positive and Gram-negative bacteria. However, the effect of the bacterial species on the chitosan antibacterial efficiency is somewhat controversial varying with the conditions and type of chitosan derivatives. In Gram-negative bacteria, the outer membrane is basically composed of lipopolysaccharides that render negative charge to the surface larger than that in Gram-positive membranes, which consist of peptidoglycan and teichoic acid. This supports the evidence observed in several studies related to a higher inhibitory effect exhibited by chitosan derivatives against Gram-negative as compared with Gram-positive bacteria.[14,21] In contrast, other studies hold that chitosan shows stronger bactericidal effects against Gram-positive because the outer membrane of Gram-negative supposes an additional barrier.[22,23] Nevertheless, other authors claim that there are no evident differences between bacterial species.[24]

Another proposed mechanism of action is based on the interaction of the positively charged chitosan with bacterial DNA that, consequently, leads to the inhibition of the messenger RNA and protein synthesis.[19,25,26] This occurs through the chitosan penetration toward the nuclei of the micro-organisms, thus the molecular weight of the chitosan seems to be crucial. In this mechanism, the chitosan must be hydrolyzed to be able to penetrate into the cell. The presence of chitosan oligomers inside bacteria, such as *Escherichia coli*, has been confirmed by confocal laser scanning microscopy.[25] Nevertheless, this proposed mechanism is controversial and has not been corroborated in the recent literature.

The third most accepted mode of action is through the chelation of metals. Chitosan exhibits an excellent chelating capacity for transition and heavy metals due to the high nitrogen content.[27,28] Chelation efficiency is higher at basic pH since the electron pair on the amine groups is available for donation to metal. In addition, the hydroxyl groups are also unprotonated at pH higher than 7–9 and the complexation occurs also *via* hydroxyl groups.[29,30] Therefore, the chitosan molecules may complex the metals surrounding bacteria in this mechanism, avoiding the flow of essential nutrients, *i.e.* needed by bacterial enzymes.[31] The available sites for complexation are, however, limited and this mode of action does not seem to be determinant.

On the basis of the reported literature, chitosan antibacterial activity is not restricted to only one mode of action; all the described mechanisms are not mutually exclusive, probably taking place simultaneous or successively.

2.2.1.2 Mechanism of Antifungal Action

The chelating properties also make chitosan an antifungal agent.[28] Several mechanisms have been proposed for its antifungal activity. Chitosan acts by interferring directly with fungal growth through interactions with the negatively charged phospholipid components of fungi membrane, causing the leakage of cellular components,[32,33] and also by binding to the fungi DNA.[10] In particular, most of the studies related to the antifungal capacity of chitosan have been associated with food and plant spoilage.[34–39] These investigations demonstrated that the presence of chitosan activated many defense responses in plants. These defense processes included accumulation of phyloalexins, pathogen-related proteins, such as chitinases, synthesis of proteinase inhibitors, formation of callose and lignifications. In general, chitosan is usually employed in plant disease control as a powerful elicitor rather than a direct antimicrobial agent.

2.2.1.3 Mechanism of Antivirus Action

The chitosan and derivatives also possess the ability to induce resistance to viral infections in plants and animals and prevent phage infection in infected cultures of micro-organisms.[40–42] Similarly to the antibacterial and antifungal activity, the antiviral capability of chitosan depends on many factors, such as the degree of deacetylation, concentration, molecular weight or chemical modifications.[43,44]

Chitosan prevents the accumulations of infectious phage progeny in infected microbial cultures. Several mechanisms have been proposed to explain the inhibition of chitosan to the replication of bacteriophages. First, chitosan can decrease the viability of cultured bacterial cells. Secondly, it can neutralize the infectivity of mature phage particles in the inoculums and/or daughter phage particles; or thirdly, it can suppress the replication of the virulent phage.[45,46]

Chitosan and derivatives are broadly used in medicine and veterinary for viral infections.[47–52] In mammals, chitosan stimulates the immune response to virus antigens, affecting the cells of the immune systems. The alteration that chitosan produces on the functional activity of the auxiliary cells involved in immune responses, such as macrophages and granulocytes, has been studied. Another important mode of action of antiviral activity in animals is the induction of the interferon synthesis, which can block virus replication. Therefore, chitosan can affect different effector mechanisms of the immune system in animals.

The ability of chitosan to suppress viral plant infections is also well known.[53–56] Several studies show that chitosan inhibits viral infections by a large variety of virus with different structures and genome expression mechanisms, in a wide number of plant species. Virus-resistant plants respond to viral inoculation by hypersensitivity, which consists in the formation of local lesions that immobilize the virus. Chitosan treatment can modulate this plant hypersensitivity decreasing the number of local lesions. On the other hand, chitosan is able to inhibit the systemic spreading of viruses in susceptible plants, avoiding the systemic viral infection. Antiviral activity of chitosan in plants is usually based on the effects that it causes in the plants inducing a wide number of resistance mechanisms,[57] for instance, initiation of defense responses that can inhibit the replication of viruses.[58]

2.2.2   Factors Affecting Antimicrobial Activity

Molecular weight

The antimicrobial activity of chitosan and derivatives is greatly influenced by the molecular weight, which is probably the main factor affecting the efficacy. However, it is difficult to establish a correlation between bactericidal activity and chitosan molecular weight based on the existing literature, which is sometimes contradictory. This is probably due to the fact that evaluation of antimicrobial activity dependence on molecular weight turns out to be rather complex, varying with several factors, such as concentration range, degree of deacetylation or micro-organisms tested.[59] Generally, glucosamine and oligomers do not show almost any growth inhibition against several micro-organisms[22] and a minimum degree of polymerization seems to be required.[18,60] To some extent, the activity increases as molecular weight does,[61] but decreases again over a certain molecular weight if this is high enough.[25] Other investigations reported, however, that an increase in molecular weight led to a decrease in the activity of chitosan[62] and even oligochitosan molecules showed better biocidal activity.[11]

Degree of deacetylation

In general, the chitosan antimicrobial activity improves as the degree of acetylation is reduced.[2,63] Higher free amine groups are present at low degrees of acetylation and, consequently, the number of protonated amine groups is increased. As a result the chitosan enhances its solubility in water, a fact that favors the interactions between chitosan and negatively charged cell walls of micro-organisms. Several studies have demonstrated the influence of degree of acetylation on the antimicrobial effectiveness against fungi, Gram-negative and Gram-positive bacteria.[63–66] Normally, the molecular weight of the chitosan affects the antimicrobial activity in a greater extent than the degree of acetylation. However, Mellegårdit *et al.* recently evaluated the influence on the antibacterial activity of the molecular weight, degree of acetylation and test organisms and no trends were revealed in the antibacterial action related to variation in the molecular weight, whereas the degree of acetylation was determinant.[67] Chitosans with a lower degree of acetylation, 16%, were more active against *Bacillus cereus*, *E. coli*, *Salmonella Typhimurium* than the more acetylated chitosans, 48%.

The pH

In relation with the degree of acetylation, the pH determines the positively charge density of chitosan and thus its antibacterial activity. At low pH, below its pK$_a$ ($\sim$6.5) the amino groups of chitosan become ionized and exhibit positive charge. Therefore, the antimicrobial effect of chitosan increases as the pH is decreased due to the higher proportion of charged amino groups.[68] Unmodified chitosan is not active at neutral pH, due to the deprotonation of the amino groups and its insolubility.[69] Many efforts, as will be discussed later, have been, however, carried out to prepare derivatives of chitosan with increasing solubility in water and antimicrobial activity at neutral or basic pH.

Temperature

Tsai and Su[68] also investigated the effect of temperature on the antimicrobial activity of chitosan against *E. coli*. The antibacterial efficacy was found to be directly proportional to temperature, increasing from 4 to 37 °C. This reduced activity at low temperatures was attributed to the decrease in the kinetic rate of the interactions between chitosan and cell surface. A similar temperature effect has been observed for other tested micro-organisms.[70]

Ionic strength

Investigations of the influence of the ionic strength on the antibacterial activity of chitosan are contradictory. Chung *et al.*[71] suggested that the activity increases with ionic strength due to the enhance solubility of chitosan and Tsai and coworkers[68] observed that the antimicrobial activity against *E. coli* decreases in the presence of sodium ions. Other studies do not reveal any effect on the activity with the addition of salt.[72] However, it is well established that the addition of divalent cations, generally, minimizes dramatically the antimicrobial activity of chitosan, probably due to the complex formation between chitosan and these divalent metal cations.[68,71,73]

2.3 Chitosan Derivatives and Composites

Chitosan presents three reactive groups on each repeat unit, *i.e.* primary (C-6) and secondary (C-3) hydroxyl groups, and the amino group on each deacetylated unit. These active groups can be easily chemically modified to improve the mechanical and physical properties as well as solubility. Water solubility in basic and neutral pH is particularly a very important factor for its antimicrobial applications. In addition, a variety of functionalities with antimicrobial properties can be introduced into the chitosan chain through chemical reactions with the reactive groups of chitosan or even graft polymerization. As an alternative to enhance its antimicrobial activity by chemical modification, the physical blending of chitosan with antimicrobial components, either other polymers or inorganic particles, such as silver or titania, has been also investigated. Although there is a large variety of chitosan modifications and derivatives reported in the literature, the most general systems are described in this section classified according to their chemical structure.

Quaternized chitosan derivatives

The quaternary ammonium chitosan is the typical kind of chitosan derivative that has attracted considerable attention. The incorporation of quaternary ammonium groups onto chitosan leads to the introduction of permanent positive charges in the polymer independently of the pH of the aqueous media, providing water solubility as well as antimicrobial activity.[74,75] The quaternary ammonium chitosan can be prepared in two different common ways. Those are, quaternization of nitrogen atoms of the amino groups as well as the incorporation of quaternary ammonium moieties on the hydroxyl group.[76] The simplest form of quaternized chitosan is *N,N,N*-trimethyl chitosan (Figure 2.3) obtained by reaction with methyl iodide, sodium iodide and sodium hydroxide.[77] This study reveals that the antibacterial activity of quaternized chitosan against *E. coli* is stronger than that exhibited by neat chitosan. The effect of the chain length of the alkyl substituent on the antimicrobial activity was evaluated introducing different alkyl groups, *i.e.* butyl, octyl and dodecyl.[78] The results against *Staphylococcus aureus* show that the longer the chain length the higher the antimicrobial activity attained. This is a consequence of the hydrophobic–hydrophobic interactions between alkyl groups and the hydrophobic interior of the bacterial membrane.

Other more recent studies introduce *N*-aryl substituents such as *N,N*-dimethylaminophenyl and pyridyl substituents by *N*-methylation and later formation of quaternary ammonium salts in the presence of sodium iodide and methyl iodide.[79] The hydrophobic affinity between the aryl groups and the bacterial cell wall seems to enhance the antimicrobial activity, in particular against *S. aureus* bacteria. Besides, more complex structures such as *N,N,N*-trimethyl *O*-(2-hydroxy-3-trimethylammonium propyl) chitosans (Figure 2.4)

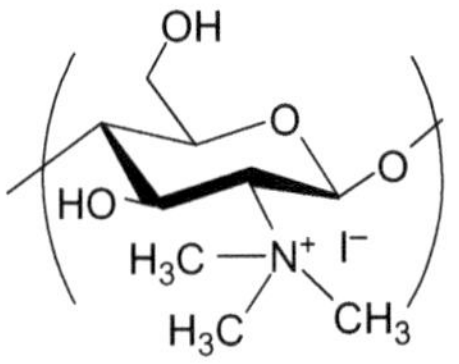

Figure 2.3   Structure of *N,N,N*-trimethyl chitosan iodide.

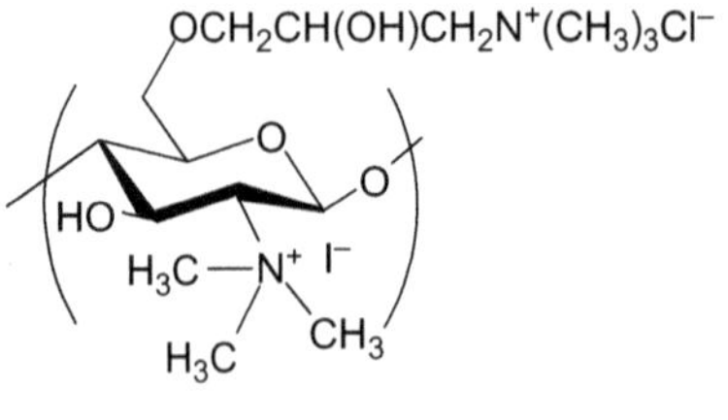

Figure 2.4   Structure   of   *N,N,N*-trimethyl   *O*-(2-hydroxy-3-trimethylammonium propyl).

with different degrees of *O*-substitution were synthesized by reacting *O*-methyl-free *N,N,N*-trimethyl chitosan with 3-chloro-2-hydroxy-propyl trimethyl ammonium chloride.[76] This quaternized chitosan exhibited enhanced antibacterial activity against *E. coli* and *S. aureus* compared with *N,N,N*-trimethyl chitosan, this increasing with a rise in the degree of substitution. There are, however, contradictory results related to the effect of quaternization, indicating that it does not always provide an enhancement of the antibacterial activity of chitosan. For instance, Chi *et al.*[80] showed that chitosan-*N*-2-hydroxypropyl trimethyl ammonium chloride was not effective against *E. coli* and *Pseudomonas aeruginosa*. In addition, it was reported that *O*-methyl free *N,N,N*-trimethyl chitosan exhibited weak antibacterial activity as compared with other non-quaternized chitosans and it even decreased as the degree of substitution increased at pH 5.5.[81] All these discrepancies found in the literature can be ascribed to the variety of factors and parameters related to chitosan and experimental conditions.

Carboxyalkyl chitosan derivatives

The carboxyalkylation of chitosan is another alternative to improve its solubility in the whole range of pH and increase the antimicrobial activity. Besides, the biocompatibility and safety for human enhance significantly.[82,83] This is the case of carboxymethyl chitosan, extensively used in biomedical applications.[84] Muzzarelli *et al.*[85,86] have reported the preparation of carboxyalkyl chitosan derivatives and worked very actively with their antimicrobial activity. The covalent immobilization of amino acid moieties, such as glycine to impart amphiphilic character was the inspiration to first develop the *N*-carboxymethyl chitosan. Its preparation was carried out by reaction of the free amino groups of chitosan with glycoxylic acid and then its reduction with sodium borohydride. The variation in concentration of the reagents allows the preparation of *N,N*-dicarboxylmethyl chitosan (Figure 2.5). Moreover, *N*-carboxypropyl and *N*-carboxybutyl chitosan were synthesized. They exhibited an improved solubility and antimicrobial activity against both Gram-negative and Gram-positive bacteria compared with chitosan.[87,88] On the other hand, other homologs

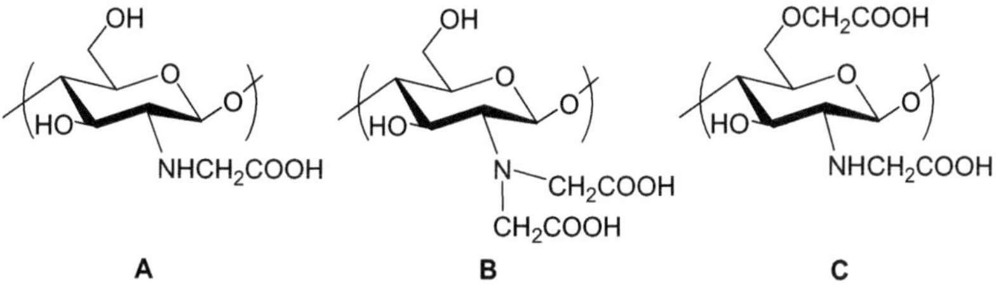

Figure 2.5 Structure of (A) *N*-carboxymethyl chitosan, (B) *N,N*-dicarboxylmethyl chitosan and (C) *N,O*-carboxymethyl chitosan.

such as *O*- and *N,O*-carboxymethyl chitosans have been also tested as anti-microbial agents.[25] It was found that the *O*-carboxymethyl chitosan was more effective than chitosan in the inhibition of the growth of *E. coli*, whereas the *N,O*-carboxymethyl chitosan showed poor antimicrobial activity. This fact was ascribed to the reduced number of free amino groups in the case of *N,O*-carboxymethyl chitosan as compared with chitosan.

Chitosan derivatives with sulfonyl groups and thio-derivatives

In general, sulfonyl groups attached to the chitosan chain provide a better solubility, resulting in an improvement of its antimicrobial action. This is the case of *N*-sulfonated and *N*-sulfobenzoyl chitosan.[89,90] High S contents provoke, however, a negative effect on the activity as a consequence of the decrease of positive charge on chitosan chains. Alternately, sulfonamide derivatives, *i.e.* 4-carboxybenzene sulfonamide chitosan showed antibacterial activity against *E. coli* and *S. aureus*.[91] Moreover, ammonium dithiocarbamate chitosan and triethylene diamine dithiocarbamate chitosan had enhanced inhibitory effects on *Fusarium oxysporum* and *Alternaria porri* fungi, which also depended on the concentration of the derivatives and fungal species.[92]

The incorporation of thiol-bearing compounds into polymers has been traditionally used to enhance properties such as the mucoadhesion or enzyme inhibitions. In recent years the study of thiolated chitosans has been extended, however, to the area of antimicrobial action. The antibacterial activity of chitosan modified with the thiol-containing 2-iminothiolane HCl and *N*-acetyl-L-cysteine was tested against a Gram-negative *E. coli* and a Gram-positive *S. aureus*.[93] It was found that the thiol moieties had a negligible effect on the antimicrobial activity of chitosan, whereas the positive charge induced by the formation of amidine groups in the 2-iminothiolane derivative was responsible for more effective disruption of the bacteria cell wall. These assessments were in agreement with other previous results.[94] However, Geisberger and coworkers[95] have recently demonstrated that chitosan-thioglycolic acid acts as a very efficient antimicrobial polymer with a broadband potential against bacteria and yeast, exhibiting superior action compared with trimethyl chitosan and carboxymethyl chitosan with higher positive zeta potential. They suggest that other modes of action apart from the electrostatic membrane interaction models must be involved.

Besides, it is well known that thioureas present antifungal, antibacterial and insecticide properties, and consequently have been also incorporated into chitosan to improve its activity. Eweis *et al.*[96] synthesized a benzoyl thiourea derivative of chitosan and demonstrated its antifungal efficacy against sugar-beet pathogens. Later, the investigation of the thiourea-modified chitosan was extended to others acyl derivatives, such as acetyl, chloroacetyl as well as benzoyl thiourea (Figure 2.6) against Gram-positive, Gram-negative bacteria and crop-threatening pathogenic fungi,[21] and all of them showed much better activity than that of native chitosan.

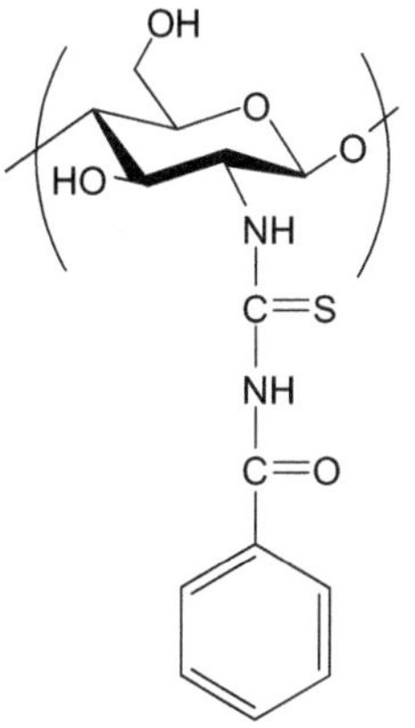

Figure 2.6 Structure of benzoyl thiourea derivative of chitosan.

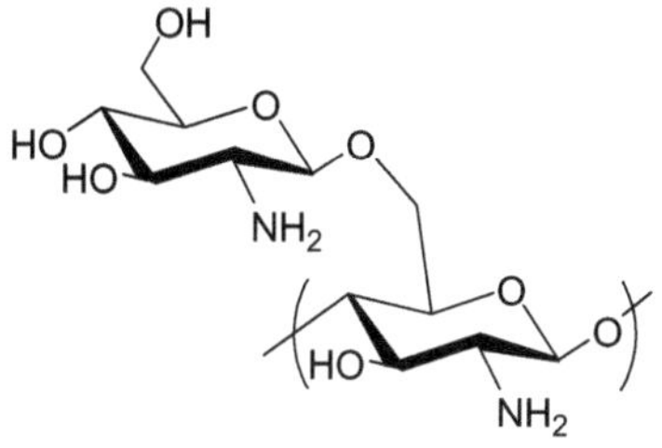

Figure 2.7 Structure of D-glucosamine-branched chitosan.

Chitosan derivatives with carbohydrates

Another exploited alternative to increase the water solubility of chitosan is the preparation of carbohydrate-branched chitosan derivatives. The antibacterial activity of 1-deoxy-1-glucit-1-yl and 1-deoxy-1-lactit-1-yl chitosan was studied against *Bacillus circulans* and *E. coli* using low concentration.[97] It was found that both significantly inhibit the growth of *B. circulans*, but they are not effective against *E. coli*. Furthermore, the antibacterial action against *E. coli* and *S. aureus* of *N*-alkylated disaccharide chitosan derivatives has been investigated by Yang *et al.*[98] They prepared water-soluble chitosan by *N*-alkylation of chitosan at the C-2 position with disaccharides including lactose, maltose and cellobiose and demonstrated that the activity is influenced by the degree of substitution and the kind of disaccharide. All the disaccharide chitosan derivatives showed higher activity than chitosan at neutral pH. The incorporation of saccharides onto the hydroxyl group at the C-6 position has been also reported (Figure 2.7).[99] The resulting D-glucosamine-branched chitosan exhibits significant growth inhibition of *Bacillus subtilis*, *S. aureus* and *Candida Albicans* compared with the native chitosan.

Chitosan composites and nanocomposites

There are a huge number of publications describing blends of chitosan with other polymers or inorganic compounds to improve their antimicrobial activity, achieve a broad antimicrobial spectrum as well as good physical properties in the case of films or scaffold fabrication where the water solubility is not a requirement. Although this topic will be discussed more indepth in the following sections concerning the applications, some of the most relevant or recent examples are described just below.

Park *et al.*[100] incorporated chitosan as an antimicrobial additive into low-density polyethylene to form films for red meat surfaces. Other composites films with antimicrobial properties based on chitosan and ethylene–vinyl alcohol copolymers (EVOH),[101] poly(vinyl alcohol),[102] poly(lactic acid),[103] cellulose[104] have been reported for food packaging applications, among others.

In addition, chitosan has been blended with others polymers that also possess antimicrobial activity. Chitosan and pectin-organic rectorite were deposited on cellulose acetate nanofibrous scaffolds for wound dressing and food packaging.[105] Pectin is also a natural and biocompatible polysaccharide that is very effective in bacterial inhibition. Therefore, the resulting nanofibrous scaffolds exhibit an improved bacterial inhibition and biocompatibility.

Lysozyme-chitosan composite films were developed for enhancing the antimicrobial properties of chitosan films[106] with a lysozyme release from the film matrix. The films with a 60 wt.% lysozyme incorporation enhanced the inhibition efficacy of chitosan films against both *Streptococcus faecalis* and *E. coli*.

Inorganic nanoparticles with antimicrobial properties can also be incorporated into chitosan to enhance its activity. Chitosan/Ag/ZnO nanoparticle composite membranes were prepared *via* a sol-cast transformation method resulting in a good and homogeneous dispersion of ZnO and Ag nanoparticles within the chitosan matrix.[107,108] The incorporation of ZnO and Ag and their respective contents had an effect on the mechanical properties and also enhanced the antibacterial activities for *B. subtilis, E. coli,* and *S. aureus, Penicillium, Aspergillus, Rhizopus* and yeast. It was also found that chitosan/Ag/ZnO films had antimicrobial activities higher than those exhibited by Chitosan/Ag and Chitosan/ZnO films, a synergetic effect being then observed.

Incorporation of titania in the obtainment of titania-chitosan nanocomposites increased the mechanical properties[109] and antimicrobial activity. Studies on the mechanical properties of chitosan-poly(*N*-vinylpyrrolidone)-TiO_2 composite materials revealed that the addition of titania significantly increased the strength. In addition, the composites exhibited great antimicrobial activity and good biocompatibility against NIH3T3 and L929 fibroblast cells.[110] Kavitha *et al.*[111] prepared different titania–chitosan nanocomposites varying the chitosan ratio by an *in situ* sol-gel method. The increase in chitosan content led to an enhancement in antibacterial activity against *S. aureus*.

2.4 Applications of Chitosan and its Derivatives in Agricultural Industries

Chitin, chitosan and their derivatives have been explored for many agricultural applications with regard to controlling plant diseases. They are biocompatible, biodegradable, well-demonstrated nontoxic for mammals, and very active against viruses, bacteria, fungi, insects and others pest. The coating with chitosan of seeds and leaves protects plants against microbial infections and also induces defense mechanisms in plants to resist diseases.[37] As an agricultural product, it can also be used as either a chemical barrier or a sticker. Moreover, chitosan has been employed as a carrier or support of fertilizers, pesticides, herbicides and insecticides for their controlled release to soil, the environmental impact of these agrochemicals being, consequently, reduced.[112] This section summarizes some of the most important developments in this field.

Chitosan and derivatives used to control plant pathogens have been extensively investigated and their efficacy depends on many parameters, as commented above (degree of deacetylation, concentration, *etc.*) and it is also related to the phatosystem. There are several proposed mechanisms of action of chitosan in reducing plant diseases. Although there is evidence showing a direct activity against pathogens,[56] it seems that chitosan is likely to act indirectly *via* enhancement of host resistance as elicitor of plant defense. Chitosan and derivatives are known to induce and modulate plant response in various plant species against different pathogens. These events include phytoalexins synthesis and accumulation, callose formation, lignification responses and the fabrication of proteinase inhibitors, activation H^+-ATPases and other series of events.[8] Besides, these biopolymers when applied on plant tissue form physical barriers that hinder the pathogen penetration and its spread to healthy tissues. Other processes involve the chelation of nutrients and minerals, preventing pathogens from accessing them. Then, these natural renewable biopolymers, chitosan and derivatives, can be utilized in a number of treatments in agriculture to control and avoid the development of diseases, thus preserving yield and quality.

Seed-coating agent

Many experimental data support the beneficial uses of chitosan in crop protection when used as coating on seeds.[113] Seed coating with chitosan compounds initially leads to antifungal and antibacterial properties at the seed surface. Subsequently, chitosan may penetrate into the seed and cell membranes and influence on the cellular metabolism of the plant.[8,114,115] Chitosan seed coatings significantly reduced root lesions in tomato plants caused by *F. oxysporum*. These studies reveal alterations in the plasma membrane of the fungi, cell-wall thickening, cytoplasm aggregation and hyphal distortion.[116]

Treating rice seeds with lower molecular weight hydrolyzed chitosan induced defense responses against *Pyricularia grisea*.[117] Others studies in rice seed coated with chitosan show that this application accelerates the germination, enhances the resistance in stress conditions, increases the yield as well as induces the immune system of plants.[118,119]

On the other hand, it is reported that the application of chitosan on peanut seeds raises the rate of germination and energy, lipase activity, and gibberellic acid and indole acetic acid levels.[120] Chitosan treatments of wheat seeds substantially improve seed germination compared with untreated cultivars by controlling seed-borne *Fusarium graminearum* infection.[121] The beneficial use of chitosan in soybean seed is found to be the protection from insects, such as agarotis, ypsilon, soybean pod borer, and soybean aphids. In addition, the treatment provokes an increase in germination, plant growth and yield.[122]

Foliar treatments

Foliar application of chitin, chitosan and derivatives to control the growth, spread and development of many diseases have been extensively reported.[8,37] For instance, the effect of foliar treatment of chitosan on the growth and yield has been investigated in okra.[123] Results reveal that chitosan concentrations of 100–125 ppm enhance the growth and development of plants and increase fruit yield in okra. Chitosan is also applied as foliar spray to barley plants and its effectiveness in inducing resistance against *Blumeria graminis f. sp. hordei* was studied.[124] After an induction phase of 3 days, the infection on primary leaf is minimized significantly, 55.5%, due to the induction of oxidative burst and deposition of phenolic compound in treated leaves that creates a hostile environment for fungal spreading. Foliar application can affect, however, the photosynthesis process. Smith and coworkers[125] studied the effect of foliar application of chitin and chitosan oligomers on photosynthesis of maize and soybean. The photosynthetic rate decreases during the first day of application but the rate did not change or even increases in some treatments on subsequent days. This increase was associated with an enhancement of the stomatal conductance and transpiration rate, whereas the intercellular concentration of CO_2 does not differ with respect to the control. On the contrary, the application of high molecular weight compounds generally decreases the net photosynthetic rate. Besides, it was reported that chitosan applied foliarly to pepper plants has an effective antitranspirant effect reducing water use of pepper plants by 26–43%.[126]

Soil amendment

Chitosan-treated soil was extensively reported to enhance plant growth and suppress soil-borne diseases. On one hand, the reduction of soil pathogens likely occurs through the elicitation of plant defense responses. On the other hand, chitosan stimulates the activity of beneficial micro-organisms in the soil, such as bacillus, fluorescent pseudomonas, actinomycetes, mycorrhiza and rhizobacteria.[127,128] For example, the substratum amendment with chitosan reduced *Fusarium acuminatum* and *Cylindrocladium floridanum* fungal infections in forest nurseries.[129] *Aspergillus flavus* was also completely inhibited in field-grown corn and peanut after soil treatment with chitosan.[130] Wild strawberry plants grown in Suppressor®, a commercial peat substrate amended with chitin-containing shellfish waste, resulted in significant plant growth effects and higher resistance against *Phytophthora fragariae* as compared with uninnoculated plants during the first 4 weeks.[128]

Controlled release

In recent years, there has been a great concern about the damage caused to the environment by chemicals used in agriculture, such as fertilizers or pesticides. Therefore, investigations in new strategies for using these compounds are of paramount interest. The application of chitosan, as matrix or complex, for the controlled release of active compounds together with the antimicrobial properties of chitosan have become an attractive alternative in agriculture for the control of viruses, bacteria and fungi.[112]

Chitosan gel beads as well as films containing entrapped herbicide atrazine and the fertilizer urea were prepared by reacting solutions of chitosan with acetic or propionic anhydride in presence of the agrochemicals.[131] The study demonstrates that prepared materials are suitable for long-term controlled release of chemicals. Chitosan-coated atrazine showed an initial rapid release of atrazine that could kill weeds present in the ground before planting the cop, followed by a constant release rate for 7 months. It was also shown regarding the chitosan-coated urea beads that the release period was prolonged related to the uncoated urea control beads. Antrazine and imidacloprid, another pesticide, have been encapsulated into carboxymethyl chitosan/bentonite composite gel to control their release in water and retard their leaching in soil. In both cases the release time was extended to 572 h for atrazine and 24 h for imidacloprid.[132] Moreover, the fabrication of chitosan microcapsules containing the water-soluble herbicide 3-hydroxy-5-methylisoxazole was reported.[133] The use of chitosan presents an advantage that its biodegradability leads to the fertilization of the soil in the ground.

2.5 Applications of Chitosan and its Derivatives in Food

The food industry is probably the major area of commercial application for chitosan and its derivatives. The antimicrobial activity and film-forming property of chitosan along with its biodegradability and biocompatibility make chitosan readily efficient in food preservation for improvement of quality and shelf life of various foods.[34,134–136] Nowadays, there is a growing concern regarding the contaminations associated with food products. Besides, many efforts are currently being carried out to address the environmental impact of discarded packaging materials. These concerns lead to the application of more restrictive safety regulation and promote investigations focused on the development of new improved systems for maintaining food properties. In this sense, there is a growing interest in consuming natural products. Thus, biomacromolecules, such as chitosan, are postulated to play a significant role in the food industry. Chitosan films especially show potential as a food-packaging material and also as an edible film and coating material for the preservation of food. Chitosan has been widely used as natural additives in food, for instance, as a color stabilization agent, emulsifier, antioxidant and gelling agents. Besides, chitosan finds application as support for the immobilization of enzymes used in food, *i.e.* in cheese production or milk clotting. Chitosan also presents dietary fiber properties because it is able to entrap intragastric oil and then precipitates in the small intestine at more neutral pH. In addition, it is reported that chitosan shows the ability to modify carbohydrate absorption. Although there is a variety of food application area as aforementioned, this present section aims at highlighting the applications focused on the improvement of food safety and shelf life, particularly those related to the antimicrobial activity of chitosan.

Food-packaging applications

Currently, many efforts have been conducted to develop innovative packages with the aim of improving quality and safety. In particular, much attention has been paid in the design of antimicrobial active systems.[137] Chitosan is a good candidate for food packaging because of its known biocompatibility, bio-degradability, antibacterial and antioxidant properties along with its film-forming ability.[138,139] Chitosan and its derivatives allow the preparation of films with moderate and successful water permeability for use in food wraps. They are good barriers for oxygen permeation, decreasing the respiration rate and, consequently, the shelf life of fruits and vegetables is prolonged by delaying oxidative, ripening and maturation processes due to reduction of ethylene and carbon dioxide production. Chitosan films are also tough, flexible, transparent and resistant to lipids. As well as its antimicrobial properties, the film properties also depend on the characteristic of chitosan. For instance, the type of acid used for film preparation has a significant effect on its properties, *i.e.* films prepared from acetic and formic acid were hard and brittle, while those obtained in lactic and citric acids were soft and stretchable.[140] The molecular weight of chitosan obviously affects the film properties, typically high molecular weights result in an increase of the strength, whereas the water vapor permeability is not significantly affected.[141] In this sense, in the food packaging industry multilayered systems are extensively used because, in general, they have better mechanical properties and barrier efficiencies. However, their fabrication is tedious, requiring many processing steps. As an alternative, in order to improve the properties of the films, such as mechanical properties after wetting or permeability, chitosan has been modified and blended with other polymers. For example the incorporation of methoxy poly(ethylene glycol)-*b*-poly(ε-caprolactone) diblock copolymer nanoparticles into chitosan enhanced the tensile strength of the films.[142] Biodegradable poly(lactic acid)/chitosan blends were prepared to improve the water vapor barrier.[143] Chitosan has been blended with EVOH, classically used in food packaging, with the objective of enhancing water-barrier properties.[144] The antimicrobial activity of chitosan is effectively expressed in aqueous systems; in film application, chitosan is, however, insoluble and only inhibits micro-organisms in direct contact.[145]

Figure 2.8 Appearance of strawberries coated with modified chitosan-based formulation containing limonene and Tween®80 emulsifier.[148]
Reprinted with permission from *Food Res. Inter.*, **44**, 1–6, Copyright 2011, Elsevier.

In this regard, a very interesting alternative is the incorporation of essential oils with antimicrobial activity into chitosan-based films.[146] The addition of essential oils not only enhances the antimicrobial activity but also reduces water permeability. Zivanovic *et al.*[147] reported that the addition of oregano essential oil into the chitosan films on processed meat decreased water permeability and increased the elasticity of the films. The incorporation of only 1–2% oregano essential oil was able to attain a 4-log reduction of *Listeria monocytogenes* and *E. coli* on bologna slices. Vu *et al.*[148] recently evaluated the antimicrobial capacities against molds and total flora isolated from strawberries coated with modified chitosan formulations with different essential oils. Red thyme (RT) and oregano extract (OE) presented strong bioactive agents, whereas limonene (LIM) and peppermint (PM) had lower antimicrobial properties. RT, PM and LIM were found to be the most efficient preservative agents in strawberries during storage at least up to 14 days at 4 °C. Formulations based on modified chitosan containing LIM in the presence of Tween®80 provided a better preservation than other formulations (Figure 2.8).

Edible-film applications

Edible films and coatings have been investigated and used for their ability of maintaining food quality, although this does not mean to totally replace the synthetic packaging films. Edible films also must exhibit selective properties to control the exchange of organic vapors, water vapor, lipids, additives and gases such as oxygen, carbon dioxide, or nitrogen.[34] For instance, the water-barrier efficiency of films is desirable to delay the surface dehydration; the control of oxygen exchange moderates the ripening of fruits, while the control of organic vapor may retain aroma compounds. Besides, edible coatings can become active films when they encapsulate or carry food additives or ingredients. That is the case of the addition of pigments or light absorbers that could reduce the effect of UV light exposition, which involves radical air reactions in foods. Concerning the antimicrobial properties of chitosan, there is a wide literature based on experiments *in vitro*. The application of chitosan as a film directly on food may involve, however, interactions between chitosan and food compounds, such as protein, fat, minerals, vitamins and salts, and consequently a change in the antimicrobial activity of chitosan. Devlieghere and coworkers[149] investigated the influence of a variety of food components on the antimicrobial action of chitosan against *Candida lambica*. They showed that starch, whey protein, and salt had a negative effect on the antimicrobial activity, whereas oil had no influence.

Although various methods are usually employed to prepare homogeneous chitosan edible coatings for food applications,[135] indeed a solution casting method is the one most employed. Agullo *et al.*[34] applied chitosan films on the surface of immature tomatoes, pears and Anquito squash from chitosan solution in acetic acid. The study was carried out during storage at 20 °C analyzing several parameters, such as weight, color, acidity or glucose and ethanol production, among others. It was stated that the food coated with chitosan shows a better quality as compared with the control samples. The same authors studied the use of chitosan as an edible coating in prepizza to evaluate its effect against *Alternaria sp*, *Penicillium sp* and *Cladosporium sp* fungi, exhibiting similar action as compared with synthetic preservatives. El Ghaouth and coworkers reported that chitosan coating was able to prolong the storage life and control decay of strawberry fruits caused by *Botrytis cinerea* and *Rhizopus sp*.[150,151] Then, other authors also proved the benefits of chitosan coating for storage of fresh strawberries.[152,153] The effectiveness of chitosan-based edible coatings was assessed in other types of food rather than fruit and vegetables. For instance, it was evaluated on model agar medium against *S. aureu*s and *L. monocytogenes* and then corroborated on real Emmental cheese (Figure 2.9).[13,154]

Chitosan coatings have been also used to extend the quality of bread, retarding the staling and reducing the micro-organism growth. The application of only 1% of chitosan in acetic acid on the surface of dough prolongs the shelf life of baguettes up to 36 h, which is much higher than the control, 12 h.[155] The storage of eggs usually encounters several problems, such as weight loss, microbial growth and interior deterioration. To solve these limitations chitosan coatings have been proposed to increase the shelf life of eggs due to the formation of a barrier for transfer of water vapor and gas from the albumen through the shell. Again the coating was carried out by dissolving chitosan in acetic acid[156] using, in some cases, glycerol as plasticizer.[157] In both cases, eggs coated with 1–3% of chitosan preserve their quality several weeks more than the untreated eggs.

Additives in foods

As aforementioned for edible films, chitosan is a human-safe biopolymer suitable for oral administration that makes it extremely interesting in the food

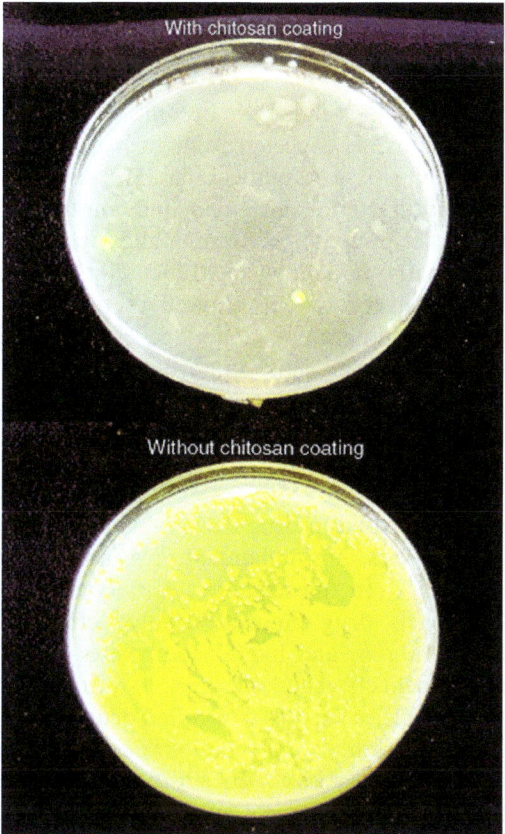

Figure 2.9 Development of *Pseudomonas aeruginosa* bacteria from Emmental cheese with and without chitosan coating inoculated on agar plate after 2 days.[154] Reprinted with permission from *J. Food Sci.*, **68**, 2788–2792, Copyright 2003, Wiley-VCH.

industry. Chitosan has been directly used as an additive in a variety of foods due to its antifungal and antibacterial properties, including meat, seafood, juices, milk, mayonnaise, jelly, *etc.*[134] For example, in apple juice, the growth of several yeasts and molds associated with food spoilage including *Zygosaccharomyces bailii*, *Schizosaccharomyces pombe* and *Saccharomyces cerevisiae* is inhibited by the addition of chitosan glutamate in a concentration ranging from 0.1 to 5 g/L.[28] The same authors, Roller and Covill[158] investigated the efficacy of chitosan glutamate against *Salmonella Enteritidis*, *Z. bailii* and *Lactobacillus fructivorans* bacteria and yeast in mayonnaise. In this case, the addition of 3 g/L of powered chitosan combined with acetic acid is enough to inactivate 10^5–10^6 of colony-forming units (CFU). Ha *et al.* studied the use of water-soluble chitosan in processed milk to reduce the bacterial and yeast spoilage that was completely inhibited during storage for 15 days at 4 and 10 °C, contrary to that observed in milk without chitosan.[159]

2.6   Applications of Chitosan and its Derivatives in Biomedical Area

Chitosan and derivative compounds have become very interesting and suitable in biomedical applications, as expected from their excellent biocompatibility and nontoxicity to living tissues, biodegradability and low immunogenicity together with their antibacterial, antifugal and antitumor activity.[84,160] In addition, this biopolymer presents gel-forming and high adsorption capacities. Over the last decade, many studies have been focused on chitosan as a potential biomaterial for drug-delivery systems as well as to create scaffolds for re-generative medicine, including areas such as gene therapy, tissue engineering and wound healing. Next, the most important biomedical applications of chitosan and its derivatives are discussed.

Drug-delivery system

Chitosan has been widely used in drug delivery as a carrier for various active agents including antibiotics, anti-inflammatories, anticancer drugs, proteins and growth factors.[161–163] In this regard, a variety of different systems are utilized such as hydrogels, drug conjugates, polyelectrolyte complexes, films, microparticles, textile fabrics, and hybrid composites. Chitosan has been ef-fectively used in drug delivery mainly because its primary amino groups that are responsible for properties such as controlled drug release through electro-static interaction, transfection, *in situ* gellation, permeation enhancement, and efflux pump inhibitory properties.[162] In addition, chemical modification of chitosan can even improve these properties. At this moment there is a product on the market as a delivery system for topical use on finger nails and toe nails, containing a chitosan derivative (hydroxypropyl chitosan) registered as Ciclopoli®. Nevertheless, there are many scientific investigations related to the use of drug-delivery systems comprising chitosan for various application sites. For example, chitosan-based delivery vehicles have been utilized for ocular therapeutics.[164] This is because first, chitosan exhibits bioadhesiveness to the ocular surfaces, and also penetration-enhancing properties across the mucosal epithelia. In addition to its pseudoelastic and viscoelastic properties desirable for ocular drug administration, the antibacterial activity is also an advan-tage.[165] The incorporation of other biomaterials into chitosan leads to the creation of new materials with improved properties. In this sense, chitosan was mixed with liposomes to create nanocomplexes that can interact with the ocular tissue and penetrate into the cells without compromising the integrity of the cell membrane.[166] Chitosan-based systems have been also used for nasal delivery, for instance, of therapeutic proteins, such as insulin[167] or even antigen vac-cines.[168] In this case, the bioavailability of nasally administered drugs is nor-mally very low with short residence time as a consequence of the low membrane permeability and secretion in nasal cavities. The use of material with mucoadhesive properties and the ability of enhancing permeation, such as chitosan might be an interesting alternative. As a local delivery system, chitosan-based compounds have also been applied to the inner ear. More specifically, chitosan-loaded neomycin was successfully injected into the ear of albino guinea pigs with effective results.[169] Concerning oral drug delivery, the direct preparation of chitosan tablets containing the drugs is limited because at pH above 6.5 chitosan precipitates. This reduces its applicability to drugs to the proximal segment of the gastrointestinal tract. In contrast, chitosan derivatives with higher cationic character do not precipitate within the pH range of the gastrointestinal tract.[170] However, in case of chitosan delivery system, most of the cases are crosslinked nano/microparticles.[171–173]

Gene therapy

In gene therapy, genetic material is used to treat or prevent disease. Currently, this technique is applied in many different illnesses such as cancer, acquired immunodeficiency syndrome (AIDS), and cardiovascular diseases. Many investigations are involved in the development of ideal gene-delivery carriers because some properties are necessary in order to transfer foreign genetic material into the human body in a successful way. First, an ideal carrier should be harmless and without provoking any immune responses. In addition, it should be able to protect the DNA until it reaches its target from the degradation by serum nucleases or from the interaction with blood components. Nowadays, the transport of DNA to cells is achieved by using viral and nonviral vectors or as naked DNA in the case of easily accessible tissues. Although viral vectors exhibit a high transfection rate and a rapid transcription, they present some drawbacks, such as toxicity, immune and inflammatory responses along with the fact that only small sequences of DNA can be inserted in the virus genome. On the other hand, nonviral vectors normally present minimal immune response, are targetable and stable. These systems include cationic polymers that can interact electrostatically with negatively charged DNA and form complexes. Of these, chitosan and derivatives have received great attention because of its biocompabitibility and low immunogenicity.[174] In general, gene-delivery systems are based on high molecular weight chitosans, which are very efficient after lung administration *in vivo*. However, they present low solubility, high viscosity and slow dissociation and thus slow release of plasmid DNA. More easily dissociated complexes were prepared by Köping-Höggard and coworkers based on highly defined oligomers.[175] Many other examples are found in the literature, for instance, in one study the *in vivo* efficacy of chitosan nanoparticles loaded with plasmid DNA for nasal mucosal immunization against hepatitis B was investigated. It seems that the prepared nanoparticles were capable of inducing systemic, mucosal and cellular immune responses upon nasal administration, therefore, with potential to be used as DNA vaccine carrier.[176]

Wound healing

Burns injuries and wounds produce locally a state of immunosuppression that is particularly susceptible to bacterial infections. *Staphylococci* and *streptococci* are the most commonly encountered pathogenic organisms acquired in superficial wounds. When bacteria proliferate in a wound not only sepsis occurs but also the processes involved in healing are seriously affected. In this context, an ideal dressing for wound-healing applications should protect wounds from infection and also promote healing. Chitin and chitosan have been extensively investigated for preventing and treating infections due to their intrinsic antimicrobial properties, and also its ability to deliver extrinsic antimicrobial agents and other compounds, such as growth factors into the injured area.[177,178] In addition, it is known that chitosan and related compounds have an effect on fibroblast activation, stimulation of cytokines production and promotion of granulation tissue formation, all these processes playing a crucial role in wound healing. These biopolymers, especially chitin, exhibit hemostatic activity and possess the ability to be bound with red blood cells, which rapidly causes bleeding to stop.[179] Another beneficial characteristic of chitosan-based dressing material is its semipermeability, preventing dehydration and contamination of the wound. Several *in vivo* (animals and clinical) works have been reported in the use of chitin, chitosan and derivatives for wound healing. Chitosan scaffolds loaded with basic fibroblast growth factor contained in gelatin microparticles were developed and evaluated in an aged mouse model.[180] It was stated that chitosan combined with growth factors accelerates wound closure in chronic ulcers as compared to control for up to 7 days. Boynuegri *et al.* tested the action of chitosan gel on periodontal regeneration in a total of 20 patients with chronic periodontitis.[181] The gel was also applied in combination with demineralized bone matrix or collagenous membrane. The radiological study reveals that all treated groups showed substantial bone fills in comparison with the nontreated control group. Despite the large number of investigations in this field, there are only few chitosan-based wound dressings at the market, *i.e.* Chitopack C® from Eisai Corporation, Ltd, only available in Japan since it is not Community European certificated nor by the Federal Drug Administration (FDA), and HemCon® from HemCon Medical Technologies Inc. However, the satisfactory results obtained in many studies in this area suggest that chitosan will be established as an important compound in the management of wounds and burns.

Tissue engineering

Tissue engineering is a multidisciplinary field involving biology, medicine, and engineering that aims to improve and replace biological function. Although the term covers a wide range of applications, much of these efforts have been encouraged by the restoring, maintaining, or enhancing tissue and organ function. Chitosan and derivatives have been postulated as very promising candidates as biomaterials for tissue engineering, first because of their aforementioned antibacterial activity, biodegradability, low immunogenicity and excellent biocompatibility.[182] Besides, many interactions with mammalian tissues are reported, presenting a positive effect for the reparation and regeneration of the tissue, including osteoblasts, fibroblasts, macrophages and keratinocytes. Chitosans can also electrostatically interact with glycosaminoglycans (heparin and heparan sulfate) and proteoglycans that it is well known that control a large number of growth factors and cytokines. Therefore, the preparation of scaffolds incorporating a chitosan–glycosaminoglycans complex may help to retain and concentrate growth factors secreted by colonizing cells.[183,184] Chitosan-based materials have been utilized in many tissues including bone, cartilage, liver, nerve, *etc.* In bone-tissue engineering chitosan has been demonstrated to promote cell growth as well as mineral rich matrix deposition by osteoblasts in culture.[185] For instance, chitosan scaffolds reinforced by calcium phosphates were fabricated in macroporous structures that also favor osteoconduction.[186] In another study, chitosan–hydroxyapatite nanocomposites were performed by *in situ* hybridization to obtain a potential material as internal fixation of bone fracture.[187] On the basis of the structural similarity of chitosan to glycosaminoglycans found in articular cartilage and liver, chitosan-based compounds have been chosen as a scaffolding material in cartilage[188] as well as liver engineering.[189] Chitosan and chitin are also suitable for nerve regeneration,[190] capable of supporting nerve cell adhesion and neurite outgrowth.

2.7 Conclusions

In recent years, the enormous population growth and environmental pollution have increased the interest of researchers in sustainable materials. Biopolymers, such as polysaccharides derived from agricultural or marine resources, are especially attractive because of their abundance, biocompatibility, and biodegradability, and, therefore, ecological safety. From all of them, chitosan is one of the most promising biopolymers not only as a result of its high reactivity but especially for its antimicrobial properties that have provided its huge potential in a wide range of field, such as biomedical, agricultural, food industry and many others, for instance, textiles or cosmetics. Although there are already several chitosan-based products on the market, additional effort and further research are needed in the development of new approaches, technologies and materials for the implementation of this type of biopolymer in the industry.

Acknowledgements

A. Muñoz-Bonilla gratefully acknowledges MINECO for her Juan de la Cierva postdoctoral contract. MINECO is fully acknowledged for financial support (MAT2010-17016 and MAT2010-19883).

References

1. L. H. Han, Y, Kimura and H. Okaauda, *Int. J. Obes. Relat. Metab. Disord.*, 1999, **23**, 174–179.
2. G. J. Tsai, S. L. Zhang and P. L. Shieh, *J. Food Prot.*, 2004, **67**, 396–398.

3. A. R. C. Duarte, J. F. Mano and R. L. Reis, *J. Supercrit. Fluids*, 2010, **54**, 282–289.
4. N. Csaba, K. M. Hoggard and M. J. Alonso, *Int. J. Pharm.*, 2009, **382**, 205–221.
5. M. M. Beppu, R. S. Vieira, C. G. Aimoli and C. C. Santana, *Science*, 2007, **30**, 126–130.
6. H. Homayoni, S. A. H. Ravandi and M. Valizadeh, *Carbohydr. Polym.*, 2009, **77**, 656–661.
7. F. Shahidi, J. K. V. Arachchi and Y. J. Jeon, *Trends Food Sci. Technol.*, 1999, **10**, 37–51.
8. E. I. Rabea, M. E.-T. Badawy, C. V. Stevens, G. Smagghe and W. Steurbaut, *Biomacromolecules*, 2003, **4**, 1457–1465.
9. S.–H. Lim and S. M. Hudson, *J. Macromol. Sci., Polym. Rev.*, 2003, **C43**, 223–269.
10. M. Kong, X. G. Chen, K. Xing and H. J. Park, *Int. J. Food Microbiol.*, 2010, **144**, 51–63.
11. M. S. Benhabiles, R. Salah, H. Lounici, N. Drouiche, M. F. A. Goosen and N. Mameri, *Food Hydrocolloid.*, 2012, **29**, 48–56.
12. M. Ignatova, K. Starbova, N. Markova, N. Manolova and I. Rashkov, *Carbohydr. Res.*, 2006, **341**, 2098–2107.
13. V. Coma, A. Martial-Gros, S. Garreau, A. Copinet, F. Salin and A. Deschamps, *J. Food Sci.*, 2002, **67**, 1162–1169.
14. Y. C. Chung, Y. P. Su, C. C. Chen, G. Jia, H. L. Wang, J. C. G. Wu and J. G. Lin, *Acta Pharmacol. Sin.*, 2004, **25**, 932–936.
15. J. Je and S. Kim, *J. Agric. Food Chem.*, 2006, **54**, 6629–6633.
16. D. H. Young and H. Kauss, *Plant Physiol.*, 1983, **73**, 698–702.
17. H. Tanga, P. Zhang, T. L. Kieft, S. J. Ryan, S. M. Baker, W. P. Wiesmann and S. Rogelj, *Acta Biomater.*, 2010, **6**, 2562–2571.
18. D. Raafat, K. von Bargen, A. Haas and H. G. Sahl, *Appl. Environ. Microbiol.*, 2008, **74**, 3764–3773.
19. N. R. Sudarshan, D. G. Hoover and D. Knorr, *Food Biotechnol.*, 1992, **6**, 257–272.
20. S. Tokura, K. Ueno, S. Miyazaki and N. Nishi, *Macromol. Symp.*, 1997, **120**, 1–9.
21. Z. M. Zhong, R. G. Xing, S. Liu, L. Wang, S. B. Cai and P. C. Li, *Carbohydr. Res.*, 2008, **343**, 566–570.
22. H. K. No, N. Y. Park, S. H. Lee and S. P. Meyers, *Int. J. Food Microbiol.*, 2002, **74**, 65–72.
23. M. S. M. Eldin, E. A. Soliman, A. I. Hashem and T. M. Tamer, *Trends Biomater. Artif. Organs.*, 2008, **22**, 121–133.
24. X. H. Wang, Y. M. Du and H. Liu, *Carbohydr. Polym.*, 2004, **56**, 21–26.
25. X. F. Liu, Y. L. Guan, D. Z. Yang, Z. Li and K. D. Yao, *J. Appl. Polym. Sci.*, 2001, **79**, 1324–1335.
26. X. Liu, L. Song, L. Li, S. Li and K. Yao, *J. Appl. Polym. Sci.*, 2007, **103**, 3521–3528.

27. R. G. Cuero, G. Osuji and A. Washington, *Biotechnol. Lett.*, 1991, **13**, 441–444.
28. S. Roller and N. Covill, *Int. J. Food Microbiol.*, 1999, **47**, 67–77.
29. E. Guibal, *Sep. Purif. Technol.*, 2004, **38**, 43–74.
30. X. Wang, Y. Du, L. Fan, H. Liu and Y. Hu, *Polym. Bull.*, 2005, **55**, 105–113.
31. A. B. V. Kumar, M. C. Varadaraj, L. R. Gowda and R. N. Tharanathan, *Biochem. J.*, 2005, **391**, 167–175.
32. Z. Guo, R. Chen, R. Xing, S. Liu, H. Yu, P. Wang, C. Li and P. Li, *Carbohydr. Res.*, 2006, **341**, 351–354.
33. J. García-Rincón, J. Vega-Pérez, M. G. Guerra-Sánchez, A. N. Hernández-Lauzardo, A. Peña-Díaz and M. G. Velázquez-Del Valle, *Pest. Biochem. Physiol.*, 2010, **97**, 275–278.
34. E. Agullo, M. S. Rodríguez, V. Ramos and L. Albertengo, *Macromol. Biosci.*, 2003, **3**, 521–530.
35. A. N. Hernández-Lauzardo, S. Bautista-Banos, M. G. Velázquez-del Valle, M. G. Méndez-Montealvo, M. M. Sánchez-Rivera and L. A. Bello-Pérez, *Carbohydr. Polym.*, 2008, **73**, 541–547.
36. M. J. Giner, S. Vegara, L. Funes, N. Martí, D. Saura, V. Micol and M. Valero, *J. Sci. Food Agric.*, 2012, **92**, 1917–1923.
37. A. El Hadrami, L. R. Adam, I. El Hadrami and F. Daayf, *Mar. Drugs*, 2010, **8**, 968–987.
38. R. K. Bai, M. Y. Huang and Y. Y. Jiang, *Polym. Bull.*, 1988, **20**, 83–88.
39. A. El Ghaouth, J. L. Smilanick, G. E. Brown, A. Ippolito, M. Wisniewski and C. L. Wilson, *Plant Dis.*, 2000, **84**, 243–248.
40. S. N. Chircov, A. V. Ilina, N. A. Surgucheva, E. V. Letunova, Y. A. Varitsev, N. Y. Tatarinova and V. P. Varlamov, *Russ. J. Plant Physiol.*, 2001, **48**, 774–779.
41. S. N. Chircov, *Appl. Biochem. Microbiol.*, 2002, **38**, 1–8.
42. W. Wang, S.-X. Wang and H.-S. Guan, *Mar. Drugs*, 2012, **10**, 2795–2816.
43. S. N. Kulikov, S. N. Chirkov, A. V. Il'ina, S. A. Lopatin and V. P. Varlamov, *Appl. Biochem. Microbiol.*, 2006, **42**, 200–203.
44. V. N. Davydova, V. P. Nagorskaia, V. I. Gorbach, A. A. Kalitnik, A. V. Reunov, T. F. Solov'eva and I. M. Ermak, *Appl. Biochem. Microbiol.*, 2011, **47**, 113–118.
45. Z. M. Kochkina and S. N. Chirkov, *Microbiology*, 2000, **69**, 258–260.
46. Z. M. Kochkina and S. N. Chirkov, *Microbiology*, 2001, **70**, 706–710.
47. W. G. Liu and K. D. Yao, *J. Control. Release*, 2002, **83**, 1–11.
48. X. W. Su, S. Zivanovic and D. H. D'Souza, *J. Food Protect.*, 2009, **72**, 2623–2628.
49. T.-S. Vo and S.-K. Kim, *Mar. Drugs*, 2010, **8**, 2871–2892.
50. R. Davis, S. Zivanovic, D. H. D'Souza and M. P. Davidson, *Food Microbiol.*, 2012, **32**, 57–62.
51. S. I. Nishimura, H. Kai, K. Shinada, T. Yoshida, S. Tokura, K. Kurita, H. Nakashima, N. Yamamoto and T. Uryu, *Carbohydr. Res.*, 1998, **306**, 427–433.

52. J. L. Dou, C. Y. Tan, Y. G. Du, X. F. Bai, K. Y. Wang and X. J. Ma, *Carbohydr. Polym.*, 2007, **69**, 209–213.
53. H. Pospieszny and J. G. Atabekov, *Plant Sci.*, 1989, **62**, 29–31.
54. H. Pospieszny and J. Giebel, *Peroxidase activity is related to the resistance against viruses induced by chitosan*. In Chitin Enzymology; R. A. A. Muzzarelli, ed. Grottammare, Atec Edizioni, 1996, vol. 2, pp. 379–383.
55. N. A. Surguchova, Y. A. Varitsev and S. N. Chirkov, *J. Rus. Phytopathol. Soc.*, 2000, **1**, 59–62.
56. S. N. Kulikov, S. N. Chirkov, A. V. Il'ina, S. A. Lopatin and V. P. Varlamov, *Prik. Biokhim. Mikrobiol.*, 2006, **42**, 224–228.
57. C. A. Ryan and E. E. Farmer, *Annu. Rev. Plant Physiol. Plant Mol. Biol.*, 1991, **42**, 651–674.
58. S. N. Chirkov, N. A. Surgucheva and I. G. Atabekov, *Dokl. Akad. Nauk*, 1995, **341**, 836–838.
59. N. Liu, X.-G. Chen, H.-J. Park, C.-G. Liu, C.-S. Liu, X.-H. Meng and L.-J. Yu, *Carbohydr. Polym.*, 2006, **64**, 60–65.
60. Y.-J. Jeon, P.-J. Park and S.-K. Kim, *Carbohydr. Polym.*, 2001, **44**, 71–76.
61. W. K. Kyung, R. L. Thomas, L. Chan and H. J. Park, *J. Food Protect.*, 2003, **66**, 1495–1498.
62. D. V. Gerasimenko, I. D. Avdienko, G. E. Bannikova, O. Y. Zueva and V. P. Varlamov, *Appl. Biochem. Microbiol.*, 2004, **40**, 253–257.
63. Y. Andres, L. Giraud, C. Gerente and P. Le Cloirec, *Environ. Technol.*, 2007, **28**, 1357–1363.
64. T. Hongpattarakere and O. Riyaphan, *Songklanakarin J. Sci. Technol.*, 2008, **30**, 1–9.
65. G. J. Tsai, W. H. Su, H. C. Chen and C. L. Pan, *Fisheries Sci.*, 2002, **68**, 170–177.
66. T. Takahashia, M. Imaia, I. Suzukia and J. Sawai, *Biochem. Eng. J.*, 2008, **40**, 485–491.
67. H. Mellegård, S. P. Strand, B. E. Christensen, P. E. Granum and S. P. Hardy, *Int. J. Food Microbiol.*, 2011, **148**, 48–54.
68. G. J. Tsai and W. H. Su, *J. Food Protect.*, 1999, **62**, 239–243.
69. Y. C. Chung, C. L. Kuo and C. C. Chen, *Bioresour. Technol.*, 2005, **96**, 1473–1482.
70. C.-Y. Chen and Y.-C. Chung, *J. Appl. Oral Sci.*, 2012, **20**, 620–627.
71. Y. C. Chung, H. L. Wang, Y. M. Chen and S. L. Li, *Bioresour. Technol.*, 2003, **88**, 179–184.
72. D. Raafat and H. G. Sahl, *Microbial Biotech.*, 2009, **2**, 186–201.
73. D. H. Young, H. Köhle and H. Kauss, *Plant Physiol.*, 1982, **70**, 1449–1454.
74. H. Tan, R. Ma, C. Lin, Z. Liu and T. Tang, *Int. J. Mol. Sci.*, 2013, **14**, 1854–1869.
75. Z. Y. Guo, R. G. Xing, S. Liu, Z. M. Zhong, X. Ji, L. Wang and P. C. Li, *Int. J. Food Microbiol.*, 2007, **118**, 214–217.
76. T. Xu, M. Xin, M. Li, H. Huang, S. Zhou and J. Liu, *Carbohydr. Res.*, 2011, **346**, 2445–2450.

77. Z. Jia, D. Shen and W. Xu, *Carbohydr. Res.*, 2001, **333**, 1–6.
78. C. H. Kim, J. W. Choi, H. J. Chun and K. S. Choi, *Polym. Bull.*, 1997, **38**, 387–394.
79. W. Sajomsang, S. Tantayanon, V. Tangpasuthadol and W. H. Daly, *Carbohydr. Polym.*, 2008, **72**, 740–750.
80. W. L. Chi, C. Q. Qin, L. T. Zeng, W. Li and W. J. Wang, *Appl. Polym. Sci.*, 2007, **103**, 3851–3856.
81. T. Xu, M. Xin, M. Li, H. Huang and S. Zhou, *Carbohydr. Polym.*, 2010, **81**, 931–936.
82. D. Fu, B. Han, W. Dong, Z. Yang, Y. Lv and W. Liu, *Biochem. Biophys. Res. Commun.*, 2011, **408**, 110–114.
83. X.-G. Chen, Z. Wang, W.-S. Liu and H.-J. Park, *Biomaterials*, 2002, **23**, 4609–4614.
84. L. Upadhyaya, J. Singh, V. Agarwal and R. P. Tewari, *Carbohydr. Polym.*, 2013, **91**, 452–466.
85. R. A. A. Muzzarelli, *Carbohydr. Polym.*, 1988, **8**, 1–21.
86. R. A. A. Muzzarelli, C. Muzzarelli, R. Tarsi, M. Miliani, F. Gabbanelli and M. Cartolari, *Biomacromolecules*, 2001, **2**, 165–169.
87. C. H. Kim and K. S. Choi, *J. Ind. Eng. Chem.*, 1998, **4**, 19–25.
88. R. A. A. Muzzarelli, R. Tarsi, O. Filippini, E. Giovanetti, G. Biagini and P. E. Varaldo, *Antimicrob. Agents Ch.*, 1990, **34**, 2019–2023.
89. C. S. Chen, J. C. Su, G. J. Tsai, *Antimicrobial effect and physical properties of sulfonated chitosan*. In Advances in Chitin Science; R. H. Chen and H. C. Chen, ed. Rita Advertising Co. Ltd., Taiwan, 1998, Vol. III, 273–277.
90. C. S. Chen, J. C. Su, G. J. Tsai, *Antimicrobial effect and physical properties of sulfonbenzoyl chitosan*. In Advances in Chitin Science; R. H. Chen and H. C. Chen, ed. Rita Advertising Co. Ltd., Taiwan, 1998, Vol. III, 278–282.
91. P. Suvannasara, K. Juntapram, N. Praphairaksit, K. Siralertmukul and N. Muangsin, *Carbohydr. Polym.*, 2013, **94**, 244–252.
92. Y. Qin, S. Liu, R. Xing, H. Yu, K. Li, X. Meng, R. Li and P. Li, *Carbohydr. Polym.*, 2012, **89**, 388–393.
93. M. M. Fernandes, A. Francesko, J. Torrent-Burgués and T. Tzanov, *React. Funct. Polym.*, 2013, **73**, 1384–1390.
94. B. Han, Y. Wei, X. Jia, J. Xu and G. Li, *J. Appl. Polym. Sci.*, 2012, **125**, E143–E148.
95. G. Geisberger, E. B. Gyenge, D. Hinger, A. Kach, C. Maake and G. R. Patzke, *Biomacromolecules*, 2013, **14**, 1010–1017.
96. M. Eweis, S. S. Elkholy and M. Z. Elsabee, *Int. J. Biol. Macromol.*, 2006, **38**, 1–8.
97. M. Yalpani, F. Johnson, L. E. Robinson, *Antimicrobial activity of some chitosan derivatives*. In Advances in Chitin and Chitosan; C. J. Brine, P. A. Sandford, J. P. Zikakis, ed. Elsevier Science Publishers Ltd., London, New York, 1992, 543–548.

98. T.-C. Yang, C.-C. Chou and C.-F. Li, *Int. J. Food Microbiol.*, 2005, **97**, 237–245.
99. K. Kurita, T. Kojima, Y. Nishiyama and M. Shimojoh, *Macromolecules*, 2000, **33**, 4711–4716.
100. S. I. Park, K. S. Marsh and P. Dawson, *Meat Sci.*, 2010, **85**, 493–499.
101. P. Fernández-Saiz, C. Soler, J. M. Lagarón and M. J. Ocio, *Int. J. Food Microbiol.*, 2010, **137**, 287–294.
102. S. Tripathi, G. K. Mehrotra and P. K. Dutta, *Int. J. Biol. Macromol.*, 2009, **45**, 372–376.
103. F. Sebastien, G. Stephane, A. Copinet and V. Coma, *Carbohydr. Polym.*, 2006, **65**, 185–193.
104. C. M. Shih, Y. T. Shieh and Y. K. Twu, *Carbohydr. Polym.*, 2009, **78**, 169–174.
105. S. Xin, X. Li, Z. Ma, Z. Lei, J. Zhao, S. Pan, X. Zhou and H. Deng, *Carbohydr. Polym.*, 2013, **92**, 1880–1886.
106. S.-I. Park, M. A. Daeschel and Y. Zhao, *J. Food Sci.*, 2004, **69**, M215–M221.
107. L.-H. Li, J.-C. Deng, H.-R. Deng, Z.-L. Liu and L. Xin, *Carbohydr. Res.*, 2010, **345**, 994–998.
108. L.-H. Li, J.-C. Deng, H.-R. Deng, Z.-L. Liu and X.-L. Li, *Chem. Eng. J.*, 2010, **160**, 378–382.
109. F. A. Al-Sagheer and S. Merchant, *Carbohydr. Polym.*, 2011, **85**, 356–362.
110. D. Archana, B. K. Singh, J. Dutta and P. K. Dutta, *Carbohydr. Polym.*, 2013, **95**, 530–539.
111. K. Kavitha, S. Sutha, M. Prabhu, V. Rajendran and T. Jayakumar, *Carbohydr Polym.*, 2013, **93**, 731–739.
112. O. Cota-Arriola, M. O. Cortez-Rocha, A. Burgos-Hernández, J. M. Ezquerra-Brauer and M. Plascencia-Jatomea, *J. Sci. Food Agric.*, 2013, **93**, 1525–1536.
113. M. E. I. Badawy and E. I. Rabea, *Int. J. Carbohydr. Chem.*, 2011, **2011**, Article ID 460381, pages 29.
114. L. A. Hadwiger, B. W. Fristensky and R. Riggleman, *Chitosan a natural regulator inplant–fungal pathogen interactions increases crop yield.* In Advances in Chitin, Chitosan and Related enzymes; G. Skjak-Braek, T. Anthonsen, P. Sanford and J. P. Zikakis, ed. Academic Press, London, 1984, pp. 227–236.
115. L. A. Hadwiger, *Plant Sci.*, 2013, **208**, 42–49.
116. N. Benhamou, P. J. Lafontaine and M. Nicole, *Phytopathology*, 1994, **84**, 1432–1444.
117. A.T. Rodriguez, M. A. Ramírez, R. M. Cárdenas, A. N. Hernández, M. G. Velázquez and S. Bautista, *Pestic. Biochem. Physiol.*, 2007, **89**, 206–215.
118. S. L. Ruan and Q. Z. Xue, *Acta Agron. Sinica*, 2002, **28**, 803–808.
119. S. Boonlertnirun, C. Boonraung and R. Suvanasara, *J. Met. Mater. Miner.*, 2008, **18**, 47–52.

120. Y. G. Zhou, Y. D. Yang, Y. G. Qi, Z. M. Zhang, X. J. Wang and X. J. Hu, *J. Peanut Sci.*, 2002, **31**, 22–25.
121. M. V. Reddy, J. Arul, P. Angers and L. Couture, *J. Agric. Food Chem.*, 1999, **47**, 1208–1216.
122. D. Zeng, S. Luo, and R. Tu, *Int. J. Carbohydr. Chem.*, 2012, **2012**, Article ID 104565, pages 5.
123. M. M. A. Mondall, M. A. Malek, A. B. Puteh, M. R. Ismail, M. Ashrafuzzaman and L. Naher, *Aus. J. Crop Sci.*, 2012, **6**, 918–921.
124. F. Faoro, D. Maffi, D. Cantu and M. Iriti, *BioControl*, 2008, **53**, 387–401.
125. W. Khan, B. Prithiviraj and D. L. Smith, *Photosynth. Res.*, 2002, **40**, 621–624.
126. M. Bittelli, M. Flury, G. S. Campbell and E. J. Nichols, *Agric. Forest Meteorol.*, 2001, **107**, 167–175.
127. A. A. Bell, J. C. Hubbard, L. Liu, R. M. Davis and K. V. Subbarao, *Plant Dis.*, 1998, **82**, 322–328.
128. J. G. Murphy, S. M. Rafferty and A. C. Cassells, *Appl. Soil Ecol.*, 2000, **15**, 153–158.
129. P. Laflamme, N. Benhamou, G. Bussiéres and M. Dessureault, *Can. J. Bot.*, 2000, **77**, 1460–1468.
130. A. J. El Ghaouth and A. Asselin, *Potential uses of chitosan in postharvest preservation of fruits and vegetables.* In Advances in Chitin and Chitosan; C. J. Brine, P. A. Sandford, and J. P. Zikakis, ed. Elsevier, Amsterdam, The Netherlands, 1992, pp. 440.
131. M. A. Teixeira, W. J. Paterson, E. J. Dunn, Q. Dunn, B. K. Li and M. F. Goosen, *Ind. Eng. Chem. Res.*, 1990, **29**, 1205–1209.
132. J. Li, J. Yao, Y. Li and Y. Shao, *J. Environ. Sci. Health B*, 2012, **47**, 795–803.
133. C. K. Yeom, Y. H. Kim and J. M. Lee, *J. Appl. Polym. Sci.*, 2002, **84**, 1025–1034.
134. H. K. No, S. P. Meyers, W. Prinyawiwatkul and Z. Xu, *J. Food Sci.*, 2007, **72**, R87–R100.
135. P. K. Dutta, S. Tripathi, G. K. Mehrotra and J. Dutta, *Food Chem.*, 2009, **114**, 1173–1182.
136. R. A. A. Muzzarelli, J. Boudrant, D. Meyer, N. Manno, M. DeMarchis and M. G. Paoletti, *Carbohydr. Polym.*, 2012, **87**, 995–1012.
137. A. Balasubramanian, L. E. Rosenberg, K. Yam and M. L. Chikindas, *J. Appl. Pack. Res.*, 2009, **3**, 193–221.
138. P. Fernández-Saiz, *Chitosan and chitosan blends as antimicrobials.* In Antimicrobial Polymers; J. M. Lagarón, M. J. Ocio and A. López-Rubio, ed. John Wiley & Son, Hoboken, New Jersey, USA, 2011, pp. 71–99.
139. C. N. Cutter, *Meat Sci.*, 2006, **74**, 131–142.
140. A. Begin and M. R. Van Calsteren, *Int. J. Biol. Macromol.*, 1999, **26**, 63–67.
141. S. Y. Park, K. S. Marsh and J. W. Rhim, *J. Food Sci.*, 2002, **76**, 194–197.
142. S. Khamhan, Y. Baimark, S. Chaichanadee, P. Phinyocheep and S. Kittipoom, *Int. J. Polym. Anal. Charact.*, 2008, **13**, 224–231.

143. E. Nugraha, S. A. Copinet, L. Tighzert and V. Coma, *J. Polym. Environ.*, 2004, **12**, 1–6.
144. P. Fernández-Saiz, M. J. Ocio and J. M. Lagarón, *Carbohydr. Polym.*, 2010, **80**, 874–884.
145. B. Ouattara, R. E. Simard, G. Piette, A. Begin and R. A. Holley, *Int. J. Food Microbiol.*, 2000, **62**, 139–148.
146. M. Abdollahi, M. Rezaei and G. Farzi, *Innov. Food Sci. Emerg. Technol*, 2012, **47**, 847–853.
147. S. Zivanovic, S. Chi and A. F. Draughon, *J. Food Sci.*, 2005, **70**, M45–M51.
148. K. D. Vu, R. G. Hollingsworth, E. Leroux, S. Salmieri and M. Lacroix, *Food Res. Inter.*, 2011, **44**, 198–203.
149. F. Devlieghere, A. Vermeulen and J. Debevere, *Food Microbiol.*, 2004, **21**, 703–714.
150. A. El Ghaouth, J. Arul, R. Ponnampalam and M. Boulet, *J. Food Sci.*, 1991, **56**, 1618–1620.
151. A. El Ghaouth, J. Arul, J. Grenier and A. Asselin, *Phytopathology*, 1992, **82**, 398–402.
152. B. M. V. Reddy, K. Belkacemi, R. Corcuff, F. Castaigne and J. Arul, *Postharvest Biol. Technol.*, 2000, **20**, 39–51.
153. C. Han, Y. Zhao, S. W. Leonard and M. G. Traber, *Postharvest Bio. Technol.*, 2004, **33**, 67–78.
154. V. Coma, A. Deschamps and A. Martial-Gros, *J. Food Sci.*, 2003, **68**, 2788–2792.
155. I. K. Park, Y. K. Lee, M. J. Kim and S. D. Kim, *J. Chitin Chitosan*, 2002, **7**, 208–213.
156. S. Bhale, H. K. No, W. Prinyawiwatkul, A. J. Farr, K. Nadarajah and S. P. Meyers, *J. Food Sci.*, 2003, **68**, 2378–2383.
157. C. Caner, *J. Sci. Food Agric.*, 2005, **85**, 1897–1902.
158. S. Roller and N. Covill, *J. Food Prot.*, 2000, **63**, 202–209.
159. T. J. Ha and S. H. Lee, *J. Korean Soc. Food Sci. Nutr.*, 2001, **30**, 630–634.
160. M. Dash, F. Chiellini, R. M. Ottenbrite and E. Chiellini, *Prog. Polym. Sci.*, 2011, **36**, 981–1014.
161. R. Riva, H. Ragelle, A. des Rieux, N. Duhem, C. Jérôme and V. Préat, *Adv. Polym. Sci.*, 2011, **244**, 19–44.
162. A. Bernkop-Schnürch and S. Dünnhaupt, *Eur. J. Pharm. Biopharm.*, 2012, **81**, 463–469.
163. J. H. Park, G. Saravanakumar, K. Kim and I. C. Kwon, *Adv. Drug Deliv. Rev.*, 2010, **62**, 28–41.
164. M. de la Fuente, M. Raviña, P. Paolicelli, A. Sánchez, B. Seijo and M. J. Alonso, *Adv. Drug Deliv. Rev.*, 2010, **62**, 100–117.
165. O. Felt, A. Carrel, P. Baehni, P. Buri and G. Gurny, *J. Ocul. Pharmacol. Ther.*, 2000, **16**, 261–270.
166. Y. Diebold, M. Jarrín, V. Sáez, E. L. S. Carvalho, M. Orea, M. Calonge, B. Seijo and M. J. Alonso, *Biomaterials*, 2007, **28**, 1553–1564.

167. X. Wang, C. Zheng, Z. Wu, D. Teng, X. Zhang, Z. Wang and C. Li, *J. Biomed. Mater. Res. B Appl. Biomater.*, 2009, **88**, 150–161.
168. M. Amidi, S. G. Romeijn, J. C. Verhoef, H. E. Junginger, L. Bungener, A. Huck-riede, D. J. Crommelin and W. Jiskoot, *Vaccine*, 2007, **25**, 144–153.
169. A. Sabera, S. P. Strand and M. Ulfendahl, *Eur. J. Pharm. Sci.*, 2010, **39**, 110–115.
170. K. Kafedjiiski, A. H. Krauland, M. H. Hoffer and A. Bernkop-Schnürch, *Biomaterials,*, 2005, **26**, 819–826.
171. S. Dhaliwal, S. Jain, H. Singh and A. Tiwary, *AAPS J.*, 2008, **10**, 322–330.
172. B. Sarmento, A. Ribeiro, F. Veiga, P. Sampaio, R. Neufeld and D. Ferreira, *Pharm. Res.*, 2007, **24**, 2198–2206.
173. S. Mitra, U. Gaur, P. C. Ghosh and A. N. Maitra, *J. Control. Release*, 2001, **74**, 317–323.
174. R. Jayakumar, K. P. Chennazhi, R. A. A. Muzzarelli, H. Tamura, S. V. Nair and N. Selvamurugan, *Carbohydr. Polym.*, 2010, **79**, 1–8.
175. M. Köping-Höggard, K. M. Varum, M. Issa, S. Danielsen, B. E. Christensen, B. T. Stokke and P. Artursson, *Gene Therapy*, 2004, **11**, 1441–1452.
176. K. Khatri, A. K. Goyal, P. N. Gupta, N. Mishra and S. P. Vyas, *Int. J. Pharm.*, 2008, **354**, 235–241.
177. T. Dai, M. Tanaka, Y. Y. Huang and M. R. Hamblin, *Expert Rev. Anti. Infect. Ther.*, 2011, **9**, 857–879.
178. A. Francesko and T. Tzanov, *Adv. Biochem. Engin./Biotechnol.*, 2011, **125**, 1–27.
179. T. H. Fischer, A. P. Bode, M. Demcheva and J. N. Vournakis, *J. Biomed. Mater. Res. A*, 2007, **80**, 167–174.
180. C. J. Park, S. G. Clark, C. A. Lichtensteiger, R. D. Jamison and A. J. W. Johnson, *Acta Biomater.*, 2009, **5**, 1926.
181. D. Boynuegri, G. Ozcan, S. Senel, D. Uç, A. Uraz, E. Oqus, B. Cakilci and B. Karaduman, *J. Biomed. Mater. Res. B Appl. Biomater.*, 2009, **90**, 461–466.
182. Y.-C. Lin, F.-J. Tan, K. G. Marra, S.-S. Jan and D.-C. Liu, *Acta Biomater.*, 2009, **5**, 2591–2600.
183. S. V. Madihally and H. W. T. Matthew, *Biomaterials*, 1999, **20**, 1133–1142.
184. A. Di Martino, M. Sittinger and M. V. Risbud, *Biomaterials*, 2005, **26**, 5983–5990.
185. Y. J. Seol, J. Y. Lee, Y. J. Park, Y. M. Lee, K. Young, I. C. Rhyu, S. J. Lee, S. B. Han and C. P. Chung, *Biotechnol. Lett.*, 2004, **26**, 1037–1041.
186. Y. Zhang and M. Zhang, *J. Biomed. Mater. Res.*, 2001, **55**, 304–312.
187. Q. Hu, B. Li, M. Wang and J. Shen, *Biomaterials*, 2004, **25**, 779–785.
188. J. K. Suh and H. W. Matthew, *Biomaterials*, 2000, **21**, 2589–2598.
189. X. H. Wang, D. P. Li, W. J. Wang, Q. L. Feng, F. Z. Cui, Y. X. Xu, X. H. Song and M. van der Werf, *Biomaterials*, 2003, **24**, 3213–3220.
190. T. Freier, R. Montenegro, H. S. Koh and M. S. Shoichet, *Biomaterials*, 2005, **26**, 4624–4632.

CHAPTER 3

Synthesis, Antimicrobial Activity and Applications of Polymers with Ammonium and Phosphonium Groups

EL-REFAIE KENAWY*[a] AND SHERIF KANDIL[b]

[a] Department of Chemistry, Polymer Research Group, Faculty of Science, University of Tanta, Tanta 31527, Egypt; [b] Department of Materials Science, Institute of Graduate Studies and Research, Alexandria University, Alexandria, Egypt
*Email: ekenawy@yahoo.com

3.1 Introduction

The worldwide spread of diseases is a great problem for modern society. Infection is one of the most serious and devastating complications faced by millions of patients annually. For example, implant-associated infections often lead to implant removal and revision surgery, imposing a huge economic burden on society.[1,2]

Numerous quite effective detergents and cleaning solutions with antimicrobial properties are frequently used in households. The major problem is that such compounds are only effective for the period of washing and shortly thereafter. Further, their frequent use might lead to the development of resistant strains of these antibiotics. In most applications it is not necessary to sterilize the whole environment, but only prevent proliferation of microbes,

RSC Polymer Chemistry Series No. 10
Polymeric Materials with Antimicrobial Activity: From Synthesis to Applications
Edited by Alexandra Muñoz-Bonilla, María L. Cerrada and Marta Fernández-García
© The Royal Society of Chemistry 2014
Published by the Royal Society of Chemistry, www.rsc.org

which form highly resistant biofilms. Therefore, developing antimicrobial coatings is mainly focused on materials that prevent the formation of such biofilms, *i.e.* that kill or repel microbes that approach and settle on the surface. This problem is evident on many materials, such as medical devices, ship hulls, and any moist environments in households, *e.g.*, showers, bathrooms, and kitchens.[3]

In order to resolve the problems of microbial contamination, researchers have intensively investigated antimicrobial materials containing various original and synthetic substances.[4] For example, inorganic antimicrobial materials carrying Ag^+ act as the most important antimicrobial substances.[5] Nevertheless, the applications of inorganic antimicrobial materials are limited because the accumulation of heavy metals may result in serious environmental problems and may be harmful to humans in the case of high metal concentration.[6] On the other hand, inorganic antimicrobial materials can form insoluble clay minerals and then easily lose their antimicrobial activity.[7] Hence, it is important to develop new antimicrobial materials with low cost and high antimicrobial activity.[8]

Polymeric antimicrobial agent or biocidal polymer holds promise to solve the problems associated with the conventional antimicrobial agent.[9–14] A biocidal polymer is one with the ability to kill micro-organisms, by acting as a source of sterilizing ions or molecules. Quaternary ammonium compounds (quats) are the essential compounds for many antimicrobials. It is known that quaternary ammonium compounds have been widely used as disinfectants. However, these suffer the disadvantages of low molecular weight compounds.[10]

After Domagk's discovery in 1935 of the biocidal properties of quats, several generations of structurally variable quats were developed.[11] The fact that quats are cationic surfactants allows the users to apply them in a variety of ways.[12] Also, the use of quaternary ammonium compounds may be a potential key driver in the emergence of antimicrobial resistance.[13]

It is now well established that when a polymer carries quaternary ammonium or phosphonium salt group, it possesses potential antimicrobial properties.[14] In this chapter we will discuss the synthesis, applications and the importance of the antimicrobial polymers with quaternary ammonium and phosphonium salts.

3.2 Why do we Need Antimicrobial Polymers?

Antimicrobial agents of low molecular weight are used for the sterilization of water, soil, as antimicrobial drugs, and as food preservatives. However, they can have the limitation of having residual toxicity even when small amounts of the antimicrobial agent are applied.[9] Current antimicrobial products, such as triclosan and silver, suffer from critical weaknesses, for instance, short active duration or high cost. Moreover, when used for clothing, such low molecular weight antimicrobial agents generally leach out from the fabrics/matrix towards the environment and to the skin of the wearers with harmful effects. Antimicrobial polymers are defined as polymers having biocidal pendant groups or biocidal inherent repeat units in the polymer chemical structure.

The development of polymers with antimicrobial activity themselves is an important area of research focused on solving the problem of contamination by micro-organisms. This alternative method of applying the antimicrobial agent avoids the inconvenience of the diffusion of the low molecular weight biocides. These are trapped through the polymeric matrix to avoid causing toxicity to the human body. Besides, antimicrobial polymers usually present longer-term activity. Also, the use of antimicrobial polymers offers promise for enhancing the efficacy of some existing antimicrobial agents and minimizing the environmental problems accompanying the use of conventional anti-microbial agents, by reducing the residual toxicity of the agents, increasing their efficiency and selectivity, and prolonging the lifetime of the antimicrobial agents. In addition, polymeric antimicrobial agents have the advantage of being nonvolatile, chemically stable, and do not permeate through skin. Therefore, they can reduce losses associated with volatilization, photolytic decomposition, and transportation of conventional low molecular weight antimicrobial agents.

The advantage of polymers used as biocides is justified in the minimization of environmental problems due to lower toxicity compared with low molecular weight compounds that are usually liquids or gases.[15] In the field of biomedical polymers, infections associated with biomaterials represent a significant challenge to the more widespread application of medical implants.

Antimicrobial polymers can be used as coatings that are nonreleasing and kill bacteria on contact.[16] Also they can be used as water-soluble polymers that could be used as disinfectants.[17] In the field of food packaging, antimicrobial polymer packaging is one of the most promising active packaging materials that have been highly effective in killing or inhibiting pathogenic micro-organisms that contaminate and spoil foods.[18,19]

Antimicrobial polymers having high molecular weight could overcome these problems, reducing or preventing the leaching out of bioactive substances. They are increasingly taken into account as a feasible alternative bioactive agent for bactericidal applications. Moreover, antimicrobial polymers are considered as an attractive way for the "nonleaching" approach in the production of bactericidal materials. This approach is interesting for many applications, in particular in the textile field, where antimicrobial polymers show several advantages with respect to low molecular weight antibacterial agents, including improved environmental stability, lack of diffusion on the wearers' skin, low skin irritation, low toxicity, good biocompatibility, low corrosion of metals and plastics, long residence time and biological activity.[20]

In the field of textiles, till now, a number of chemicals have been employed to impart antibacterial activity to textiles. These chemicals include inorganic salts, organometallics, iodo-phors (substances that slowly release iodine), phenols and thiophenols, onium salts, antibiotics, heterocyclics with anionic groups, nitro compounds, ureas and related compounds, formaldehyde derivatives, and amines.[21] However, many of these chemicals when used in the field of textiles are toxic to humans and cannot easily degrade in nature. Antimicrobial polymers can minimize these problems.[13,15,16,22,23]

3.3 Polymers with Antimicrobial Activity

Due to the interest and the importance of the subject of the polymers with antimicrobial activities, there have been number of reviews in the field of the chemistry and applications of antimicrobial polymers, which have considered definite classes of antimicrobial polymers and targeted specific applications.[9,23]

In this chapter, we will survey most of the reported polymers with antimicrobial activity having quaternary ammonium and phosphonium salts.

3.3.1 Polymers with Quaternary Ammonium Salts

Polymers with quaternary ammonium and phosphonium groups are probably the most explored kind of polymeric biocides. It is generally accepted that the mechanism of the bactericidal action of the polycationic biocides involves destructive interaction with the cell wall and/or cytoplasmic membranes.[10] Since most bacterial cell walls are negatively charged, and most antimicrobial polymers are positively charged there is an interaction between the positively charged polymers and the negatively charged cell wall leading to destruction of the cell wall and, consequently, to cell death. Polymeric materials may interact more effectively with the cell of Gram-positive bacteria as their polyglycane outer layer is sufficiently loosely packed to facilitate deep penetration of the polymer chain inside the cell to interact with the cytoplasmic membrane. They may also kill the micro-organism without penetration inside the cell by disrupting the cell wall. Figure 3.1 depicts a comparison between the antimicrobial-polymer-treated *Staphylococcus aureus* and the untreated ones.[24]

The results of the study showed that, with a growing number of phosphonium units in the polymers, the antimicrobial activity was found to rise, due to the increase in the interaction between the cationic species with the cell wall and membrane, which might lead to morphological cell changes, and even cell wall perforation. Consequently, the use of these polymers leads to K^+ leakage out of the microbial cells.

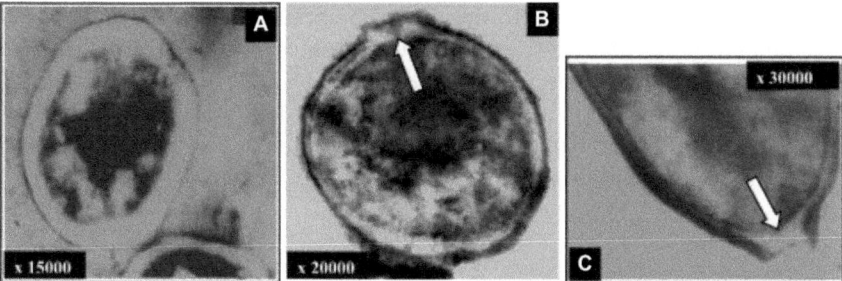

Figure 3.1 Electron scanning micrograph of [A] Normal: Control *Staphylococcus aureus* living cell in which completely normal cell membrane can be found. [B] and [C] *Staphylococcus aureus* cells, treated with polymeric quaternary phosphonium salt.

3.3.1.1 Polyquaternary Ammonium Based on Synthetic Polymers

Ikeda and Tazuke synthesized and investigated the antimicrobial activity of homopolymers of polyacrylates with quaternary ammonium and phosphonium salts.[25] When the molecular weight of the polymer was lower than 5×10^4 Da, the antimicrobial property rose with the increase of the molecular weight, while the antibacterial activity decreased sharply with the increase of the molecular weight of the polymer when it exceeded 1.2×10^5 Da. The discrepancy of antimicrobial properties was explained on the basis of the permeability through the cell wall. The molecular weight dependence of poly(trialkyl vinylbenzyl ammonium chloride) (Figure 3.2) against *S. aureus* was evaluated by Ikeda and Tazuke.

Recently, Nonaka and coworkers synthesized methacryloyloxyethyl trialkyl phosphonium chlorides/*N-iso*propylacrylamide copolymers. They found that the antibacterial activity of the copolymers against *Escherichia coli* rose with increasing the alkyl chain length in the phosphonium groups in the copolymer.[26,27]

Rationalization of the parabolic relationship between antibacterial properties and alkyl chain lengths has been debated. It has been attributed to (1) dual binding sites on the surface for which the relative binding affinities at each site differ for long and short alkyl substituents or (2) different aggregation behavior for long and short hydrophobes.[28]

Kenawy *et al.* reported the synthesis of crosslinked copolymers based on the polymerization reaction of vinylbenzyl chloride either with 2-chloroethyl vinyl ether or with methyl methacrylate, using divinylbenzene as a crosslinker.[29]

The crosslinked copolymers were further modified by quaternization with triphenylphosphine and triethylamine (Figure 3.3). The antimicrobial activities of the modified copolymers were evaluated against various micro-organisms (*S. aureus*, *E. coli*, *Bacillus subtilis*, *Aspergillus flavus*, *Fusarium oxysporum*, and *Candida albicans*). It was found that the diameter of the inhibition zone varied according to the active group in the copolymer and also to the examined micro-organism. In general, the copolymers showed antimicrobial activity against all the tested micro-organisms. However, the compound with the triphenylphosphonium salt of the modified copolymer was the most effective one against both the bacteria and fungi species. A concentration of 20 mg/mL of triphenylphosphonium salt of the modified copolymer killed 100% of *C. albicans*. However, this polymer activity toward *A. flavus* was lower and left 44% of the organisms surviving at 20 mg/mL.

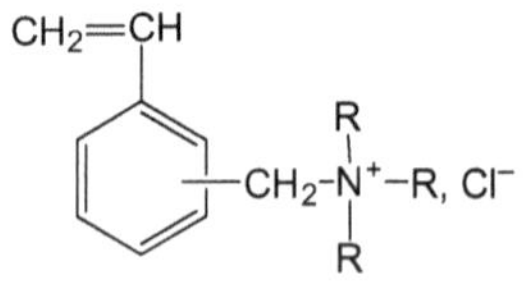

Figure 3.2 Poly(trialkyl vinylbenzyl ammonium chloride).

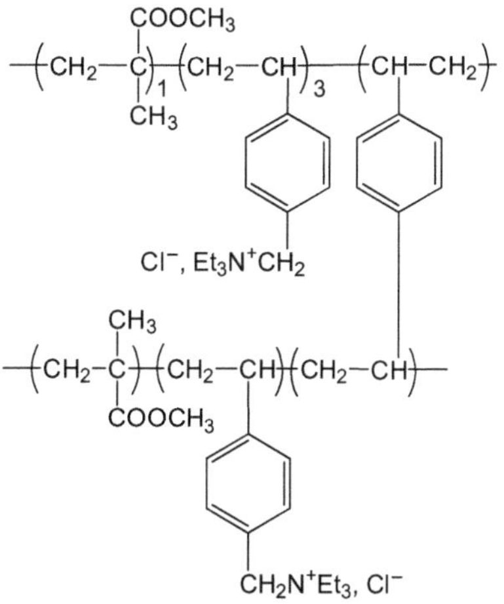

Figure 3.3 Crosslinked copolymer based on vinylbenzyl chloride with methyl methacrylate further quaternized with triethylamine.

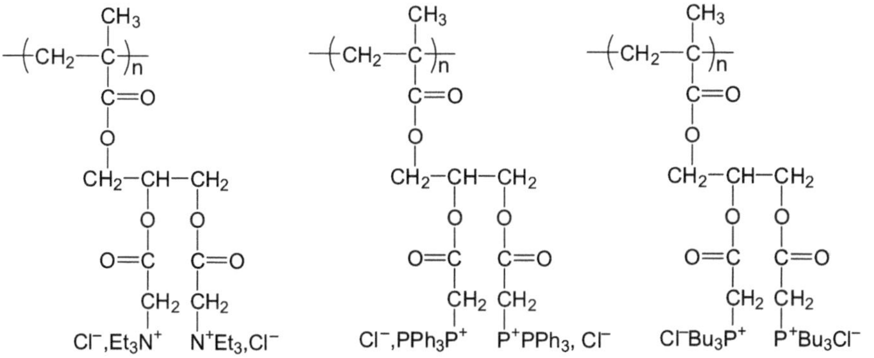

Figure 3.4 Glycidyl methacrylate polymers having quaternary ammonium and phosphonium groups.

Modified glycidyl methacrylate polymers having quaternary ammonium and phosphonium groups were prepared.[30] Three different antimicrobial polymers were prepared (Figure 3.4), and the antimicrobial activities of these polymers were tested against Gram-negative bacteria (*E. coli, Pseudomonas aeruginosa, Shigella* sp., and *Salmonella typhae*), Gram-positive bacteria (*B. subtilis* and *Bacillus cereus*), as well as the fungus *Trichophytun rubrum*. The tested polymers showed significant antimicrobial activity against Gram-negative bacteria and the fungus after 24 h contact time, whereas these polymers were less active against Gram-positive bacteria. It was reported that the tributyl phosphonium salt was able to kill 100% of the fungus *T. rubrum* at a concentration of 10 mg/mL and a contact time of 24 h. The polymer with triphenyl phosphonium salt was able to kill 100% *P. aeruginosa* at a lower concentration of 5 mg/mL;

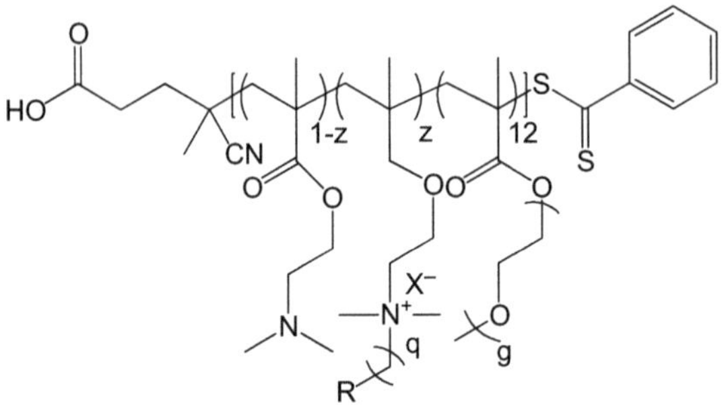

Figure 3.5   Poly(quaternary ammonium) salts containing different functionalities from functional halides.

the same concentration of polymer with ammonium salt was able to kill 100% *P. aeruginosa* at the same contact time.

New methacrylate monomers containing pendant quaternary ammonium moieties based on 1,4-diazabicyclo-[2.2.2]-octane were synthesized by Dizman *et al.*[31] The polymers exhibited good antimicrobial activities against *S. aureus* and *E. coli*. The activity increased as the *N*-alkyl chain length augmented from four to six carbons.

Venkataraman *et al.* synthesized well-defined poly(ethylene glycol) polymers with tertiary amines from readily available commodity monomers 2-(dimethylamino)ethyl methacrylate and oligo(ethylene glycol) methyl ether methacrylate by reversible addition–fragmentation chain-transfer polymerization (Figure 3.5). By employing a simple and efficient postpolymerization functionalization strategy, the tertiary amines were quaternized to result in cationic polymers.[32]

Minimum inhibitory concentration (MIC), *i.e.* the polymer concentration to completely inhibit the bacterial growth, was found in these systems to be dependent on both the nature of functional group and the hydrophobicity of the polymer. Evaluating the series of antimicrobial polymers obtained by the quaternization of the poly[2-(dimethylamino)ethyl methacrylate-*co*-oligo(ethylene glycol) methyl ether methacrylate] with the functional halides, X–(CH$_2$)q–R, both the length of the alkyl spacer (q) and the chemical functionality (R) were found to significantly influence the minimum inhibitory concentration against the Gram-positive bacteria *B. subtilis*. Of the polymers tested, the polymers with the short alkyl spacer were found to have lower MIC. The best polymer was the one with the methyl substituent, having a MIC value of about 20 mg/L.

The antimicrobial activities of these cationic polymers were determined against Gram-positive bacteria *B. subtilis*. Amongst the functional groups, both the alkyl and the alcohol groups were found to be effective, with MIC values in the range of 20–80 mg/L. The hemolytic properties of polymers were analyzed against mouse red blood cells. The polymers with a short alkyl or hydroxyl group still retained strong antimicrobial activity. The overall hydrophobicity of

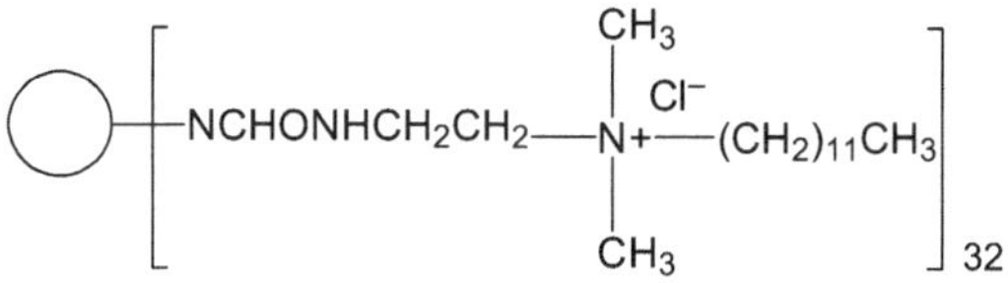

Figure 3.6 Quaternary dendrimer ammonium salt.

the polymer influenced its hemolytic behavior. These polymers can be promising antimicrobial agents.

Charles *et al.* modified poly(amidoamine) dendrimer into quaternary ammonium salts using tertiary amines with different chain lengths: dimethyldodecyl amine, dimethylhexyl amine, and dimethylbutyl amine using an efficient synthetic route (Figure 3.6).[33] The antimicrobial activity of these dendrimer ammonium salts against *Staphylococcus* and *E. coli* bacteria was examined using the disc diffusion method. It was found that the quaternary ammonium salt prepared with the dimethyldodecyl amine exhibits antimicrobial efficacy against *Staphylococcus* and *E. coli* bacteria.

3.3.1.2   Polyquaternary Ammonium Based on Natural Polymers

Hydroxypropyltrimethyl ammonium chloride chitosan was synthesized with different degrees of substitution (6%, 18% and 44%) of quaternary ammonium by reacting chitosan with glycidyl trimethyl ammonium chloride.[1] The antibacterial activities of these polymers were tested *in vitro* against *S. aureus*, Methicillin-resistant *S. aureus*, and *Staphylococcus epidermidis*. Mouse fibroblasts and bone marrow derived stromal cells (hMSCs) were used to investigate the biocompatibility of the hydroxypropyl trimethyl ammonium chloride. The results show that the antibacterial activities of the hydroxypropyl trimethyl ammonium chloride with 18% or 44% substitution were significantly higher against all tested bacteria.

Quaternary *N*-[2-(*N,N,N*-tri-alkyl ammoniumyl and 2-pyridiniumyl) acetyl] derivatives of chitosan polymer, chito-oligomer, and glucosamine (monomer) were synthesized for the purpose of investigating the structure activity relationship on the antibacterial effect.[34] In order to obtain chitosan derivatives with the bulky *N,N*-dimethyl-*N*-dodecyl- and *N,N*-dimethyl-*N*-butyl side chains, three steps were needed, starting from 3,6-*O*-di-*tert*butyldimethylsilyl chitosan as the key intermediate. The quaternary ammoniumyl acetyl derivatives of glucosamine were synthesized from glucosamine or tetra-*O*-acetylglucosamine. *N,N,N*-trimethyl chitosan was used as a reference compound for investigating antibacterial activity. Clinical Laboratory Standard Institute protocols were used to determine MIC for activity against clinically important Gram-positive strains *S. aureus*, and Gram-negative strains of *E. coli*, *P. aeruginosa* and *Enterococcus faecalis*. In general, the *N*-(2-(*N,N*-dimethyl-*N*-dodecyl ammoniumyl) acetyl) derivatives of chitooligomer and glucosamine monomer were more active against bacteria than derivatives with shorter alkyl chains. In contrast, the *N*-(2-(*N*,*N*-dimethyl-*N*-dodecyl ammoniumyl) acetyl) derivatives of chitosan were less active than derivatives with *N*-(2-*N*,*N*,*N*-tri-metylammoniumyl)acetyl or *N*-(2-(*N*-pyridiniumyl)acetyl) quaternary moiety. *N*,*N*,*N*-trimethyl chitosan was the most active compound in this study.

Fan *et al.* reported the preparation of cationic derivatives of pectin by reacting pectin with 3-chloro-2-hydroxypropyltrimethylammonium chloride in the presence of sodium hydroxide[14] (Figure 3.7). The chemical structures of the derivatives were characterized by using elemental analysis, FT-IR, and ^{13}C NMR spectroscopies. *In vitro* antimicrobial activity assessment exhibited by quaternized pectin showed pronounced inhibitory effect against the three bacteria (*S. aureus*, *E. coli*, and *B. subtilis*). The improved functionalities of the derivative might be explained by its polycationic characteristics.

Holappa *et al.* reported a synthetic procedure for the preparation of water-soluble quaternary chitosan *N*-betainates (Figure 3.8).[35,36] Chitosan *N*-beta-inates were prepared having various degrees of substitution and structurally uniform molecular structure, thus structure–activity relationships and the effect of degree of quaternization can be exactly determined. The results reported by the authors revealed that the antimicrobial activity of chitosan *N*-betainates increased with a decreasing degree of substitution in acidic conditions, which suggests that the positive charge has to be situated on the amino group in the chitosan backbone if one wishes to achieve efficient antimicrobial activity.

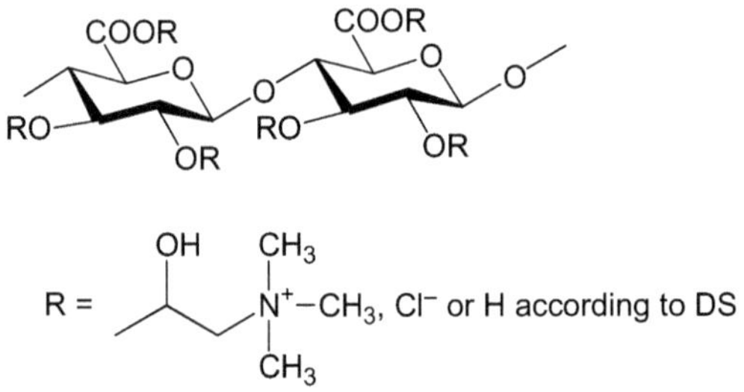

Figure 3.7   Quaternization of pectin.

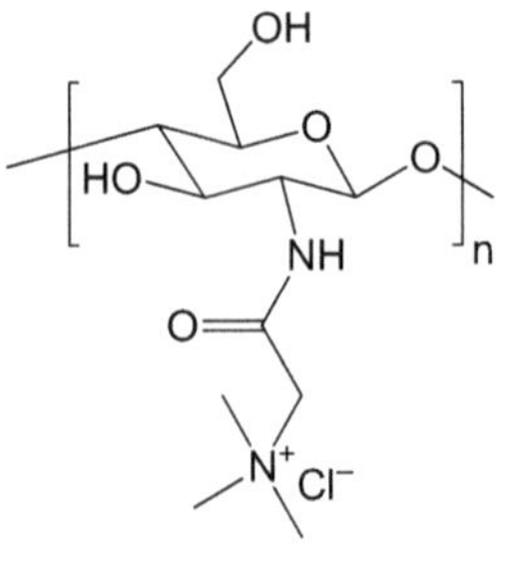

Figure 3.8   Quaternary chitosan *N*-betainate chloride.

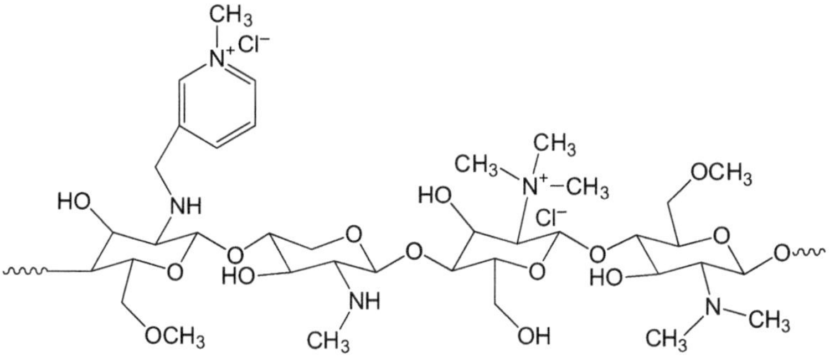

Figure 3.9   Methylated *N*-(3-pyridylmethyl) chitosan derivatives.

They concluded that just the introduction of a quaternary ammonium moiety into chitosan is not sufficient to obtain antimicrobial action, but the key issue is the optimal positioning of the positive charge in relation to the polymer backbone.

Fu *et al.* reported the synthesis of three water-soluble chitosan derivatives bearing double functional groups by means of 2,3-epoxypropyltrimethyl ammonium chloride and benzaldehyde as modifiers through formation of Schiff base, reduction, *N*-methylation and *O*-quaternization.[22] The synthesized chitosan derivatives were applied to cotton fabrics together with citric acid as the crosslinking agent to evaluate their use as durable antimicrobial textile finishing agents. The optimal finishing conditions were investigated and the antibacterial activities of finished fabrics were tested with a "Shake-Flask Test". The produced fabric showed strong antimicrobial activities and fairly good durability. The antibacterial efficiency of *S. aureus* and *E. coli* was more than 99% and 96%, respectively, when crosslinked with citric acid (14%, o.w.f) through a two-bath process. Among chitosan derivatives the *O*-quaternized-*N*,*N*-diethyl-*N*-benzyl ammonium chitosan chloride exhibited the best response. Antimicrobial activity using *O*-quaternized-*N*-chitosan Schiff bases finishing fabric was still more than 75% antimicrobial activity after 20 consecutive home launderings.

Sajomsang *et al.* reported the synthesis of methylated *N*-(3-pyridylmethyl) chitosan chloride with various degrees of *N*-substitution, degrees of quaternization (DQ), and molecular weights by single methylation of *N*-(3-pyridylmethyl) chitosan with iodomethane (Figure 3.9).[37] The bactericidal activity of these polymers was determined by using minimum inhibitory concentration and minimum bactericidal concentration methods against *E. coli* and *S. aureus* bacteria compared with the *N*,*N*,*N*-trimethyl chitosan chloride in accordance with the standard method of the National Committee for Clinical Laboratory Standards. It was found that the bactericidal activity was dependent on the chemical structure of quaternary ammonium moiety and the molecular weight, at a constant degree of quaternization. The results revealed that the *N*,*N*,*N*-trimethyl ammonium group has higher bactericidal activity than the *N*-methyl-pyridinium group at a similar degree of quaternization and molecular weight.

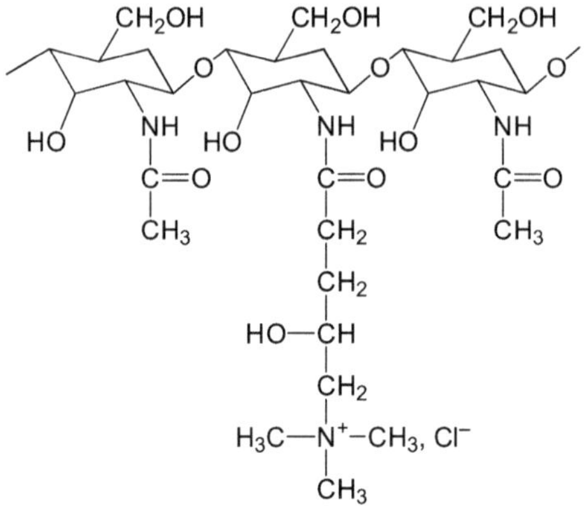

Figure 3.10 Chito-oligosaccharide with quaternary ammonium functionality.

Kim *et al.* synthesized chito-oligosaccharide with quaternary ammonium functionality. The introduction of a quaternary ammonium group to chito-oligosaccharide has been accomplished by coupling of glycidyl trimethy-lammonium chloride to chito-oligosaccharide in aqueous solution without using an additional catalyst (Figure 3.10). Its antimicrobial activity was evaluated against *Streptococcus mutans*, which is a principal etiological agent of dental caries in humans.[38] The resulting polymer increased the inhibition to above 80% against *S. mutans* after 5 h, whereas the chito-oligosaccharide showed the growth inhibition of about 10%. It was found that the anti-microbial activity of the chito-oligosaccharide could be considerably enhanced by the introduction of quaternary ammonium functionality.

3.3.2   Polymers with Quaternary Phosphonium Salts

3.3.2.1   *Polyquaternary Phosphonium Based on Synthetic Polymers*

Kanazawa and coworkers prepared poly(tributyl (4-vinylbenzyl) phosphonium chloride) (Figure 3.11) and investigated the molecular weight dependence of their antimicrobial activities against *S. aureus* in saline solution. The results revealed that the antibacterial properties increased as molecular weight rose from 1.6×10^4 to 9.4×10^4 Da.[39]

A new class of antimicrobial polymers consisting of poly(phenylene oxide) was synthesized by Chang *et al.*[40] and their antimicrobial activities were investigated. The modified poly(phenylene oxide) was accomplished by selective bromination of poly(phenylene oxide) followed by quaternization reactions with various tertiary phosphines (Figure 3.12).

The antimicrobial activities of these quaternized polymers were tested against Gram-positive bacteria (*S. Epidermidis*) and Gram-negative bacteria (*E. coli*). The triphenylphosphonium-modified polymer showed excellent

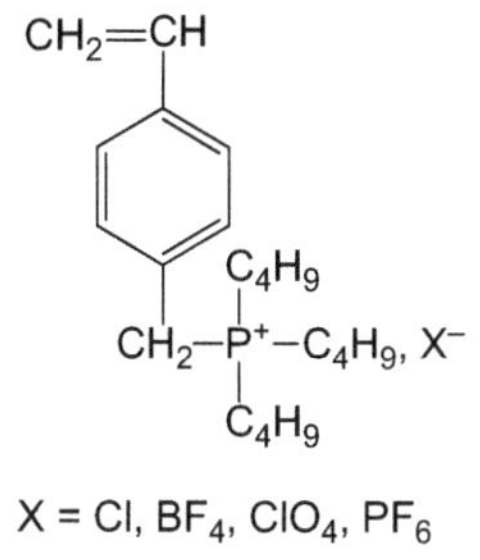
CH2=CH
C4H9
CH2–P+–C4H9, X−
C4H9
X = Cl, BF4, ClO4, PF6

Figure 3.11 Poly(tributyl (4-vinylbenzyl) phosphonium chloride).

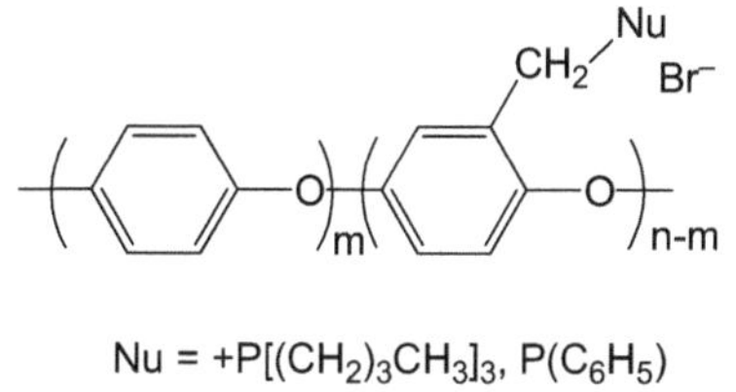
Nu
CH2
Br−
O
m
O
n-m
Nu = +P[(CH2)3CH3]3, P(C6H5)

Figure 3.12 Quaternized poly(phenylene oxide).

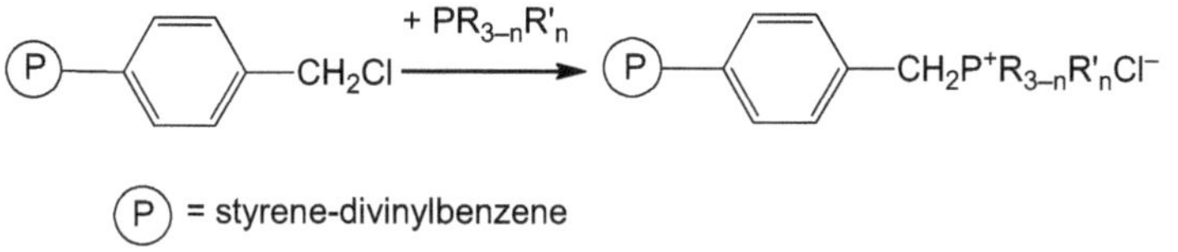
P
CH2Cl
+ PR3–nR'n
CH2P+R3–nR'nCl−
P = styrene-divinylbenzene

Figure 3.13 Quaternary phosphonium salts of styrene–divinylbenzene copolymers.

antibacterial activity against both types of bacteria. Generally, the thermal stability of phosphonium-modified bromonated poly(phenylene oxide) was superior to that of the ammonium analog, and the increase in the functionalization of the polymer backbone resulted in an improved antimicrobial activity.

Popa *et al.* synthesized quaternary phosphonium salts that were grafted on styrene–divinylbenzene copolymers as macromolecular supports by polymer-analogous reactions (Figure 3.13). The products were proved to have antibacterial activity against *S. aureus*, *E. coli* and *P. aeruginosa*. A great advantage was that the active species grafted on insoluble carriers could be used in repeated cycles, only needing sterilization prior to use.[41]

Anthierens *et al.* synthesized alkyne-containing poly(butylene adipate) and functionalized the polymer through the copper-catalyzed azide-alkyne "click" reaction with a quaternary phosphonium group (Figure 3.14).[42] The resulting functionalized polyester was tested for antimicrobial activity both in dispersion by a dynamic shake flask method, and on the surface. In both cases the polyester was found to significantly reduce the cell counts of *E. coli*.

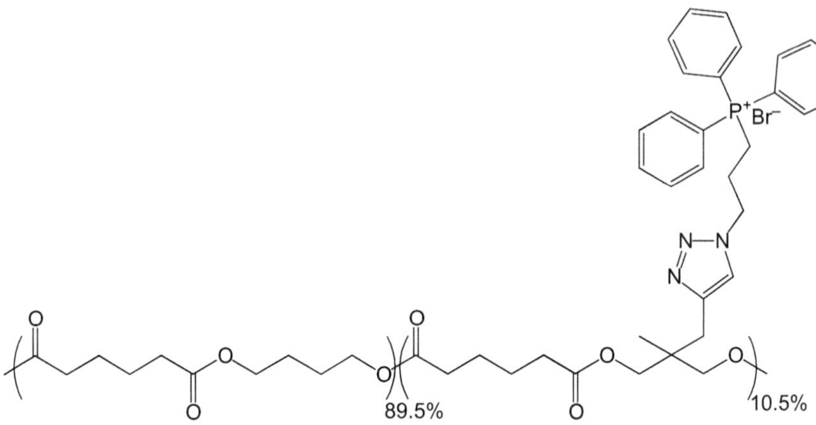

Figure 3.14   Alkyne-functionalized poly(butylene adipate).

The results showed that the antimicrobial activity was caused by the co-valently attached quaternary phosphonium groups and not by the freely available residues or the unreacted alkyne groups. The presented polymer can, after coating or lamination on a carrier material, show applications as an antimicrobial packaging film for food or medical purposes.

Kenawy *et al.* synthesized glycidyl methacrylate polymers having quaternary phosphonium groups[30] (see polymeric structure on right side of Figure 3.4), and their antimicrobial activities were tested against Gram-negative bacteria (*E. coli*, *P. aeruginosa*, *Shigella* sp., and *S. typhae*), Gram-positive bacteria (*B. subtilis* and *B. cereus*), as well as the fungus *T. rubrum*. These polymers exhibited significant antimicrobial activity against Gram-negative bacteria and the fungus after 24 h contact time, whereas they were less active against Gram-positive bacteria. It was reported that the tributyl phoshonium salt was able to kill 100% of the fungus *T. rubrum* at a concentration of 10 mg/mL and a contact time of 24 h. The polymer with triphenylphosphonium salt was able to kill 100% *P. aeruginosa* at a lower concentration of 5 mg/mL.

3.3.2.2   Polyquaternary Phosphonium Based on Natural Polymers

Chun Li *et al.* prepared new antimicrobial natural rubber by its chemical modification. Quaternary phosphonium salt (QPs) groups were covalently immobilized onto the molecular chains of epoxidized natural rubber using halogenated acid as an intermediary.[43] The results showed that the ring-opening epoxidized natural rubber and the quaternary phosphonium salt of the modified epoxidized natural rubber were successfully prepared (Figure 3.15). The structures of the modified rubbers were confirmed, while the antimicrobial activity of the quaternary phosphonium salt of the modified epoxidized natural rubber was investigated. The proliferation of micro-organisms tested can be

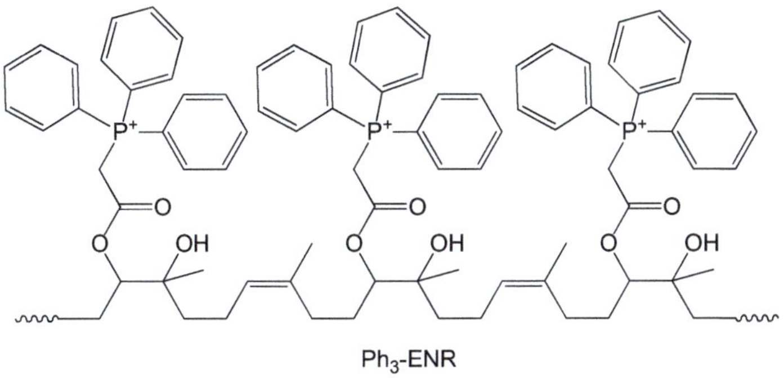

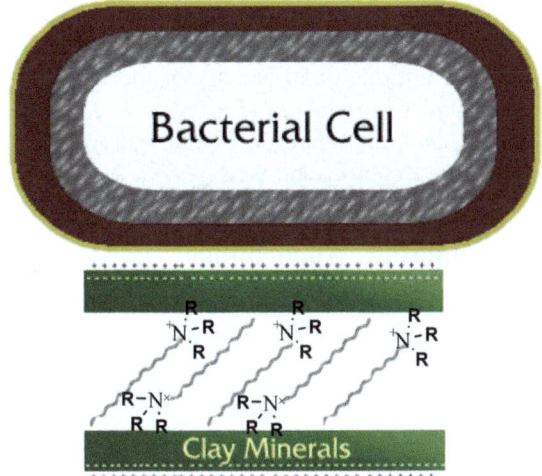

Ph$_3$-ENR

Figure 3.15 Triphenyl phosphine epoxidized natural rubber.

Figure 3.16 The schematic diagram of the antimicrobial activity of tetradecyl tributyl phosphonium bromide -clay minerals (redrawn from ref. 8).

effectively inhibited by using quaternary phosphonium salt modified epoxidized natural rubber. Meanwhile, the modified epoxidized natural rubber illustrated better antimicrobial activity towards Gram-negative *E. coli* than Gram-positive *S. aureus*. The described method provided potential applications in preparing permanent antimicrobial rubbers.

Ting Wu *et al.* reported different clay minerals intercalated by quaternary phosphonium salt (tetradecyl tributyl phosphonium bromide, TDTB) and evaluated the antimicrobial activity of these organo-clay minerals on *E. coli* and *S. aureus*.[8] Figure 3.16 shows a schematic presentation of the clay intercalated with TDTB.

As reported by the authors, the antimicrobial activity of those organo-clay minerals depended on three factors; these are the releasing amount of tetradecyl tributyl phosphonium bromide; surface charge; and particle size of organo-clay minerals. The results established that antimicrobial activity was the synergic effect of these three factors. Among the four organo-clay minerals, clay-tetra-decyl tributyl phosphonium bromide exhibited the highest antimicrobial activity because of its releasing amount of organic antimicrobials.

3.4 Examples of Applications of Polymers with Ammonium and Phosphonium Groups

3.4.1 Water Treatments

Crosslinked antimicrobial polymers for antimicrobial treatment were reported by Kenawy *et al.*[29] The crosslinked copolymers were synthesized based on copolymerization of vinylbenzyl chloride either with 2-chloroethyl vinyl ether, or methyl methacrylate using the divinylbenzene as crosslinker. The crosslinked copolymers were further modified by quaternization with triphenylphosphine and triethylamine. The antimicrobial activity of the modified copolymers was tested against various micro-organisms (*S. aureus, E. coli, B. subtilis, A. flavus, F. oxysporum* and *C. albicans*). The copolymers showed good antimicrobial activity against the tested micro-organisms.

3.4.2 Dental Composites Applications

Beyth *et al.* reported the preparation of dental composites containing quaternary ammonium polyethylenimine nanoparticles.[19] The structure of the synthesized nanoparticles of quaternary ammonium polyethylenimine is that depicted in Figure 3.17.

   The dental composite was prepared by adding the synthesized polymer to three commercial composite resins. The antibacterial activity was tested with *S. mutans*. Antimicrobial tests using *S. mutants* showed that these polyethyleneimine nanoparticles when incorporated in dental composite resins at a concentration as low as 1%, exhibited a strong antibacterial effect against the tested bacteria. Composite resin materials that were incorporated with

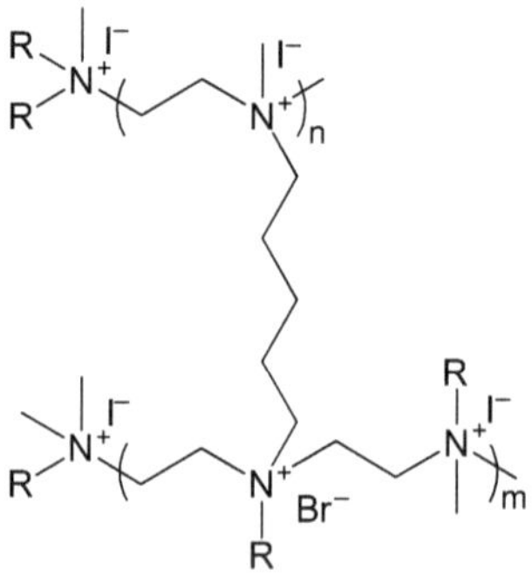

Figure 3.17   Quaternary ammonium polyethylenimine.

polyethyleneimine nanoparticles, maintained antibacterial properties over 1 month without leaching out and with no alteration of the original mechanical properties. For composite resin restorations, incorporation of antibacterial nanoparticles may prevent biofilm formation and secondary caries.

3.4.3 Food Packaging

Aliphatic polyester poly(butylene adipate), which was functionalized with a quaternary phosphonium group by means of the copper-catalyzed azide-alkyne "click" reaction, has been investigated as a potential polymer for antimicrobial packaging material.[42] This reaction has the great advantage of being carried out under mild conditions, thus the properties and chain length of the polyester backbone are not altered. The functionalized polyester was able to significantly reduce cell counts of *E. coli*, both in dispersion and directly on the surface.

3.4.4 Textiles

The growth of microbes on textiles during use and storage negatively affects the wearer as well as the textile itself.[44] Hsu and Klibanov synthesized a photosensitive hydrophobic polycationic salt starting from branched polyethylenimine.[45] Positive charges were maximized by methylating the nucleophilic amino groups into quaternary ammonium groups, yielding the final photosensitive *N*-alkyl-polyethylenimine. Plain cotton fabric was then dipped into a solution of the polymer in dichloroethylene and dried in the dark. The polymer can be covalently bonded to the cotton fabric by means of UV light. Antimicrobial efficiency of the coated-fabric was determined against *E. coli* and *S. aureus* in phosphate-buffered saline at 250 rpm at 37 °C for 2 h with *E. coli* and at room temperature for 4 h with *S. aureus*.

3.4.5 Anticoagulant

Fan *et al.* synthesized quaternary ammonium chitosan sulfates with diverse degrees of substitution by reacting quaternary ammonium chitosan with an uncommon sulfating agent ($N(SO_3Na)_3$) that was prepared from sodium bisulfite ($NaHSO_3$) through reaction with sodium nitrite ($NaNO_2$) in a homogeneous aqueous system. Its anticoagulation activity *in vitro* was determined by an activated partial thromboplastin time assay, a thrombin time assay and a prothrombin time assay. The results reported by the authors proved that anticoagulation assays of quaternary ammonium chitosan sulfates were significantly prolonged activated partial thromboplastin time and thrombin time, and they demonstrated that the introduction of sulfate groups into the quaternary ammonium chitosan structure, obviously improved its anticoagulant activity. The study showed that its anticoagulant properties strongly depended on its degrees of substitution, concentration and molecular weight.[46]

3.4.6 Lowering Cholesterol

In the last several decades intensive, scientific, and clinical evidences have been accumulated, and demonstrated the relationship between elevated blood cholesterol levels and an increased risk for the development of coronary heart disease.[47] This coronary heart disease, which is linked to high levels of cholesterol or triacylglycerols carrying lipoproteins especially low-density lipoproteins (LDL)-bound cholesterol, comes under the general phenomena of hyperlipoproteinemias. The drugs of first choice for treatment of hyperlipoproteinemias are the bile acid sequestrants, (cholestyramine and colesevelam hydrochloride) (Figures 3.18 and 3.19).[47–49] Cholestyramine and colesevelam hydrochloride are two examples of FDA approved commercial bile acid sequestrants that are used as pharmaceutical agents for the reduction of elevated levels of plasma cholesterol.

The bile acid sequestrants act as anion exchange resins, binding bile acids in the lumen of the small intestine. Bile acid sequestrants are able to interrupt the enterohepatic circulation of bile acids. This results in an increased hepatic synthesis of bile acids from cholesterol because bile acids suppress the microsomal hydroxylase that catalyzes the rate-determining step in the conversion of cholesterol to bile acids.

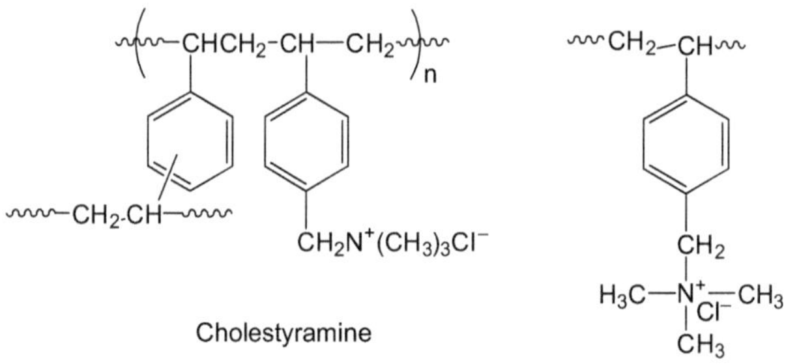

Figure 3.18   Structure of cholestyramine.

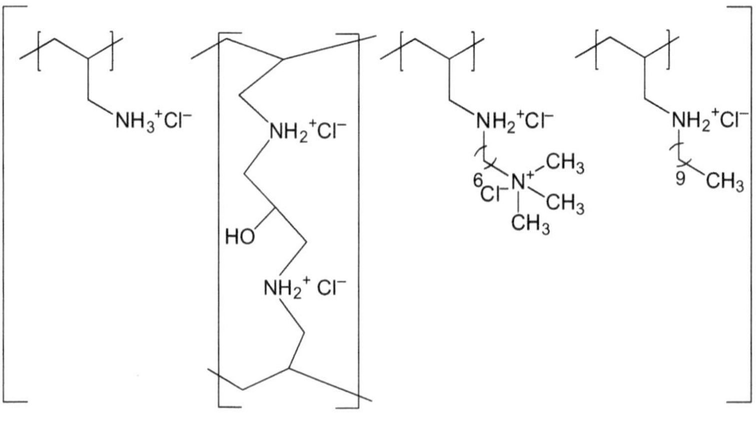

Figure 3.19   Structure of Colesevelam hydrochloride.

The bile acid sequestrant agents are the oldest and safest choice in the treatment of patients with elevated levels of low-density lipoprotein cholesterol; these drugs are not absorbed by intestines.[50,51] Because bile acid sequestrant is highly positively charged macromolecules, it can bind with the negatively charged bile acids. Also, they are not absorbed and after binding with bile acid they are excreted in the stool.

Cameron *et al.* investigated the synthesis of polystyrene-*b*-poly(*N*,*N*,*N*-trimethyl ammonium ethylene acrylamide chloride) by employing both polystyrene-*b*-poly(*tert*-butyl acrylate) and its hydrolyzed derivative, polystyrene-*b*-poly-(acrylic acid) as starting materials, and coupling them with *N*,*N*-dimethyl ethylene diamine, followed by quaternization (Figures 3.20 and 3.21). These polymers aggregate as potential matrices for bile acid sequestering in the human gut.[52]

Huval *et al.* prepared a series of amine-functionalized polymers based on polyether backbones by the chemical modification of poly(epichlorohydrin) and poly(2-chloroethylvinyl ether) (Figure 3.22).[53] The amine functional

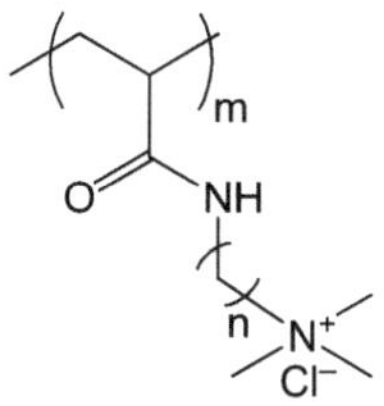

Figure 3.20 Chlorotrimethylammonium-bearing repeat unit for poly(acrylamide) (QPDA-*n*) with an alkyl spacer where *n* is typically 2–12.

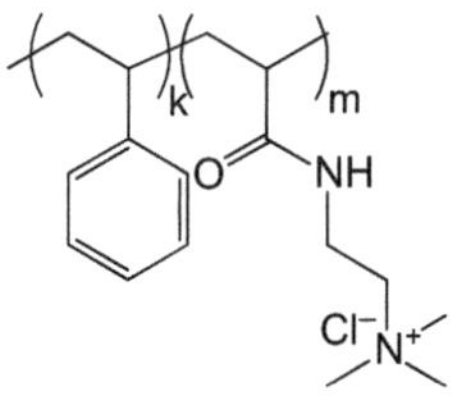

Figure 3.21 Polystyrene-*b*-poly(*N*,*N*,*N*-trimethylammonium ethylene acrylamide chloride) (PS-*b*-PTMEACl).

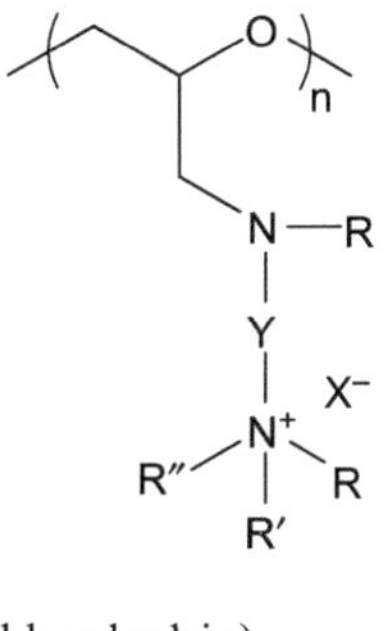

Figure 3.22 Modified poly(epichlorohydrin).

polyethers exhibited promising bile acid sequestration properties during *in vivo* experiments using hamsters as animal models, providing a novel approach for treating hypercholesterolemia. Some of these polymers show efficacy superior to commercially available bile acid sequestrants. The results suggest that these novel polyammonium gels may be useful as cholesterol-lowering agents.

References

1. Z. X. Peng, L. Wang, L. Du, S. R. Guo, X. Q. Wang and T. T. Tang, *Carbohydr. Polym.*, 2010, **81**, 275–283.
2. S. M. Kurtz, E. Lau, J. Schmier, K. L. Ong, K. Zhao and J. Parvizi, *J. Arthroplasty*, 2008, **23**, 984–991.
3. J. C. Tiller, C. Sprich and L. Hartmann, *J. Control. Release*, 2005, **103**, 355–367.
4. E.-R. Kenawy, S. S. Al-Deyab, N. O. Shaker, B. M. El-Sadek and H. B. Abeer Khattab, *J. Appl. Polym. Sci.*, 2009, **113**, 818–826.
5. D. Zhao, J. Zhou and N. Liu, *Appl. Clay Sci.*, 2006, **33**, 161–170.
6. H. He, D. Yang, P. Yuan, W. Shen and R. L. Frost, *J. Colloid Interf. Sci.*, 2006, **297**, 235–243.
7. Y. Zhou, M. Xia, Y. Ye and C. Hu, *Appl. Clay Sci.*, 2004, **27**, 215–218.
8. T. Wu, A.-G. Xie, S.-Z. Tan and X. Cai, *Colloid Surf. B: Biointerfaces*, 2011, **86**, 232–236.
9. E.-R. Kenawy, S. D. Worley and R. Broughton, *Biomacromolecules*, 2007, **8**, 1359–1384.
10. E.-R. Kenawy, F. I. Abdel-Hay, A. E.-R. R. El-Shanshoury and M. H. El-Newehy, *J. Polym. Sci. Part A: Polym. Chem.*, 2002, **40**, 2384–2393.
11. G. Domagk, *Deutsche Medizinische Wochenschrift*, 1935, **61**, 829–832.
12. A. Skrzypczak, B. Brycki, I. Mirska and J. Pernak, *Eur. J. Med. Chem.*, 1997, **32**, 661–668.
13. S. Buffet-Bataillon, P. Tattevin, M. Bonnaure-Mallet and A. Jolivet-Gougeon, *Int. J. Antimicrob. Ag.*, 2012, **39**, 381–389.
14. L. H. Fan, M. Cao, S. Gao, W. P. Wang, K. Peng, C. Tan, F. Wen, S. X. Tao and W.G. Xie, *Carbohydr. Polym.*, 2012, **88**, 707–712.
15. B. L. Rivas, E. D. Pereira, M. A. Mondaca, R. J. Rivas and M. A. Saavedra, *J. Appl. Polym. Sci.*, 2003, **87**, 452–457.
16. S. B. Lee, R. R. Koepsel, S. W. Morley, K. Matyjaszewski, Y. Sun and A. J. Russell, *Biomacromolecules*, 2004, **5**, 877–882.
17. R. Adelmann, M. Mennicken, D. Popescu, E. Heine, H. Keul and M. Moeller, *Eur. Polym. J.*, 2009, **45**, 3093–3107.
18. E. Salleh, I. I. Muhamad and N. Khairuddin, *Asian Chitin J.*, 2007, **3**, 55–68.
19. N. Beyth, I. Yudovin-Farber, R. Bahir, A. J. Domb and E. I. Weiss, *Biomaterials*, 2006, **27**, 3995–4002.
20. A. Varesano, C. Vineis, A. Aluigi and F. Rombaldoni, in *Science against Microbial Pathogens: Communicating Current Research and Technological*

Advances, ed. A. Méndez-Vilas, Microbiology Book Series, Formatex Research Center, Badajoz, Spain, 2011, Number 3, 99.

21. S. H. Lim and S. M. Hudson, *Carbohyd. Res.*, 2004, **339**, 313–319.
22. X. Fu, Y. Shen, X. Jiang, D. Huang and Y. Yan, *Carbohydr. Polym.*, 2011, **85**, 221–227.
23. A. Muñoz-Bonilla and M. Fernandez-Garcia, *Prog. Polym. Sci.*, 2012, **37**, 281–339.
24. E.-R. Kenawy and Y. A.-G. Mahmoud, *Macromol. Biosci.*, 2003, **3**, 107–116.
25. T. Ikeda and S. Tazuke, *Polym. Prep.*, 1985, **26**, 226–227.
26. T. Nonaka, Hua, T. Ogata and S. Kurihara, *J. Appl. Polym. Sci.*, 2003, **87**, 386–393.
27. Y. Uemura, I. L Moritake, S. Kurihara and T. Nonaka, *J. Appl. Polym. Sci.*, 1999, **72**, 371–378.
28. C. Z. Chen, N. C. Beck-Tan, P. Dhurjati, T. K. Van Dyk, R. A. LaRossa and S. L. Cooper, *Biomacromolecules*, 2000, **1**, 473–480.
29. E.-R. Kenawy, F. I. Abdel-Hay, A. Abou El-Magd and Y. Mahmoud, *React. Funct. Polym.*, 2006, **66**, 419–429.
30. E.-R. Kenawy, F. I. Abdel-Hay, A. R. El-Shanshoury and M. H. El-Newehy, *J. Control. Release*, 1998, **50**, 145–152.
31. B. Dizman, M. O. Elasri and L. J. Mathias, *J. Appl. Polym. Sci.*, 2004, **94**, 635–642.
32. S. Venkataraman, Y. Zhang, L. Liu and Y.-Y. Yang, *Biomaterials*, 2010, **31**, 1751–1756.
33. S. Charles, N. Vasanthan, D. Kwon, G. Sekosan and S. Ghosh, *Tetrahedron Lett.*, 2012, **53**, 6670–6675.
34. O. V. Rúnarsson, J. Holappa, C. Malainer, H. Steinsson, M. Hjálmarsdóttir, T. Nevalainen and M. Másson, *Eur. Polym. J.*, 2010, **46** 1251–1267.
35. J. Holappa, T. Nevalainen, P. Soininen, M. Elomaa, R. Safin, M. Másson and T. Järvinen, *Biomacromolecules*, 2005, **6**, 858–863.
36. J. Holappa, M. Hjálmarsdóttir, M. Másson, Ö. Rúnarsson, T. Asplund, P. Soininen, T. Nevalainen and T. Järvinen, *Carbohydr. Polym.*, 2006, **65** 114–118.
37. W. Sajomsang, U. R. Ruktanonchai and P. Gonil and Ch. Warin, *Carbohydr. Polym.*, 2010, **82**, 1143–1152.
38. J. Y. Kim, J. K. Lee, T. S. Lee and W. H. Park, *Int. J. Biol. Macromol.*, 2003, **32** 23–27.
39. A. Kanazawa, T. Ikeda and T. Endo, *J. Polym. Sci. Part A: Polym. Chem.*, 1993, **31**, 1441–1450.
40. H.-I. Chang, M.-S. Yang and M. Liang, *React. Funct. Polym.*, 2010, **70**, 944–952.
41. A. Popa, C. M. Davidescu, R. Trif, Gh. Ilia and S. Iliescu and Gh. Dehelean, *React. Funct. Polym.*, 2003, **55**, 151–158.
42. T. Anthierens, L. Billiet, F. Devlieghere and F. Du Prez, *Innov. Food Sci. Emerg. Technol.*, 2012, **15**, 81–85.
43. C. Li, Y. Liu, Q.-Y. Zeng and N.-J. Ao, *Mater. Lett.*, 2013, **93**, 145–148.
44. Y. Gao and R. Cranston, *Text. Res. J.*, 2008, **78**, 60–72.

45. B. B. Hsu and A. M. Klibanov, *Biomacromolecules*, 2011, **12**, 6–9.
46. L. Fan, P. Wu, J. Zhang, S. Gao, L. Wang, M. Li, M. Sha, W. Xie and M. Nie, *Int. J. Biol. Macromol.*, 2012, **50**, 31–37.
47. P. Zarras, *J. Polym. Sci. Part A: Polym. Chem.*, 2004, **42**, 701–707.
48. C. Luca, S. Dragan, V. Barboiu and M. Dima, *J. Polym. Sci. Part A: Polym. Chem.*, 1980, **18**, 449–451.
49. C. C. Huval, S. R. Holmes-Farley, W. H. Mandeville, R. Sacchiero and P. K. Dhal, *Eur. Polym. J.*, 2004, **40**, 693–701.
50. R. W. Mahley and T. P. Bersot, In Goodman & Gilman's *The Pharmacological Basis of Therapeutics*, 10th edn pp 971. A. Goodman Gilman, T. W. Rall, A. S. Nies, and P. Taylor, ed.; McGraw-Hill, New York, 2001.
51. P. Zarras and O. Vogl, *Prog. Polym. Sci.*, 1999, **24**, 485–516.
52. N. S. Cameron, A. Eisenberg and G. R. Brown, *Biomacromolecules*, 2002, **3**, 116–123.
53. C. C. Huval, M. J. Bailey, S. R. Holmes-Farley, W. H. Mandeville, K. Miller-Gilmore, R. J. Sacchiero and P. K. Dhal, *J. Macromol. Sci. Part A-Pure Appl. Chem.*, 2001, **38**, 1559–1565.

CHAPTER 4

Water-Soluble Antimicrobial Polymers for Functional Cellulose Fibres and Hygiene Paper Products

HUINING XIAO*[a,c] AND LIYING QIAN[b]

[a] Department of Chemical Engineering, University of New Brunswick, Fredericton NB Canada E3B 5A3; [b] State Key Laboratory of Pulp & Paper Engineering, South China University of Technology, Guangzhou, China; [c] School of Environmental Science and Engineering, North China Electric Power University, Baoding, China 071003
*Email: hxiao@unb.ca

Lessons learned from the spreading of *Escherichia coli* in various events have made it clear that fatal disease can be created due to the bacterial contamination without proper protection, including packaging. The maintenance of hygiene in various environments is of great importance. This has created an urgent need to develop antimicrobial packaging materials, both paper- and polymer-based, for daily uses. To render commonly used materials or articles antimicrobials, a common approach is to utilise antimicrobial agents, which, however, might not be strongly bonded to the substrates and could leach or wash off. To ensure the strong bonding of antimicrobial agents to substrates, a polymer such as starch can serve as a carrier. More importantly, antimicrobial starch could be readily applied in papermaking processes as a functional wet-end additive, thus leading to the hygiene paper products that are promising for various applications. The development of water-soluble antimicrobial polymers

RSC Polymer Chemistry Series No. 10
Polymeric Materials with Antimicrobial Activity: From Synthesis to Applications
Edited by Alexandra Muñoz-Bonilla, María L. Cerrada and Marta Fernández-García
© The Royal Society of Chemistry 2014
Published by the Royal Society of Chemistry, www.rsc.org

in recent years is the main focus of this chapter. The antimicrobial polymers with multiple functions such as wet-strength enhancing are also particularly valuable for some paper products like tissues and kitchen towels. As a further extension, cyclodextrin-based polymer has also been developed in the past decade. The cationic branches ensure the high water solubility, whereas the cyclodextrin as core could accommodate appropriate antibiotics for various applications.

4.1   Water-Soluble Polymers for Rendering Cellulose Fibres Antimicrobial

4.1.1   Guanidine-Based Polymers

Guanidine-based polymer is a series of polymers bearing guanidine groups, which has been extensively used as an antimicrobial agent due to its high water solubility, wide spectrum antimicrobial activity, excellent biocide efficiency and nontoxicity. Guanidine salt, biguanidine salt, cyanamide, dicyandiamide and other similar small molecules could act as monomers to impart the anti-microbial guanidio group to polymer to obtain guanidine-based polymers. The guanidio group is believed to be the effective antimicrobial group in guanidine-based polymer for its cationic charge. The strong positive charges in the guanidine-based polymer facilitate the rapid binding to the negative charged membrane of bacteria and displacing the presence of Mg^{2+} and Ca^{2+} ions simultaneously. This binding is also to the lipopolysaccharide and peptidoglycan components of the cell wall, which can change the phospholipids environment of the membrane and cause the loss of its biofunction, or destroy the cytoplasmic membrane and lead to the death of the bacteria.[1] Polyhexamethylene guanidine hydrochloride (PHGH) is a kind of typical guanidine-based polymers synthesised from the condensation polymerisation of hexamethylenediamine and guanidine hydrochloride. It is a mixture of polymer chains with various lengths, different molecular structures and comparatively low molecular weights. PHGH are often reckoned as an oligomer with only several repeat units and its molecular weight is difficult to characterise using classical methods for its high charge density. The detailed molecular structures in the condensed PHGH are shown in Figure 4.1 and seven types of structures are found. Among them, three types are linear, which are terminated with amine groups and/or guanidine groups; and the other four are branched or cyclic. The branched structures result from the branched reaction of the imine group of guanidinium and the cyclic structures are formed from the intramolecular reactions between the terminal amines. The contents of branched and cyclic molecules are increased by prolonging the reaction time and elevating the reaction temperature. As an effective antimicrobial agent, guanidine-based polymer has attracted extensive interest for rendering paper products antimicrobials, which require a high level of hygiene, such as hospitals, pharmaceutical units and food factories. It can also be tailor-made as novel additives by combining with azetidinium ring (AZR) or other functional

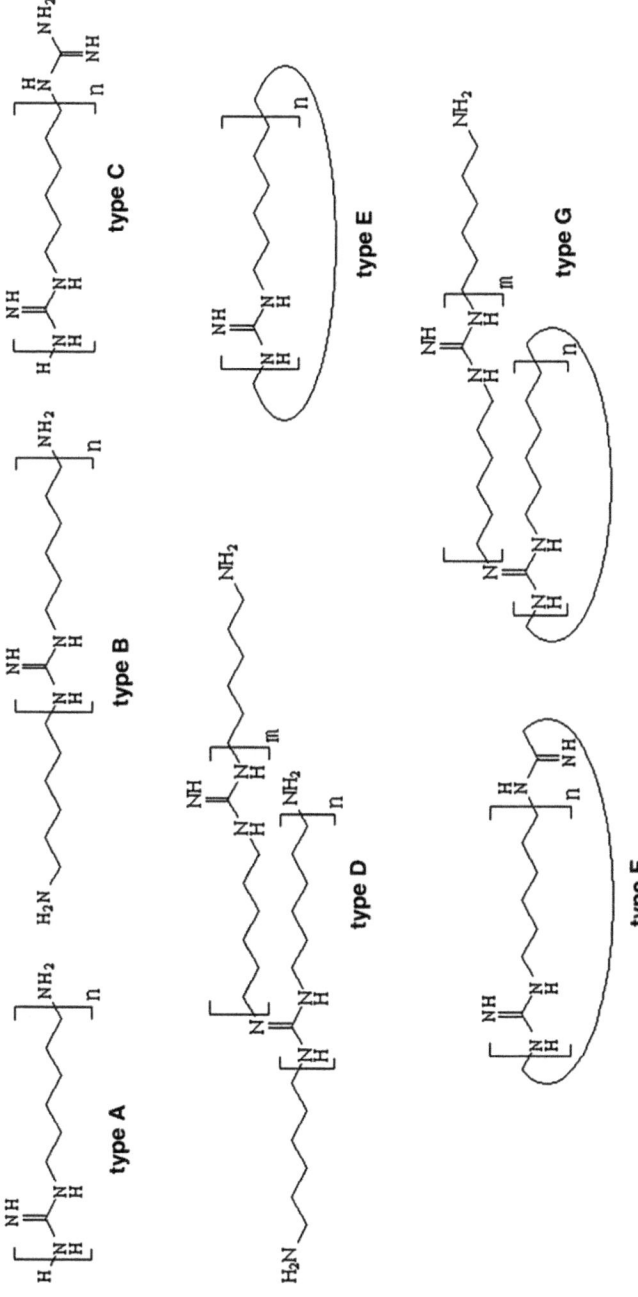

Figure 4.1 Molecular structures in the PHGH characterised using an electrospray ionisation time-of-flight mass spectrometry (ESI-TOF-MS). Reprinted from *Mater. Sci. Eng. C*, **29**, 1776–1780. Copyright (2009) with permission from Elsevier.

groups for specific applications such as tissue, kitchen towel due to its high reactivity.

4.1.2   Gemini Surfactant and Polymer

Quaternary ammonium salts (QAS) have been the most widely used antimicrobial agents owing to their excellent cell membrane penetration properties, low toxicity, good environmental stability, nonirritation, low corrosivity and extended residence time.[2] But the use of the conventional QAS in some fields has been limited due to the developed microbial resistance against them. Recently, the gemini surfactants containing the QAS groups has attracted considerable research interest not only for the higher antibacterial activities but also for the reactive functional groups in the chain. In particular, gemini surfactants with ester or amide spacers possess excellent antimicrobial activities. The reactive gemini surfactants can also be used as a macromonomer to copolymerise with other monomers to obtain the antimicrobial polymer.[3] Scheme 4.1 shows the synthesis route of gemini surfactant monomers containing a double bond and its copolymerisation with other monomers.[4] Herein, a convenient and efficient monoamidation of alkylenediamines with the quaternary ammonium gemini surfactants of L-lysine ester was employed and the yields of N-acylalkylenediamines were thus very high. Finally, the L-lysine quaternary ammonium gemini surfactants were obtained from the other amino group left on the N-acylalkylenediamines easily linked to acrylic chloride. The resulting monomers potentially provide a new strategy for developing antimicrobial polymers with extremely high activities owing to the polycationic structures, which facilitate the penetration of polymer chains into cell membranes, thus improving the pathogen deactivation. For example, the gemini surfactants could be effortlessly copolymerised with N-(3-aminopropyl) methacrylamide hydrochloride (APMA) (Scheme 4.1) using a conventional free-radical polymerisation. The antibacterial and antifungal activities of these gemini surfactants and its copolymers are shown in the Table 4.1 and both of them possess the excellent inhibition activity against the bacteria and fungi.[4]

4.2   *In Situ* Grafting of Antimicrobial Polymer

4.2.1   Grafting onto Cellulose Fibre

It is important to develop an appropriate way to graft the antimicrobial monomers or polymers onto the surface of cellulose fibre, *i.e.* incorporating antimicrobial agents with cellulose fibre through covalent bonding, which is critical in the textile industry for durable application. The advantage of this approach is that the antimicrobial agents would not leach out from the substrates to ensure the long-term inhibition activity of cellulose fibre. Different types of photosensitisers have been covalently grafted on cotton fabric using a 1,3,5-triazine derivative as the linker to prepare the materials for anti-microbial applications.[5] Chitosan was also crosslinked with cotton fabrics by

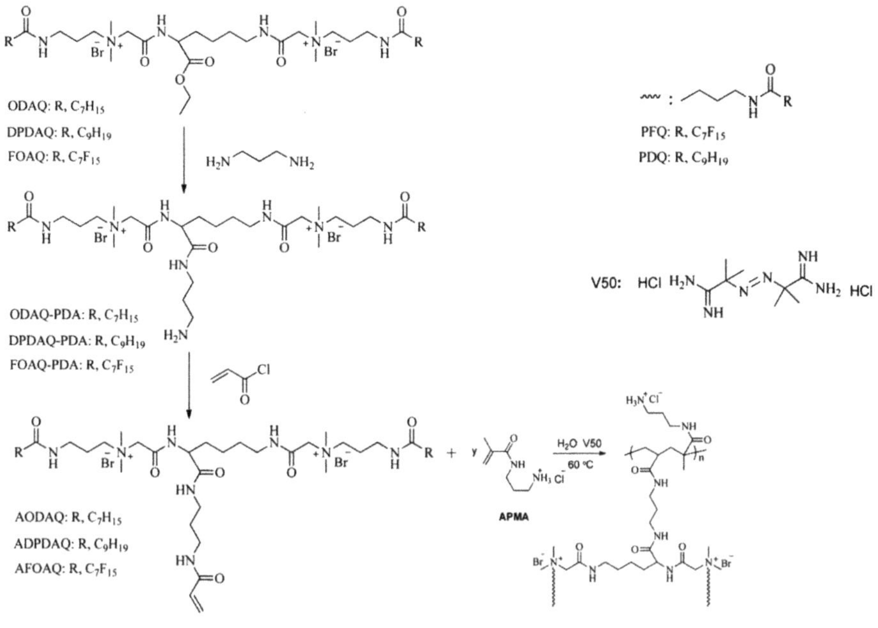

Scheme 4.1 Synthetic route of gemini surfactant monomers containing a double bond and its copolymerisation with APMA.
Reprinted from *Polym. Chem.* **3**, 907–913. Copyright (2012) with permission from the Royal Society of Chemistry.

Table 4.1 Minimal inhibitory concentration results of gemini surfactant monomers, their corresponding copolymers and commercial reference *n*-dodecyl-trimethylammonium bromide (DTAB) ($\mu molL^{-1}/g\ mL^{-1}$).
Reprinted from *Polym. Chem.* **3**, 907–913. Copyright (2012) with permission from the Royal Society of Chemistry.

Sample	*Staphylococcus aureus*	*Escherichia coli*	*Candida albicans*	*Aspergillus niger*
DTAB	250 (77.1[a])	250 (77.1[a])	500 (154.2[a])	250 (77.1[a])
ADPDAQ	7.81 (7.88[a])	15.63 (15.8[a])	250 (252.2[a])	62.5 (63.1[a])
AFOAQ	3.91 (5.84[a])	31.25 (46.7[a])	15.63 (23.4[a])	62.5 (93.4[a])
PDQ	74.0[a]	148.0[a]	–	–
PFQ	54.0[a]	54.0[a]	–	–

[a]The unit is $\mu g\ mL^{-1}$.

epicholorohydrin (EP) to provide antimicrobial properties and good dyeability to the fabrics.[6] The *in situ* graft copolymerisation of varieties of monomers onto cellulose has been carried out by different techniques, such as irradiation with ultraviolet light, gamma rays, plasma ion beams, atom-transfer radical polymerisation, enzymatic grafting and by ceric (IV) ion initiation methods. Enzymatic grafting is an ecofriendly approach to meet the growing consumers' expectation of higher hygiene standards and safer products together with environment protection concerns.[7] Cellulose fibres have been grafted with laccases and natural phenols in order to develop antimicrobial properties. Syringaldehyde (SA), acetosyringone (AS) and *p*-coumaric acid (PCA) were enzymatically grafted onto unbleached flax fibres to give the antimicrobial properties to the fibre against *Staphylococcus aureus*, *Klebsiella pneumoniae* and *Pseudomonas aeruginosa*. The ceric (IV) ion initiation offers great advantages of forming radicals at the cellulose backbone through a single-electron-transfer process to promote grafting of monomers onto cellulose. As redox systems, a ceric ion–cellulose complex is initially formed as a result of one electron transfer, then ceric ion is reduced to cerous ion and a free radical is created on the cellulose backbone. Acrylic acid (AA), acrylonitrile (AN),[8] guanidine polymer[9] and β-cyclodextrin (β-CD) derivatives loaded with benzoic acid[10,11] were grafted onto cellulose fibre using ceric ammonium nitrate as an initiator.

Prior to PHGH being grafted onto cellulose fibre, the unsaturated double bonds were first introduced by reacting PHGH with glycidyl methacrylate (GMA); and then the modified PHGH was *in situ* copolymerised onto cellulose (Scheme 4.2). The copolymerisation was confirmed by the attenuated total reflectance-Fourier transform infrared spectroscopy (ATR-FTIR) spectra and energy-dispersive X-ray spectroscopy. Figure 4.2(1) shows the IR spectra of cellulose fibres and PHGH grafted fibres. It was confirmed the introduction of PHGH from the absorption peak at 1633 cm^{-1} due to imino groups (from

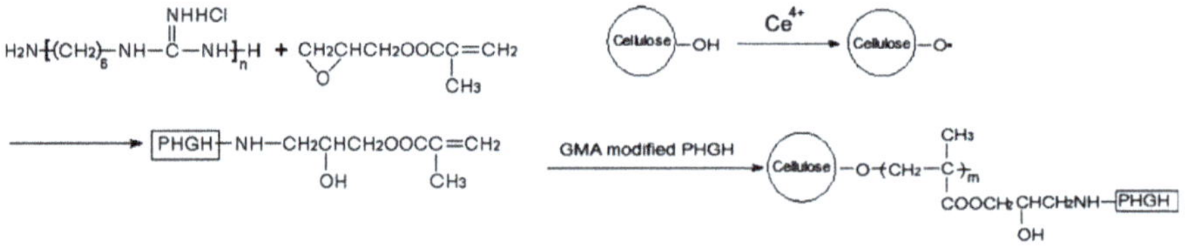

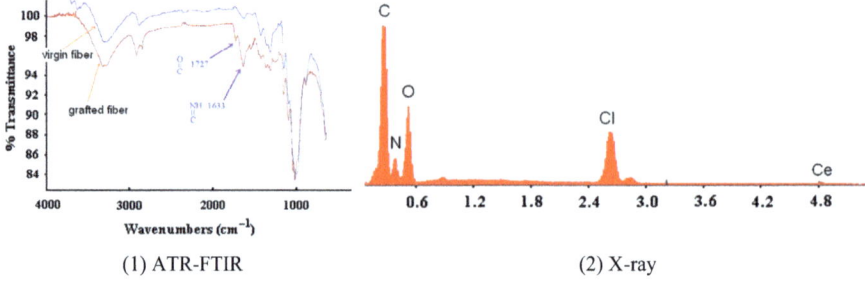

(1) GMA modified PHGH   (2) graft copolymerisation

Scheme 4.2   *In situ* copolymerisation of PHGH onto cellulose.
Reprinted from *Carbohyd. Polym.*, **69**, 688–696. Copyright (2007) with permission from Elsevier.

(1) ATR-FTIR   (2) X-ray

Figure 4.2   Characterisation of PHGH grafted cellulose fibre.
Reprinted from *Carbohyd. Polym.*, **69**, 688–696. Copyright (2007) with permission from Elsevier.

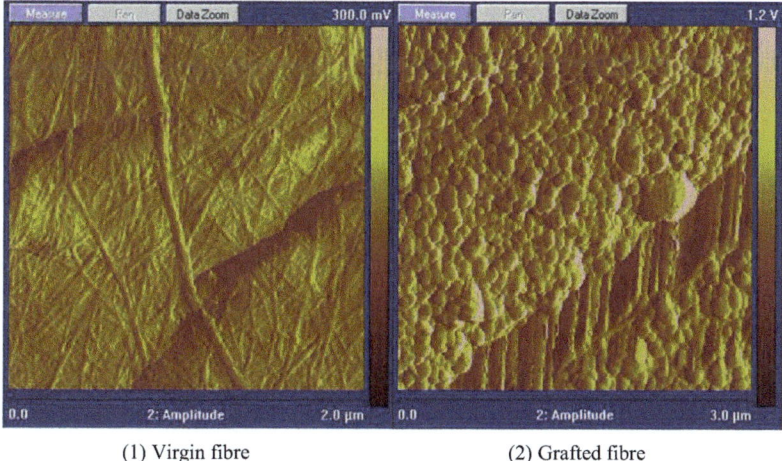

(1) Virgin fibre   (2) Grafted fibre

Figure 4.3   Morphology of virgin and grafted fibres.
Reprinted from *Carbohyd. Polym.*, **69**, 688–696. Copyright (2007) with permission from Elsevier.

PHGH) and at 1727 cm^{-1} due to carboxyl groups (from GMA). Figure 4.2(2) shows the energy dispersive X-ray spectroscopy of PHGH grafted fibres. The peaks of elements N, and Cl are attributed to the grafted PHGH. It should be noted that the peak location of nitrogen is quite near to those of carbon and oxygen, thus creating the overlapped peaks with nitrogen to some extent.

Atomic force microscopy (AFM) is an effective tool for revealing the morphology of the modified fibres and identifying the location of the grafted polymers. The fibrillar structure of the virgin fibre surface can be seen from the Figure 4.3. The random orientation of the cellulose microfibrils indicates they belong to the primary layer of the fibre. After graft copolymerisation, the surface of fibres appears to be covered with the granules consisted of grafted polymer and the linearly oriented part composed of crystal cellulose microfibrils.

4.2.2 Grafting onto Starch

Starch, a natural polysaccharide, has been extensively used in the paper industry to increase the dry strength and retain fines. The antimicrobial monomers or polymers can also be grafted onto starch to obtain the modified starch that could readily be added into the papermaking process as wet-end additives. The resulting antimicrobial paper products could be used in hospitals, pharmaceutical production units, *etc*. The antimicrobial agents with low molecular weight (MW) such as Ag ions or unmodified guanidine polymer can desorb easily from the fibre surface, or leach out during storage, thus impairing their antimicrobial performance.[12] Therefore, the key objective was to covalently bond antimicrobial polymers with relatively low MW onto the backbone of starch (as a carrier) in an attempt to enhance their retention on

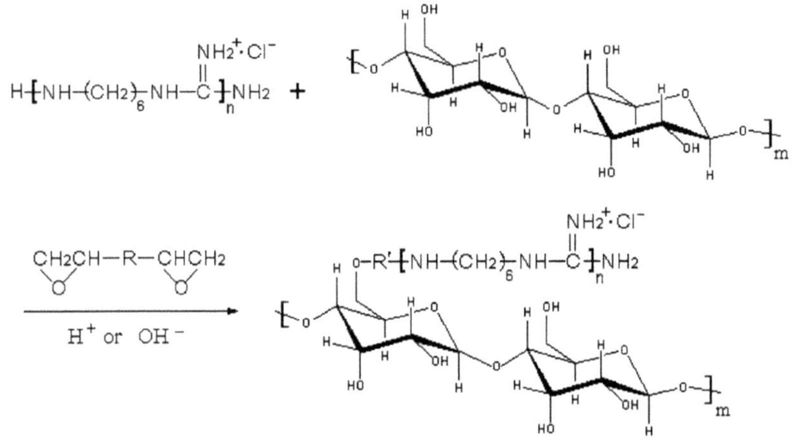

Scheme 4.3 Synthesis of guanidine polymer modified starch.
Reprinted from *Cellulose*, **15**, 609–618. Copyright (2008) with kind
permission from Springer Science and Business Media.

cellulose fibres. Furthermore, this kind of modified starch can be readily
adopted and applied to the papermaking industry without extra capital
investment or much change of existing manufacture process. Finally, the
antimicrobial modified starch is a multifunctional additive, which is desirable
for papermaking industries to develop innovative products.

The branched antimicrobial starch can be prepared by coupling PHGH onto
starch based on the hydroxyl groups of starch, and glycerol diglycidyl ether
(GDE) is used to couple the guanidine polymer to starches (shown in
Scheme 4.3).[13] The reactivity between guanidine polymer and amylose or
amylopectin should not be significantly different because the modification was
based on the hydroxyl groups of the starch. The crosslinking might occur over
the reaction, which can be eliminated by carefully controlling the reaction
conditions such as reaction time, temperature, pH and so on. FT-IR and ^{1}H-NMR
confirmed the grafting of PHGH on starch *via* GDE coupling and the coupling
efficiency could be calculated based on the results of C/N element analysis.

The guanidine-polymer-modified starch is cationic charged, contributed
from the guanidine side chains. As a result, the modified starch could readily
adsorb onto the surface of cellulose fibre *via* electrostatic association; and retain
in the paper product. However, the adsorption process of guanidine-modified
starch might be different from the commercial ones for the various molecular
structures (shown in Figure 4.4).[14] The cationic part of guanidine-modified
starch originates from the guanidine side chain with an average molecular
weight about 2600 (corresponding to about 13 positive-charged groups per side
chain). The guanidine-modified starch can adsorb onto the recycled fibre even
more than the virgin fibre and ultimately achieve the similar antimicrobial
performance and increase the strength properties of papers made from different
fibres. This is important to eliminate the impairment of recycled fibre from
impurities in pulp and the hornification phenomenon from the recycling.

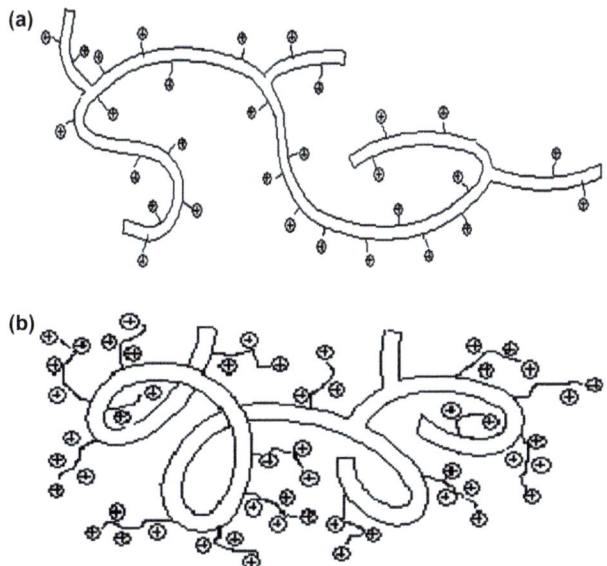

Figure 4.4 Molecular structure of cationic starches (a) Ordinary cationic starch, (b) Guanidine-modified starch.
Reprinted from *Cellulose*, **15**, 619–629. Copyright (2008) with kind permission from Springer Science and Business Media.

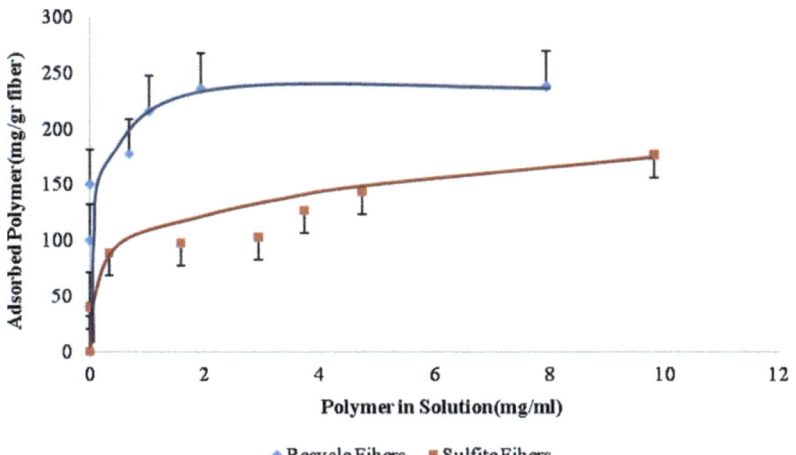

Figure 4.5 Adsorption isotherms of PHGH-modified starch on recycled and sulfite fibres.
Reprinted from *J. Biomat. Sci. Polym. Ed.*, **21**, 1359–1370. Copyright (2010) with permission from Taylor & Francis Ltd.

Because of the shortage of virgin fibres and environmental concerns, the application of recycled fibres is steadily increasing in paper industries, which raises the strong demands for the recycled fibres with improved quality.[15,16]

Figure 4.5 shows the adsorption isotherms of the starch with pendant PHGH chains on recycled and sulfite fibres. The driving force for such an adsorption is the electrostatic attraction between the cationic groups of PHGH-modified starch and anionic groups of fibres.[17] The antimicrobial polymer adsorbed more on recycled fibres than on sulfite fibres because of the higher surface charge densities and specific surface areas of recycled fibres. Meanwhile, the adsorption conditions including temperature, ionic strength and pH can influence the adsorption process and intrinsic electrostatic interaction between guanidine-modified starch and cellulose fibre.

4.3 Wet-Strength Antimicrobial Polymer Systems for Hygiene Paper Products

4.3.1 Modified Guanidine Polymer Containing Azetidinium Ring Group (E-PHDGC)

Wet-strength agents are necessary for making a number of paper products, such as tissue, paper towels, kitchen paper, food wrapper and bank notes.[18,19] Meanwhile, antimicrobial agents are also required in these products to protect human beings from being attacked by bacteria or virus. Therefore, functional polymer with synergistic wet-strength and antimicrobial activity is very important for these kinds of hygiene paper products. The four-membered AZR in polyamideamine-epichlorohydrin (PAE), a widely used wet-strength additive in the papermaking industry, is believed to be the effective group for developing the wet strength of paper.[20] From the point of polymer structure and design, the macromolecules bearing both guanidio group and AZR group are expected to be wet-strength antimicrobial additives for cellulose paper.[21] The polymer was normally obtained by two steps: prepolymer was first synthesised by condensation from monomers with various ratios, and then EP was introduced to form a four-membered ring (AZR) structure. Hexamethylene diamine (HMDA), diethylenetriamine (DETA) and adipic acid (AA) are the most common monomers that can be selected to polymerise with guanidine hydrochloride (GH) by polycondensation to obtain the prepolymer. Scheme 4.4 shows the synthesis of wet-strength antimicrobial polymer from the monomers like HMDA, DETA and GH.[22] After introducing EP, prepolymer could also be crosslinked *via* the secondary amine of DETA and HMDA; and this side-reaction should be minimised by controlling the reaction conditions. The wet-strength antimicrobial polymer was proved to be a mixture with various molecular structures for the side reactions in both steps. Both inhibition and the wet-strength enhancing abilities of the polymer are determined by the monomer ratios.

For the polymer shown in Scheme 4.4, minimum inhibition concentration (MIC) was lowered with higher HMDA content at the same amount of guanidio groups, though guanidine was generally reckoned to be valid to inhibit bacteria. Both molecular chain length and molecular stiffness of HMDA are larger than those of DETA, which suggests HMDA is behaving like a rigid

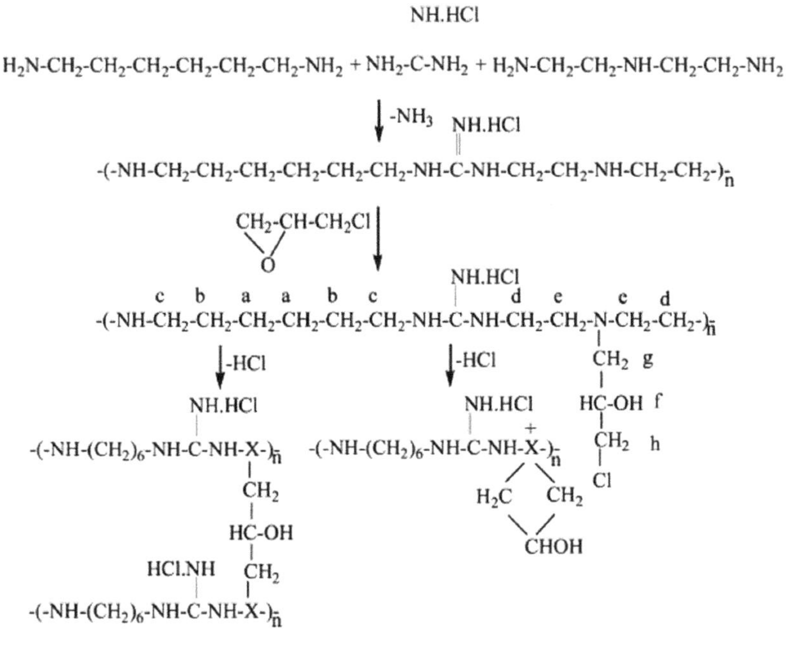

NH.HCl

H$_2$N-CH$_2$-CH$_2$-CH$_2$-CH$_2$-CH$_2$-CH$_2$-NH$_2$ + NH$_2$-C-NH$_2$ + H$_2$N-CH$_2$-CH$_2$-NH-CH$_2$-CH$_2$-NH$_2$

$\downarrow$ -NH$_3$ NH.HCl

-(-NH-CH$_2$-CH$_2$-CH$_2$-CH$_2$-CH$_2$-CH$_2$-NH-C-NH-CH$_2$-CH$_2$-NH-CH$_2$-CH$_2$-)$_n$

CH$_2$-CH-CH$_2$Cl $\downarrow$
$\diagdown$O$\diagup$

    c   b   a   a   b   c   NH.HCl   d   e   e   d

-(-NH-CH$_2$-CH$_2$-CH$_2$-CH$_2$-CH$_2$-CH$_2$-NH-C-NH-CH$_2$-CH$_2$-N-CH$_2$-CH$_2$-)$_n$

$\downarrow$ -HCl $\downarrow$ -HCl   CH$_2$ g

NH.HCl   NH.HCl   HC-OH f

-(-NH-(CH$_2$)$_6$-NH-C-NH-X-)$_n$ -(-NH-(CH$_2$)$_6$-NH-C-NH-X-)$_n$ CH$_2$ h

         CH$_2$ Cl

         HC-OH   H$_2$C CH$_2$

HCl.NH CH$_2$ CHOH

-(-NH-(CH$_2$)$_6$-NH-C-NH-X-)$_n$

a Note: X is $-(-CH_2-CH_2-N-CH_2-CH_2-)-$, and the side chain (or four-member ring) was connected to $-N-$.

Scheme 4.4 Synthesis of Modified Guanidine-Based Polymer with AZR group (E-PHDGC).
Reprinted from *ACS Appl. Mater. Interf.*, **3**, 1895–1901. Copyright (2011) with permission from the American Chemical Society.

rod. The rigidity and hydrophobicility enhanced by HMDA might facilitate the attachment of guanidine compounds or polymers to bacteria cells and further strengthen the attraction between polymer and bacteria apart from the electrostatic association between the cationic guanidio group and the anionic charge on bacterial surface. With increasing HMDA content, the polymer becomes more rigid and easier to destroy the cell membrane, so MIC to *E. coli* (ATCC11229) was as low as 8 ppm for the polymer without DETA. Antimicrobial activities of the polymer to bacteria can be visually observed by revealing the morphology of bacteria with AFM. Figure 4.6 presents the morphology of fresh *E. coli* and treated ones by modified guanidine polymer with AZR group.[23] For fresh *E. coli*, the surface membrane was structured and integrated, and there were no indentations and grooves on cell surface. An air-drying process prior to AFM observation hardly dehydrated the bacterial cells and there were no leaked residues around them. After being treated with the guanidine polymer, *E. coli* showed different patterns at lower and higher concentrations than MIC. Bacterial cells could maintain the intact shape after 30 min in the guanidine polymer solution when the polymer concentration was lower than MIC. There were no obvious indentations observed on the surface,

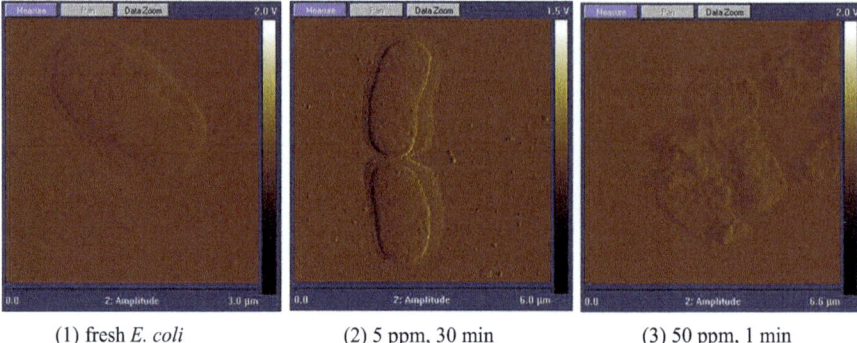

| (1) fresh *E. coli* | (2) 5 ppm, 30 min | (3) 50 ppm, 1 min |

Figure 4.6 Morphology of *E. coli* treated with modified guanidine polymer with AZR group.
Reprinted from *Polymer*, **49**, 2471–2475. Copyright (2008) with permission from Elsevier.

whereas a small amount of leakage was found around the entire membrane. The bacterial cells collapsed and the cell membrane was completely destroyed after being exposed to the polymer solution at a higher concentration (50 ppm) for 1 min, resulting in hardly distinguishing individual bacterial cells. Comparing AFM images of fresh *E. coli* and treated *E. coli*, it could be concluded that the antimicrobial mechanism of modified guanidine polymer with AZR group was to destroy the cell membrane of *E. coli* and induce the leakage of intracellular components from bacterial cells.

4.3.2 E-PHDGC/Carboxymethyl Cellulose Polyelectrolyte Complex

Polyelectrolyte complex (PEC), formed by cationic and anionic polyelectrolytes, has attracted great interest because of its unique properties and various applications. PEC could be water-soluble complex or water-insoluble nanoparticles, depending on the types of polyelectrolytes, charge ratio, ionic concentration, *etc*. The properties of water-soluble polymers can be changed or enhanced after the formation of PEC. It was reported that polyelectrolyte complexes of PAE and carboxymethyl cellulose (CMC) could improve the wet strength of a handsheet more efficiently than PAE alone. Modified guanidine polymer with AZR group can also form the polyelectrolyte complex with an anionic polyelectrolyte to increase the synergistic effect, especially the wet-strength enhancing ability. By forming complex, wet-strength antimicrobial polymer could be retained in handsheets effectively, thus improving strength properties *via* forming homo- and heterocrosslinking with fibres.[24] Wet/dry strength ratio of handsheets with PEC in various addition modes is several times higher than that of handsheets with the wet-strength antimicrobial polymer alone. Isothermal adsorption of wet-strength antimicrobial polymer forming PEC with anionic CMC was proved to be more than triple compared with the polymer alone system. Essentially, modified guanidine polymer with AZR group (E-PHDGC) is a kind of cationic polyelectrolyte with high charge density and low molecular weight, which resulted in the low adsorption amount on the cellulose fibre; anionic polyelectrolyte can form the bridging among the cationic polymer chains, thus enhancing the adsorption, which was also influenced by adsorption time, temperature, fibre consistency and salt concentration.

In order to reveal the characteristics of the polyelectrolyte complex, layer-by-layer (LBL) film can be self-assembled on a solid surface by alternate adsorption of cationic and anionic polyelectrolytes. The morphology, roughness and thickness of LBL film reveal the assembly mode of the PEC, which is affected by MW and charge density of polyelectrolytes, charge ratio of two polyelectrolytes, ionic strength, pH, *etc.* With the increasing of layer numbers of LBL film, the surface roughness showed different patterns for low and high molecular weight CMC. For low MW CMC, the roughness increased as the layer numbers of the assembly rose. For high MW CMC, however, the roughness was increased sharply with the first few layers and then decreased with more layers assembled. This might be attributed to the long chains of high-MW CMC, which facilitate the formation of loops when adsorbing onto the wafer surface treated with cationic antimicrobial-wet-strength polymer, resulting in significant roughness increase when building initial polyelectrolyte multilayer. AFM images of polyelectrolyte multilayer films (Figure 4.7) also revealed that the cationic antimicrobial-wet strength polymer and anionic CMC formed the complex with micrograins shape at the first stage of the layer assembly, regardless of MW of CMC.[25] However, as more layers were built up, the morphological changes appeared to be influenced by CMC molecular weights. For low-MW CMC, particle-like aggregates occurred at a high number of assembly layers. In contrast, for high-MW CMC, similar aggregates were observed at a low number of assembly layers; and subsequent layers might match each other or adsorb into the "valley" between "hills",[26] thus decreasing surface roughness.

4.3.3 Chitosan/Tripolyphosphate Nanocapsule

The natural polymer, chitosan, has been employed by the papermaking industry, but when used as a wet-strength agent with a high molecular weight, it showed relatively low antibacterial activity. In an attempt to increase antimicrobial properties, while maintaining high wet strength, guanidine-modified chitosan was prepared since it has been demonstrated that guanidine polymers exhibited higher antimicrobial activities than chitosan. Meanwhile, a nanocapsule system formed by chitosan with a crosslinking agent is also an effective way to deliver drug or antimicrobial agents. Chitosan is readily soluble in an acidic environment due to protonation of the amine groups. The resultant positive charge makes it possible to prepare nanocapsules by ionotropic gelation with multivalent anions, such as tripolyphosphate (TPP).[27] Many factors

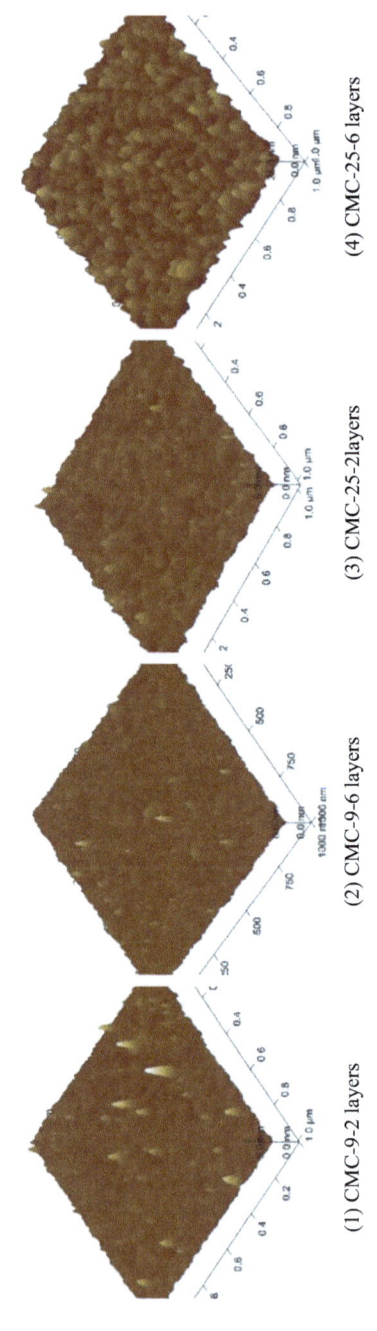

(1) CMC-9-2 layers   (2) CMC-9-6 layers   (3) CMC-25-2layers   (4) CMC-25-6 layers

Figure 4.7 Morphology of LBL film assembled by wet-strength antimicrobial polymer and CMC (CMC-9: MW = 90 000; CMC-25: MW = 250 000).

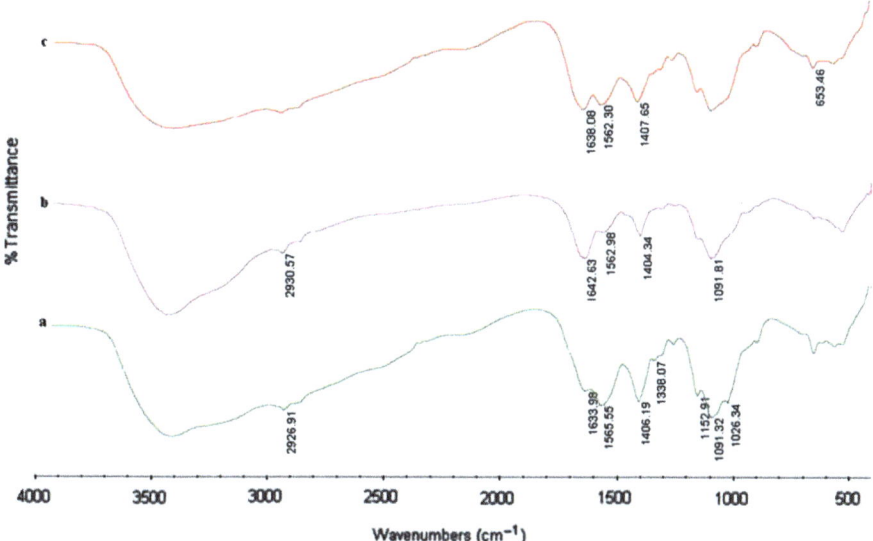

Figure 4.8 FTIR spectra of native chitosan (a), chitosan–PHGH prepolymer complex (b), chitosan–PHGH complex (c).

have important impacts on the characteristics of the nanocapsule, such as chitosan concentration, MW and crosslinking agent/chitosan ratio.

The guanidine polymers can also be wrapped into the chitosan/TPP nano-capsule to develop novel wet-end additives with synergistic antimicrobial and wet strength for the production of specialty value-added papers. The modified guanidine polymer can be mixed with chitosan in an acetic acid aqueous solution in the existence of sodium tripolyphosphate to form chitosan–guanidine nanoparticles. Figure 4.8 shows the FTIR spectra of chitosan and chitosan–PHGH complexes.[28] In comparison with the spectra of native chitosan, some changes in the FTIR spectra of chitosan–PHGH prepolymer and chitosan–PHGH nanoparticles can be observed. The major changes in the FTIR spectra for the nanocapsules are that the bands at 1633 cm^{-1} were shifted to a higher wave number and the intensity of the stretching of carbonyl amide band strengthened, and that the intensity of the bands of N–H bending vibration was weakened and shifted to 1562 cm^{-1}. It could be concluded from the FTIR analysis that the modified guanidine polymer was introduced into chitosan/TPP nanocapsules. The modified guanidine polymer can be also released slowly from the chitosan/TPP nanocapsules shown in Figure 4.9.[29] The released amount increased within 24 h, and levelled off after 120 h. The release rate and amount of PHGH from the nanocapsules was faster than that of PHGHE due to its lower MW. The initial increase in release within 24 h was attributed to the guanidine polymer chains dissociating from the external surface of nano-particles. The remainder belonged to guanidine polymer chains located inside of the nanocapsules. The slow release of guanidine polymers from the nano-capsules ensures the high retention of antimicrobial polymers in paper and

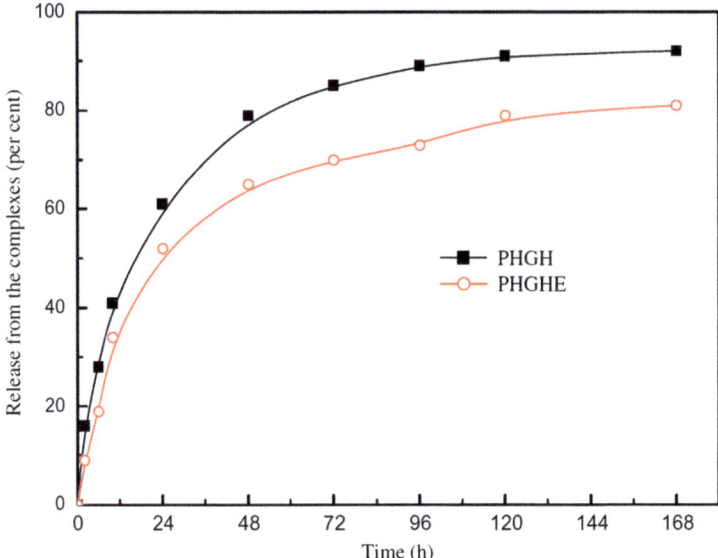

Figure 4.9   Release curves of modified guanidine polymer from the chitosan/TPP
nanocapsules (chitosan to guanidine was 2 : 1).
Reprinted from *Bioresource Technol.*, **101**, 5693–5700. Copyright (2010)
with permission from Elsevier.

potentially enhances the long-term effectiveness of the polymer in inhibiting
bacterial growth. The chitosan/TPP nanocapsules loaded with guanidine
polymer exhibit an excellent inhibition activity to *E. coli* and *S. aureus* and
extraordinary wet-strength-enhancing ability at proper dosage and chitosan/
guanidine polymer ratio.

4.4   Cationic Branched Polymers as Functional Additives for Antimicrobial Paper

4.4.1   Cyclodextrin-Based Polymer/Antibiotics (CPβCD) Complex as Functional Additives

Cyclodextrins (CDs) are a series of torus-shaped cyclic oligosaccharides consisting of six to eight glucose units linked by α-1, 4 bonds. β-CD, which consists
of seven glucose units, is the most important and widely used one among them.
The internal hydrophobic cavities in the CDs can host the guest molecules to
form the inclusion complexes, which can improve the properties of the guest
compound or enable release under a controlled manner. Several influencing
factors that control the incorporation of a guest compound in the cavity of the
cyclodextrin include steric effect, release of high-energy water, and hydrophobicity. In order to overcome the drawbacks of β-CD, such as poor water
solubility and the small size of the cavity, cyclodextrin-based polymers were

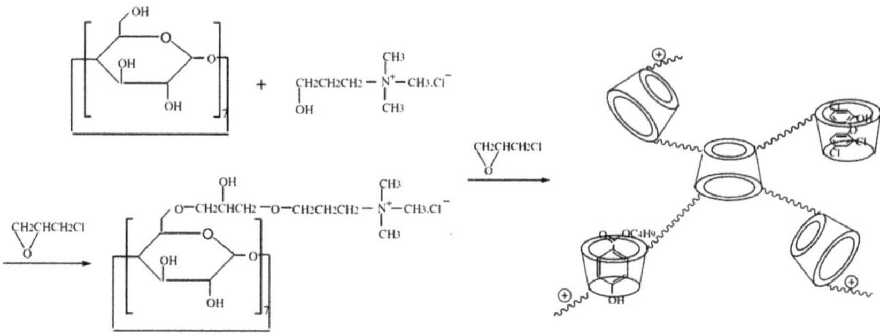

Scheme 4.5 Synthesis of cationic β-CD based polymer and its inclusion with (a) triclosan and (b) butylparaben.
Reprinted from *Cellulose*, **16**, 309–317. Copyright (2009) with kind permission from Springer Science and Business Media.

developed due to their high solubility in water and capability to include relatively larger guest molecules by the cooperation of two adjacent β-CD moieties on a polymer chain.[30–34] It was found that cationic cyclodextrin-based polymers of relatively high molecular weight and low cationic charge density exhibited good drug inclusion and dissolution abilities.[35,36] The antibiotics, such as triclosan and butylparaben, with similar size to the β-CD cavity can be easily included to form the complex with antimicrobial activity. By forming the complexes with cationic β-CD-based polymers, the solubility of antibiotics can be significantly improved and the cationic charge can help the adsorption of the antibiotics complexes onto the anionic cellulose fibre surfaces. The typical scheme of synthesising cationic β-CD-based polymer and forming a host–guest complex with antibiotics is shown in Scheme 4.5.[37] During the reaction, EP plays an important role as crosslinker for imparting cationic charge to β-CD and polymerising to cationic polymer. Five β-CD molecules were contained in one cationic polymer averagely according to the molecular weight of cationic polymer measured by gel permeation chromatography (GPC). The cationic β-CD-based polymer was highly water-soluble because of the introduced cationic charges compared with the poor solubility of β-CD. There are various methods to obtain the complex of cationic β-CD-based polymer and antibiotics; and grinding is a simple and convenient one. The hygiene paper products modified with the antimicrobial complexes could be prepared *via* adsorption or spraying. It was found that the inhibition effect of the spraying is better than that of the adsorption (see the results in Table 4.2).[37] The shaking flask method is also an effective way to assess the antimicrobial activity of paper products. The effects of antibiotic concentration and contacting time are listed in Table 4.3.[38] As can be seen, the growth inhibition raised with the increase in antibiotics' concentration or contacting time. The antimicrobial activity of triclosan is higher than that of butylparaben, whereas butylparaben exhibited a faster inhibition effect than triclosan because the smaller molecular size, which facilitates the release of butylparaben from the complex and promotes the deactivation of the bacteria in a relatively shorter contact time at

Table 4.2   Antimicrobial activities of the complex with cationic β-CD based polymer and antibiotics by diffusion method. Reprinted from *Cellulose*, **16**, 309–317. Copyright (2009) with kind permission from Springer Science and Business Media.

Sample	Antibiotics content[b] (% wt)	Diameter of inhibition zone (mm)[c] (against E. coli 11229)	Diameter of inhibition zone (mm)[c] (against Salmonella)
Blank	0	0	0
Troclosan/CPβCD	0.5	6.8	7.2
Troclosan/CPβCD	1.0	24.8	22.6
Troclosan/CPβCD[a]	0.5	–	23.0
Butylparaben/CPβCD	0.4	5.0	6.2
Butylparaben/CPβCD	1.0	7.3	9.3
Butylparaben/CPβCD[a]	0.5	–	12.3

[a]Samples were prepared by spraying antimicrobial agent on the paper
[b]The antimicrobial agent content means triclosan or butylpharaben weight percentage based on wood fibers (o.d.).
[c]The diameter of inhibition zones equals to the diameter difference between the ring and the paper sheet.

Table 4.3   Antimicrobial activities against *E. coli* ATCC 11229 of the complex with cationic β-CD-based polymer and antibiotics by shaking flask method. Reprinted from *Macromol. Rapid Commun.*, **28**, 2244–2248. Copyright (2007) with permission from John Wiley and Sons.

Contacting time[b]	Growth inhibiton[a]		Concentration[c]	Growth inhibiton[a]	
	%			%	
min	Butylparaben	Triclosan	wt.-%	Butylparaben	Triclosan
0.5	50.00 ± 5.62	24.58 ± 5.21	0.03	4.54 ± 1.85	23.18 ± 4.67
1	61.86 ± 6.85	37.71 ± 4.33	0.06	9.09 ± 2.76	50.23 ± 8.26
3	95.76 ± 0.51	61.02 ± 8.96	0.125	54.55 ± 6.48	68.06 ± 5.82
10	100^d	95.76 ± 0.63	0.25	81.82 ± 5.25	87.09 ± 3.56
30	100	99.63 ± 0.52	0.5	99.85 ± 0.21	99.73 ± 0.38
60	100	100	1	100	100

[a]The number of colonies of blank control was 2.2×10^6 CFU·ml^{-1}.
[b]The concentration of antibiotics was fixed at 0.5 wt.-% (based on the weight of cellulose fibers).
[c]Contacting time was fixed at 1 h.
[d]100% of growth inhibition indicates all the bacteria were killed and there were no colonies in the agar dishes atter incubation. Therefore, the standard deviations of such samples are not listed in the table.

appropriate concentrations. The antimicrobial mechanism of triclosan or butylparaben and CDβCPs complexes is different from that of guanidine polymers, which tend to destroy the cell membrane and eventually deactivate the bacteria according to the AFM. The role of triclosan is to block the synthesis of lipids and inhibit the enzyme enoyl-acyl carrier protein reductase,[39] while butylparaben may be linked to the mitochondrial depolarisation depletion of cellular ATP (adenosine triphosphate) through uncoupling of oxidative phosphorylation.[40]

4.4.2   Ciprofloxacin-Containing Copolymer

The advantage of the antibiotics included in the cationic β-CD-based polymer could be control released from paper product and inhibit the bacteria sustainably. On the other hand, the majority of the antimicrobial agents steadily permeating out from the paper product gives rise to poisonous influences on the human body and weakened inhibition activity. Covalent bonding of antimicrobial pharmacophores to polymeric papermaking additives would reduce or eliminate the permeation problems, which can be done either by polymerisation of monomers possessing antimicrobial activity or by direct chemical anchoring of drug moieties onto functional groups of the ordinary polymers.

Fluoroquinolones are a widely used class of antimicrobials that inhibit bacterial replication by their action on DNA gyrase. One of such compounds is the ciprofloxacin (CPF) that is an antibiotic with broad-spectrum activity[41] and can be covalently linked to papermaking additives due to the presence of reactive amine and carboxylic acid functional groups.[42] Polyacrylamide (PAM), a synthetic polymer with widespread applications in the papermaking industry has been utilised as an intensifier, retention aid, filtration aid or flocculant according to its molecular weight. Diallyl dimethyl ammonium chloride (DADMAC) and other cationic charged monomers could be also copolymerised with acrylamide (AM) to obtain the cationic polyacrylamide (CPAM) to enhance the adsorption of the polymer onto cellulose fibre. Scheme 4.6 shows the synthesis of the tripolymer containing CPF that was bonded covalently as a pendant to the polymer backbone *via* a free-radical

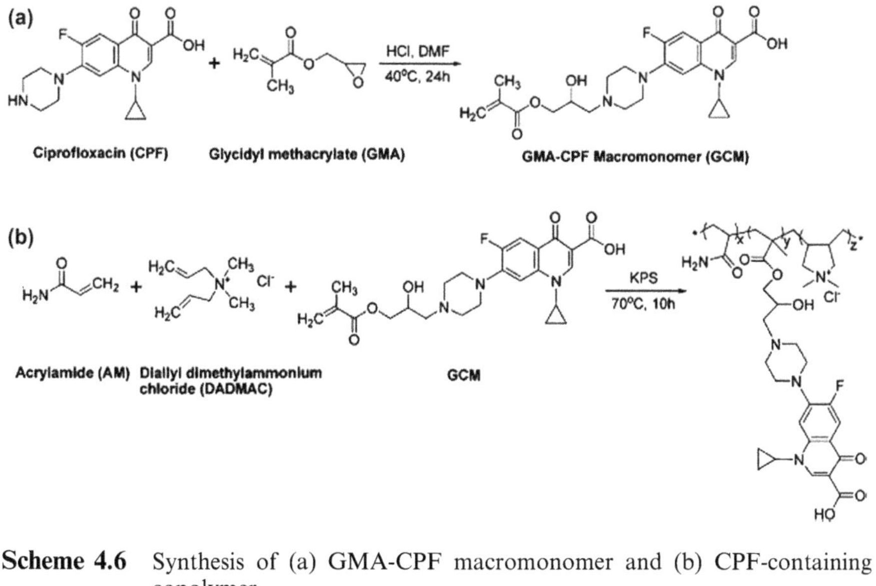

Scheme 4.6   Synthesis of (a) GMA-CPF macromonomer and (b) CPF-containing copolymer.
Reprinted from *J. Biomat. Sci. Polym. Ed.*, **23**, 1115–1122. Copyright (2012) with permission from Taylor & Francis Ltd.

Table 4.4   Minimum inhibitory concentration (MIC) of polymer against
              E. coli. Reprinted from *J. Biomat. Sci. Polym. Ed.*, **23**, 1115–
              1122. Copyright (2012) with permission from Taylor & Francis Ltd.

	AM feed content (molar ratio, %)	GCM feed content (molar ratio, %)	CPF content (wt%)[a]	MIC (μg/ml)
P(AM-DAD)	70	0	0	500
Tripolymer 1	70	5	4.62	250
Tripolymer 2	70	10	8.27	125
Tripolymer 3	70	20	12.53	62.5
Tripolymer 4	60	30	21.10	15
Tripolymer 5	65	30	17.61	15
P(DAD-GCM)	0	10	15.23	30
CPF	–	–	–	7
GCM	–	100	69.98	7

[a]Measured by UV spectrophotometry at 277 nm.

polymerisation.[43] First, GMA reacted with ciprofloxacin to generate a reactive
GMA-CPF macromonomer (GCM) by a ring-opening reaction. After that,
copolymerisation of GCM, DADMAC and acrylamide was carried out *via*
free-radical polymerisation using potassium persulfate (KPS) as an initiator.
Various monomer ratios generate a series of tripolymers with various molecular
weights, charge density and antimicrobial activity. The tripolymers are thermally stable up to 200 °C much higher than the temperature in the drying
section of a paper machine. Practically, CPF was incorporated into the tripolymer in a low content as an antimicrobial moiety to obtain excellent inhibition
effects (shown in Table 4.4).[43] The GCM as well as the pure drug ciprofloxacin
has a similar MIC value (around 7 μg/mL) against *E. coli*, indicating a very
effective antibacterial activity. With the increase of CPF content in the tripolymer, the MIC values of the antimicrobial polymers decreased, which indicated the enhanced antibacterial activities. The antimicrobial activity of CPF,
which is one of the second-generation quinolones, depends on the bicyclic
heteroaromatic pharmacophore and the nature of the peripheral substituents.
The ciprofloxacin-containing copolymers would have excellent antibacterial
activity either against Gram-positive or against Gram-negative bacteria
(*E. coli*) because Gram-negative ones have more complicated cell walls of
bilayer structure than Gram-positive bacteria.

4.5   Conclusions

Typical water-soluble antimicrobial polymers for functional cellulose fibres and
hygiene paper products are summarised in this chapter. Due to the low reactivity and heterostructure of cellulose fibres, appropriate functional polymers
rendering cellulose fibres and paper products antimicrobial should contain
highly reactive groups and/or be cationically charged apart from the high inhibition efficiency, broad spectrum antimicrobial activity and nontoxicity.
Antimicrobial polymers with highly reactive groups such as guanidine-based

polymer are readily grafted onto the cellulose fibres or papermaking additives; moreover, they can also be copolymerised with other functional groups to obtain the novel additives for papermaking. Cationic charged antimicrobial polymers not only improve the water solubility but also enhance the interaction between polymer and bacteria or cellulose fibre, therefore increasing the inhibition ability and retention efficiency. Meanwhile, the cationic charged polymer could form polyelectrolyte complex or nanocapsule to ensure the high retention in paper products. Guanidine-based polymer is especially suitable for endorsing the antimicrobial to cellulose fibre and paper products due to its high reactivity and cationic charge. Gemini surfactant and polymer are novel antimicrobial agents with high antibacterial activities and the reactive functional groups for paper products.

References

1. D. Wei, Q. Ma, Y. Guan, F. Hu, A. Zheng, X. Zhang, Z. Teng and H. Jiang, *Mater. Sci. Eng. C*, 2009, **29**, 1776–1780.
2. T. Hong and H. Xiao, *Tetrahedron Lett.*, 2008, **49**, 1759–1761.
3. D. Mingming, L. Jiehua, F. Xiaoting, Z. Jian, T. Hong, G. Qun and F. Qiang, *Biomacromolecules*, 2009, **10**, 2857–2865.
4. Y. Zhang, M. Ding, L. Zhou, H. Tan, J. Li, H. Xiao, J. Li and J. Snow, *Polym. Chem.*, 2012, **3**, 907–913.
5. C. Ringot, V. Sol, M. BarriÒre, N. Saad, P. Bressollier, R. Granet, P. Couleaud, C. Frochot and P. Krausz, *Biomacromolecules*, 2011, **12**, 1716–1723.
6. S.-H. Lee, M.-J. Kim and H. Park, *J. Appl. Polym. Sci.*, 2010, **117**, 623–628.
7. A. Fillat, O. Gallardo, T. Vidal, F. I. J. Pastor, P. DÚaz and M.B. Roncero, *Carbohyd. Polym.*, 2012, **87**, 146–152.
8. W. Dahou, D. Ghemati, A. Oudia and D. Aliouche, *Biochem. Eng. J.*, 2010, **48**, 187–194.
9. Y. Guan, H. Xiao, H. Sullivan and A. Zheng, *Carbohyd. Polym.*, 2007, **69**, 688–696.
10. D. Ghemati, A. Oudia, D. Aliouche and S. Lamouri, *Appl. Biochem. Biotech.*, 2009, **159**, 532–538.
11. M. H. Lee, K. J. Yoon and S.-W. Ko, *J. Appl. Polym. Sci.*, 2000, **78**, 1986–1991.
12. Z. Ziaee, L. Qian, Y. Guan, P. Fatehi and H. Xiao, *J. Biomater. Sci.-Polym. Ed.*, 2010, **21**, 1359–1370.
13. Y. Guan, L. Qian, H. Xiao and A. Zheng, *Cellulose*, 2008, **15**, 609–618.
14. Y. Guan, L. Qian, H. Xiao, A. Zheng and B. He, *Cellulose*, 2008, **15**, 619–629.
15. A. Hernadi and I. Lele, *Papiripar*, 2004, **48**, 97–101.
16. L. Qian, B. He, Y. Huang and H. Zhan, *Prog. Paper Recycl.*, 2005, **14**, 19–23.

17. N. Maximova, J. Laine and P. Stenius, *Pap. Puu-Pap. Tim.*, 2005, **78**, 176–182.
18. J. Gigac, M. Fiserova and Z. Osvaldik, *Wood Res.*, 2005, **50**, 73–83.
19. T. Saito and A. Isogai, *Ind. Eng. Chem. Res.*, 2007, **46**, 773–784.
20. T. Obokata, M. Yanagisawa and A. Isogai, *J. Appl. Polym. Sci.*, 2005, **97**, 2249–2255.
21. L. Qian, Y. Guan, B. He and H. Xiao, *Mater. Lett.*, 2008, **62**, 3610–3612.
22. L. Qian, H. Xiao, G. Zhao and B. He, *ACS Appl. Mater. Interf.*, 2011, **3**, 1895–1901.
23. L. Qian, Y. Guan, B. He and H. Xiao, *Polymer*, 2008, **49**, 2471–2475.
24. T. Obokata and A. Isogai, *Colloids Surf. A*, 2007, **302**, 525–531.
25. B. He, L. Qian, X. Li, H. Xiao, *Proceedings of the 63rd Appita Annual Conference*, 289–294, Melbourne, VIC, Australia, April 19–22, 2009.
26. P. M. Mendes, M. N. Belgacem, C. A. V. Costa and A. P. Costa, *Cell. Chem. Technol.*, 2003, **37**, 439–451.
27. H. Jonassen, A.-L. Kjøniksen and M. Hiorth, *Biomacromolecules*, 2012, **13**, 3747–3756.
28. L. Qian, X. Li, S. Sun and H. Xiao, *J. Biobased Mater. Bio.*, 2011, **5**, 219–222.
29. S. Sun, Q. An, X. Li, L. Qian, B. He and H. Xiao, *Bioresource Technol.*, 2010, **101**, 5693–5700.
30. J. Li and H. Xiao, *Tetrahedron Lett.*, 2005, **46**, 2227–2229.
31. J. Li, H. Xiao, Y. S. Kim and T. L. Lowe, *J. Polym. Sci. Polym. Chem.*, 2005, **43**, 6345–6354.
32. Z. Guo, X. Chen, X. Zhang, J. Xin, J. Li and H. Xiao, *Tetrahedron Lett.*, 2010, **51**, 2351–2353.
33. H. Xiao and N. Cezar, *J. Colloid Interf. Sci.*, 2005, **283**, 406–413.
34. J. Li, P. R. Modak and H. Xiao, *Colloids Surf. A*, 2006, **289**, 172–177.
35. J. Li, H. Xiao, J. Li and Y. P. Zhong, *Inter. J. Pharma.*, 2004, **278**, 329–342.
36. J. Li, Z. Guo, J. Xin, G. Zhao and H. Xiao, *Carbohyd. Polym.*, 2010, **79**, 277–283.
37. L. Qian, Y. Guan, Z. Ziaee, B. He, A. Zheng and H. Xiao, *Cellulose*, 2009, **16**, 309–317.
38. Y. Guan, L. Qian and H. Xiao, *Macromol. Rapid Commun.*, 2007, **28**, 2244–2248.
39. H. J. Buschmann and E. Schollmeyer, *J. Cosmet. Sci.*, 2002, **53**, 185–191.
40. X. Lu and Y. Chen, *J. Chromatogr. A.*, 2002, **955**, 133–140.
41. Y. Xue, Y. Guan, A. Zheng and H. Xiao, *Colloids Surf. B*, 2013, **101**, 55–60.
42. R. Sripriya, M. S. Kumar, M. R. Ahmed and P. K. Sehgal, *J. Biomater. Sci. Polym. Ed.*, 2007, **18**, 335–351.
43. Y. Xue, Y. Guan, A. Zheng, H. Wang and H. Xiao, *J. Biomater. Sci. Polym. Ed.*, 2012, **23**, 1115–1122.

CHAPTER 5

Polymer-Based Synthetic Mimics of Antimicrobial Peptides (SMAMPs) – A New Class of Nature-Inspired Antimicrobial Agents with Low Bacterial Resistance Formation Potential

FRANZISKA DORNER AND KAREN LIENKAMP*

Albert-Ludwigs-Universität Freiburg, Department of Microsystems Engineering (IMTEK) and Freiburg Institute for Advanced Studies (FRIAS), Freiburg, Germany
*Email: lienkamp@imtek.uni-freiburg.de

5.1 Introduction

5.1.1 Why are SMAMPs a Promising Complement to Traditional Antibiotics?

Humankind used unselective biocides like sulfur dioxide, ethanol and copper long before bacteria as the underlying cause for infections had been discovered. Today, the aim of antibacterial treatment is to selectively eradicate or inhibit a

RSC Polymer Chemistry Series No. 10
Polymeric Materials with Antimicrobial Activity: From Synthesis to Applications
Edited by Alexandra Muñoz-Bonilla, María L. Cerrada and Marta Fernández-García
© The Royal Society of Chemistry 2014
Published by the Royal Society of Chemistry, www.rsc.org

pathogen without harming the patient. In other words, researchers are looking for compounds with selective toxicity. Antibiotics commonly used in clinical settings fulfill these requirements. The mechanism of action of antibiotics varies. However, they typically aim at defined cell receptors. This may, for example, lead to prevention of crosslinking of peptidoglycan in the bacteria's cell wall, or the inhibition of protein and nucleic acid biosynthesis in the bacteria, and results in the death or growth inhibition of the bacterial pathogen.[1] However, a critical issue when using antibiotics is the growing number of acquired resistance of certain bacteria against these drugs. This makes the treatment of infections increasingly difficult, especially when immunocompromised patients are involved.

Acquired resistance is a natural phenomenon. Each bacterial population contains less-sensitive variants that are still viable when the more-sensitive species are already killed by the antibiotic. These robust variants may react by slight cellular modifications that enhance their viability. The more often an antibiotic is applied, the higher the probability of an acquired resistance. Common resistance mechanisms are: the production of inactivating enzymes in the bacteria; changes in the receptor molecules to which the antibiotic binds; modification of the cell membrane permeability by changing the structure of membrane channels, or other transmembrane transport mechanisms; enhancing the activity of efflux pumps that carry antibiotics out of the cell; the overexpression of the antibiotic target molecule; or alteration of the target metabolism.[1] Today, the natural phenomenon of resistance is aggravated by many patients' lack of compliance to medical advice, the needless use of antibiotics when other therapies are possible, and the irresponsible use of antibiotics in livestock and dairy farming. As a result, multiresistant bacteria are nowadays found in most hospitals and many other public places in the western world.[2-4] Originating from either the patient, medical staff, or visitors, they spread through the hospital environment, which then provides a continuous supply of pathogens.[5-8] 25 000 people in Europe and about 90 000 people in the USA annually die from infections with multiresistant bacteria.[2,9] Thus, the need for new antibiotic substances as back-up compounds is ever increasing. So is the pressure to find antibacterial substances featuring new modes of action. Antimicrobial peptides (AMPs) and their synthetic mimics (SMAMPs) are such substance classes. As we will show in this chapter, some of these compounds have shown great antimicrobial activity and simultaneous cell compatibility *in vitro*. They feature an unspecific, yet cell-selective mechanism of action, which makes them less prone to cause bacterial resistance. In combination with strict hygiene protocols, AMPs and SMAMPs may therefore have the potential to help contain hospital-acquired bacterial infections.

5.1.2 Natural Antimicrobial Peptides (AMPs) – Structure and Mechanisms of Action, Cell Selectivity, Quantification of Antimicrobial Activity, and Bacterial Resistance in AMPs

Antimicrobial peptides are natural substances with broad-spectrum antimicrobial activity. They can be found in almost all existing organisms: from

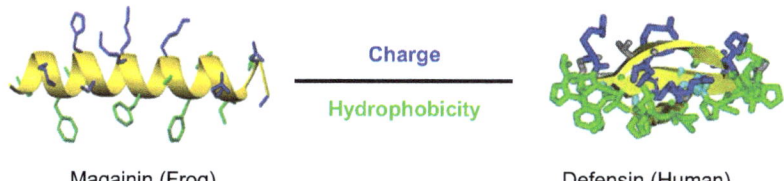

Figure 5.1   Two antimicrobial peptides – the frog-derived magainin and one of the human defensins. The hydrophobic (green) residues of the amino acids are located on one side of the molecule, while the cationic, hydrophilic (blue) residues are on the opposite side. This facial amphiphilicity enables the particular mode of action of these antimicrobial molecules.
Picture modified with permission from *Adv. Polym. Sci.*, 2013, **251**, 141–172. Copyright 2010, Springer Verlag.

eubacteria, protozoa, algae and fungi, to plants and all the animal kingdom.[10] Previous reviews on AMPs exist, most notably the ones by Brogden and Brogden[10,11] and Zasloff.[12] For the readers' convenience, we will reiterate some of the important facts presented in these publications, to which we refer for further details. The special feature of AMPs is their facially amphiphilic molecular structure: the amino acid sequence of AMPs is arranged in such a way that the charged cationic residues come to lie on one face of the molecule, while the hydrophobic amino acid residues are placed on its opposite face.* This principle is illustrated in Figure 5.1 for the frog AMP magainin (left, alpha-helix backbone) and the human AMP defensin (right, beta-sheet backbone).

AMPs were already isolated in the 1950s, and their impressive antibacterial activity, in combination with a much lower toxicity to cells of the host organism, was reported on a phenomenological level. Their special structural features – cationic charge and low molecular weight – were reported later in the 1960s/1970s.[11,13–15] However, it took until the 1980s/1990s for the field to mature, and to yield detailed studies about the mode of action of these unique molecules.[11,16–24] In addition to facial amphiphilicity, the typical characteristics of cationic AMPs are that they have 2–59 amino acids per molecule, a high content of the basic amino acids lysine and arginine, a high content of hydrophobic amino acids like (phenyl)alanine, tryptophane and leucine, and a ratio of hydrophobic to charged amino acids ranging from 1:1 to 1:2.[11] AMPs with highly defined secondary structure were typically more active than those with less formal arrangements.[11]

There are several theories about the mechanisms of action of AMPs. However, the details are still not fully understood. The scientific community seems to agree that many AMPs form pores through bacterial membranes. The first step of interaction between the AMP and the bacteria is AMP localization at the cell membrane. Because AMPs are cationic, and bacterial cells have a net negative surface charge, it is assumed that this adhesion step proceeds *via*

---

*As reviewed by Brogden, there are also neutral and anionic antimicrobial peptides. (see K. A. Brogden, *Nature Reviews Microbiology*, 2005, **3**, 238–250). However, we restrict our discussion to cationic AMPs for this review.

electrostatic attraction.[11] Indeed, both Gram-positive and Gram-negative bacteria bear a net negative charge on their outer barrier (due to layers of lipopolysaccharide and peptidoglycan, respectively). However, it is not yet fully understood how AMPs move from these negatively charged cellular envelopes to the actual cell membrane, which also has a net negative surface potential.[11] Most likely, they form initially a polyelectrolyte complex with peptidoglycan and/or lipopolysaccharide. This complexation helps AMPs to attach to the cell, and later to migrate or diffuse to the actual cell membrane. Once AMPs reach the membrane, they are embedded with their cationic face into the hydrophilic region of the lipid bilayer, which leads to membrane thinning.[11,25,26] From this initial orientation parallel to the cell membrane (at low concentrations) they then rearrange to a perpendicular orientation at higher local AMP concentrations. This results in insertion of their hydrophobic face through the hydrophobic lipid part of the membrane.[11] There are several theories as to how this insertion takes place in detail. The most important ones are the barrel–stave model, the carpet model, and the toroidal pore model.[11] These are illustrated in Figure 5.2.

In the barrel–stave model,[27] AMPs first attach parallel to the pathogen membrane, aggregate, and then insert into the membrane bilayer. They thus form a hydrophilic channel thought the bacterial membrane that resembles a barrel made of AMP staves (Figure 5.2a). In the carpet model,[27] the AMPs attach to the membrane in parallel orientation and form a carpet-like structure. At a critical local AMP concentration, this AMP carpet disrupts the membrane, forms micelles with the membrane pieces that have been torn out in the process, and creates a hole in the remaining membrane (Figure 5.2b). In the toroidal pore model,[27] the AMPs attach to the membrane, insert through it, and induce a local change in the membrane curvature, so that a pore is formed without tearing out parts of the membrane (Figure 5.2c).[11] Many analytical techniques, most notably different microscopic methods and studies with phospholipid model vesicles,

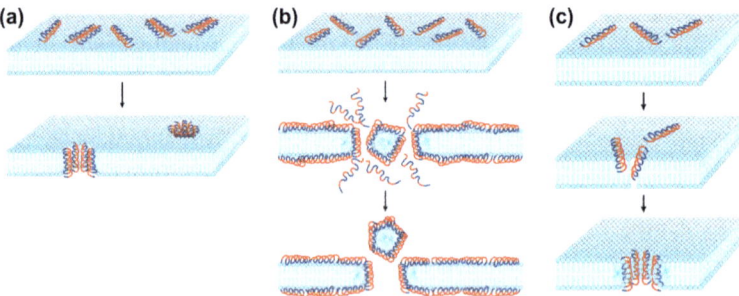

Figure 5.2   Mechanisms of AMP-induced pore formation in bacterial membranes. After surface attachment, the insertion of the AMPs through the bacterial membrane could proceed *via* several mechanisms. (a) Barrel–stave model; (b) carpet model; (c) toroidal pore model; red: hydrophilic part, blue: hydrophobic part of the AMPs.
Picture modified with permission from *Nature Rev. Microbiol.*, 2005, **3**, 238–250. Copyright 2005, Nature Publishing Group.

have shown that each group of AMPs may behave differently, so that these and other models may be considered as descriptions of different AMP/bacteria combinations.[11] The timescale of bacterial killing can vary from instantaneous to 90 min, with prolonged cell-damaging activity of up to 8 h.[11]

Because mammalian cell membranes mainly consist of zwitterionic phospholipids, they are overall charge neutral. As a result, there is no electrostatic driving force for the attachment of AMPs to these cells. AMPs can thus selectively target the negatively charged bacteria membranes in the presence of the overall neutral mammalian cells. This selective toxicity can be quantified by determining the selectivity index (also called "the selectivity") of a substance. It is defined as the ratio of a parameter that quantifies cell compatibility, and a parameter that determines antimicrobial activity. In many publications from the field of AMPs and SMAMPs, the selectivity is the ratio of the hemolytic concentration against erythrocytes (HC_{50}), and the minimum inhibitory concentration (MIC) against bacteria. The HC_{50} value is the substance concentration at which 50% of a defined amount of erythrocytes are lysed, and the MIC is the lowest concentration at which a certain percentage (typically 90% or 100%) of a bacterial population is killed or growth inhibited. Details of these assays are given elsewhere.[28] The MIC, and consequently also the selectivity, depends of course on the bacterial strain. To give some numbers, the frog peptide magainin has a selectivity of about 10, whereas the selectivity of human defensins ranges in the 100s.

Due to the unspecific mechanism by which AMPs interact with bacteria, resistance formation to AMPs is much slower than towards antibiotics with more specific bacterial cell targets. For example, a simple modification of a target receptor in the bacteria may render a traditional antibiotic inactive, while the bacteria have to modify their entire cell membrane chemistry to become resistant to AMPs. Yet resistance in AMPs has been observed: *Staphylococcus aureus* can change its surface charge by shifting positively charged amino acids from the cytoplasm to the cell wall.[29–31] Other bacteria, like *Klebsiella pneumoniae*, can form capsules that limit AMP access to their membranes.[32] However, the process of AMP resistance formation is relatively slow. AMPs have been around for presumably hundreds of thousands of years, yet resistance to them is much less pronounced that resistance to our fairly recent antibiotic drugs.

Thus, AMPs have the combined properties of high antimicrobial activity, good cell compatibility and low potential for resistance formation. So far, however, AMP applications are scarce, because AMPs are also prone to enzymatic degradation, and can only be obtained in limited quantities synthetically or from natural sources. These disadvantages of otherwise very promising compounds stimulated research towards the more easily accessible synthetic mimics of antimicrobial peptides (SMAMPs).

5.1.3 From AMPs to Synthetic Mimics of Antimicrobial Peptides (SMAMPs)

Soon after the first systematic studies of natural AMPs came out, artificial AMP-like substances with similar structural features were synthesized. This had two aims. First, the ability to systematically tune the properties of artificial AMP analogs would enable researchers to understand the essential structural motifs for antimicrobial activity and cell selectivity. Secondly, AMPs have several disadvantages that researchers hoped to overcome: they are difficult to obtain in large quantities, have a short half-life *in vivo* (due to hydrolysis or degradation), and tend to fail under physiological conditions in animal models.[10] Artificial systems, on the other hand, promised to be obtainable in large quantities. Specific structural variations would improve their stability and possibly enhance their activity, which would be beneficial for real-life applications. It was hoped that this would lead to a broader range of applications as drugs or antimicrobial materials.[10]

The evolution of the field from AMPs to polymer-based synthetic mimics of antimicrobial peptides (SMAMPs) has been previously described.[33–41] We will therefore only briefly recapitulate the most important steps in the development of these substances. The most intuitive starting point was the development of AMP-like peptides from natural α-amino acids, but with unnatural amino acid sequences.[42–45] A list of review papers on this topic is given by Brogden and Brogden.[10] Next, besides α-amino acids, peptides based on β-amino acids were studied.[24,42–44,46–48] These had defined secondary structures, which at that time seemed to be a crucial prerequisite for antimicrobial activity. Later, researchers discovered that peptides of L- and D-amino acids, which did not have formal rigid backbones, were still highly active.[49–51] From there, the field developed in two directions: some groups worked with structural units that were still rigid, but less complicated than the α-helix or β-sheet of peptides, such as aromatic oligomers.[52–54] Others directly moved to molecules without any rigid structural element, *i.e.* to synthetic polymers including poly(methacrylates), nylon-3-like polymers, and poly(norbornenes).[55–63] In the following, we will concentrate on the field of synthetic polymer-based SMAMPs, and in particular on these three classes of polymers, which have been studied in most detail compared to other systems.

5.2 Polymer-Based Synthetic Mimics of Antimicrobial Peptides

5.2.1 What is the Difference between Polymer-Based SMAMPs and Previously Reported Antimicrobial Polymers?

Research goes through generational cycles. As a result, research areas that were once extensively studied, and then abandoned, become topical again. In some cases, this is caused by progress in neighbouring fields that open up new possibilities, in others it is driven by new or improved analytical techniques. Research in antimicrobial polymers is such a case. Already in the 1970–1990s, polymers with inherent antimicrobial activities were studied.[64–71] While excellent pioneering work was published at that time, these polymer systems and their biological activity were described on a phenomenological level, but not fully understood or correlated with physical properties. On the one hand, this was because the conceptual tie between such polymers and AMPs had not been established. On the other hand, controlled polymerization techniques were not accessible, and precise structural control could not be obtained. This made it difficult to assess structure–property relations. Also, many of these systems were merely designed as biocides, and their biocompatibility with higher organisms was not studied and optimized. Therefore, most of these polymers cannot be considered as SMAMPs, as they lack the selectivity for bacteria over mammalian cells, and in most cases do not contain the structural features that are prerequisite for such selectivity.

It goes without saying that polymer systems that consist of an antimicrobial component (silver ions, antibiotics, *etc.*) and an otherwise inactive polymeric carrier are also not SMAMPs. In the following, we will exclusively discuss polymer families that were developed with the aim of obtaining antimicrobially active, yet cell selective substances, and that are thus true synthetic mimics of antimicrobial peptides.

5.2.2 Poly(acrylic) SMAMPs

Poly(acrylics) constitute an important polymer family in many areas of basic science and industry. It is therefore no surprise that this class was chosen as one of the first synthetic platforms for polymer-based SMAMPs. Many of the building blocks and monomers necessary for poly(acrylic) SMAMPs are easily accessible, as they are either commercially available or can be obtained in a few synthetic steps. In particular, SMAMPs based on poly(methacrylates)[55,72–75] and poly(acrylamides)[76,77] were investigated in detail. The biggest contributions in the development of poly(acrylic) SMAMPs came from the labs of Kuroda and DeGrado. Already in 2005, DeGrado and coworkers designed SMAMPs based on amphiphilic poly(methacrylate) derivatives.[55] These SMAMPs were obtained by free radical polymerization using a chain transfer agent to control the degree of polymerization, as shown in Figure 5.3a.[55] Starting from the hydrophobic butyl methacrylate, Kuroda and coworkers synthesized an extensive library of poly(methacrylate) SMAMPs and supplemented their experimental results with theoretical calculations. A summary of the most important structural variations of poly(methacrylate)-based SMAMPs is shown in Figure 5.3b.

It was already known from the AMP literature that molecular weight and the ratio of hydrophobic to hydrophilic molecule parts are important parameters to achieve control over the antimicrobial activity and cell selectivity. Therefore, DeGrado and Kuroda varied both parameters in their first study on the antimicrobial behavior of random amphiphilic poly(methacrylates).[55] Three series of polymers were obtained, with high (7900–10 100 g/mol), medium (4500–6000 g/mol), and low (1300–1900 g/mol) molecular weights. (Co)polymers with molar ratios of 0% to about 60% of butyl repeat units were studied.

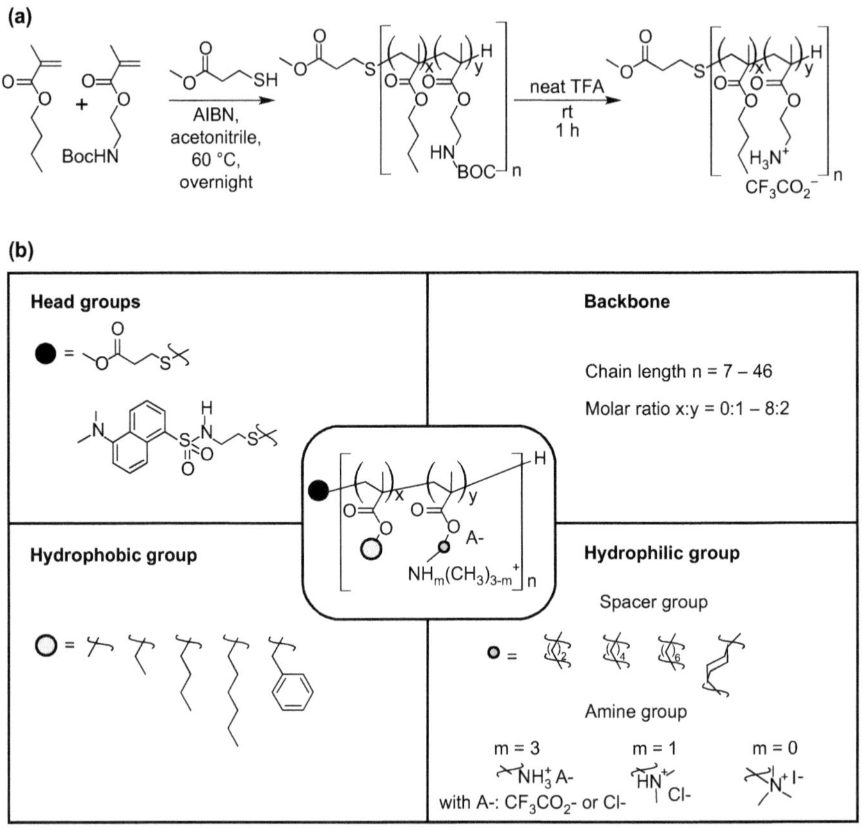

(a)

(b)

Figure 5.3 (a) Synthesis of a poly(methacrylate) based SMAMP by free-radical polymerization of butyl methacrylate and *N*-(*tert*-butoxycarbonyl) aminoethyl methacrylate. AIBN was used as initiator and methyl 3-mercaptopropionate as chain-transfer agent. The *tert*-butoxycarbonyl (Boc) protective group was removed by trifluoroacetic acid; (b) Structural variants in the poly(methacrylate) based SMAMP family. Center: General SMAMP structure consisting of a poly(methacrylate) backbone to which a head group (black dot), a hydrophobic group (big light gray dot), and a hydrophilic group with different spacer lengths (small dark grey dot) are attached. Polymer chain lengths, monomer molar ratios, head groups, and the hydrophilic and hydrophobic groups were varied as indicated.

As shown by the MIC and HC_{50} data of these polymers,[†] increasing the butyl content had two effects: it raised the hemolytic activity in each series, as well as the antimicrobial activity against *Escherichia coli*. However, this trend did not continue linearly. At a certain critical hydrophobicity, the MIC value did not

---

[†]Authors differ in the way they define and determine MIC and HC_{50} for their SMAMPs, as there is not yet consensus and/or a standardised procedure to determine these parameters. Therefore, we report these biological data as defined by the authors in the original publication, and will attempt comparisons between polymer classes only when these parameters are defined similarly.

decrease further, but leveled into a plateau. This was due to the reduced water solubility of the butyl-rich copolymers, which caused aggregation or precipitation, and made them biologically inaccessible. It was also found that the high molecular weight polymer series was more hemolytic than the two low molecular weight series. The most active polymer of the study had an MIC of 16 mg/mL. However, that polymer also had an HC_{50} in the same range and therefore a low selectivity. The best polymer in the study had a selectivity of about three. The authors attributed this to poor control over the average molecular weight (*i.e.* high polydispersity), and no control over the repeat unit sequence along the polymer backbone.[55] These experimental results were compared to theoretical data from molecular dynamics simulations of poly(methacrylate) SMAMPs in contact with lipid membranes.[72] As revealed by these simulations, the primary amine groups on the SMAMP trigger the first interaction between the SMAMP and the lipid bilayer through contacts with the phosphates of the lipid head groups. The butyl groups were then inserted into the hydrophobic part of the model membrane.[72] This led to partitioning of the SMAMPs in the lipid bilayer, with the SMAMP primary amine groups in the lipid head group region, and the SMAMP butyl groups in the lipid tail region. In the light of these findings, the authors concluded that the hydrophobic groups cause the antimicrobial activity, because they cause the actual membrane rupture. The primary amine groups, which are partially protonated under physiological conditions, are responsible for the selectivity of these polymers, because they will preferentially attach to negatively charged bacterial membranes and not to the zwitterionic mammalian cell membranes.[72]

In another study, Kuroda and coworkers investigated the impact of hydrophobicity on the antimicrobial activity and hemolytic properties of poly(methyl methacrylate) SMAMPs in more detail. They obtained five series of copolymers with different ratios of the hydrophilic aminoethyl methacrylate repeat unit and different hydrophobic R-substituted methacrylate repeat units (R = methyl, ethyl, butyl, hexyl and benzyl, as illustrated in Figure 5.3b). In addition to varying the ratio of the hydrophobic to hydrophilic groups, they also investigated different molecular weight ranges.[73] As in the previous study, the antimicrobial activity against *E. coli* in each polymer series was low for the copolymers with low content of the hydrophobic repeat unit. With increasing amount of hydrophobicity, the MIC decreased, and then leveled into a plateau for the polymers with the highest hydrophobicity. When plotted against their mass concentration, these polymers did not strongly depend on molecular weight in the range observed (*ca.* 1200–8700 g/mol).[73] When the molecular weight of the polymers was taken into account, *i.e.* when the data were plotted as molar concentration, the antimicrobial activity of the polymers against *E. coli* showed a small dependence on the chain length, with somewhat higher activity for the longer polymer chains. However, the HC_{50} values remained low, resulting in selectivities of only up to 10.

Based on this body of data, Kuroda and coworkers attempted a correlation of the HC_{50} values with the overall hydrophobicity of the SMAMPs.[73] The results are shown in Figure 5.4. They compared the HC_{50} data with either the

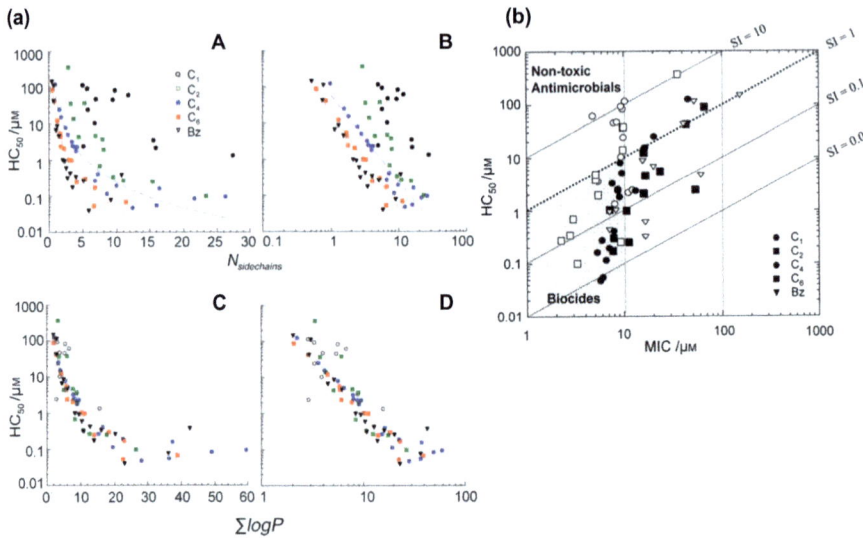

Figure 5.4   (a) HC$_{50}$ of poly(methacrylate) SMAMPs with different hydrophobic groups (C$_1$ to C$_6$ and benzyl) plotted *vs.* the number of polymer side chains ($N_{\text{side chains}}$) (panels A and B), and $\sum \log P$ (panels C and D). $N_{\text{side chains}}$ is the molar fraction of hydrophobic groups of each polymer, multiplied by the degree of polymerization; *P* is the calculated partition coefficient between octanol and water for the poly(methacrylate)-based SMAMPs with different hydrophobic groups; (b) Relationship between HC$_{50}$ and MIC of poly(methacrylate) SMAMPs with different side chains (C$_1$ to C$_6$ and benzyl). The selectivity index, defined as HC$_{50}$/MIC, is plotted as guide lines. Polymers with an MIC lower than 10 µM and a selectivity smaller than 1 were classified as biocides. Polymers with higher selectivities and MICs lower than 10 µmol/L were classified as nontoxic antimicrobials.[73] Reprinted with permission from *Chem. Eur. J.* 2009, **15**, 1123–1133; Copyright 2009, Wiley-VCH.

number of carbon atoms in the polymer side chain (Figure 5.4a, panels A and B), or the calculated octanol/water partition coefficient of model systems (Figure 5.4a, panels C and D).[73] It was then assumed that the polymer chains may exist in three different states during a hemolysis experiment: as collapsed polymer chains, as free polymer chains in solution, and as membrane-bound or inserted polymer chains. These states were assumed to be in mutual equilibrium. HC$_{50}$ was then defined as the sum of the polymer concentrations in all states, which, together with a numerically modeled partition coefficient for the different side chains, led to the fit in Figure 5.4a, A and B. When summing over the partition coefficients of all side chains, a plot of HC$_{50}$ *vs.* $\sum \log P$ could be created, where *P* is the partition coefficient (Figure 5.4a, panels C and D).

While some scaling of HC$_{50}$ with overall hydrophobicity was observed in these plots, these quantification attempts remained rather on an empirical level, and a correlation with solid physical properties was unfortunately not yet possible. Though somewhat crude, this model was a first quantitative attempt to correlate the experimentally determined biological properties with molecular features and mechanisms, and thus it went far beyond the previous empirical, synthesis-focused property optimization approaches.

Several of Kuroda's poly(methacrylate) SMAMPs were also equipped with fluorescent dansyl groups and exposed to model membranes that mimicked either red blood cells or bacteria. Surprisingly, fluorimetry data showed that the hydrophilic poly(ethanolamine methacrylate) homopolymer had a higher affinity for neutral membranes than for negative ones.[73] The more hydrophobic SMAMPs did not show this large preference, but had enhanced binding to all model membranes. Overall, the data showed that binding to lipid model membranes depended both on the polymer hydrophobicity and the chemical structure of the lipid.[73] Kuroda and coworkers also categorized their SMAMPs according to their selectivity as either biocidal or nontoxic antimicrobials (Figure 5.4b). Polymers with MICs below 10 µmol/L and selectivities below 1 were classified as biocides, while those with selectivities above 10 were categorized as nontoxic antimicrobials, *i.e.* SMAMPs.

The most promising members of the series of poly(methyl methacrylate) and poly(butyl methacrylate) SMAMPs were also tested against various clinically relevant bacteria strains, where some showed broadband antimicrobial activity, although the selectivity of the most active SMAMPs was low.[38] To demonstrate that SMAMPs do not form resistance in *E. coli*, this bacterium was exposed 21 times to sublethal doses of a poly(methyl methacrylate) and a poly(butyl methacrylate) SMAMP, respectively. Two standard antibiotics were used as controls. While this procedure increased the MICs of the antibiotics ciprofloxacin and norfloxacin by a factor of 250–500, the MICs of the two SMAMPs remained unchanged.[38] In further tests in the presence of physiological amounts of sodium chloride, the MIC of these SMAMPs increased by a factor of 2–4. Adding bivalent salts at a concentration of 2 mmol/L caused an increase in the MIC of 2–4 times for the methyl SMAMPs, but not for the butyl SMAMPs. It was also demonstrated that these SMAMPs caused outer-membrane permeability in *E. coli* at concentrations below the MIC, but not inner-membrane damage. This may be a hint that SMAMPs get trapped in the outer membrane and do not reach the inner membrane of Gram-negative bacteria.[38] This is in line with the results of model experiments on *E. coli* using poly(oxanorbornene) SMAMPs (see below),[78] and may indicate that other mechanisms besides membrane permeation may play a role in SMAMP activity.

Some representative polymers of the poly(methyl methacrylate-*co*-aminoethyl methacrylate) series and the poly(butyl methacrylate-*co*-aminoethyl methacrylate) series (all with low MICs and moderate to high hemolytic activities) were studied in more detail by dye-leakage experiments and several assays using human red blood cells: the classic hemolysis assays, a hemagglutination assay, erythrocyte count, and an osmoprotection assay, Figure 5.5).[79] For the methyl copolymer series, the thus obtained data show that the critical value for hemagglutination (C_{agg}) scales inversely with polymer charge – the higher the aminoethyl methacrylate repeat unit content, the lower C_{agg}, *i.e.* the more hemaglutionation. Hemagglutination is an important, yet often neglected aspect in SMAMP research, as it may lead to thrombosis. At the

(a)

Sample	MIC (μg/mL)	HC$_{50}$ (mg/mL)	C$_{agg}$ (mg/mL)
P$_0$	500 ± 0	>2000	125 ± 0
PM$_{10}$	500 ± 0	>2000	250 ± 0
PM$_{28}$	500 ± 0	>2000	250 ± 0
PM$_{47}$	63 ± 0	>2000	2000 ± 0
PM$_{63}$	16 ± 0	114 ± 7	not observed
PB$_{27}$	16 ± 0	13 ± 2	not observed
magainin-2	125 ± 0	>250	not observed
melittin	13 ± 0	2 ± 0.2	not observed

(b)

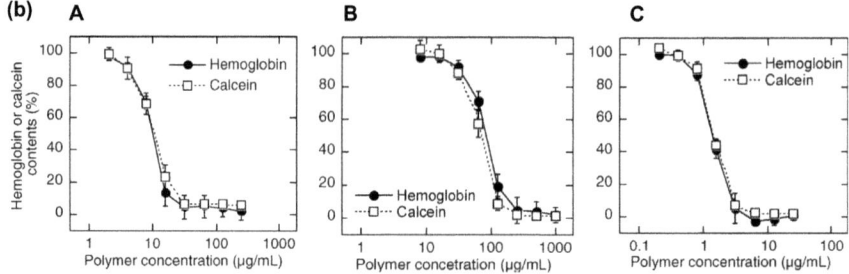

Figure 5.5 (a) MIC against *E. coli*, HC$_{50}$ and hemagglutionation concentration (C$_{agg}$) for selected poly(methacrylate) SMAMPs: P0 = poly(aminoethyl methacrylate), PM10 = 10% methyl comonomer, 90% aminoethyl comonomer, *etc.*, PB27 = 27% butyl comonomer, 73% aminoethyl comonomer. The AMPs magainin-2 and mellitin were used as controls. (b) Hemolysis and calcein leakage from erythrocytes *vs.* concentration (PB27, PM63 and melittin).[79] Modified/reprinted with permission from *Biomacromolecules* 2011, **12**, 260–268; Copyright 2011, American Chemical Society.

same time, the hemolytic concentration of these polymers was high. Thus, highly charged SMAMPs may be unsuitable for biomedical applications in spite of good HC$_{50}$ data, if C$_{agg}$ is much below HC$_{50}$. In dye-leakage experiments with calcein-loaded erythrocytes, it was shown that the concentration at which 50% of dye was released was in good agreement with the previously determined HC$_{50}$ data (Figure 5.5b). Using erythrocyte cell counts, it was demonstrated that all hemoglobin was released from one cell once that cell was sufficiently damaged, *i.e.* at higher polymer concentrations, the number of damaged cells increased. However, it was not fully understood why only a fraction of the cells releases hemoglobin at a given time.[79] Osmoprotection experiments were also conducted. Without adding osmotically active species to the sample, the SMAMP-exposed erythrocytes underwent lysis due to a higher osmotic pressure inside the cell. Adding osmoprotectants like sucrose and poly(ethylene gylcol) suppressed this, which demonstrated that the SMAMPs formed nanopores or other nanosized lesions at the erythrocyte membrane.[79]

Another subset of these SMAMPs with different hydrophobicities was studied using sum frequency generation vibrational spectroscopy.[80] This surface-sensitive IR-related spectroscopic technique allowed the interaction of SMAMPs with model lipid membranes to be studied. It was confirmed that SMAMPs with high MIC (125 µg/mL to 500 µg/mL) and HC_{50} (> 2000 µg/mL) were not membrane active at a given concentration, while the more hydrophobic SMAMPs with low MIC (15.6–62.5 µg/mL) and low HC_{50} (31.9 µg/mL and 55.1 µg/mL) interacted with the phosphatidylglycerol model membranes. A parallel alignment to the membranes was observed.[80]

Kuroda's group also investigated how changing the cationic, hydrophilic group of the SMAMPs affected their antimicrobial and hemolytic properties.[74] They synthesized poly(methacrylate) SMAMPs with primary amine, tertiary amine and quaternary ammonium groups (Figure 5.3b). These were combined with methyl methacrylate and butyl methacrylate as hydrophobic repeat units. In the poly(methyl methacrylate) series, the polymers with a primary amine group where the most active ones, with MIC values as low as 16 µg/mL, and excellent selectivities up to 125 for the copolymer with 53% primary amine repeat units.[74] The SMAMPs with tertiary amine groups were less antimicrobially active and had selectivities of only up to 8. The polymers with quaternary ammonium groups were inactive. These findings are in accordance with Kuroda's previous simulation data, where the role of the primary amine group as a trigger for further SMAMP–membrane interaction was identified.[72] The corresponding poly(butyl methacrylate) copolymers had excellent MIC values for both the primary and tertiary amine groups. Here, very little difference between the nature of the amine group and the resulting antimicrobial activity was observed. The polymers carrying the primary amine groups were slightly more toxic, with HC_{50} values in the range of 14–200 µg/mL, compared to 20–1600 µg/mL for the tertiary amine-butylmethacrylate copolymers. This resulted in better selectivities (up to 65) for the tertiary amine polymer series.[74] The authors also investigated the effect of pH on MIC and HC_{50} data. Overall, it was concluded that polymers with *N*-protonated groups are better candidates to obtain nontoxic SMAMPs.[74] Such polymers are pH sensitive and may be deprotonated during their interaction with the bacterial membrane, which may be important for their ability to penetrate bacteria.[74] As in the previous paper with DeGrado,[73] a numeric correlation between biological properties and polymer structure was attempted. For some polymers, a somewhat linear correlation between the log(selectivity) and the fraction of hydrophobic groups was obtained.[74] To further investigate the role of cationic groups in the membrane disruption process, Kuroda and coworkers labeled copolymers containing different cationic structures with fluorescent dansyl groups. Using fluorimetry, they could thus quantify the partitioning of their copolymers between the hydrophobic and the hydrophilic parts of a zwitterionic lipid membrane.[81] In combination with ^{31}P solid-state NMR experiments, they concluded that primary amines had greater membrane-disrupting abilities than quaternary ammonium salts.

Kuroda and coworkers also synthesized a set of copolymers where the spacer length between the poly(methacrylate) backbone and the cationic charge was gradually increased from 2-aminomethylene methacrylate to 6-aminohexylene methacrylate (Figure 5.3b).[82] The key idea of this polymer design was to enable the side chains to move more freely and to adopt peptide-like secondary conformations. The MIC data showed that the homopolymers with a spacer length of two and four carbon atoms were inactive against bacteria, while the homopolymer with a 6-carbon atom spacer had an MIC of 10 µg/mL. The authors note that "elongation of spacer length enhances the activity of the cationic homopolymers without incorporation of hydrophobic comonomers".[82] An alternative interpretation would be that by increasing the hydrophobicity of the polymer *through a longer alkyl chain spacer*, these molecules become more active. A similar effect was observed in a family of poly(norbornene imide) polymers.[83] The data interpretation for the copolymers derived from these homopolymers (2-aminomethylene methacrylate, 4-aminobutylene methacrylate or 6-aminohexylene methacrylate copolymerized with either ethyl methacrylate or butyl methacrylate) is also difficult. It is impossible to differentiate between effects of the spacer as a structural feature that provides a higher level of flexibility and the effect of an overall change in molecule hydrophobicity, because changing the spacer length alters both. Even when comparing the high antimicrobial activity of the 6-aminohexylene polymers with the lower activity of the corresponding cyclohexane derivatives, the argument of equal hydrophobicity (due to the same number of carbon atoms) is difficult to hold up, as these two functional groups have very different hydrophobic moments due to the different molecular surface area of hexyl compared with cyclohexyl. The same difficulties arise in the interpretation of the HC_{50} data. Additional high-performance liquid chromatography (HPLC) measurements that quantify the polymer hydrophobicity may help to identify polymers with equal hydrophobicity, which might then allow conclusions to be reached about the impact of spacer flexibility. Overall, the reported copolymers span a wide range of selectivities,[‡] with the best ones having selectivities of 63 and 53, respectively. In bactericidal kinetics experiments, some of Kuroda's polymers killed 99.99% bacteria within 30 min. Again, it is difficult to attribute differences in these kinetics measurements to spacer versus general hydrophobicity effects.[82] In simulation experiments that modeled these polymers, Kuroda showed that the SMAMP polymer backbones were oriented parallel to a model lipid membrane. The simulated end-to-end distances of the respective polymers revealed that the hydrophobic polymers were more stretched than the hydrophilic ones. The hydrophobic arms were directed into the membrane core, and the amine groups towards the membrane/water interface.[82] This is in line with the assumed facially amphiphilic mechanism for SMAMP activity and previous simulation results.[72]

The effect of increasing backbone hydrophilicity was studied by comparing the more hydrophilic poly(methacrylamides) with their more hydrophobic poly(methacrylate) homologues.[76] Also, this backbone change improved the chemical stability of the polymers in basic aqueous solutions. Two series of

---

[‡]HC_{50}/MIC as defined by the authors.

copolymers were synthesized: poly(butyl methacrylamide-*co*-aminopropyl methacrylamide) and the corresponding hexyl homologues. These polymers showed a maximum in the MIC (against *E. coli* and *S. aureus*) *vs.* hydrophobicity curves, which means that adding hydrophobicity to the polymers first made them less antimicrobially active, before an improvement of the MIC was observed.[76] This is unusual and in contrast to the results of most other SMAMP series, which typically go through a minimum with increasing hydrophobicity, or level into a plateau. The exact reason for this anomalous behavior was not explained. The hexyl homologues were more active than the butyl homologs.[76] This underlines that poly(methacrylamides) are more hydrophilic than the analogous poly(methacrylates) – the corresponding hexyl methacrylate copolymers would be already insoluble and thus inactive. Also, the HC_{50} values of the poly(methacrylamides) were much higher than those of the corresponding poly(methacrylates), which is a further indication for the higher overall hydrophilicity of the amide system. Titration data showed that at pH 7.4, the fraction of protonated repeat units was smaller for the poly(methacrylates) than for the corresponding poly(methacrylamides).[76] Since the activity of the poly(methacrylates) is higher than that of the poly(methacrylamides), the authors conclude that the presence of neutral amine groups enhances SMAMP activity, and that by choosing groups with different pKa, SMAMP activity could be further tuned.[76] This is an interesting hypothesis that deserves further detailed study.

The authors also tested SMAMP cytotoxicity by measuring the metabolic activity (IC_{50}) of human epithelial HEp-2 cells, which can be correlated with cell viability, using the XTT colorimetric assay.[76] Overall, the metabolic activity was significantly reduced at concentrations as low as 20 µg/mL. These values were much lower than the HC_{50} values reported for this polymer system, which indicates that HC_{50} is a good screening tool, but not a quantitative indicator for cell toxicity and therefore it should not be used as the only indicator when considering clinical application potential. The best polymer of the poly(methacrylamide) series, the poly(aminopropyl methacrylamide) homopolymer, had a selectivity (HC_{50}/MIC) of > 64 for *E. coli* and an amazing selectivity of > 640 for *S. aureus*. Unfortunately, its IC_{50} value for HEp-2 cells was very low, resulting in an IC_{50}/MIC value of only <1.3. The authors also found that, while hemolysis usually increases with increased hydrophobicity and decreased charge, the opposite was true for the IC_{50} values.[76] This is in agreement with the known fact that polycations are usually cell toxic. Kuroda here pointed out the general predicament of the field: too much charge causes toxicity to cells, too much hydrophobicity causes hemolysis. Thus, an assay that would easily allow quantifying "the hydrophobicity" *vs.* charge of a new SMAMP would be highly desirable. So far, however, all attempts to do so (HPLC, determination of octanol/water partition coefficients, *etc.*) have not been very convincing. The data obtained from dye-leakage experiments with vesicles that model *S. aureus*, *E. coli* and erythrocytes mirrored the data from the MIC and HC_{50} experiments of the poly(methacrylamide) polymers. When looking at the dye-leakage kinetics in more detail, it was found that hydrophobic polymers caused rapid and complete leakage (associated with rapid membrane rupture), while the more cationic polymers caused more gradual leakage.[76] The latter was particularly true with the more anionic model vesicles and was interpreted as a more modulated interaction between SMAMPs and lipids,[76] which is in our opinion consistent with a polyelectrolyte complex formation prior to membrane rupture.

One weakness of many antimicrobial polymers, at least from an in-patient applications point of view, is their lack of biodegradability. To solve this problem, Kuroda and coworkers demonstrated the incorporation of degradable groups into poly(acrylates) by simultaneous chain growth and step-growth polymerization of *tert*-butyl acrylate and 3-butenyl 2-chloropropionate.[84] While the polydispersity of the copolymers thus obtained was quite broad, given the sensitivity of SMAMP activity to molecular weight, this approach is an important step towards degradable poly(acrylic) SMAMPs with tunable properties.

In summary, it was shown that poly(acrylic) SMAMPs can be easily synthesized and have interesting biological properties. This makes them promising candidates for clinical applications.

5.2.3   Nylon-3-Based Polymer SMAMPs

Another synthetic platform that was used to obtain SMAMPs with tunable properties are poly(amides), more specifically the nylon-3 or poly(β-lactam) family. Comparing the three well-studied polymer SMAMP platforms (poly(acrylics), poly(norbornenes) and nylon-3), the nylon-3 backbone is the most AMP like, and certainly the most hydrophilic backbone. This is a great advantage when designing SMAMPs, as the extent to which the molecular properties can be tuned is larger for a more hydrophilic backbone. In consequence, polymers with superb selectivities for bacteria over mammalian cells were obtained using this approach.[40,60,85,86] After having worked extensively on peptide-based SMAMPs, Gellman and coworkers published the first paper on nylon-3-based polymer SMAMPs in 2007. The rationale behind this work was to test the hypothesis that such polymers could self-organize when in contact with bacterial membranes.[60] As it was believed that the helical conformation of peptide-SMAMPs was induced by contact to biomembranes, Gellman and coworkers proposed that a random polymer of cationic and hydrophobic subunits would also self-organize in an irregular, yet globally amphiphilic structure when in contact with such membranes (Figure 5.6). The fact that these molecules were highly antimicrobially active, yet cell selective, supports this hypothesis.[60] The nylon-3 polymers were obtained by anionic ring-opening polymerization (AROP) of β-lactams, as shown in Figure 5.7a. Since then, Gellman and coworkers have synthesized a whole family of nylon-3 SMAMPs, which are summarized in Figure 5.7b.

The general structure of the nylon-3 polymers is shown in the center box. In detailed studies, the chain length,[85,87] the structure of the hydrophobic and cationic repeat units,[40,60,85,87–89] and their molar ratio[85,87,88] was varied. The

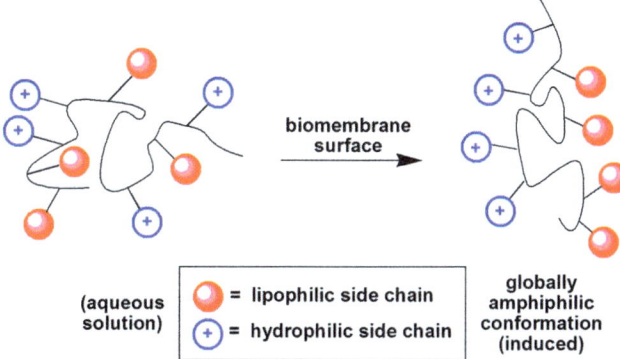

Figure 5.6   Hypothesis explaining the activity of polymeric SMAMPs that do not
have rigid, AMP-like polymer backbones. When in contact with a
biomembrane, they adopt a globally amphiphilic, yet irregular conformation (compared to the structural precision of α-helices or β-sheets).[40]
Reprinted with permission from *J. Am. Chem. Soc.* 2009, **131**, 9735–9745;
Copyright 2009, American Chemical Society.

polymers were also equipped with different *N*- and *C*-termini to study the effect
of hydrophobicity variation at these positions.[86]

In the first series of nylon-3 SMAMPs, with cyclohexyl as hydrophobic unit,
the ratio of the hydrophobic and the hydrophilic group was gradually varied. It
was observed that increasing the proportion of the hydrophobic repeat unit
increased the antimicrobial activity (decreased MIC values) and the hemolytic
activity (decreased MHC values).[§] This is in line with reports on other SMAMP
families.[55,56,63] An optimum selectivity of 32 (MHC/MIC, which does not
correspond numerically to selectivities expressed as HC_{50}/MIC, see footnote)
was obtained for the nylon-3 copolymer with 63% cationic repeat units and
37% hydrophobic cyclohexyl repeat units.[60] This selectivity is in the range of
performance of the natural host defense peptides tested in the same panel.
When the chain length of this 63% polymer was varied from 7–58 repeat units,
the MIC remained mostly unaffected, with a little loss of activity against
S. aureus. This could be due to the inability of these polymers to penetrate the
peptidoglycan cell wall.[60] Additionally, the polymers became substantially
more hemolytic at chain lengths above 30 repeat units (Figure 5.8a).[60]

In dye-leakage studies, it was shown that the nylon-3 copolymer with 60%
cationic repeat units permeabilized anionic vesicles that mimic the composition
of Gram-negative and Gram-positive bacteria.[85] When exposed to overall
charge-neutral vesicles that mimic mammalian cells, no dye leakage was

---

[§]Gellman and coworkers use the following metrics: MIC for antimicrobial activity and MHC, or
minimum hemolytic concentration, for the hemolytic activity. MHC is generally much lower than
HC_{50}. Therefore, these values are not comparable. Consequently, MHC/MIC selectivities are
numerically much lower than HC_{50}/MIC selectivities, although the molecules may be equally
selective for bacteria over mammalian cells. Consequently, care must be taken when comparing
these values.

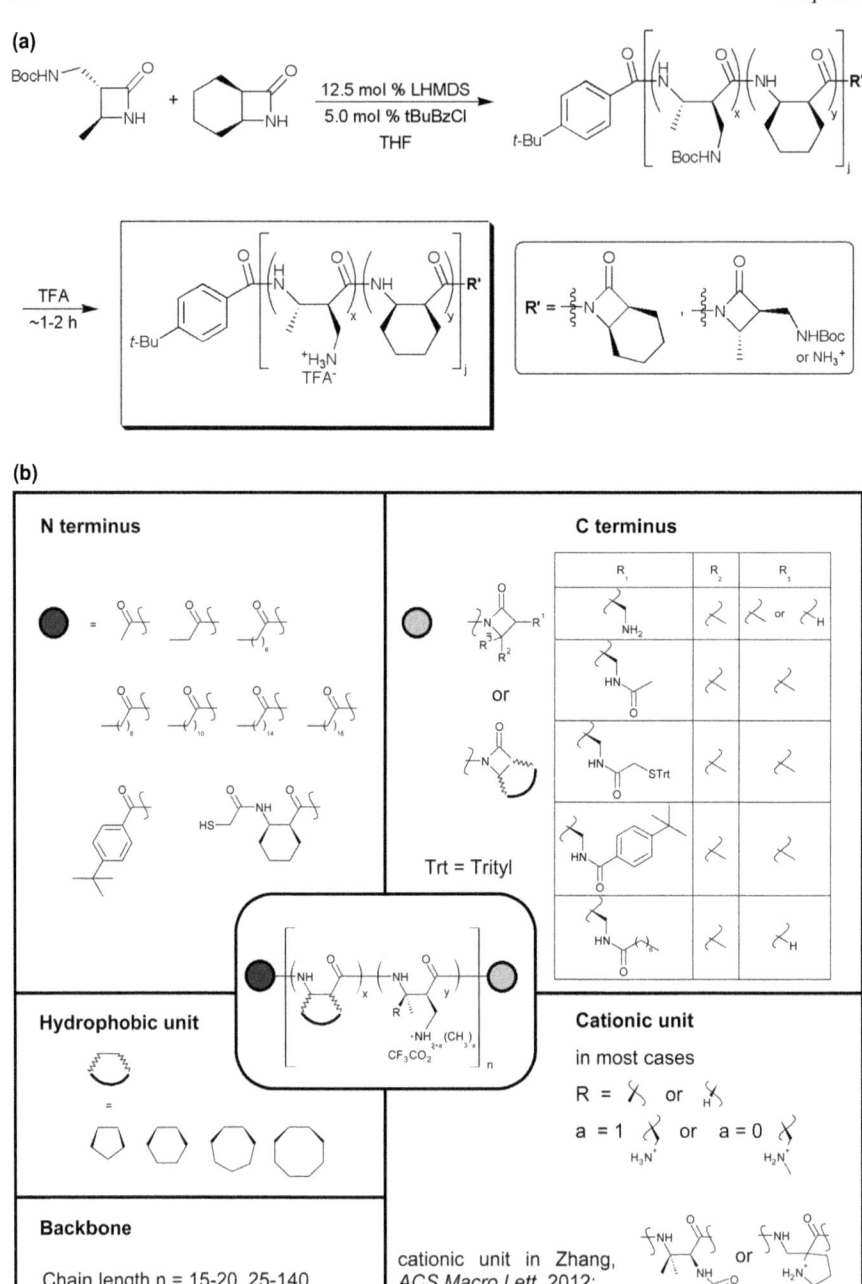

Figure 5.7 (a) Example for the synthesis of a nylon-3-based SMAMP by ring-opening polymerization with LiN(SiMe₃)₂ as catalyst.[60] Modified with permission from *J. Am. Chem. Soc.* 2007, **129**, 15474–15476; Copyright 2007, American Chemical Society; (b) Structure of poly(β-lactams) (= nylon-3)-based SMAMPs consisting of an N-terminal group (black dot), a C-terminal group (grey dot), and various hydrophobic and cationic groups.

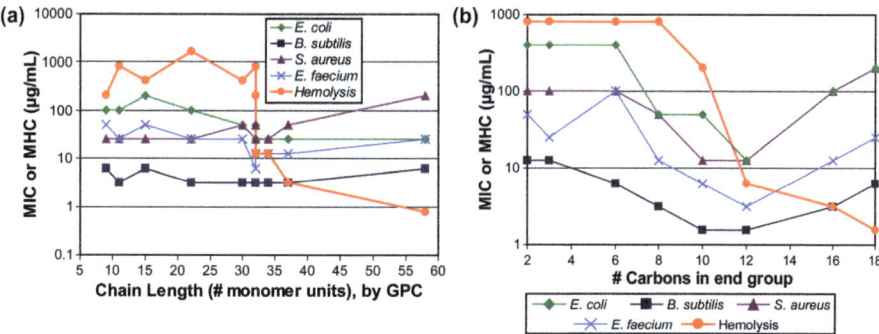

Figure 5.8 Biological properties (MIC against various bacteria and minimum hemolytic concentration (MHC)) of a series of nylon-3 SMAMPs with 63% cationic groups and (a) different polymer chain lengths; (b) different numbers of carbon atoms in the acyl tail at the polymer N-terminus.[40] Reprinted with permission from *J. Am. Chem. Soc.* 2009, **131**, 9735–9745; Copyright 2009, American Chemical Society.

observed.[85] Isothermal calorimetry (ITC) measurements confirmed that there was polymer binding to the anionic bacteria-mimicking vesicles, but not to the charge-neutral ones.[85] Using differential scanning calorimetry, Gellman and coworkers showed that the addition of the 60% nylon-3 polymers to mixed lipid bilayers caused a phase separation of the anionic lipids from the zwitterionic lipids in the bilayer.[85] This demonstrated that nylon-3 based SMAMPs may have another mechanism of action besides mere disruption or pore formation by insertion into bacterial membranes. They can induce phase boundaries in lipid bilayers, and defects at these phase boundaries further de-stabilize the bacterial membrane. Further ITC studies indicated strong binding between the outer negatively charged components of bacterial membranes, *i.e.* lipopolysaccharides (in Gram-negative bacteria) and peptidoglycane (in Gram-positive bacteria). This may cause a displacement of Ca^{2+} in these cell wall/membrane parts that may compromise their overall function. Using chromogenic receptor molecules, it was further shown that the nylon-3 SMAMP could permeabilize the inner membrane of *E. coli* bacteria even at low concentrations, while high polymer concentrations block transport mechanisms through the outer membrane.[85] The latter event can kill the bacteria even when there is no rupture of the inner membrane of Gram-negative bacteria. Com-bined, these studies show that the antimicrobial activity of these polymers may be a combination of different mechanisms that act simultaneously on different parts of the cell wall or membrane.

Next, the nylon-3 polymer with 63% cationic repeat units and chain lengths of 27–35 repeat units was functionalized with different alkyl groups (C_2 to C_{18}) at the *N*-terminus (Figure 5.7b).[40] The impact of this modification on the biological data is shown in Figure 5.8b. As in the case of increasing the side-chain hydrophobicity in other systems,[63,90] a modest hydrophobicity increase improved the antimicrobial activity, while too much hydrophobicity reduced it, presumably due to the beginning of molecular insolubility or aggregation. The optimum antimicrobial activity occurred at a chain length of 12 at the

N-terminus. An increased hemolysis was already observed for C_{10}, so that an optimal selectivity was obtained for a C_8 chain.[40] In a different set of experiments, the cyclohexyl group was replaced by cycloheptyl and cyclooctyl as hydrophobic moiety. This decreased the MIC values relative to the cyclohexyl series. In both series, a minimum in the MIC *vs.* % cationic lactam curve was observed. At that point, adding more hydrophobicity increased the MIC values again, possibly due to the beginning insolubility or aggregation of these polymers.[40] This was investigated in more detail by determining the critical aggregation concentration (CAC) of these SMAMPs. The more hydrophobic the SMAMPs, the lower were their CACs. Consequently, after reaching the minimum MIC at a moderate hydrophobicity, the MIC of these polymers increased again because the SMAMP aggregates were not bioavailable. As pointed out by the authors, this could be due to the inability of the aggregates to permeate the bacterial membranes. Interestingly, aggregate formation does not affect the MHC, most likely because erythrocytes do not have a protective cell wall like bacteria, and are thus accessible for aggregated and nonaggregated SMAMPs.

As discussed, the AROP approach by which the nylon-3 SMAMPs were obtained provided good control over the functional groups at the *N*-terminus of the polymers. The *C*-terminal position, on the other hand, could not be as easily modified. Therefore, Gellman and coworkers modified the functional groups at this position by a block-copolymer-like synthesis. When the hydrophobic and hydrophilic β-lactam monomers that constituted the main part of the polymer (the "first block") were consumed, they added one equivalent of a different monomer to the still active polymer chain to form the "second block".[86] On average, this yields a second block with one repeat unit, and thus one functional group, at the *C*-terminus. When different functionalities are thus attached to the *N*- and *C*-termini, these polymers can be considered as heterotelechelic, and are attractive building blocks for various applications. However, as pointed out by the authors, it could not be clearly demonstrated that the functionalization at the *C*-terminus led to exactly one functional end-group per polymer chain. Indeed, it is more likely that a mixture of none, one and more end-groups per polymer chain was obtained. To check whether the *C*- and *N*-termini of the polymer were chemically equivalent, the authors compared the biological activity of *N*-acetyl/*C*-*p*-(*tert*-butyl)benzamide terminated SMAMPs with the inversely modified analog, *i.e.* with an *N*-(*p*-*tert*-butyl)-benzamide/*C*-acetyl-terminated SMAMP, that was otherwise identical. Both polymers had relatively similar MIC data, but the MHC of the SMAMP with the *C*-*p*-(*tert*-butyl)benzamide (25 and 3.13 µg/mL, respectively) was drastically lower than the MHC for the *C*-acetyl SMAMP (800 µg/mL). The biological data thus show that adding hydrophobicity at the *C*-terminus has a more drastic effect than adding "the same amount" of hydrophobicity at the *N*-terminus. However, is it really the same amount? The authors assume that there is more or less one functional group at the *C*-terminus, and argue that the deviation in the biological data is because the *N*-termini and the *C*-termini are not chemically equivalent (which had been observed with α-peptides).[86] They then hypothesize that the difference between the *C*- and *N*-termini is due to a gradient distribution of repeat units from the *N*- to the *C*-terminus, with less-hydrophobic groups at the *N*-terminus than at the *C*-terminus.[86] However, since the *C*-terminus is not well defined in their polymers, this is a bold interpretation of the data. It is true that the *N*-termini and *C*-termini are not chemically equivalent, because the *N*-termini bear exactly one functional group per chain, while the *C*-termini of the entire polymer sample represent a polydisperse distribution of chain ends. One could thus argue that even a small fraction of polymers with two or more *C*-*p*-(*tert*-butyl)benzamide end-groups (say, 10%) could drastically decrease the MHC. Indeed, when the reagent amount for inserting the *C*-*p*-(*tert*-butyl)benzamide end-group was increased (*i.e.* when the probability of having more doubly *C*-*p*-(*tert*-butyl)benzamide functionalized polymers is higher), the MHC dropped further from 25 to 3.13 μg/mL.[86] Also, a previous report showed that the modification of the *N*-terminus with C_8 or C_{12} chains, *i.e.* a difference of just 4 carbon atoms per polymer chain, led to a decrease in the MHC of two orders of magnitude, while the MIC was much less affected.[40] This emphasized that even a small fraction of polymer chains with a hydrophobic blob at the chain end (formed by two or more *C*-*p*-(*tert*-butyl)benzamide moieties per polymer chain) could cause the observed differences in the MHC of these inversely *C*/*N*-substituted SMAMPs. Of course, a combination of the gradient theory proposed by the authors and the end-group polydispersity theory is also possible. However, it won't be possible to quantify the impact of each of these molecular imprecisions until a synthetic method that introduces one single *C*-terminal end-group becomes available.

In another series of experiments, Gellman and coworkers changed the nature of the cationic group. They synthesized polymers with different secondary amine structures and evaluated the effect of this modification on the hemolytic and antibacterial properties (Figure 5.7b). While the primary amine polymer had MICs of 1.56–12.5 μg/mL against four different bacteria, and an MHC of 12.5 μg/mL, the methyl-substituted secondary amine showed much less activity against *E. coli* (MIC = 100 mg/mL) and was almost inactive against *S. aureus* (MIC = 400 μg/mL). The MHC, however, rose to 100 μg/mL. The polymer with the more complex secondary amine was inactive against *S. aureus* and *Enterococcus faecium*, while its MHC was high. The corresponding polymer with a cyclic secondary amine was inactive against all four bacteria and non-hemolytic (MHC = 800 μg/mL).[91] One reason for the decreased activity of these polymers may be that they are less charged under physiological conditions than the corresponding primary amines. As Palermo and Kuroda have shown, this has a crucial effect of the SMAMPs' ability to attach to bacterial membranes, and thus on their activity.[74]

In summary, the effect of hydrophobic modifications, chain-length variations, and modification of the cationic group on the biological properties of nylon-3 polymer are consistent with the results obtained for other SMAMP families. In addition, the detailed biophysical studies that were conducted with these polymers added many interesting details that advanced the overall understanding of SMAMP interaction with bacteria and mammalian cells.

5.2.4 Poly(norbornene)-Based Synthetic Mimics of Antimicrobial Peptides

The above-described families of nylon-3 and poly(acrylic) polymers had promising to very good properties in terms of antimicrobial activity and hemocompatibility, and are interesting materials in many respects. Still, they do not reach the same level of structural precision that AMPs have, as they are obtained by statistical copolymerization of one monomer carrying the hydrophilic group, and a second monomer carrying the hydrophobic group. This means that there is no control over the repeat unit sequence, and thus over the distribution of hydrophilic and hydrophobic groups along the polymer chain. However, any repeat unit sequence that is not strictly alternating will lead to hydrophilic and hydrophobic "blobs" along the polymer chain. Consequently, the local hydrophobic/hydrophilic balance, *i.e.* the local amphiphilicity, of nylon-3 and poly(acrylic) SMAMPs is disturbed. While this clearly does not impede their ability to adopt a global facially amphiphilic conformation when in contact with membranes, it may limit the optimization of the system. It is known that the presence of hydrophobic blobs, for example at a chain end, will decrease the HC_{50} values. This then limits the maximum selectivity that can be obtained.

This inherent restriction is not present in the case of poly(norbornene) SMAMPs. Here, two functional groups (*i.e.* the hydrophobic and the hydrophilic group) can be attached to the same repeat unit, so that the resulting polymer is not only globally, but locally facially amphiphilic. Poly(norbornenes) are obtained by ring-opening metathesis polymerization (ROMP). This allows excellent control over the polymer molecular weight and polydispersity, and precise chain-end functionalization by functionalized initiators and quenching reagents. However, the poly(norbornene) platform also has its imperfections: poly(norbornenes) as a backbone are overall more hydrophobic than poly(acrylates) or nylon-3 polymers, which restricts the amount of hydrophobicity that can be added to each repeat unit before the polymer precipitates in aqueous solution. Additionally, stereoisomers are created in the ROMP step, where the double bond can have either *cis* or *trans* configuration. Both may limit the overall property optimization of a given polymer series. There are currently two synthetic platforms to obtain poly(norbornene) SMAMPs, the imide platform and the ester platform. The general monomer and polymer synthesis for these structures is shown in Figure 5.9. In the imide approach (Figure 5.9a), three different functional groups can be put on one repeat unit. Typically, R1 and R2 are hydrophobic groups and contribute to the overall hydrophobicity per repeat unit. The second functionality, the charged group, is incorporated into R3. The imide approach has the advantage that R1/R2 and R3 are truly on opposite faces of the molecule, as in the case of natural AMPs. However, the synthesis of such monomers is quite involved. Every variation in R1 or R2 requires a separate set of monomer precursors, *i.e.* every monomer has to be synthesized from scratch. The ester approach (Figure 5.9b) has the advantage that it is a modular, construction kit-like platform. To obtain a series of polymers with tunable

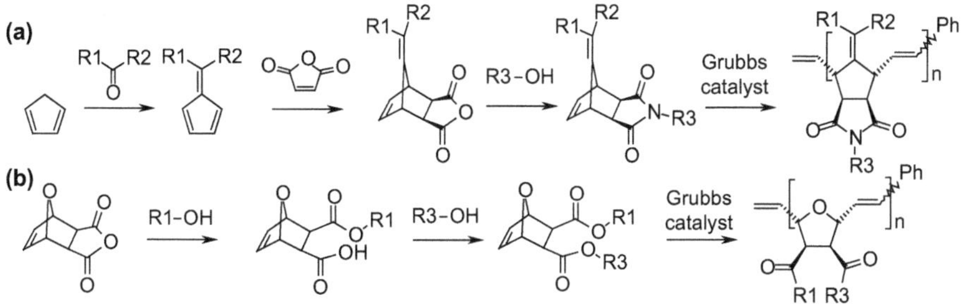

Figure 5.9 Synthesis of poly(norbornene) monomers and polymers using ring-opening metathesis polymerization (R1, R2 = hydrophobic group, R3 = protected form of the hydrophilic, cationic group). (a) Imide platform, (b) ester platform.[34]
Reprinted with permission from *Chem. Eur. J.* 2009, **15**, 11784–11800; Copyright 2009, Wiley-VCH.

properties, one has simply to decide which functional group will remain constant, and which one will be variable. The constant group is then introduced as R1, and the final monomers can be obtained from the same intermediate by introducing different R3s in the last step of monomer synthesis. While easier synthesis is a huge benefit, one has to bear in mind that the ester approach yields "less perfect" structures than the imide approach. In the ring-opening step of the anhydride precursor (Figure 5.9b), R1 can open either to the left or the right, yielding a pair of stereoisomers. These perpetuate through the polymer synthesis and may influence the property optimization of the system.

A summary of all currently known poly(norbornene)-based SMAMPs is given in Figure 5.10. The effect of systematically varying the structural parameters in these SMAMPs on their biological properties is discussed in the following sections.

The effect of increased hydrophobicity on the biological properties of poly(norbornenes) has been investigated in great detail. Several series of poly(norbornene) homo- and copolymers with gradually increasing hydrophobicity have been synthesized.[34,56,63,83,92] The first poly(norbornene) series based on the imide platform is shown in Figure 5.10 (poly1 to poly4).[56] This series consists of polymers with facially amphiphilic repeat units. The hydrophilic, cationic group was attached to the imide and was kept constant. The functional group on the opposite face of the molecule, attached to the poly(norbornene) backbone, was made successively more hydrophobic.[56] As shown in Figure 5.11a, the antimicrobial activity of the poly1 to poly4 series first improved with increasing hydrophobicity, but then went through a minimum, with maximum activity for poly3, and lower activity for poly4, presumably due to the poor solubility of poly4. The hemolytic activity of these polymers also increased with increasing hydrophobicity from poly2 to poly4. Since poly3 had the lowest MIC$_{90}$, and poly2 the highest HC$_{50}$, copolymers from these two facially amphiphilic repeat units were synthesized in an attempt to merge the beneficial properties of both polymers. The biological data for the

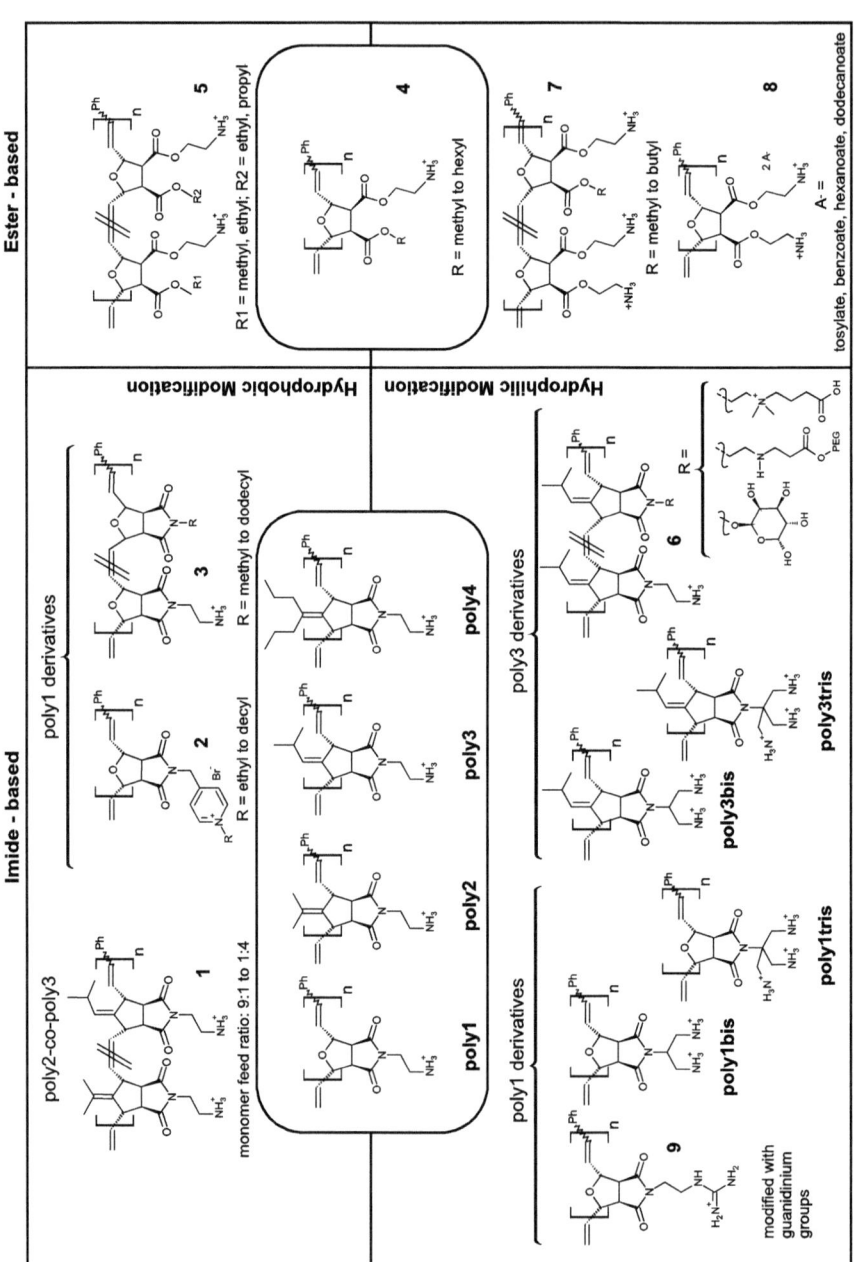

Figure 5.10 Library of poly(norbornene)-based SMAMPs. Left: poly(norbornene-imide) platform, right: poly(oxonorbornene-diester) platform. The polymers in the center boxes are the parent polymer series. Hydrophobic modifications are shown in the top boxes, hydrophilic modifications are in the bottom boxes.

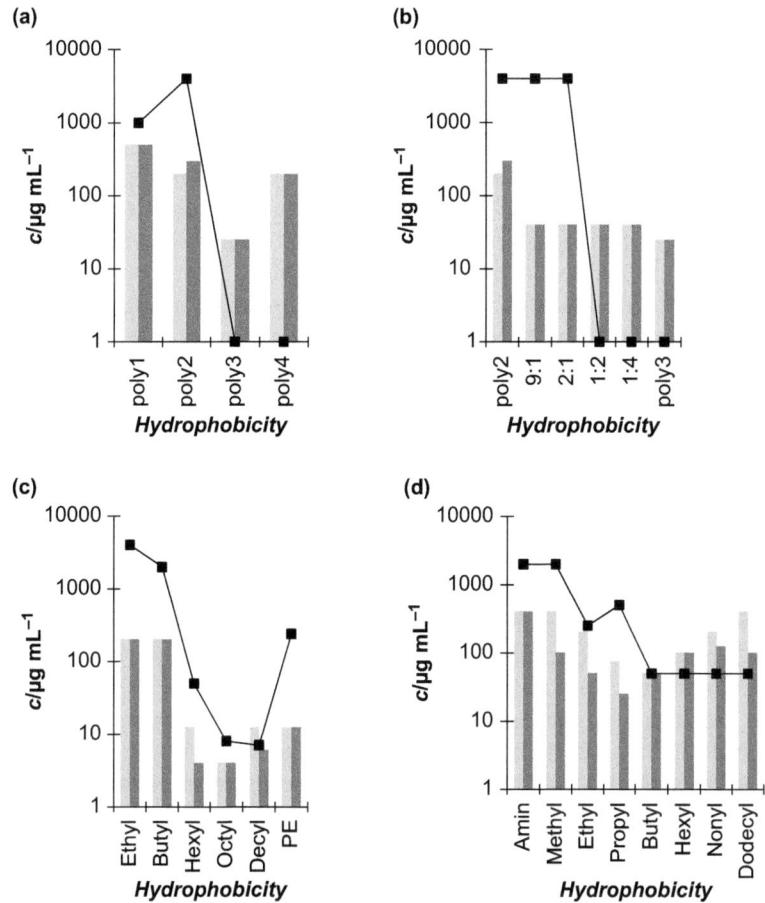

Figure 5.11 Effect of increasing the hydrophobicity on the biological properties of several series of poly(norbornene imide) SMAMPs. The biological data are plotted as concentration (MIC_{90} or HC_{50}, respectively) *versus* increasing hydrophobicity. Light gray columns = MIC_{90} against *E. coli*; dark gray columns = MIC_{90} against *B. subtilis* (for a to c) or *S. aureus* (d); black squares = HC_{50} of human erythrocytes. (a) poly1 to poly4, (b) copolymers of poly2 and poly3 (series 1 in Figure 5.10), (c) pyridinium-based homopolymers (series 2 in Figure 5.10), (d) statistical copolymers based on poly1 (series 3 in Figure 5.10).[34,56,63,83,92]

Data from *Chem. Eur. J.* 2009, **15**, 11784–11800; modified with permission; Copyright 2009, Wiley-VCH.

resulting copolymers are shown in Figure 5.11b. As can be seen from these data, for monomer feed ratios from 2:1 to 9:1, the resulting copolymers are nonhemolytic. Surprisingly, they stayed antimicrobially active even when the polymer contained only 10% of the antimicrobially active cationic repeat unit. This led to selectivities as high as 100 against both Gram-negative *E. coli* and Gram-positive *Bacillus subtilis* bacteria.[56]

To understand the mechanism of action of these polymers, the authors performed dye-leakage experiments with model membranes. They investigated the lysis of neutral cholesterol:phosphatidyl choline vesicles (as a mimic for erythrocytes) and of negatively charged phosphatidylserine:phosphatidyl choline vesicles (as a model for bacterial membranes) when exposed to structurally different SMAMPs.[92] They found that the lysing properties of poly2, poly3 and the poly2–3 copolymers were in good agreement with their antibacterial activities, poly2 being inactive and the other two polymers showing marked dye-leakage.[56,92] In the most comprehensive physical study on ROMP-based SMAMPs to date, HPLC, dye-leakage studies, light-scattering, ITC and fluorescence microscopy were combined to investigate the mechanism of polymer-membrane interactions of the poly1 to poly4 series in more detail.[93] A linear correlation between HPLC elution times and the alkyl side-chain length from poly1 to poly4 was observed, which proved the intuitive assumption that the polymer hydrophobicity increased with increasing alkyl chain length. The dye-leakage data of these polymers also follow their MIC trend, *i.e.* the inactive poly1 does not lyse vesicles, whereas poly2 and poly3 are increasingly membrane active.[93] Dynamic light scattering was used to monitor the effect of SMAMP addition on the hydrodynamic radius of lipid vesicles as a function of time. While the radii of vesicles exposed to poly1 remained unaltered, those exposed to poly2 and poly3 grew significantly over time. This is another indication that poly1 is not membrane active, whereas poly2 and poly3 clearly are, although the light-scattering studies do not capture the significant difference in the MIC_{90}s of these polymers (200 *vs.* 25 µg/mL). It is also not clear whether the vesicle growth is due to aggregation or vesicle fusion. Consequently, this effect was further studied by fluorescence microscopy on dye-labeled vesicles and stained bacteria cells.[93] When the vesicles that are routinely used for dye-leakage studies, with a diameter of about 200 nm, were exposed to poly3, giant fluorescent vesicles appeared. The stained *E. coli* bacteria aggregated, as has been observed with other SMAMPs.[90] These aggregation phenomena highlight that SMAMPs are not just very complicated detergents. When added to vesicles, detergents would just dissolve the membranes, instead of causing vesicle fusion or aggregation. Also, detergents do not have the ability to differentiate between cells. ITC revealed that, while no binding interaction between the vesicles and poly1 or poly4, respectively, was observed, there was a strong binding event between the vesicles and poly2 or poly3.[93] Fitting the data with modeling software showed that the binding between the vesicles and these SMAMPs is entropically favorable. The overall free enthalpy of binding was about the same for both poly2 and poly3. However, marked differences were observed in the binding stoichiometry – the ratio of vesicle lipids to ammonium groups of the polymer was 0.4 for poly2, and 1.06 for the more active poly3. From this, the authors drew the conclusion that membrane rupture, and thus antimicrobial activity, necessitates a minimum amount of SMAMP molecules attached to the membrane. Once that threshold of molecules per vesicle is passed, membrane rupture may occur.

In another polymer series, Eren *et al.* investigated the effect of increasing the hydrophobicity of the alkyl spacer length that is attached to the imide group on the biological properties. Their polymer series was based on the relatively hydrophilic poly1. The imide group was substituted with pyridinium groups on ethyl to dodecyl spacers, and a phenylethyl spacer (Figure 5.10, series 2). The biological data for these molecules are shown in Figure 5.11c:[83] first, the MIC_{90} continuously increases with longer alkyl spacers, and goes through a minimum for R = octyl. The HC_{50} follows this trend, probably due to the surfactant-like nature of the long alkyl chains with the quaternary pyridinium. Thus, the overall selectivity of these polymers remained low.[83] In dye-leakage experiments, Eren *et al.* investigated the effect of their polymers on phosphatidyl choline vesicles (erythrocyte mimics) and *E. coli* extract, and found a good correlation between the membrane-disruptive properties of these polymers and their HC_{50} and MIC_{90} data, respectively.[83]

Another series of poly(norbornene) SMAMPs with gradually increasing hydrophobicity was based on statistical poly(norbornene) copolymers. In this series, a hydrophilic monomer was copolymerized with a hydrophobic monomer, much like in the case of poly(methacrylate) SMAMPs. The hydrophilic monomer (the poly1 monomer) remained constant, while the chain length of the R group of the imide was gradually increased (series 3 in Figure 5.10). The idea of this synthetic design, which gave up the facial amphiphilicity of the poly(-norbornene) repeat unit, was to see if one could obtain a polymer series with tunable antimicrobial properties without the need to go through the tedious synthetic procedures of the poly1–4 series.[57] The biological data for the statistical copolymer are shown in Figure 5.11d. Due to the high structural similarity between these polymers and the poly1–4 series, high selectivities were also expected. However, while the new polymers (series 3 in Figure 5.10) followed the general trends that had been observed before (a minimum value for the MIC_{90}, and HC_{50} values that decreased with increasing hydrophobicity), their overall selectivities remained much lower. The maximum selectivity (HC_{50}/MIC_{90}) was only 20, and a deviation from the 1 : 1 monomer feed ratio did not improve it. It was concluded that the segregation of the functional groups onto two different repeat units, which caused hydrophobic "blobs" in the statistical polymer, had an unfavorable effect on the hemolytic activity of the polymer.

Another homopolymer series with varying hydrophobicity was obtained from the ester-based synthesis platform (Figure 5.9b, series 4 in Figure 5.10). The hydrophobic residue R of these polymers was systematically changed from R = methyl to hexyl. The MIC_{90} and HC_{50} values for this series are summarized in Figure 5.12a. As can be seen from these data, the trends observed in the antibacterial and hemolytic activities of the ester homopolymers fit well in the general picture: antimicrobial activity first increases from methyl to propyl, at which point a minimum is reached, and then decreases again due to poor polymer solubility and aggregation of the more hydrophobic homologues. They thus become unavailable to interact with the bacteria membrane, *i.e.* inactive. Also, as for the other poly(norbornene) series, the HC_{50} values were highest for the most hydrophilic polymers and then decrease significantly as the polymers become more hydrophobic. The homopolymer with R = ethyl was broadband active and had the best selectivity ($HC_{50}/MIC_{90} = 28$ for both

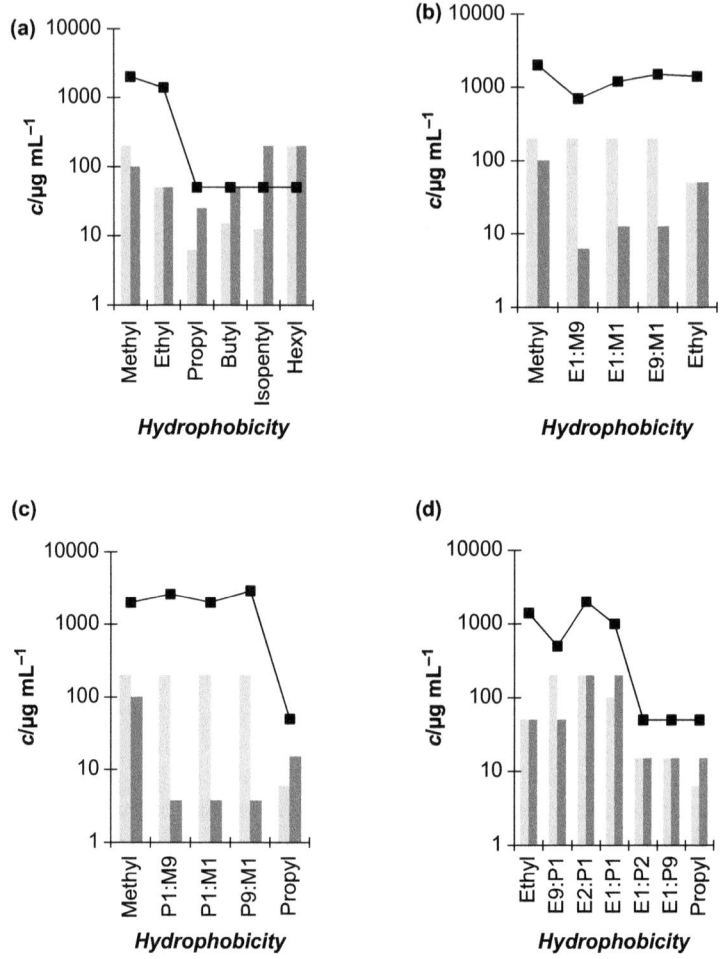

Figure 5.12 Effect of increasing hydrophobicity throughout several series of poly(oxonorbornene diester) SMAMPs on their biological properties. The biological data are plotted as concentration (MIC_{90} or HC_{50}, respectively) *versus* increasing hydrophobicity. Light gray columns = MIC_{90} against *E. coli*; dark gray columns = MIC_{90} against *S. aureus*; black squares = HC_{50} of human erythrocytes. (a) Series 4 in Figure 5.10, (b) to (d) series 5 in Figure 5.10 (methyl–ethyl, methyl–propyl, and ethyl–propyl copolymers).[34]
Data from *Chem. Eur. J.* 2009, **15**, 11784–11800; modified with permission; Copyright 2009, Wiley-VCH.

E. coli and *S. aureus*). The corresponding methyl polymer had the highest HC_{50} value, while the propyl polymer was the most active one. To obtain better selectivities, the facially amphiphilic monomers with R = methyl, ethyl and propyl were polymerized pairwise at different ratios to fine-tune the hydrophilic/hydrophobic balance. Three copolymer series (Series 5 in Figure 5.10, with R1/R2 = methyl/ethyl, methyl/propyl and ethyl/propyl) were thus obtained. The biological data from these copolymers are summarized in Figures 5.12b–d.

Copolymerization of the ethyl with the propyl comonomer gave only little improvement of the overall antimicrobial activity, with the expected decrease in the HC_{50} with increasing hydrophobicity, and low to moderate selectivity. However, in the case of the methyl–ethyl and methyl–propyl copolymers, a drastic improvement of the HC_{50} was observed, with values > 2000 mg/mL. Simultaneously, some of these polymers had antimicrobial activities as low as 3.75 µg/mL against the Gram-positive *S. aureus*. This resulted in a selectivity > 500, which is on the order of magnitude of the natural human defensins. Even at a feed ratio of only 1 : 9 of the active to the nonhemolytic repeat unit, these high selectivities were observed. Unexpectedly, however, these polymers were only weakly active against the Gram-negative *E. coli*. They were therefore termed "doubly selective" SMAMPs: they were not only selective for bacteria over mammalian cells, but preferentially killed one kind of bacteria over another. The "double selectivity" of the ester-based copolymers was unexpected because they can be considered as the direct analog of the poly2–poly3 copolymers, which are active against both Gram-negative and Gram-positive bacteria. Both copolymer families consist of facially amphiphilic repeat units. Their structural difference is that the two functional groups of the poly2–poly3 copolymers are spatially separated from each other by the polymer backbone, whereas those on the ester-based poly(oxonorbornenes) are on the same side of the molecule. This may hinder their self-organization at the bacterial membrane and thus reduce their activity. In dye-leakage experiments, Lienkamp *et al.* studied polymers from series 4 on *S. aureus* mimicking cardiolipin vesicles. They found a good correlation between the MIC_{90} for *S. aureus* and the vesicle leakage for the methyl to butyl homopolymers.[94] For the hexyl polymer of series 4 (Figure 5.10), discrepancies between dye-leakage experiments and biological data were observed. While this hydrophobic polymer was inactive in the MIC experiment, it caused significantly dye leakage. The reason for this is the low solubility of hydrophobic polymers at the comparatively high concentrations of the MIC experiment.[94] The SMAMP concentration in the dye-leakage experiments is usually one or two orders of magnitude lower than in the MIC experiments, as these experiments are understood to be much more sensitive.[95] Thus, if a polymer has poor solubility in aqueous media, it will seem less active in the MIC experiment, but it will still be active in the dye-leakage experiment, causing a discrepancy in the results from the two methods. It has been previously pointed out by Gellman and coworkers that this discrepancy is not observed when the membrane activity of a SMAMP against erythrocyte-mimicking vesicles is compared to hemolysis data.[40] As erythrocytes are not surrounded by a cell wall, they may still interact with aggregated hydrophobic SMAMPs, and are therefore lysed.

In summary, adding hydrophobicity to poly(norbornene) SMAMPs increases their antimicrobial activity up to a point at which they become insoluble or aggregate. At the same time, it makes them more hemolytic. Copolymerization of antimicrobially active with nonhemolytic monomers yields poly(norbornene) SMAMPs with improved selectivities, however, their broadband antimicrobial activity may suffer. While successive structural simplification in going from AMPs *via* foldamers to SMAMPs showed that many design features, such as the rigid nature of the backbone, were not essential for obtaining high selectivities for bacteria over mammalian cells, it was shown with the segregated SMAMPs that facial amphiphilicity on the repeat unit level was critical.[90]

Another hypothesis in SMAMP design was that the hemolytic activity of an active but toxic polymer could be reduced by copolymerizing the respective monomer with a hydrophilic comonomer. This approach is discussed in the following section. As a first example, Colak *et al.* modified poly3, the most active and hemolytic polymer in the poly1 to poly4 series, with nonionic and zwitterionic hydrophilic repeat units.[96] Unlike the above-described approach of tuning the hydrophilic/hydrophobic balance by shorter or longer alkyl chains, these additional hydrophilic repeat units (with R = sugar residue, zwitterionic side chain, or poly(ethylene glycol) side chain) dilute the overall charge density of the polymer (series 6 in Figure 5.10). The biological activity of the resulting polymers is shown in Figures 5.13a–c, respectively.[96] The data indeed showed that gradually making poly3 more hydrophilic reduced its hemolytic activity; however, the dilution of the active amine group also rendered these molecules increasingly inactive. Therefore, the selectivities of these polymers remained low. It is assumed that the charge density of those molecules is simply not high enough to trigger attachment to bacterial membranes; therefore, only little preferential activity towards bacteria over mammalian cells was observed.

A more direct approach of making SMAMPs more hydrophilic is to increase the charge density per repeat unit, which does not cause as much structural alteration as introducing a hydrophilic comonomer. It was already known from the AMP literature that the positively charged group is an important design feature to obtain biological activity, which made this strategy promising. Two polymer series with systematic variation of this parameter were synthesized, one of them imide-based (poly1bis and -tris, and poly3bis and -tris, respectively, see Figure 5.10),[97] and the other ester-based (series 7 in Figure 5.10).[78,94] For the imide-based series, poly1 and poly3 were taken as starting points. Structurally alike polymers carrying two and three charges per repeat unit were obtained, and their biological properties were compared with those of their parent compounds.[97] The biological data for these polymers are reported in Figure 5.13d. The hydrophobic poly3, which is active and toxic, became drastically less hemolytic and more active against *E. coli* as the charge doubled. Interestingly, this polymer had a selectivity of 140 for *E. coli*, but the selectivity for *S. aureus* was only 12. Thus, poly3bis had the opposite Gram-selectivity of the above-reported doubly selective SMAMPs. Further addition of charge (poly3tris) did not improve the biological properties, and the Gram-selectivity also vanished. In the poly1-based series, the hemolytic activity of the already hydrophilic poly1 did not improve upon addition of charge. However, the polymer became more active against *S. aureus* bacteria, which resulted in selectivity for Gram-positive

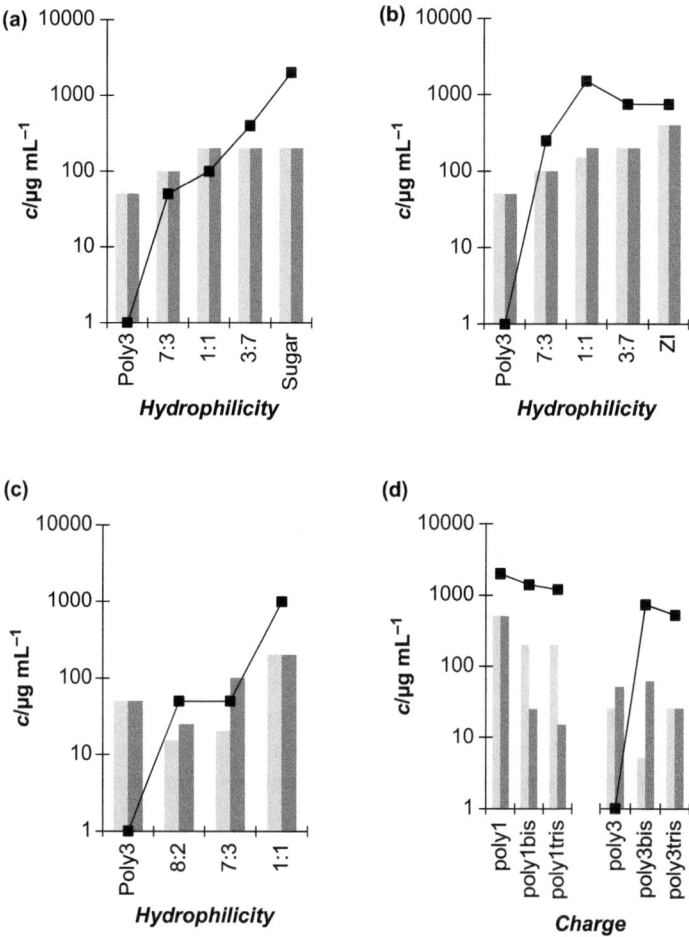

Figure 5.13 Effect of increasing the hydrophilicity of poly(norbornene imide) SMAMPs on their biological properties. The biological data are plotted as concentration (MIC_{90} or HC_{50}, respectively) versus increasing hydrophilicity. Light gray columns = MIC_{90} against *E. coli*; dark gray columns = MIC_{90} against *S. aureus*; black squares = HC_{50} of human erythrocytes. Data from[96,97] and (a) Series 6 in Figure 5.10, R = Sugar, (b) series 6, R = zwitterion, (c) series 6, R = PEG,[96] (d) poly1 and poly3 with modified charge density, left: poly1 derivatives, right: poly3 derivatives.[97]

Data from *Chem. Eur. J.* 2009, **15**, 11784–11800; modified with permission; Copyright 2009, Wiley-VCH.

over Gram-negative bacteria for both poly1bis and poly1tris. These results showed that the net effect of adding charge (and thus hydrophilicity) on the biological properties depends on the overall hydrophobicity of the polymer series. In dye-leakage studies of these polymers, good agreement of the leakage from vesicles mimicking red blood cells, *E. coli* and *S. aureus* with the respective biological data of poly1bis/tris and poly3bis/tris was observed.[97]

In another study of the charge-hydrophobicity balance and its correlation with biological activity, the poly(oxonorbornene diester) platform was used. Four series of copolymers with a doubly charged repeat unit, and a singly charged repeat unit with a variable R group were synthesized (series 7 in Figure 5.10, R = methyl to butyl).[78,94] The thus-obtained polymer series had different overall hydrophobicities between each series, but an equal number of charges at a given monomer feed ratio (*e.g.* M1 : D1, E1 : D1, P1 : D1 and B1 : D1 all consist of a 1 : 1 ratio of diamine to alkyl repeat units, and have increasing hydrophobicity from methyl (M) to butyl (B)). Different monomer feed ratios allowed the charge density to be continuously varied across the series. The biological properties of those polymers are summarized in Figure 5.14.

As these data indicate, all the methyl and ethyl copolymers are nonhemolytic, whereas the propyl and butyl copolymers become hemolytic with high alkyl comonomer content. Thus, the properties of those latter two series are dominated by the hydrophobicity of those R groups. It was found with complementary methods that the hydrophobicity of the monoamine-methyl homopolymer closely resembled that of the diamine homopolymer, while the ethyl to butyl homopolymers were significantly more hydrophobic. Thus, the methyl copolymers were identified as the most suitable model system to study the effect of increasing charge density at approximately constant overall hydrophobicity. Indeed, the properties of this polymer series (Figure 5.14) are very similar to those of the poly1 derivatives (Figure 5.13d): with increasing charge, the hemolytic activity is only slightly affected; however, the activity against *S. aureus* dramatically improves, making these polymers also doubly selective. When charge is reduced in the methyl-diamine series (from M9:D1 to methyl in Figure 5.14a), there is a sudden jump in the MIC from 4 to 100 µg/mL. The same is found in the poly1 derivatives when going from one to two charges per repeat unit. These findings, together with AMP literature data[98] led to the postulation that there is a specific charge threshold that needs to be exceeded to obtain decent activities against *S. aureus*. Rather than a certain number of charges per repeat unit, this charge threshold has to be understood as a minimum charge density, or charge per unit volume, and the exact threshold number of charges per repeat unit will be slightly different for each SMAMP series, depending on the molecular volume of the repeat units. On the molecular level, this postulated charge threshold translates into a minimum charge density that is necessary to trigger successful attachment of the SMAMP to the bacterial membrane. Once enough charge is present to enable this attachment, the overall hydrophobicity of the molecule will determine to what extent the SMAMP is able to penetrate the bacterial membrane, and thus whether it is active or not.

To investigate the reason for the double selectivity of the methyl-diamine copolymers and the diamine homopolymer (MIC$_{90}$s for *S. aureus* = 15 µg/mL, for *E. coli* > 200 µg/mL),[63] Lienkamp *et al.* studied the behavior of the diamine homopolymer (polymer 8 in Figure 5.10, with trifluoroacetate counterions, $M_n \sim 3000$ g/mol), towards both cardiolipin (*S. aureus* mimic) and phosphatidylethanolamine:phosphatidylglycerol (PE:PG, *E. coli* mimic)

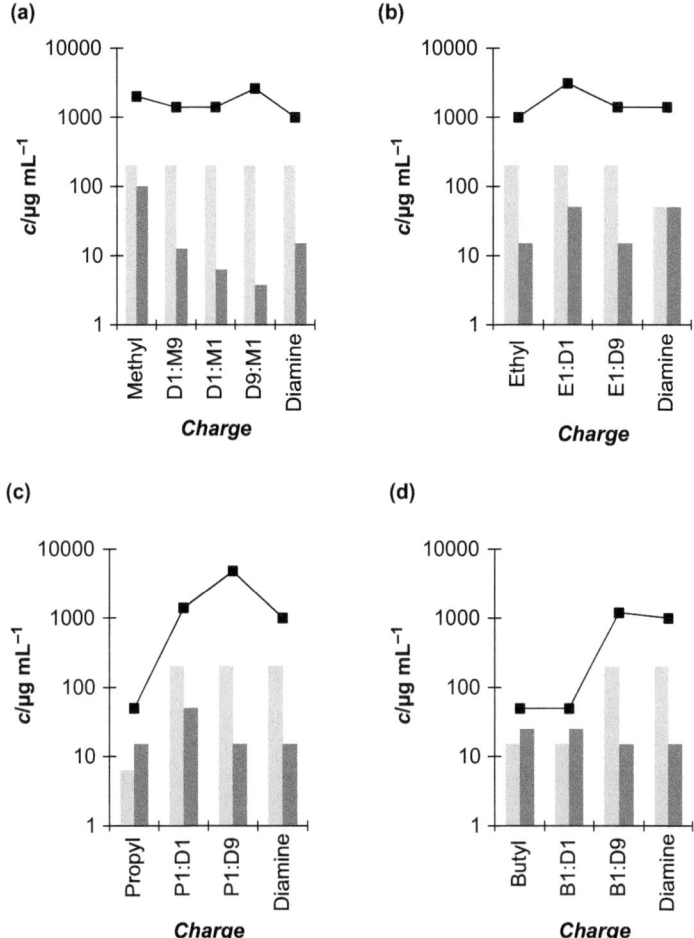

Figure 5.14 Effect of increasing the charge and hydrophilicity of poly(oxonorbornene diester) SMAMPs (series 7 in Figure 5.10) on their biological properties. The biological data are plotted as concentration (MIC_{90} or HC_{50}, respectively) versus increasing charge. Light gray columns = MIC_{90} against *E. coli*; dark gray columns = MIC_{90} against *S. aureus*; black squares = HC_{50} of human erythrocytes. (a) methyl copolymers, (b) ethyl copolymers, (c) propyl copolymers, (d) butyl copolymers.[34]
Data from *Chem. Eur. J.* 2009, **15**, 11784–11800; modified with permission; Copyright 2009, Wiley-VCH.

vesicles.[78] Surprisingly, although the MIC_{90}s of the 3000 g/mol sample of polymer 8 were dramatically different for *E. coli* and *S. aureus*, this polymer caused almost identical leakage for both vesicle types (Figure 5.15a). This demonstrated that the differences in lipid composition of the two bacteria were not responsible for the observed differences in MIC_{90}. When dye-leakage studies fail to model cell–SMAMP interactions, this may indicate that other

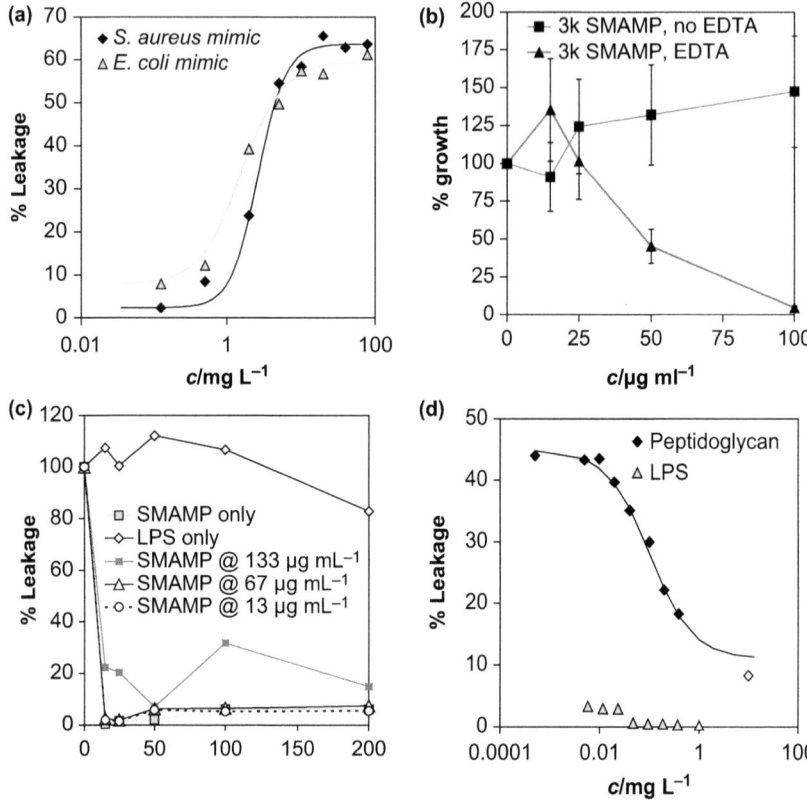

Figure 5.15   (a) Dye-leakage percentage versus SMAMP concentration of *E. coli* and *S. aureus* mimicking vesicles when exposed to polymer 8 (Figure 5.10); (b) MIC of polymer 8 (Figure 5.10) against regular *E. coli* cells (no EDTA) and *E. coli* cells with EDTA-damaged outer membrane; (c) MIC experiment of polymer 8 (Figure 5.10) on *S. aureus* in the presence of LPS; (d) binding of polymer 8 (Figure 5.10) with LPS and peptidoglycan.[78] *Chem. Eur. J.* 2009, **15**, 11784–11800; modified with permission; Copyright 2009, Wiley-VCH.

components of the cell structure are important, which are not adequately modeled by the simple bilayer membrane. Examples for such structures are the peptidoglycan cell wall of Gram-positive bacteria, or the double-membrane structure of Gram-negative bacteria including the lipopolysaccharide layer (LPS), which surrounds the outer membrane of Gram-negative bacteria (*E. coli*). To investigate which part of *E. coli* was responsible for the loss of SMAMP activity, two series of modified MIC experiments were conducted. First, *E. coli* cells were exposed to ETDA, which selectively damaged their outer membranes. These membrane-compromised bacteria had a lower MIC than regular, intact *E. coli* cells (Figure 5.15b). On the other hand, the MIC$_{90}$ of *S. aureus* in the presence of added LPS extract was the same as that without additional LPS, meaning that there is no irreversible binding between the

SMAMP and LPS (Figure 5.15c). Thus, it was concluded that it is the second lipid bilayer, and not the additional LPS layer that reduces the SMAMP activity. Binding to lipids in the outer plasma membrane may reduce the SMAMP concentration at the plasma membrane to such an extent that the SMAMP becomes inactive against *E. coli* bacteria.[78] Since it was observed that the diamine polymer with a molecular mass of 3000 g/mol (polymer 8 in Figure 5.10) was active against *S. aureus*, while the corresponding 50 000 g/mol polymer was inactive, it was considered that the peptidoglycan cell wall of Gram-positive organisms was an impenetrable barrier for large SMAMPs. This could be either due to SMAMP binding, or due to the inability of large molecules to diffuse or reptate through the peptidogylcan layer. Dye-leakage experiments on *S. aureus*-mimicking cardiolipin vesicles, to which peptidoglycan extract had been added, led to the conclusion that irreversible binding to peptidoglycan does not occur on the timescale of the experiment (Figure 5.15d, SMAMP-peptidoglycan binding reduces leakage much less than SMAMP-LPS binding), but that the large SMAMPs cannot penetrate the peptidoglycan mesh and thus do not reach the plasma membrane.[78]

The effect of counterion exchange on the biological properties of the SMAMPs 8 (Figure 5.10) was also studied.[78] The hydrophilic counterions of the most hydrophilic ester-based polymer (polymer 8 in Figure 5.10) were exchanged by hydrophobic organic counterions (*e.g.* hexanoate and tosylate). While the original idea was that exchanging these counterions would impart hydrophobicity onto the polymer and makes it more active, it was found that ion exchange of this polymer completely eliminated its antibacterial activity. Using dye-leakage studies, it was found that the ion-exchanged polymers were not membrane active, unlike the parent polymer 8. This revealed that the protonated primary amine groups of these polymers and the organic counterions formed such a tight ion pair that the overall positive charge of the polymer was masked.[78] While this meant that exchange of the counterions does not provide an additional handle to tune the antibacterial activity of SMAMPs, these findings agreed with previous studies using inorganic counterions that correlated the tightness of the ion pair to the antibacterial activities of the polymer.[67]

The previously presented poly(norbornene) SMAMP data all referred to samples with a molecular weight of roughly 3000 g/mol, although most of the studies mentioned investigated two or more molecular weights of each polymer type. In this section, we summarize how molecular weight affects SMAMP properties. Ilker *et al.* found only a weak molecular weight dependence for their poly3 compound, which at that time did not seem significant.[56] In the case of the poly2 and poly4 series, no trend was observed as those polymers were in the inactive regime. Eren's low and high molecular weight polymers ($M_n \sim 3,000$ g/mol and 10,000 g/mol, respectively) all had similar antibacterial and hemolytic activities, whether they were in the active or inactive regime.[83] For the "segregated" copolymers, Gabriel *et al.* found that the high molecular weight polymers ($M_n \sim 10\,000$ g/mol) were less active than the low molecular weight polymers ($M_n \sim 3000$ g/mol) by a factor of 2–8.[90] Some of them were also slightly more hemolytic. The same general trend was found by Lienkamp *et al.* for the higher molecular weight ester-based polymers (series 4 in Figure 5.10, $M_n \sim 10\,000$ g/mol). Compared with their $M_n \sim 3000$ g/mol analogs, these polymers were generally less active against *E. coli*, with the exception of the $10\,000$ g/mol propyl polymer, which was surprisingly active against that bacterial type. More notably, they were all inactive against *S. aureus*.[63] Similarly, the diamine homopolymer (polymer 8 with TFA counterions, Figure 5.10) showed a systematic decrease in activity against *S. aureus* with molecular weight, together with inactivity against *E. coli* at all molecular weights. This leads to the hypothesis that these particular polymers get stuck in the peptidoglycan layer of Gram-positive bacteria at higher molecular weights. To investigate the molecular weight effect in more detail, especially in the low molecular weight region, a series of oligomers from the propyl homopolymer (series 4 in Figure 5.10, R = propyl) was prepared. These data showed that the molecular-weight dependence for both the hemolytic and antibacterial activities is highly nonlinear and different for each bacterial species involved. In the present case, Oligo 1 (average number of repeat units = 1) was selective for *S. aureus* over *E. coli*, while Propyl_3k (average number of repeat units = 9–10) shows the opposite tendency. It is thus very difficult to draw general conclusions concerning the dependency of biological activity on molecular weight. In most cases, the general rule seems to hold that, when the molecular weight is above a certain threshold value, the polymers become inactive, as discussed above. Below this threshold, however, it is not possible to predict which molecular weight will give the best activities and selectivities, as this strongly depends on the overall hydrophobicity of the particular polymer studied.

Recently, an unusual SMAMP has been discovered (polymer 9 in Figure 5.10). While most of the previously discussed ROMP-based SMAMPs had protonated primary amine groups as the positively charged moiety, this polymer contained guanidinium groups. It had broad-spectrum antibacterial activity against both Gram-negative (*E. coli* and *Serratia marcescens*, $MIC_{90} = 6$ µg/mL and 50 µg/mL, respectively) and Gram-positive (*S. aureus* and *B. subtilis*, $MIC_{90} = 12$ µg/mL for both) bacteria. Together with a remarkably low hemolytic activity ($HC_{50} = 1500$ µg/mL), this yields a selectivity for *E. coli* over red blood cells of 250, which is the highest selectivity so far observed for a broad-spectrum SMAMP. Using bactericidal kinetics studies, it was shown that this polymer caused a 5 log reduction in less than 60 min at four times the MIC_{90}, meaning that the polymer is indeed bactericidal and not just bacteriostatic.[62] Comparative dye-leakage studies with poly1, poly3 and polymer 9 showed that, in spite of its low MIC_{90} value, polymer 9 did not lyse model membranes. Similarly, while the active poly3 caused membrane damage and cell aggregation, as observed in fluorescence microscopy experiments, this SMAMP was membrane inactive. These are clear indications that the antimicrobial activity of this polymer is not due to extensive membrane damage, as is the case for the other SMAMPs.

In summary, a large variety of poly(norbornene) SMAMPs was obtained. By carefully tuning the overall hydrophobicity and charge density of these molecules, polymers with tailor-made properties, from inactive/nonhemolytic *via* active/nonhemolytic to active/toxic, were obtained. As the biological properties of a SMAMP result from the interplay of many parameters, it is not yet possible to predict the exact properties of such molecules from their mere chemical structure. However, as demonstrated above, the effect of certain design features, such as charge and hydrophobicity, on the properties across a polymer series is meanwhile quite well understood.

5.3 Conclusion

Polymer-based synthetic mimics of antimicrobial peptides (SMAMPs) not only emulate antimicrobial peptides (AMPs) structurally, but also capture their key biological properties. Several series of poly(methacrylate)-, nylon-3- and poly(norbornene)-based SMAMPs were synthesized. The most optimized structures in each series have a high selectivity for bacteria over mammalian cells, which is of the order of magnitude of the cell selectivity of AMPs. Combined with their improved hydrolytic stability, low tendency to cause bacterial resistance, and easy synthetic availability, these polymers are excellent candidates for biomedical applications. By systematic structural variation of these polymers, several design rules for SMAMPs became apparent. Also, many biophysical studies have investigated the interaction of SMAMPs with biomembranes (bacteria or erythrocytes) and artificial model membranes (lipid vesicles). These studies revealed many mechanistic details and led to a number of theories on how SMAMPs kill bacteria. While this body of data may not yet tell the complete story of SMAMP–membrane interactions, the following conclusions can be drawn:

- A rigid polymer backbone that directs the hydrophobic groups to one side of the molecule, and the hydrophilic groups to the opposite side, is not required to simultaneously obtain high antimicrobial and low hemolytic activity. It is hypothesized that SMAMPs can self-organize to a globally amphiphilic conformation when they are in contact with biomembranes.
- A minimum amount of positive charge is necessary for antibacterial activity; otherwise the polymer cannot bind to the bacterial membrane. The charged group seems to trigger the attachment of the SMAMP to the bacterial membrane. Primary amine groups seem to make the SMAMP more active than secondary or tertiary amines, because they are usually protonated under physiological conditions. Unlike permanently charged quaternary ammonium groups, they are also pH-sensitive and may be deprotonated during their interaction with the bacterial membrane. This may be important for the SMAMPs' ability to penetrate bacteria. Sufficient charge is also necessary for cell selectivity, because the charged SMAMP will preferentially bind to the more negatively charged bacterial membranes and not to the zwitterionic mammalian cell membranes. Polymers with *N*-protonated groups also appear to be better candidates to obtain nontoxic SMAMPs than permanently charged quaternary ammonium groups.

- Sufficient hydrophobicity is required for insertion through the bacterial lipid membrane, as well as consecutive membrane disruption or permeabilization. If the hydrophobicity is too low, the SMAMP becomes membrane inactive. Adding more hydrophobic groups increases the antimicrobial activity, but too much hydrophobicity causes insolubility or aggregation of the SMAMP in aqueous media. This decreases the antimicrobial activity, because the aggregates are not able to permeate the bacterial membranes. Interestingly, aggregate formation does not affect hemolysis, most likely because erythrocytes do not have a protective cell wall like bacteria, and are thus accessible for aggregated and non-aggregated SMAMPs.
- Most polymeric SMAMPs are antimicrobially active because they disrupt bacterial membranes rather than forming well-defined, discrete pores. However, there is also evidence for other mechanisms: for example, it was shown that some SMAMPs caused outer membrane permeability in *E. coli*, but no inner membrane damage. In other works, the bacteria were growth inhibited even though the inner membrane was still intact. This may be evidence that SMAMPs get trapped in the outer membrane, which could disturb the bacterial metabolism and thus cause antimicrobial activity. Nylon-3-based SMAMPs were shown to induce phase boundaries in lipid bilayers, which destabilized the bacterial membrane. They also bound to lipopolysaccharides and peptidoglycan, and could block transport mechanisms through the outer membrane of *E. coli*.[85]
- Some SMAMPs caused hemagglutination at concentrations well below the HC_{50}. This indicates that SMAMPs need to be studied in more detail, and by methods other than just the HC_{50} assay, to judge their biocompatibility, if they are intended for medical applications. However, HC_{50} remains a valuable screening tool to screen SMAMP toxicity.

Overall, the many SMAMP molecules that have been synthesized so far cover virtually all the structural modifications that are possible. It is, of course, feasible to choose yet another polymer backbone and repeat the exercise of optimizing the system by variation of the well-known parameters. While there may be practical, application- or patent-driven reasons to do so, this will probably not increase our understanding on how SMAMPs interact with cells and organisms. Unless a novel synthetic approach promises new structural precision, such as the alternating polymers by Sampson and coworkers,[99] research should now focus on a deeper mechanistic understanding of SMAMP action, and on a more thorough characterization of their *in vitro* and *in vivo* toxicity, to bring these interesting molecules into clinical applications.

Acknowledgements

Karen Lienkamp's lab is funded by the Emmy-Noether-Program of the German Research Foundation (DFG, grant ID LI1714/5-1) and the Freiburg Institute for Advanced Studies (FRIAS). Cofunding of the Studienförderwerk

Klaus Murmann der Stiftung der Deutschen Wirtschaft (Franziska Dorner) is also gratefully acknowledged. A part of this work is based on an earlier review that was coauthored by Prof. G. N. Tew, University of Massachusetts Amherst, Amherst/MA, USA, and has been modified with permission from *Chem. Eur. J.* 2009, **15**, 11784; Copyright 2009, Wiley-VCH. Wibke Hartleb is gratefully acknowledged for proof-reading the manuscript.

References

1. D. F. H. Hahn, S. H. E. Kaufmann, U. Ullmann, *Medizinische Mikrobiologie und Infektiologie*, Springer, Berlin, 2001.
2. CDC, 1999. http://www.cdc.gov.
3. E. Klein, D. L. Smith and R. Laxminarayan, *Emerg. Infect. Dis.*, 2007, **13**, 1840–1846.
4. K. Okuma, K. Iwakawa, J. D. Turnidge, W. B. Grubb, J. M. Bell, F. G. O'Brien, G. W. Coombs, J. W. Pearman, F. C. Tenover, M. Kapi, C. Tiensasitorn, T. Ito and K. Hiramatsu, *J. Clin. Microbiol.*, 2002, **40**, 4289–4294.
5. K. Page, M. Wilson and I. P. Parkin, *J. Mater. Chem.*, 2009, **19**, 3819–3831.
6. S. J. Dancer, *J. Hosp. Infect.*, 2004, **56**, 10–15.
7. A. R. B. Dietze, C. Wendt and H. Martiny, *J. Hosp. Infect.*, 2001, **49**, 255–261.
8. G. P.-B. J. M. Boyce, C. and T. K. Chenevert, *Infect. Control. Hosp. Epidemiol.*, 1997, **18**, 622–627.
9. EARS-Net 2011. http://www.ecdc.europa.eu/en/activities/surveillance/EARS-Net/Pages/index.aspx.
10. N. K. Brogden and K. A. Brogden, *Int. J. Antimicrob. Agents*, 2011, **38**, 217–225.
11. K. A. Brogden, *Nature Rev. Microbiol.*, 2005, **3**, 238–250.
12. M. Zasloff, *Nature*, 2002, **415**, 389–395.
13. J. G. Hirsch, *J. Exp. Med.*, 1956, **103**, 589–611.
14. H. I. Zeya and J. K. Spitznagel, *Science*, 1963, **142**, 1085–1087.
15. D. Friedberg, I. Friedberg and M. Shilo, *Infec. Immunity*, 1970, **1**, 311–318.
16. R. I. Lehrer, A. Barton, K. A. Daher, S. S. L. Harwig, T. Ganz and M. E. Selsted, *J. Clin. Invest.*, 1989, **84**, 553–561.
17. T. Ganz, M. E. Selsted and R. I. Lehrer, *Eur. J. Haematol.*, 1990, **44**, 1–8.
18. B. L. Kagan, M. E. Selsted, T. Ganz and R. I. Lehrer, *Proc. Natl. Acad. Sci. USA*, 1990, **87**, 210–214.
19. V. N. Kokryakov, S. S. L. Harwig, E. A. Panyutich, A. A. Shevchenko, G. M. Aleshina, O. V. Shamova, H. A. Korneva and R. I. Lehrer, *FEBS Lett.*, 1993, **327**, 231–236.
20. R. I. Lehrer, A. K. Lichtenstein and T. Ganz, *Annu. Rev. Immunol.*, 1993, **11**, 105–128.
21. T. Ganz and R. I. Lehrer, *Pharmacol. Ther.*, 1995, **66**, 191–205.
22. R. E. W. Hancock and R. Lehrer, *Trends Biotechnol.*, 1998, **16**, 82–88.
23. O. Shamova, K. A. Brogden, C. Zhao, T. Nguyen, V. N. Kokryakov and R. I. Lehrer, *Infect. Immun.*, 1999, **67**, 4106–4111.

24. M. Zasloff, *Proc. Natl. Acad. Sci. USA*, 1987, **84**, 5449–5453.
25. H. W. Huang, *Biochemistry*, 2000, **39**, 8347–8352.
26. F.-Y. Chen, M.-T. Lee and H. W. Huang, *Biophys. J.*, 2003, **84**, 3751–3758.
27. K. Matsuzaki, O. Murase, N. Fujii and K. Miyajima, *Biochemistry*, 1996, **35**, 11361–11368.
28. J. Rennie, L. Arnt, H. Z. Tang, K. Nusslein and G. N. Tew, *J. Ind. Microbiol. Biotechnol.*, 2005, **32**, 296–300.
29. A. Peschel, R. W. Jack, M. Otto, L. V. Collins, P. Staubitz, G. Nicholson, H. Kalbacher, W. F. Nieuwenhuizen, G. Jung, A. Tarkowski, K. P. van Kessel and J. A. van Strijp, *J. Exp. Med.*, 2001, **193**, 1067–1076.
30. A. Peschel, M. Otto, R. W. Jack, H. Kalbacher, G. Jung and F. Gotz, *J. Biol. Chem.*, 1999, **274**, 8405–8410.
31. S. A. Kristian, M. Durr, J. A. G. Van Strijp, B. Neumeister and A. Peschel, *Infect. Immun.*, 2003, **71**, 546–549.
32. M. A. Campos, M. A. Vargas, V. Regueiro, C. M. Llompart, S. Alberti and J. A. Bengoechea, *Infect. Immun.*, 2004, **72**, 7107–7114.
33. J. A. Robinson, *Curr. Opin. Chem. Biol.*, 2011, **15**, 379–386.
34. K. Lienkamp and G. N. Tew, *Chem. Eur. J.*, 2009, **15**, 11784–11800.
35. K. Lienkamp, *Nachr. Chem.*, 2011, **59**, 719–723.
36. K. Lienkamp, A. E. Madkour and G. N. Tew, *Adv. Polym. Sci.*, 2013, **251**, 141–172.
37. G. N. Tew, R. W. Scott, M. L. Klein and W. F. De Grado, *Acc. Chem. Res.*, 2010, **43**, 30–39.
38. I. Sovadinova, E. F. Palermo, M. Urban, P. Mpiga, G. A. Caputo and K. Kuroda, *Polymers (Basel, Switz.)*, 2011, **3**, 1512–1532.
39. E. F. Palermo and K.-I. Kuroda, *Appl. Microbiol. Biotechnol.*, 2010, **87**, 1605–1615.
40. B. P. Mowery, A. H. Lindner, B. Weisblum, S. S. Stahl and S. H. Gellman, *J. Am. Chem. Soc.*, 2009, **131**, 9735–9745.
41. B. Mensa, Y. H. Kim, S. Choi, R. Scott, G. A. Caputo and W. F. DeGrado, *Antimicrob. Agents Chemother.*, 2011, **55**, 5043–5053.
42. E. A. Porter, X. Wang, H. S. Lee, B. Weisblum and S. H. Gellman, *Nature*, 2000, **404**, 565.
43. D. Liu and W. F. DeGrado, *J. Am. Chem. Soc.*, 2001, **123**, 7553–7559.
44. R. F. Epand, T. L. Raguse, S. H. Gellman and R. M. Epand, *Biochemistry*, 2004, **43**, 9527–9535.
45. J. A. Patch and A. E. Barron, *J. Am. Chem. Soc.*, 2003, **125**, 12092–12093.
46. Y. X. Chen, C. T. Mant, S. W. Farmer, R. E. W. Hancock, M. L. Vasil and R. S. Hodges, *J. Biol. Chem.*, 2005, **280**, 12316–12329.
47. H. S. Won, S. J. Jung, H. E. Kim, M. D. Seo and B. J. Lee, *J. Biol. Chem.*, 2004, **279**, 14784–14791.
48. Y. Hamuro, J. P. Schneider and W. F. DeGrado, *J. Am. Chem. Soc.*, 1999, **121**, 12200–12201.
49. Y. Shai and Z. Oren, *Peptides*, 2001, **22**, 1629–1641.
50. J. Hong, Z. Oren and Y. Shai, *Biochemistry*, 1999, **38**, 16963–16973.
51. Z. Oren and Y. Shai, *Biochemistry*, 1997, **36**, 1826–1835.

52. G. N. Tew, D. Liu, B. Chen, R. J. Doerksen, J. Kaplan, P. J. Carroll, M. L. Klein and W. F. DeGrado, *Proc. Natl. Acad. Sci. USA*, 2002, **99**, 5110–5114.
53. D. H. Liu, S. Choi, B. Chen, R. J. Doerksen, D. J. Clements, J. D. Winkler, M. L. Klein and W. F. DeGrado, *Angew. Chem., Int. Ed.*, 2004, **43**, 1158–1162.
54. H. Tang, R. J. Doerksen and G. N. Tew, *Chem. Commun.*, 2005, 153–1539.
55. K. Kuroda and W. F. DeGrado, *J. Am. Chem. Soc.*, 2005, **127**, 4128–4129.
56. M. F. Ilker, K. Nusslein, G. N. Tew and E. B. Coughlin, *J. Am. Chem. Soc.*, 2004, **126**, 15870–15875.
57. L. Arnt and G. N. Tew, *J. Am. Chem. Soc.*, 2002, **124**, 7664–7665.
58. L. Arnt and G. N. Tew, *Langmuir*, 2003, **19**, 2404–2408.
59. L. Arnt, K. Nuesslein and G. N. Tew, *J. Polym. Sci., Part A: Polym. Chem.*, 2004, **42**, 3860–3864.
60. B. P. Mowery, S. E. Lee, D. A. Kissounko, R. F. Epand, R. M. Epand, B. Weisblum, S. S. Stahl and S. H. Gellman, *J. Am. Chem. Soc.*, 2007, **129**, 15474–15482.
61. V. Sambhy, B. R. Peterson and A. Sen, *Angew. Chem., Int. Ed.*, 2008, **47**, 1250–1254.
62. G. J. Gabriel, A. E. Madkour, J. M. Dabkowski, C. F. Nelson, K. Nusslein and G. N. Tew, *Biomacromolecules*, 2008, **9**, 2980–2983.
63. K. Lienkamp, A. E. Madkour, A. Musante, C. F. Nelson, K. Nusslein and G. N. Tew, *J. Am. Chem. Soc.*, 2008, **130**, 9836–9843.
64. T. Ikeda, A. Ledwith, C. H. Bamford and R. A. Hann, *Biochim. Biophys. Acta, Biomembr.*, 1984, **769**, 57–66.
65. T. Ikeda, H. Hirayama, H. Yamaguchi, S. Tazuke and M. Watanabe, *Antimicrob. Agents Chemother.*, 1986, **30**, 132–136.
66. N. Kawabata and M. Nishiguchi, *Appl. Environ. Microbiol.*, 1988, **54**, 2532–2535.
67. A. Kanazawa, T. Ikeda and T. Endo, *J. Polym. Sci., Part A: Polym. Chem.*, 1993, **31**, 3031–3038.
68. A. Kanazawa, T. Ikeda and T. Endo, *J. Polym. Sci., Part A: Polym. Chem.*, 1993, **31**, 2873–2876.
69. A. Kanazawa, T. Ikeda and T. Endo, *J. Polym. Sci., Part A: Polym. Chem.*, 1993, **31**, 335–343.
70. S. D. Worley and G. Sun, *Trends Polym. Sci. (Cambridge, UK)*, 1996, **4**, 364–370.
71. E. F. Panarin, M. V. Solovskii and O. N. Ekzemplyarov, *Khim.-Farm. Zh.*, 1971, **5**, 24–26.
72. I. Ivanov, S. Vemparala, V. Pophristic, K. Kuroda, W. F. DeGrado, J. A. McCammon and M. L. Klein, *J. Am. Chem. Soc.*, 2006, **128**, 1778–1779.
73. K. Kuroda, G. A. Caputo and W. F. DeGrado, *Chem. Eur. J.*, 2009, **15**, 1123–1133.
74. E. F. Palermo and K. Kuroda, *Biomacromolecules*, 2009, **10**, 1416–1428.
75. B. Findlay, G. G. Zhanel and F. Schweizer, *Antimicrob. Agents Chemother.*, 2010, **54**, 4049–4058.
76. E. F. Palermo, I. Sovadinova and K. Kuroda, *Biomacromolecules*, 2009, **10**, 3098–3107.

77. L. C. Paslay, B. A. Abel, T. D. Brown, V. Koul, V. Choudhary, C. L. McCormick and S. E. Morgan, *Biomacromolecules*, 2012, **13**, 2472–2482.
78. K. Lienkamp, K.-N. Kumar, A. Som, K. Nuesslein and G. N. Tew, *Chem. Eur. J.*, 2009, **15**, 11710–11714.
79. I. Sovadinova, E. F. Palermo, R. Huang, L. M. Thoma and K. Kuroda, *Biomacromolecules*, 2011, **12**, 260–268.
80. C. W. Avery, E. F. Palermo, A. McLaughlin, K. Kuroda and Z. Chen, *Anal. Chem.*, 2011, **83**, 1342–1349.
81. E. F. Palermo, D.-K. Lee, A. Ramamoorthy and K. Kuroda, *J. Phys. Chem. B*, 2011, **115**, 366–375.
82. E. F. Palermo, S. Vemparala and K. Kuroda, *Biomacromolecules*, 2012, **13**, 1632–1641.
83. T. Eren, A. Som, J. R. Rennie, C. F. Nelson, Y. Urgina, K. Nusslein, E. B. Coughlin and G. N. Tew, *Macromol. Chem. Phys.*, 2008, **209**, 516–524.
84. M. Mizutani, E. F. Palermo, L. M. Thoma, K. Satoh, M. Kamigaito and K. Kuroda, *Biomacromolecules*, 2012, **13**, 1554–1563.
85. R. F. Epand, B. P. Mowery, S. E. Lee, S. S. Stahl, R. I. Lehrer, S. H. Gellman and R. M. Epand, *J. Mol. Biol.*, 2008, **379**, 38–50.
86. J. Zhang, M. J. Markiewicz, B. P. Mowery, B. Weisblum, S. S. Stahl and S. H. Gellman, *Biomacromolecules*, 2012, **13**, 323–331.
87. R. Liu, K. S. Masters and S. H. Gellman, *Biomacromolecules*, 2012, **13**, 1100–1105.
88. M. T. Dohm, B. P. Mowery, A. M. Czyzewski, S. S. Stahl, S. H. Gellman and A. E. Barron, *J. Am. Chem. Soc.*, 2010, **132**, 7957–7967.
89. R. Liu, K. Z. Vang, P. K. Kreeger, S. H. Gellman and K. S. Masters, *J. Biomed. Mater. Res., Part A*, 2012, **100A**, 2750–2759.
90. G. J. Gabriel, J. A. Maegerlein, C. F. Nelson, J. M. Dabkowski, T. Eren, K. Nusslein and G. N. Tew, *Chem. Eur. J.*, 2009, **15**, 433–439.
91. J. Zhang, M. J. Markiewicz, B. Weisblum, S. S. Stahl and S. H. Gellman, *ACS Macro Letters*, 2012, **1**, 714–717.
92. M. F. Ilker, H. Schule and E. B. Coughlin, *Macromolecules*, 2004, **37**, 694–700.
93. G. J. Gabriel, J. G. Pool, A. Som, J. M. Dabkowski, E. B. Coughlin, M. Muthukumar and G. N. Tew, *Langmuir*, 2008, **24**, 12489–12495.
94. K. Lienkamp, A. E. Madkour, K.-N. Kumar, K. Nuesslein and G. N. Tew, *Chem. Eur. J.*, 2009, **15**, 11715–11722.
95. A. Som, S. Vemparala, I. Ivanov and G. N. Tew, *Biopolymers*, 2008, **90**, 83–93.
96. S. Colak, C. F. Nelson, K. Nusslein and G. N. Tew, *Biomacromolecules*, 2009, **10**, 353–359.
97. Z. M. Al-Badri, A. Som, S. Lyon, C. F. Nelson, K. Nusslein and G. N. Tew, *Biomacromolecules*, 2008, **9**, 2805–2810.
98. M. Pasupuleti, B. Walse, B. Svensson, M. Malmsten and A. Schmidtchen, *Biochemistry*, 2008, **47**, 9057–9070.
99. A.-R. Song, S. G. Walker, K. A. Parker and N. S. Sampson, *ACS Chem. Biol.*, 2011, **6**, 590–599.

CHAPTER 6

Prevention of Hospital and Community Acquired Infections by Using Antibacterial Textiles and Clothing

GANG SUN

University of California, Davis, USA
Email: gysun@ucdavis.edu

6.1 Introduction

Transmissions of infectious diseases in healthcare facilities, particularly these drug-resistant species, have been an important and urgent challenge to the infection-control community. As an example, *Staphylococcus aureus* (SA) is the number one source of hospital-associated infections, and the most dangerous SA infections in hospitals are caused by its drug-resistant strain such as methicillin-resistant *Staphylococcus aureus* (MRSA).[1] A recent study[2] revealed that MRSA infections in US academic medical facilities have doubled during the years of 2003–2008. MRSA cases acquired through healthcare facilities are named as hospital-acquired infections (HAI). However, the MRSA infections are not only occurring in hospitals and have been spread to many public places including schools, excise rooms and gyms. Those MRSA cases found from the community facilities are called community-acquired infections (CAI). Community-acquired infections are often distinguished from the hospital-acquired

RSC Polymer Chemistry Series No. 10
Polymeric Materials with Antimicrobial Activity: From Synthesis to Applications
Edited by Alexandra Muñoz-Bonilla, María L. Cerrada and Marta Fernández-García
© The Royal Society of Chemistry 2014
Published by the Royal Society of Chemistry, www.rsc.org

diseases by the types of organisms that affect patients who are recovering from a disease or injury.[3] Both HAI and CAI are transmitted by surface contacts of person to devices and person to person. Most of the transmissions are linked to textile materials, such as bed linens, towels, dresses, and surgical uniforms.[4,5]

More recently, a super germ called carbapenem-resistant *Enterobacteriaceae* or CRE, which is resistant to the last line of antibiotics, has been found in 41 states in the US since it was first reported in 2001. The CRE is mostly transmitted among those elderly and with weak immune systems by surface contacts, but so far only occur in hospitals or nursing homes, rather than in the community. The transmission pattern seems to be related to all medical devices, and hand hygiene plays a key role in prevention of the transmission, similar to the cases of MRSA. According to the facts and patterns, the transmission of CRE is very obviously related to all medical used materials, including textiles as well.

In fact, micro-organisms can be transmitted from sources onto surfaces of materials by contacts and aerosol deposition and can survive on the surfaces for weeks and even months in hospital environments.[5,6] During the course of survival on the surfaces of materials, the germs can undergo several stages of growth leading to formation of biofilms if enough time is given,[7] which are described in Figure 6.1. The biofilms formed on surfaces of materials cannot be easily eradicated by spraying regular biocides, except for the use of chlorine bleach or a solution of glutaldehyde.[7] The micro-organisms on the surfaces of the materials can directly transmit diseases by contacts or serve as an aerosol source at any stage of growth, which is particularly possible for flexible textiles. Realizing the above paths of microbes, antimicrobial functions on the materials have been considered as a potential tool in reducing transmission of diseases, assuming that the function could reduce the population of micro-organisms.

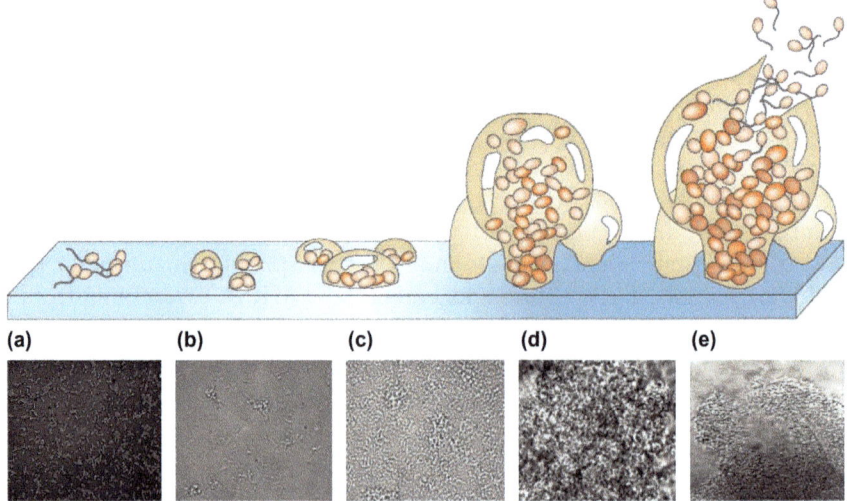

(a)   (b)   (c)   (d)   (e)

Figure 6.1   Formation of biofilms on surfaces of materials,[7] with permission from Nature Publishing Group.

Due to such a speculation and urgent need in control of infectious diseases many antimicrobial polymers, coating, fibers and fabrics have been developed in the recent decades.[8,9] However, the use of these new functional materials has not resulted in significant reduction of MRSA infections for years. There is a need to carefully review the functions of the antimicrobial textiles and their applications in the medical arena so as to improve the infection control and reduce MRSA or CRE transmissions within or outside healthcare settings.

6.2 Transmission Paths and Textiles

People in hospital environments including healthcare workers and patients are potential carriers of SA and MRSA, and experimental results[10] have revealed that SA lives on human bodies in pretty high rates regardless of age, gender, and profession (Figure 6.2). Nasal carriage of SA has been the most important

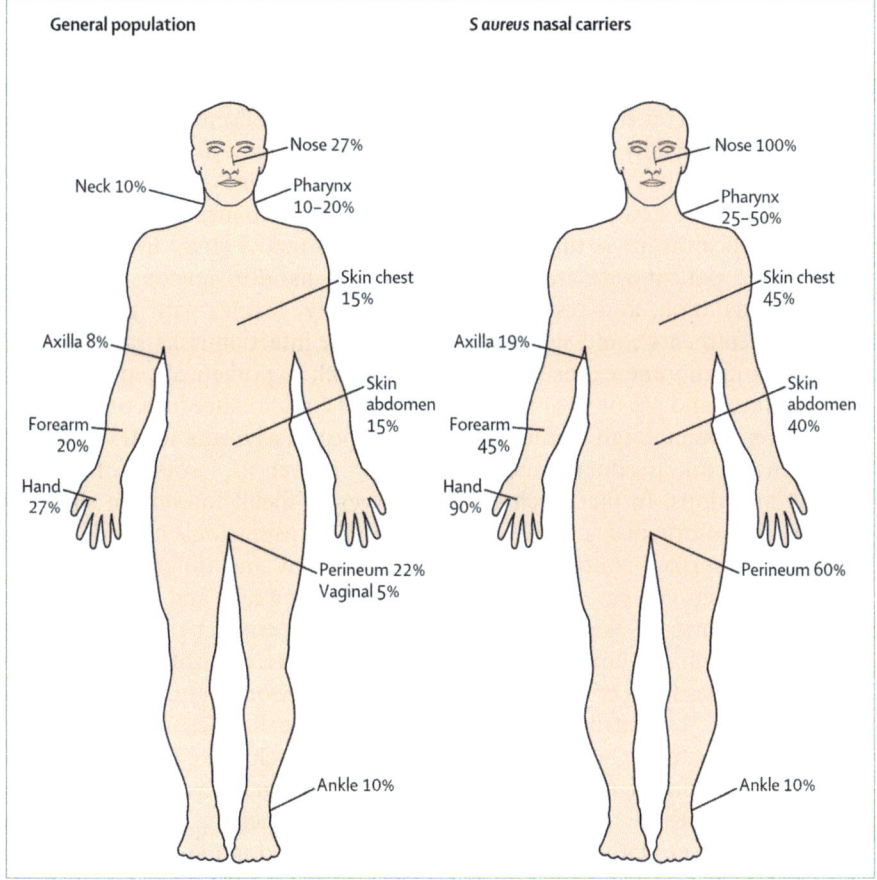

Figure 6.2 *S. aureus* carriage rates per body site in adults,[10] with permission from Elsevier.

source of the disease with 100% rate on all human SA carriers. Different investigations on healthcare workers have revealed that nurses were more often found as MRSA nasal carriers than physicians and nonpatient contact healthcare workers,[11,12] while resident orthopaedic surgeons have higher nasal carriage rates of *S. aureus* and similar carriage rate of MRSA to a special patient group.[13] More recently, female healthcare workers with children residing status have higher risk in carrying SA than other professionals.[14] These facts just confirmed that healthcare workers are sources of pathogens, and with these moving sources available the infection control is indeed a complicated and difficult mission to accomplish in hospitals.[15]

Meanwhile, pathogenic micro-organisms carried by human bodies could be transmitted by hand touching noses and then textiles or other surfaces, possibly spreading them to all areas.[16] Eventually, hand hygiene has been considered critical in infection control,[17] however, it only has limited effect since the bacteria reservoirs are always available for the hands. If the reservoirs of pathogens always exist in human bodies, the live bacteria will be continuously provided and supplied to be transmitted from the human bodies. Furthermore, micro-organisms could survive on medical textiles for weeks and months and can contribute to secondary transmission from these materials.[18–20] Again, since most common transmission paths of micro-organisms from surfaces to surfaces and materials to materials in indoor environments are through surface contacts and aerosol depositions,[4,5,21] both hand hygiene and environmental cleaning could be and have been considered as effective measures in infection controls in healthcare settings.[17,22] An integrated model study in a room occupied by a patient with transmissible paths of hand-to mucous membrane contact, inhalation, and respiratory droplet spray, textiles with proper antibacterial treatments could significantly reduce the infection risks *via* the hand-to-mucous membrane exposure pathway,[23] which hypothetically proved that clean surfaces and use of antimicrobial materials could reduce infections of SA.

However, even though textiles have been suspected as media for transmitting pathogens, textile products were not clinically proven as a source of directly causing infections. In fact, such evidence is very difficult to confirm since infectious micro-organisms are everywhere and transmissions caused by the aerosol and surface contact are difficult to prevent and differentiate. Since micro-organisms can survive on textile materials for weeks and even months,[5] during the course of existence on the surfaces, any contact or touch on the contaminated areas could cause transmissions. Aerosolization of micro-organisms on surfaces of medical textiles, such as uniforms and bed linens has been reported,[24,25] contributing to transmission of the diseases from surfaces of surviving media to the air and other surfaces, such as clothing materials. Hand touches and surface contacts will bring the micro-organisms from textiles to textiles, materials to human, and from person to person.[16] Overall, these processes are interrelated and intercontaminating, making the spreading of the diseases into a chain reaction of transmissions (Figure 6.3).

According to the transmission paths of pathogens (Figure 6.3), antibacterial functions on surfaces of medical devices and textile materials that can provide

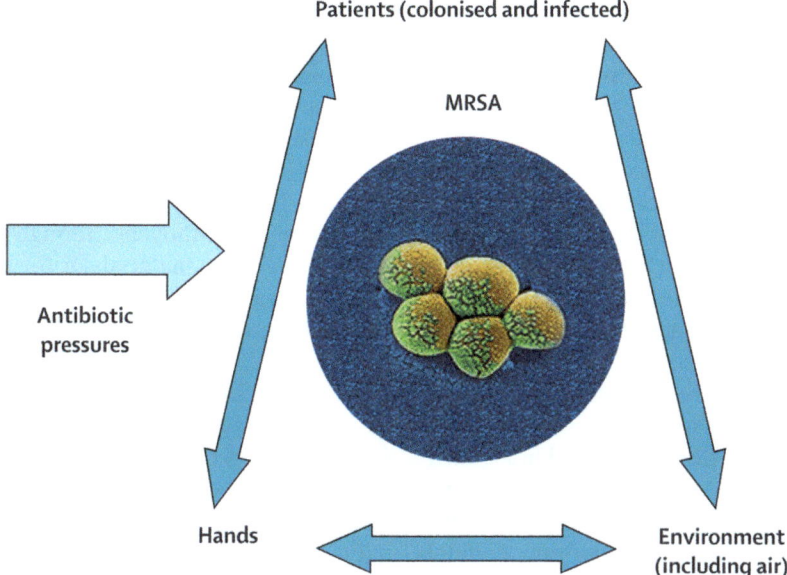

Figure 6.3 Dynamic transmission cycle of MRSA,[5] with permission from Elsevier.

absolute clean surfaces should be able to cut off the chain reactions of transmissions and, consequently, reduce the cases of infections,[5,26] and such functions have been considered as a necessary protective measure on all medical
textiles including uniforms, dresses, bed linens, curtains, and furniture.[27]
During the entire life and repeated uses, the antibacterial medical textiles
should rapidly and continuously kill any pathogens upon contact so as to
provide absolutely clean surfaces. Thus, the killing power, speed, durability and
effectiveness of the functions against a range of micro-organisms, such as
bacteria, viruses, and even spores, would be critical and very important to be
evaluated, and such functions are naturally different from the needs for consumer products. In addition, medical textiles should provide other necessary
properties per different products. For example, uniforms and patient dresses,
bed linens, and surgical gowns and drapes are basically different in performance. Thus, a fundamental understanding of application requirements, necessary performance and desired antibacterial functions of the product is
necessary. Here, the main focus of the chapter is on antimicrobial functions of
medical textiles.

6.3 Different Antimicrobial Functions

Antimicrobial function is a broad claim of the functions on materials that could
generally inhibit or eliminate growth of micro-organisms. If textiles could
eliminate growth of micro-organisms, the speed of the materials to completely
kill the micro-organisms on the surfaces is related to the efficacy in prevention

of transmission of the micro-organism. A long contact time to completely in-activate the pathogens could leave more opportunities for potential transmis-sion during the course of the function, which is not considered as perfect antimicrobial properties. Rapid elimination and inactivation of entire micro-organisms on the surfaces of textile are the mostly desired functions on medical textile and devices. Based on the analysis, regular antimicrobial textiles that could only inhibit growth of micro-organisms during a long contact time are not enough in prevention of crosstransmission of pathogens, though such functions are effective in reducing biodegradation of the materials or for odor-control applications. Thus, it is absolutely necessary to differentiate the anti-microbial functions based on the power of the functions, as well as the intended applications.

The microbial inhibiting functions have been defined as *biostatic functions*, which are quite useful in textile conservation and some hygienic application such as odor control.[28] Under such a circumstance, growth rates and numbers of micro-organisms on the textile materials can be significantly reduced, however, live bacteria or pathogens could still be found on their surfaces. Al-though the infection rate could be reduced, the residual micro-organisms still could cause transmission of infections at certain level. Thus, the prevention of infectious diseases, such as MRSA in hospitals, could not be achieved by using such materials as medical textiles. In fact, clinical studies also proved that the textiles incorporated with quaternary ammonium salt structures were able to reduce but were unable to completely eliminate all MRSA on the textiles.[29]

The function of rapid inactivation of a broad spectrum of micro-organisms can be called the biocidal or antibacterial function. Textiles possessing such functions should be able to rapidly eliminate live micro-organisms on surfaces of textiles, and, consequently, the use of such textiles could cut off the chain of disease transmission through contacts and aeresolization. Such an anti-microbial function should be the strongest among all antimicrobial claims and thus the ideal one for medical textiles, particularly for biologically protective products. Biocidal textiles could ensure complete disinfection of diseases on the surfaces of the materials, preventing transmissions of infectious diseases caused by surface contacts on textile products. In reality, most antimicrobial textiles currently available on the market do not have this function, and development of such biologically protective materials is the long-term goal for textile scientists.

The biocidal textiles have the strongest antimicrobial effects, but also bring concerns when they are used as consumer products. The biocidal func-tions of clothing materials could create a very clean environment surrounding a human body, which is harmful for the human immune system. Thus, only textiles that absolutely need such functions should have biocidal functions, which include all medical textiles and biological protective clothing. More specifically, healthcare worker uniforms, patient dresses, linens, curtains, covers, wipes, wraps, and surgical gowns and drapes need strong biocidal functions.

6.4 Measurements of Biocidal Functions on Textiles

Antimicrobial functions on textiles can be evaluated qualitatively and quantitatively by following several organizational and industrial standard methods. In the United States, the most widely employed quantitative and qualitative methods are the American Association of Textile Chemist and Colorist (AATCC) test method 100 (quantitative) and 147 (qualitative).[30] The American Society for Testing and Materials (ASTM) international test standard ASTM E2149 is also used by many researchers and companies.[31] Due to the fact that micro-organisms exposed onto textiles could transmit through several ways, including by hand touching, aerosolization, and penetration through textiles during the survival, ideal test methods that are employed in evaluating antimicrobial function should provide results that can predict potential reductions of the transmissions.

The AATCC test method 147 uses fabric strips placed onto agar plates that are inoculated by micro-organisms and then the distance of cleared edges of the fabric sample to the nearest growth zone of the micro-organisms is measured. The method is simple and effective in evaluating antimicrobial functions on textiles that are provided by biocides, which will be released during the testing or use. But it is not so effective in evaluating nonleaching antimicrobial agents on textiles and cannot provide information on how quick and powerful are the functions. In other words, it is not applicable to evaluating biocidal functions on textiles.

The AATCC test method 100 is the most widely used quantitative test protocol for evaluating antimicrobial functions on textiles with use of both either Gram-positive or Gram-negative bacteria. Variation factors of running the tests include sample weight, size, and surface area, amount and bacterial concentration of inoculums, and contact time. Complete absorption of the inoculums by the fabric samples should be ensured during the test. After a proper contact time, which could be as short as a few minutes or as long as many hours, the inoculated sample will be neutralized by a sterilized solution with volume known and then after thorough mixing, an aliquot of the dilution will be taken out and undergoes serial dilutions. Surviving organisms are recovered and incubated for 18–24 h on agar plates. The number of surviving bacteria counted as colony forming units (CFU). The results should report sample size, bacterial concentrations in inoculating fabrics, contact time, and percentage or log reduction of the micro-organisms in comparison to an untreated control sample.[30]

This test method could measure antimicrobial functions provided by nonleaching surface bound biocides, so-called contact kill, and also by leaching biocides on textiles, noncontact kill. The advantages of this method come from the quantitative results – containing both speed and power (log reduction) against individual micro-organism, which is different from the results obtained from AATCC test method 147. However, the AATCC test method 100 protocol may have difficulty in detecting biocidal functions on hydrophobic fabrics due to poor surface contact between inoculums and surfaces of biocides on the fibers.

ASTM International test method ASTM E2149 is a quantitative screening test, generally known as the "Shake Flask Method". A fabric sample in any shape with a weight of 1 gram is immersed in 50 mL of an inoculated bacterial solution in a flask, which is placed on a wrist action shaker. A flask containing the bacterial solution but without any fabric sample serves as one control, and another flask with an untreated fabric is used as another control. After an exposure of 1–24 h, an aliquot of the solution from the flask is placed onto nutrient agar plates and then incubated for 18–24 h. The viable number of colonies are counted, and a reduction is calculated using the known initial bacterial concentration and the final count after exposure to the test sample or is calculated *versus* an untreated control.

This test method can be employed in screening many irregular solid antimicrobial products, such as films and textiles under dynamic conditions, but may not be the best to represent the contact kill functions provided by the materials. The testing conditions of soaking and shaking materials in bacterial solutions do not simulate the transmission patterns of pathogens by medical devices and textiles. Thus, the results are unable to represent antimicrobial functions of medical textiles in hospital environments. In addition, the detected kill speed is restricted by, at least, one hour shaking time.

The above-mentioned test protocols for antimicrobial functions on textiles could measure mostly concerned contact and leaching antimicrobial properties, which may be able to represent prevention of transmission of diseases by contact and aerosolization if the materials could provide complete kills. However, in a study on the evaluation of wet bacterial penetration through antimicrobial textile, it was found that another test method, ISO 22610, might be able to better reflect both contact kill and prevention on wet penetration of micro-organisms.

ISO 22610 is a test method to determine the resistance to wet bacterial penetration for surgical drapes, gowns and clean air suits, used as medical devices, for patients, clinical staff and equipment. This test uses a device shown in Figure 6.4.

The procedure is as follows: on top of an agar plate on the holder (5) place the testing fabric, inoculated polyurethane (PU) film and a cover high-density polyethylene (HDPE) film in sequence and fasten them with the tenter rings. Apply the finger (weigh 3N) on the top of the assembly. Switch to a new plate after every 15 min run and repeat four times using the same sample assembly. At the end of the test, turn the side of the fabric sample in contact with the inoculate to a new agar plate and cover HDPE film again to run the finger for 15 min. Incubate all six agar plates at 36 °C for 48 h. The results should show the expected plate for penetration (EPP) using the following technique:

$X1$, $X2$, $X3$, $X4$, $X5$ are the numbers of colonies on the five plates from one of five runs of a covering material. Z is the plate count from the inverted test specimen.

$T = Z + X1 + X2 + X3 + X4 + X5$

$CUM1 = X1/T$

$CUM2 = X2 + X1/T$

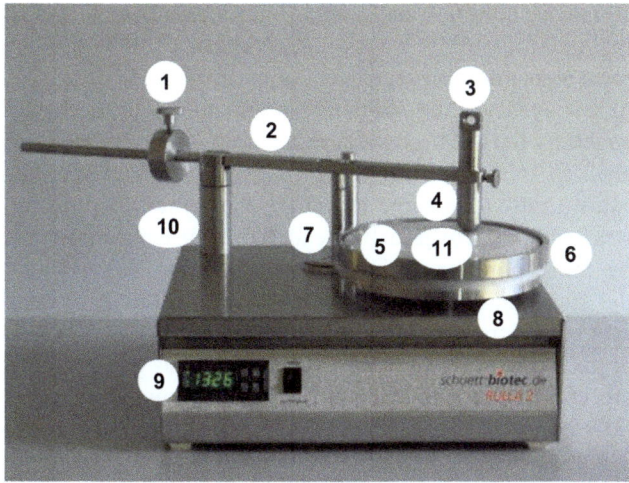

Figure 6.4 Rolla II bacteria wet penetration tester (Shown in the figure: (1) Counter-weight; (2) Balancing arm with finger; (3) Dynamometer attachment point; (4) Stainless steel finger; (5) Petri dish holder; (6) Tenter frame/double steel ring; (7) Excenter; (8) Turntable; (9) Timer; (10) Ball bearing; (11) Petri dish–fabric sample–cover PU and HDPE films.)

$$CUM3 = X3 + X2 + X1/T$$
$$CUM4 = X4 + X3 + X2 + X1/T$$
$$CUM5 = X5 + X4 + X3 + X2 + X1/T$$
$$EPP = 6 - (CUM1 + CUM2 + CUM3 + CUM4 + CUM5)$$

If biocidal fabrics are tested with this method, a control sample without biocidal functions should be conducted.

6.5 Antimicrobial Textiles and Biocidal Functions

In recent years, developments of antimicrobial textiles have attracted a lot of attention from researchers and, consequently, have resulted in many products with the claims of antimicrobial functions.[29,32,33] Antimicrobial functions of textiles are created by incorporating proper antimicrobial agents onto the materials. Antimicrobial agents suitable for textiles are quite limited, including quaternary ammoniums salts,[29,34] heavy-metal ions (Cu and Ag) or metal nanoparticles (Ag),[32,33,35,36] halamines,[37] others,[38] and photoinduced bio-cides.[39,40] However, among these available technologies, only a handful of them are able to provide so defined biocidal functions that are needed on medical textiles.

6.5.1 Quaternary Ammonium Salts and Similar Compounds

Quaternary ammonium salts are a group of ionic compounds containing cationic quaternary ammonium nitrogen atoms connected with four alkyl or aryl groups and anionic counterions, such as chloride or bromide. Among the four alkyl groups, one preferably is a long alkyl chain group containing more than eight hydrocarbons and also serves as the hydrophobic group. The quaternary ammonium salt agents inactivate micro-organisms with the hydrophobic long chain approaching the cell lipid membrane and quaternary positive charged site breaking the cell membrane resulting in expiration of the micro-organisms. This mechanism may take some time to execute the functions, and the active cationic sites could be covered by dead bacterial cells so as to have limited power. The most popular antimicrobial quaternary ammonium salt is tri-methoxysilyl quaternary ammonium chloride, an Environmental Protection Agency (EPA) registered biocide and has been widely used in many consumer and institutional products. The fabrics treated by this group of chemicals could demonstrate effective antimicrobial functions following the shaking jar test (ASTM E2149) by soaking a sample in a bacterial broth. But other methods, such as the directly agar overlay method, a direct contact test could not reveal the functions.[34] Similar quaternary ammonium salt products with some structural differences are available on the market. A major drawback of this group of biocides is that the antimicrobial cationic sites can be poisoned by anionic surfactants if laundered improperly. A recent clinical study demon-strated the use of a new quaternary ammonium salt as biocides on medical scrubs could reduce bacteria populations on the materials but did not provide complete elimination of micro-organisms, MRSA, Vancomycin-resistant *Enterococcus* (VRE) and Gram-negative rod (GNR) (Table 6.1).[29] The testing protocol is as follows: Scrubs were cultured before and after working shifts by thirty healthcare workers, and the cultures were collected with a collection swab by a 15-s up and down rub. Each swab was inoculated into enrichment broths to increase the isolation rate of *S. aureus*. After incubation, the broths were serially diluted and plated onto Trypticase soy agar to determine CFU.[29] The differences of log CFU were quite significant, particularly for MRSA, indi-cating the antimicrobial functions. But, the residual live CFUs of MRSA were also significantly high. Such a result is consistent with the mechanism of

Table 6.1 Comparison of difference in apparel mean log colony-forming unit (CFU) count overall.[29]

Mean log CFU count on overall samples					
	Study (samples, n)	*Control (sampltes, n)*	*Difference*	*SE of difference*	*P*
MRSA					
Leg cargo pocket	6.71 (12)	11.84 (16)	5.13	1.1493	0.0002
Abdominal area	7.54 (25)	11.35 (25)	3.81	1.23	0.0056
VRE					
Leg cargo pocket	0 (0)	12.68 (1)	12.68	NA	NA
Abdominal area	12.68 (1)	12.27 (5)	0.41	2.8917	0.9013
GNR					
Leg cargo pocket	4.41 (1)	13.02 (1)	8.61	NA	NA
Abdominal area	9.14 (3)	10.36 (2)	1.22	3.4376	0.7569

antimicrobial functions and potential outcomes offered by biocides on the textiles, not possessing rapid and powerful biocidal ability.

6.5.2 Silver and Copper

Several heavy metals, such as silver and copper, can provide antimicrobial functions and also have been the most popular biocides employed in textiles and medical devices in recent years.[32,33,35,36,41] Silver has been well known for its antimicrobial functions and has a superior safe record to human as silverware has been used for thousand years. Silver can be applied onto surfaces of fibers and textiles in different forms, such as ions, nanoparticles, and even plating. However, due to safety concerns on nanoparticles, silver nanoparticles have not been approved for use in textiles. Regardless of which type of silver is used, the silver ion is responsible for providing antimicrobial functions. The generally recognized mechanism is reaction between silver ions and sulfhydryl groups in proteins. The silver-treated textiles could provide effective antimicrobial functions by leaching out silver ions and by direct contact. A clinical study of using silver-treated clothing in dermal wound sites showed reduced *S. aureus* CFU in comparison to control samples, again only reduction of CFU not elimination of the micro-organism.[41]

Copper as an antimicrobial agent has received enormous attention since the US EPA approved copper-containing alloy products with the antimicrobial claims in 2008. Copper alloys have been applied onto surfaces of many medical devices to provide antimicrobial functions.[35] The efficacy of copper in control of growth of micro-organisms was proven clinically.

6.5.3 Halamine Chemistry

Halamine is a structure of chlorine or bromine covalently connected to imide, amide and amine nitrogen. Halamine compounds are biocides but also can stabilize active chlorine in water,[42,43] thus, water-soluble halamines have been widely used as water disinfectants in swimming pools and even in municipal water treatment. Due to the excellent biocidal functions and safety record as disinfectants in swimming pools, halamine structures have been incorporated onto solid materials, such as polymers and textiles.[28,37]

Halamine polymers and treated textiles all demonstrated powerful and speedy biocidal functions against a broad spectrum of micro-organisms. Halamine polymers have been developed into solid drinking-water disinfectants and received EPA approval in 2009. Halamine-treated cotton fabrics have been widely used in dish cloths, incontinence pads, wipers and mops, medical uniforms, and bed linens. Biocidal functions of halamine treated cotton and cotton/polyester fabrics were proven excellent, both in killing speed and log reduction (Table 6.2).[37]

Halamine could release free chlorine in aqueous solutions (eqn (6.1)), which provides disinfection function, and could serve as biocides on solid surfaces (eqn (6.2)). Chemically bonded halamine structures kill micro-organisms

Table 6.2 Biocidal functions of halamine-treated cotton fabrics.[37]

Micro-organism	Contact time	Log reduction Cotton	Cotton/PET (35/65)
E. coli	2 min	6	6
Staphylococcus aureus (SA)	2 min	6	6
Salmonella choleraesuis	2 min	7	6
Shigella	2 min	6	7
Candida albicans	2 min	6	6
Brevibacterium	2 min	8	8
Pseudomonas aeruginosa	2 min	6	6
Methicillin-resis. SA (MRSA)	2 min	3	6
Vancomycin-resis. Enterococcus (VRE)	2 min	6	6

*AATCC test method 100. Both fabrics were plain woven treated by 6% 1,3-dimethylol-5,5-dimethyl hydantoin.

mostly by a contact oxidation reaction, rapidly oxidizing sulfide bonds or sulfhydral groups in proteins. In addition, chlorine (Cl^+) atoms on halamine bonds can quickly react with hydrogen atoms on peptide bonds in proteins, which could lead to expiration of micro-organisms as well. Such antimicrobial functions are quick and powerful without generating resistance from micro-organisms, a potential technology that could meet the requirement for prevention of transmission of MRSA in hospitals. The biocidal functions of the halamine-treated textiles can be repeatedly recharged in chlorine bleach solution, a reverse reaction shown in eqn (6.2). Such rechargeable biocidal functions is mostly meaningful for reusable medical textiles such as doctors' uniforms, nurses and patients' dresses, linens, and wipes and mops.

$$\text{>N-Cl} + H_2O \rightleftharpoons \text{>N-H} + Cl^+ + OH^- \tag{6.1}$$

$$\text{>N-Cl} \underset{\text{Bleach}}{\overset{\text{Kill bacteria}}{\rightleftharpoons}} \text{>N-H} \tag{6.2}$$

Based on the biocidal functions of halamine textiles, contact biocidal functions of the fabrics were compared with the ability of prevention of bacterial wet penetration of the fabrics. In this study, pure cotton fabrics were treated with 4% 1,3-dimethylol-5,5-dimethyl hydantoin (DMDMH), 4% 3-methylol-2,2,5,5-tetramethyl-4-imidazolidin-4-one (MTMIO), and combinations of both compounds in different ratios with a total concentration of 4%. Such a combination incorporated varied amounts of very lethal imide, amide and very stable amine halamines onto the fabrics.[44] More 3-methylol-2,2,5,5-tetramethyl-4-imidazolidin-4-one (MTMIO) on the fabrics, more stable amine halamines and slow biocidal functions exist on the fabrics (Figure 6.5).[44] At the meantime, slow biocidal functions also lead to more bacteria wet penetrations through the fabrics (Figure 6.6). Such a result further proved that prevention of transmission of micro-organism requires rapid and powerful biocidal functions on medical textiles.

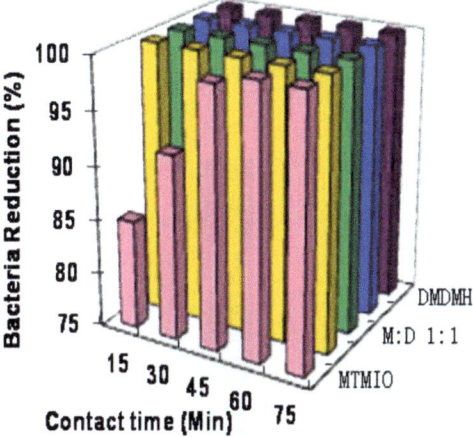

Figure 6.5 Biocidal functions of halamine fabrics with different activities against *E. coli.* M = MTMIO; D = DMDMH.

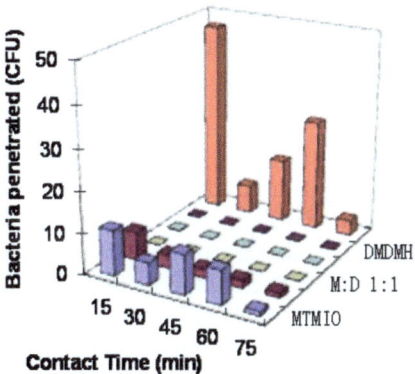

Figure 6.6 Wet bacteria (*E. coli*) penetration through biocidal halamine fabrics M = MTMIO; D = DMDMH.

6.5.4 Light-Induced Biocides

In recent years, some photosensitive agents have shown proper antimicrobial functions on surfaces of polymers and textiles when exposed to UVA and daylight.[45–47] Titanium dioxide nanoparticles were coated onto textile surfaces and the materials could demonstrated antimicrobial or self-cleaning functions under light exposure.[45] Benzophenone is an organic photosensitizer, several compounds containing the benzophenone group have been incorporated onto textiles, and the products exhibited the light-induced antimicrobial functions under UVA (365 nm) light.[46] With structural similarity, some anthraquinone compounds also revealed antimicrobial functions after being exposed to the UVA light.[40] The anthraquinone derivatives could be incorporated onto

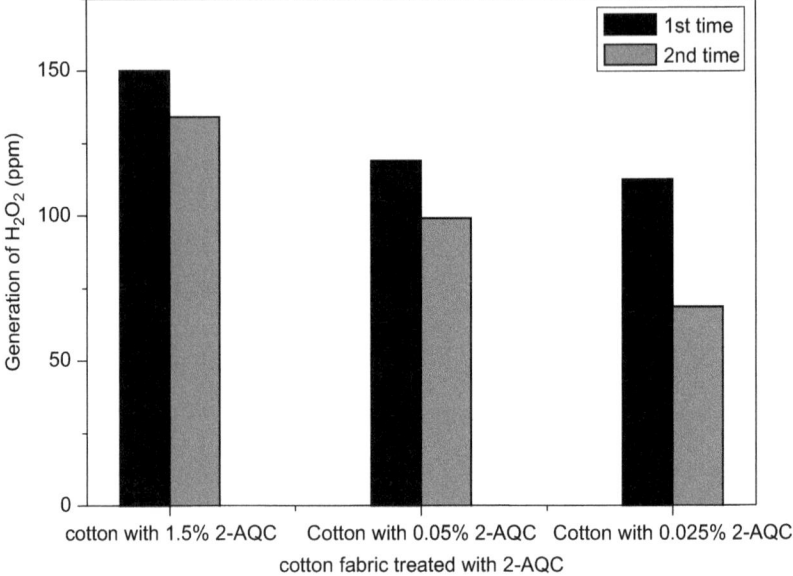

Figure 6.7 Generation of H_2O_2 by cotton treated with different concentration of 2-AQC.[47]

cotton, protein and nylon fibers by using textile-dyeing processes, and the products showed proper light-induced biocidal functions.

The photoinduced antimicrobial functions on both benzophenone and anthraquinone incorporated fabrics are due to radicals generated by these light-active agents on fabrics, which also lead to formation of several reactive oxygen species (ROS) including hydroxyl radicals, singlet oxygen and hydrogen peroxide. As an example, after one hour UVA (365 nm) exposure, 2-anthra-quinone carboxylic acid (2-AQC) treated cotton fabrics showed a significant amount of H_2O_2 in aqueous solution (Figure 6.7),[47] which could provide antimicrobial functions on the materials even after the light is shut off. Hydrogen peroxide is a disinfectant widely used in textile laundry and surface cleaning. Thus, the light-induced antimicrobial agents might generate more interest from researchers and manufacturers of medical textiles.

6.6 Safety of Biocidal Textiles

The above discussions have differentiated the biostatic and biocidal functions on textile products. Biocidal textiles are considered as a potentially effective tools in prevention of transmission of diseases in hospitals and should be used as personal protective gear in healthcare facilities. Thus, biocidal medical textiles should be developed and widely used in hospitals to reduce the transmission of MRSA and any infection caused by contact and crosstransmission of pathogens through textile materials. However, such biocidal functions are only necessary for medical protections in healthcare facilities but should not be incorporated onto consumer products.

Because biocides are intended to kill micro-organisms, many biocidal products pose significant risk to human health and welfare. The use of biocidal textiles could have significant adverse effects on healthy skin environment if skin flora-good micro-organisms are killed by the clothing. Leached biocides could be a threat to the natural environment as well. Generally, dermal toxicity and sensitivity of nonleaching antimicrobial textiles should be evaluated. Biocides that could be leached out from textiles should have a comprehensive toxicity evaluation. For example, the toxicological database of trimethoxysilyl quaternary ammonium salt is comprised of acute toxicity studies, a subchronic dermal toxicity study, one subchronic oral study in rats, one developmental toxicity study in rats, and six mutagenicity studies. Additional results of avian acute toxicity and dietary studies and from the freshwater invertebrate acute toxicity studies characterized it as practically nontoxic to birds but also considered it highly toxic to freshwater invertebrates in acute studies. The trimethoxysilyl quaternary ammonium salts are classified as being moderately toxic to coldwater fish species.

Halamine structures provide rapid biocidal functions on textiles and leach-free chlorine ions into air and water, making them ideal for healthcare protective textiles. Halamine structures are generally safe to humans due to the use as swimming pool and drinking-water disinfectants, but the halamine biocidal textiles should still go through various toxicology and safety evaluations. At the meantime, halamine structures have some limitations in applications, including stability against dry heat and UV exposure, as well as compatibility with color and other treatments on textiles. Overall, halamine-treated textile have the potential to reduce transmission of MRSA in healthcare facilities due to the biocidal functions.

6.7　Ideal Medical Use Textiles

Medical textiles are potential carriers and growth media of infectious diseases in hospitals. The use of biocidal textiles should be able to reduce the transmission of infectious diseases, such as MRSA, depending on the proper power and speed of the functions. Ideal biocidal textiles should provide very quick inactivation against a broad spectrum of micro-organisms upon contact and prevent any potential transmission of the germ through contact, aerosolization and penetration through the textiles. The biocidal textiles should be nontoxic to humans and the environment, stable, durable and rechargeable if the textiles are reusable. We anticipate the reduction of infections of MRSA in hospitals due to the use of environmentally friendly biocidal medical textiles in the future.

References

1. F. R. DeLeo and H. F. Chambers, *J. Clin. Invest.*, 2009, **119**, 2464–2474.
2. M. Z. David, S. Medvedev, S. F. Hohmann, B. Ewigman and R. S. Daum, *Infect. Control Hosp. Epidemiol.*, 2012, **33**, 782–789.

3. S. Stefani, D. R. Chung, J. A. Lindsay, A. W. Friedrich, A. M. Kearns, H. Westh and F. M. MacKenzie, *Inter. J. Antimicrob. Ag.*, 2012, **39**, 273–282.
4. S. Fijan and S. Š. Turk, *Int. J. Env. Res. Public Health*, 2012, **9**, 3330–3343.
5. S. J. Dancer, *Lancet Infect. Dis.*, 2008, **8**, 101–113.
6. J. M. Bartley and R. N. Olmsted, *Clin. Microbiol. Newsl.*, 2008, **30**, 113–117.
7. D. Davies, *Nature Rev. Drug Discov.*, 2003, **2**, 114–122.
8. D. R. Monteiro, L. F. Gorup, A. S. Takamiya, A. C. Ruvollo-Filho, E. R. d. Camargo and D. B. Barbosa, *Int. J. Antimicrob. Agents*, 2009, **34**, 103–110.
9. N. Stobie, B. Duffy, S. J. Hinder, P. McHale and D. E. McCormack, *Colloids Surf. B. Biointerfaces*, 2009, **72**, 62–67.
10. H. F. L. Wertheim, D. C. Melles, M. C. Vos, W. van Leeuwen, A. van Belkum, H. A. Verbrugh and J. L. Nouwen, *Lancet Infect. Dis.*, 2005, **5**, 751–762.
11. B. P. Suffoletto, E. H. Cannon, K. Ilkhanipour and D. M. Yealy, *Ann. Emergency Med.*, 2008, **52**, 529–533.
12. M. C. Elie-Turenne, H. Fernandes, José, R. Mediavilla, M. Rosenthal, B. Mathema, A. Singh, Tiffany R. Cohen, Kimmerle A. Pawar, H. Shahidi, Barry N. Kreiswirth and Edwin A. Deitch, *Infect. Control Hosp. Epidemiol.*, 2010, **31**, 574–580.
13. R. Schwarzkopf, R. C. Takemoto, I. Immerman, J. D. Slover and J. A. Bosco, *J. Bone Jt. Surg. (Am.)*, 2010, **92**, 1815–1819.
14. K. Olsen, M. Sangvik, G. S. Simonsen, J. U. E. Sollid, A. Sundsfjord, I. Thune and A.-S. Furberg, *Epidemiol. Infect.*, 2013, **141**, 143–152.
15. W. C. Albrich and S. Harbarth, *Lancet Infect. Dis.*, 2008, **8**, 289–301.
16. L. S. Munoz-Price, K. L. Arheart, J. P. Mills, T. Cleary, D. DePascale, A. Jimenez, Y. Fajardo-Aquino, G. Coro, D. J. Birnbach and D. A. Lubarsky, *Am. J. Infect. Control*, 2012, **40**, e245–e248.
17. D. J. Weber, W. A. Rutala, M. B. Miller, K. Huslage and E. Sickbert-Bennett, *Am. J. Infect. Control*, 2010, **38**, S25–S33.
18. J. Blanchard, *AORN J.*, 2009, **89**, 409–411.
19. S. Galvin, A. Dolan, O. Cahill, S. Daniels and H. Humphreys, *J. Hosp. Infect.*, 2012, **82**, 143.
20. M. Maclean, S. J. MacGregor, J. G. Anderson, G. A. Woolsey, J. E. Coia, K. Hamilton, I. Taggart, S. B. Watson, B. Thakker and G. Gettinby, *J. Hosp. Infect.*, 2010, **76**, 247–251.
21. D. L. Butler, Y. Major, G. Bearman and M. B. Edmond, *J. Hosp. Infect.*, 2010, **75**, 137–138.
22. S. Khodavaisy, M. Nabili, B. Davari and M. Vahedi, *J. Prev. Med. Hyg.*, 2011, **52**, 215–218.
23. M. Nicas and G. Sun, *Risk Anal.*, 2006, **26**, 1085–1096.
24. T. Shiomori, H. Miyamoto, K. Makishima, M. Yoshida, T. Fujiyoshi, T. Udaka, T. Inaba and N. Hiraki, *J. Hosp. Infect.*, 2002, **50**, 30–35.
25. A. P. Sergent, C. Slekovec, J. Pauchot, L. Jeunet, X. Bertrand, D. Hocquet, L. Pazart and D. Talon, *Orthop. Traumatol. Surg. Res.*, 2012, **98**, 441–445.

26. D. Cardo, P. H. Dennehy, P. Halverson, N. Fishman, M. Kohn, C. L. Murphy and R. J. Whitley, *Am. J. Infect. Control*, 2010, **38**, 671–675.
27. S. J. Dancer, *Eur. J. Clin. Microbiol. Infect. Dis.*, 2011, **30**, 1473–1481.
28. G. Sun and S. D. Worley, *J. Chem. Edu.*, 2005, **82**, 60–64.
29. G. M. L. Bearman, A. Rosato, K. Elam, K. Sanogo, M. P. Stevens, C. N. Sessler and R. P. Wenzel, *Infect. Control Hosp. Epidemiol.*, 2012, **33**, 268–275.
30. AATCC, http://www.aatcc.org/testing/methods/.
31. ASTM, http://www.astm.org/.
32. D. Hamilton, A. Foster, L. Ballantyne, P. Kingsmore, D. Bedwell, T. J. Hall, S. S. Hickok, A. Jeanes, P. G. Coen and V. A. Gant, *J. Hosp. Infect.*, 2010, **74**, 62–71.
33. A. Mariscal, R. M. Lopez-Gigosos, M. Carnero-Varo and J. Fernandez-Crehuet, *Eur. J. Clin. Microbiol. Infect. Dis.*, 2011, **30**, 227–232.
34. P. R. Murray, A. C. Niles and R. L. Heeren, *J. Clin. Microbiol.*, 1988, **26**, 1884–1886.
35. J. O'Gorman and H. Humphreys, *J. Hosp. Infect.*, 2012, **81**, 217–223.
36. C. L. Dupont, G. Grass and C. Rensing, *Metallomics*, 2011, **3**, 1109–1118.
37. G. Sun, X. Xu, J. R. Bickett and J. F. Williams, *Ind. Eng. Chem. Res.*, 2001, **40**, 1016–1021.
38. S. J. O'Hanlon and M. C. Enright, *Int. J. Antimicrob. Agents*, 2009, **33**, 427–431.
39. K. H. Hong and G. Sun, *Carbohydr. Polym.*, 2011, **84**, 1027–1032.
40. N. Liu, G. Sun and J. Zhu, *J. Mater. Chem.*, 2011, **21**, 15383–15390.
41. A. Gauger, M. Mempel, A. Schekatz, T. Schäfer, J. Ring and D. Abeck, *Dermatology*, 2003, **207**, 15–21.
42. D. E. Williams, E. D. Elder and S. D. Worley, *Appl. Environ. Microbiol.*, 1988, **54**, 2583–2585.
43. S. D. Worley, D. E. Williams and R. A. Crawford, *Crit. Rev. Environ. Control*, 1988, **18**, 133–175.
44. L. Qian and G. Sun, *J. Appl. Polym. Sci.*, 2004, **91**, 2588–2593.
45. K. Qi, W. A. Daoud, J. H. Xin, C. L. Mak, W. Tang and W. P. Cheung, *J. Mater. Chem.*, 2006, **16**, 4567–4574.
46. K. H. Hong and G. Sun, *Carbohydr. Polym.*, 2008, **71**, 598–605.
47. N. Liu and G. Sun, *Ind. Eng. Chem. Res.*, 2011, **50**, 5326–5333.

CHAPTER 7

Synthesis, Structure and Antimicrobial Activities of Polymeric and Nonpolymeric Silver and other Metal Complexes

KENJI NOMIYA,* NORIKO CHIKARAISHI KASUGA AND AKIHIKO TAKAYAMA

Department of Chemistry, Faculty of Science, Kanagawa University, Hiratsuka, Kanagawa 259-1293, Japan
*Email: nomiya@kanagawa-u.ac.jp

7.1 Introduction

Concerning the antimicrobial activities of silver materials, we have had several interesting issues: (1) how different are the antimicrobial behavior and modes of action of aqueous silver(I) ions (Ag^+), silver nanoparticles and metallic silver (both zerovalent silver with different sizes) and silver(I) complexes, (2) how can we control the antimicrobial behavior of silver materials, (3) how can we design the antimicrobial-active silver(I) complexes by using appropriate ligands, and (4) how different are the antimicrobial mechanisms of silver(I) complexes and other metal complexes, and so on. In particular, we are interested in the first issue, *i.e.* the difference in antimicrobial behavior or modes of action among aqueous silver ions (Ag^+) and silver(I) complexes.

RSC Polymer Chemistry Series No. 10
Polymeric Materials with Antimicrobial Activity: From Synthesis to Applications
Edited by Alexandra Muñoz-Bonilla, María L. Cerrada and Marta Fernández-García
© The Royal Society of Chemistry 2014
Published by the Royal Society of Chemistry, www.rsc.org

From such viewpoints, in this chapter, we review recent reports of antimicrobial activities of aqueous silver(I) ions as $AgNO_3$ (Section 7.2), of silver nanoparticles (Section 7.3) plus reports on the synthesis, structures and antimicrobial activities of polymeric/nonpolymeric silver(I) complexes (Section 7.4.1). Nonpolymeric complexes include mononuclear silver(I) complexes and silver(I) cluster complexes such as dinuclear, tetranuclear and polynuclear clusters and so on. We also review the structure and antimicrobial activities of other polymeric and nonpolymeric metal complexes (Section 7.4.2).

The structures of polymeric and nonpolymeric metal complexes described here are based on X-ray crystallography or single-crystal X-ray structure analysis. The antimicrobial activities are usually based on data evaluated by minimum inhibitory concentration (MIC), only if comments or notes are not described. In this chapter, the numbering of ligands and complexes is defined by a combination of the section number and the numbers of ligands and/or complexes themselves without chapter number 7 such as, for example, **4-1-1-L1** for the sulfadiazine ligand and **4-1-1-C1** for the silver(I) sulfadiazine complex, both described in Section 7.4.1.1.

In Section 7.4.1, silver(I) complexes include (1) classification based on the coordinating donor atoms of ligands (N, S, O, P, C atoms and so on) (Sections 7.4.1.2, 7.4.1.3, 7.4.1.4 and 7.4.1.6) and (2) classification based on appropriate ligands such as sulfadiazine (Section 7.4.1.1), amino acids (Section 7.4.1.5), coumarin (Section 7.4.1.7), thiosemicarbazone (Section 7.4.1.8) and hinokitiol (Section 7.4.1.9). Polymeric materials are found to be mostly silver(I) complexes with heterocyclic nitrogen atoms such as imidazole, triazole, tetrazole, and amino acid ligands such as histidine, arginine, aspartic acid, asparagine, *etc.* and their phosphine derivatives. On the other hand, the monomeric, mononuclear silver(I) complexes have been realized as *N*-heterocyclic carbene (NHC)-silver(I) complexes, coumarin-silver(I) complexes and so on. Silver(I) cluster complexes, with and without antimicrobial activities, have also been realized, such as the $K_{12}[Ag_8(2\text{-mba})_{10}]$ complex (**4-1-3-C3**), $[Ag(2\text{-Hmna})]_6$ complex (**4-1-3-C6**), $[Ag(pfbt)(PPh_3)]_6$ complex (**4-1-3-C11**), $[H_4Ag_8(2\text{-mpa})_6(PPh_3)_6]$ complex (**4-1-3-C13**), $[Ag(hino)]_2$ complex (**4-1-9-C1**), $[Ag(thiosemicarbazone)]_4$ (**4-1-8-C2**) complex and so on.

In general, as to the phosphine ligands, most polymeric and nonpolymeric silver(I) phosphine complexes, but not all (Section 7.4.1.4), show no antimicrobial activity (Sections 7.4.1.2–7.4.1.5), while gold(I) phosphine complexes show selective activities against Gram-positive bacteria (Section 7.4.2.1). Polymeric silver(I) phosphine complexes are cited here even if no activity is shown. However, the mononuclear silver(I) complexes without antimicrobial activity are not cited.

In Section 7.4.2, other metal complexes such as gold(I) (7.4.2.1), zinc(II) (7.4.2.2), copper(II), zinc(II), iron(II) (7.4.2.3), copper(II), zinc(II), nickel(II), antimony(III), bismuth(III) (7.4.2.4) and magnesium, manganese(II), nickel(II), zinc(II), molybdenum(VI), tungsten(VI), copper(II), palladium(II), iron(III), titanium(IV), hafnium(IV), indium(III), bismuth(III), antimony(III), *etc.* (7.4.2.5), *i.e.* metal complexes except silver(I), are described, which are formed with appropriate ligands such as phosphine (7.4.2.1), pyrithione (7.4.2.2), chitosan (7.4.2.3), Schiff base (7.4.2.4) and hinokitiol (7.4.2.5). Most complexes, except chitosan complexes with polymeric ligands, are non-polymeric, mono- and dinuclear metal complexes showing antimicrobial activities.

Representation of abbreviations of several neutral ligands should be noted. In particular, the number of protons has been erroneously described and confused in previous papers, such as histidine (H_2his),[1,2] acetylglycine (Hacgly),[3] pyrrolidon-carboxylic acid (Hpyrrld).[1,4,5] Herein, we use their correct forms.

7.2 Antimicrobial Activities of Aqueous Silver(I) Ion

The antimicrobial activity of silver materials (metallic silver and the silver(I) ion) has been known for years. Since ancient times, human beings have used silver empirically as dishes and as a water preservative.[6–12] The most widely documented uses are prophylactic treatment of burns and water disinfection. The toxicity of silver materials to human cells is considerably lower than that to bacteria.[13] However, the mechanisms by which silver kills cells are not well known. Information on resistance mechanisms is apparently contradictory and even the chemistry of silver(I) ions in such systems is poorly understood.

The silver(I) ion is known to be highly reactive and can readily bind with a variety of negatively charged ions such as inorganic anions, proteins, RNA and DNA, and also with soft donor atoms such as sulfur and phosphorus. It is difficult to know whether strong binding reflects toxicity or detoxification; some sensitive bacterial strains have been reported as accumulating more silver than corresponding resistant strains, and in others the reverse apparently occurs. In several cases resistance has been shown to be plasmid mediated. Plasmids are reported to be difficult to transfer, and can also be difficult to maintain, as we have found. Attempts to find biochemical differences between resistant and sensitive strains have met with limited success; differences are subtle, such as increased cell surface hydrophobicity in resistant *Escherichia coli*.

Some of the problems are due to defining conditions in which resistance can be observed.[6] The silver(I) ion has been shown to bind to components of cell-culture media, and the presence of chloride is necessary to demonstrate resistance. The silver species used must also be considered. This is usually water-soluble $AgNO_3$, which readily precipitates as AgCl. The clinically preferred compound is the highly insoluble silver sulfadiazine (Section 7.4.1.1), which does not cause hypochloraemia in burns. It has been suggested that resistant bacteria are those unable to bind silver(I) ion more tightly than does chloride. It may be that certain forms of insoluble silver are taken up by cells, as has been found for nickel. There is evidently a subtle interplay of solubility and stability that should reward further investigation.

Some evidence on the antibacterial mechanism of the silver(I) ion has been presented by some researchers. Feng and coworkers combined electron micro-copy and X-ray microanalysis to investigate $AgNO_3$-treated Gram-negative

E. coli and Gram-positive *Staphyrococcus aureus*.[14] They found that silver and sulfur elements in electron-dense granules either surrounding the cell wall or inside the cells caused DNA to lose its replication ability, and that the protein became inactivated after treatment with the silver(I) ion.

Holt and Bard studied the behavior of *E. coli* upon the addition of ≤ 10 μM $AgNO_3$ using electrochemical techniques.[15] This group found that the addition of ≤ 10 μM $AgNO_3$ to suspended or immobilized *E. coli* results in stimulated respiration before death, signifying the uncoupling of respiratory control from ATP synthesis. This is symptomatic of the interaction of the silver(I) ion with enzymes of the respiratory chain. The use of $[Fe(CN)_6]^{3-}$ (hexacyanidoferrate(III) ion) as an alternative electron acceptor to oxygen suggests that the silver(I) ion inhibits at a low potential point of the chain, possibly the nicotinamide adenine dinucleotide dehydrogenase (NADH) stage. Potassium concentration and the presence of glucose affected the uptake rate. Measurement of silver(I) ion uptake by immobilized cells showed that approximately 60% of the silver(I) ion taken up is transported into the cell and the remaining 40% binds to the outside of the cell.

Park and coworkers studied the antibacterial effect and action mechanism of a silver(I) ion in solution that was electrically generated for *S. aureus* and *E. coli* by analyzing the growth, morphology, and ultrastructure of the bacterial cells following treatment with the silver ion solution.[16] Bacteria were exposed to the silver(I) ion solution for various periods of time, and the antibacterial effect of the solution was tested using the conventional plate count method and flow cytometric (FC) analysis. Transmission electron microscopy (TEM) showed considerable changes in the bacterial cell membranes upon silver(I) ion treatment, which might be the cause or consequence of cell death. They concluded that the results suggested that silver ions may cause *S. aureus* and *E. coli* bacteria to reach an active and nonculturable state and eventually die.

Wang and coworkers investigated the influence of silver ions on *E. coli* and *Staphylococcus epidermidis* using atomic force microscopy (AFM).[17] Both single cells and cell communities were visualized, and the nanoscale ultrastructure images of these two bacteria strains before and after stimulation with silver ions were obtained. The results showed that in the case of *E. coli* after treatment with silver ions, vesicles appeared on the cell walls, and the size of the vesicles became larger with an increase in incubation time. However, in the case of *S. epidermidis* after treatment with silver ions, irregular and deep grooves appeared on the cell walls, and the cytoplasm membrane shrank and became separated from the cell wall. The significant differences in cell wall changes between *E. coli* and *S. epidermidis* after treatment with silver ions were related to their structural characters.

There are some solid materials for antimicrobial purposes, such as silver-doped zeolite, ceramics and fibers and they are used in many places. However, the details of the antimicrobial mechanisms of solid materials are more difficult to study and remain to be elucidated. Most of them show bactericidal activities against micro-organisms, which are due to active silver species. Tsuchido and coworkers proposed that the generation of a reactive oxygen species induced by silver material damaged/killed bacteria cells and that this was the mechanism of silver-doped zeolite.[18]

7.3 Antimicrobial Activity of Silver Nanoparticles (AgNPs)

Metal nanoparticles have been the focus of attention since 2000 because particles of 1–100 nm show unique properties due to their large surface area and quantum-size effects compared with those of bulk metal materials. Among them coinage metal nanoparticles are of interest because they are expected to be useful for catalysis, biosensing, drug delivery, nanodevice fabrication and so forth.[19–23]

Silver nanoparticles (AgNPs) in different formulations and with different shapes and sizes exhibit various antimicrobial activities. However, the mechanisms of antimicrobial activities of silver(I) ions and AgNPs, and their toxicity to human tissues are not fully characterized. The efficiency of the antibacterial activities of metallic silver and the silver(I) ion is well known. Since the silver(I) ion prevents the growth of various micro-organisms at low concentration, it is difficult to elucidate which silver species are effective as mechanisms of antimicrobial activities. Many researchers believe that slow release of the silver(I) ion from various materials prevents the increase of micro-organisms. Is the hydrated silver(I) ion the only active species for antimicrobial activities of silver materials? Recently, some evidence that AgNPs serve as the active species in the antimicrobial mechanism has been reported.[19]

Stable AgNPs were prepared and their shape and size distribution were characterized by particle analyzer and TEM study by Jeong and coworkers.[24] The antimicrobial activities of AgNPs were investigated against yeast, E. coli, and S. aureus. In these tests, Müller Hinton agar plates were used and AgNPs of various concentrations were supplemented in liquid systems. As a result, yeast and E. coli were inhibited at a low concentration of AgNPs, whereas the growth-inhibitory effects on S. aureus were mild. The free-radical generation effect of Ag nanoparticles on microbial growth inhibition was investigated by electron spin resonance (ESR) spectroscopy. These results suggested that AgNPs can be used as effective growth inhibitors against various micro-organisms, making them applicable to diverse medical devices and antimicrobial control systems.

Lara and coworkers challenged different drug-resistant pathogens of clinical importance (multidrug-resistant Pseudomonas aeruginosa, ampicillin-resistant E. coli O157:H7 and erythromycin-resistant Streptococcus pyogenes) with a suspension of AgNPs.[25] By means of a luciferase-based assay, it was determined that AgNPs (1) inactivate a panel of drug-resistant and drug-susceptible Gram-positive and Gram-negative bacteria, (2) exert their antibacterial activity through a bactericidal rather than bacteriostatic mechanism, and (3) inhibit the bacterial growth rate from the time of first contact between the bacteria and the nanoparticles. Additionally, strains with a phenotype resistant to AgNPs were developed and used to explore the bactericidal mode of action of AgNPs.

Through the Kirby–Bauer test, it was shown that the general mechanism of bactericidal action of AgNPs is based on inhibition of cell-wall synthesis, protein synthesis mediated by the 30s ribosomal subunit, and nucleic acid synthesis. The data of Lara and coworkers[25] suggested that AgNPs are effective broad-spectrum biocides against a variety of drug-resistant bacteria, which makes them a potential candidate for use in pharmaceutical products and medical devices that may help to prevent the transmission of drug-resistant pathogens in different clinical environments.

Sondi and Salopek-Sondi used scanning transmission electron microscopy (STEM) to show that treated *E. coli* cells were damaged by the formation of "pits" in the cell wall of the bacteria, while AgNPs were found to accumulate in the bacterial membrane.[26] As mentioned by other researchers, they also reported that the size, shape and composition of silver nanoparticles can have a significant effect on their antimicrobial efficiency. Morones and coworkers used high-angle annular dark-field (HAADF) STEM to ascertain that AgNPs act primarily in three ways against Gram-negative bacteria: (1) nanoparticles, mainly in the range of 1–10 nm, are attached to the surface of the cell membrane and drastically disturb its proper functions, like permeability and respiration; (2) they are able to penetrate inside the bacteria and cause further damage by possibly interacting with sulfur- and phosphorus-containing compounds such as DNA; (3) nanoparticles release silver(I) ions, which make an additional contribution to their bactericidal effect.[27] The potential use of AgNPs to control pathogens, with emphasis on their action against pathogenic bacteria, their toxicity and possible mechanisms of action, was reviewed by Duran *et al.*[28]

It seems that several processes are involved in the antimicrobial activity of AgNPs, *i.e.* silver(I) ions are released from the AgNPs, the nanomechanical action of AgNPs against micro-organism membranes, and the formation of other chemically active species, such as the hydroxyl radical by AgNPs.

7.4   Antimicrobial Behavior of Silver and other Metal Complexes

7.4.1   Polymeric and Nonpolymeric Silver Complexes

7.4.1.1   *Polymeric Silver(I) Sulfadiazine Complex*

Silver sulfadiazine **4-1-1-C1**, a complex made from Ag[I] and the sulfadiazine ligand **4-1-1-L1** (N1-pyrimidin-2-ylsulfonilamide or 2-sulfanilamidopyrimidine), is used clinically as an antimicrobial and antifungal agent. It is an insoluble polymeric compound, which releases Ag[I] ions slowly, and is applied topically as a cream to prevent bacterial infections in cases of severe burns. The slow release of antimicrobial Ag[I] ions from inorganic or organic polymer matrices[29,30] is of industrial importance. If the slow release is true, the antimicrobial action of **4-1-1-C1** would be the same as that of the Ag[I] ion itself.

4-1-1-L1

4-1-1-C1

As related metal complexes, Fox and coworkers[31] have described the metal (zinc, cerium and cobalt) sulfonamides as antibacterial agents in topical therapy. The apparent antibacterial efficacy of zinc sulfadiazine and cerium sulfadiazine and the metabolic role of other trace metals suggested that the sulfonamide salts of these metals might be of therapeutic value, so various sulfonamide salts were prepared and studied *in vitro* and *in vivo*. Only zinc sulfathiazole and zinc sulfamethoxazole were as effective as zinc sulfadiazine in animal studies. Only cobalt sulfadiazine appeared comparable to zinc and cerium sulfadiazine in healing burn wounds in rats. It is not yet known whether other metal sulfadiazines have this attribute.

Cook and Turner[32] have described the crystal and molecular structures of the 1:1 silver sulfadiazine complex, **4-1-1-C1**. Each Ag was coordinated to 3 sulfadiazine molecules in a distorted tetrahedron of 3 N atoms and 1 O. The molecules were also linked by sheets of NH · · · OS bonds.

Baenziger and Struss independently reported the crystal structure of **C1**: the nitrogen atoms of the pyrimidine ring coordinate to two different silver atoms to form polymeric chains extending through the crystal.[33] Each silver atom in this chain is also coordinated to one oxygen atom from the sulfonyl group of the **4-1-1-L1** molecule in the chain. A second identical chain (related by center of symmetry) is joined to the first chain by the coordination of the silver atom from each chain to the imido nitrogen atom of the **4-1-1-L1** molecule in the other chain. In addition, the silver atom in one chain is only 2.916(1) Å from the symmetry-related silver atom in the other chain. The double-stranded chains are further hydrogen bonded by the amine hydrogen atoms and sulfonyl oxygen atoms to form planar sheets of the double-stranded chains. If the Ag–Ag interaction is taken into account, the coordination about silver becomes a distorted trigonal bipyramid.

The mechanism of Ag[I] cytotoxicity is unknown. Cell-wall damage may be important, and it has been shown that Cys150 in phosphomannose isomerase, an essential enzyme for the biosynthesis of cell walls of *Candida albicans*, is an Ag[I] target in this organism.[30,34]

Mastrolorenzo and coworkers have reported the antifungal activity of silver and zinc complexes of sulfadrug derivatives incorporating arylsulfonylureido moieties.[35,36] Two well-known antimicrobial sulfonamides, *i.e.* sulfadiazine (**4-1-1-L1**) and sulfamerazine, were reacted with arylsulfonyl isocyanates, affording several new arylsulfonylureido derivatives. These compounds were subsequently used as ligands (in the form of conjugate bases, as sulfonamide anions) for the preparation of metal complexes containing silver and zinc. The newly synthesized complexes, unlike the free ligands, proved to act as effective antifungal agents against several *Aspergillus* and *Candida* species, some of them showing activities comparable to ketoconazole. The mechanism of antifungal action of these complexes seems to be different from that of the azole antifungals acting as lanosterol-14-alpha-demethylase inhibitors. The levels of sterols assayed in the fungi cultures treated with these new antifungals were equal in the absence or in the presence of the tested compounds. This is in strong contrast to similar experiments in which ketoconazole has been used as an antifungal, when drastically reduced ergosterol amounts could be detected. Thus, it is probable that the inhibition of phosphomannose isomerase, a key enzyme in the biosynthesis of yeast cell walls, imparts antifungal activity to the new metal complexes reported here.

It should also be noted that there is a clinical report showing that some strains were susceptible to antibiotic but resistant to **4-1-1-C1**.[37]

7.4.1.2 Polymeric Silver(I)–Nitrogen Bonding Complexes and their Phosphine Derivatives

$_\infty$[Ag(imd)] (**4-1-2-C1**), [Ag(Himd)$_2$]NO$_3$ (**4-1-2-C2**), $_\infty$[Ag(1,2,3-triz)] (**4-1-2-C3**), $_\infty$[Ag(1,2,4-triz)] (**4-1-2-C4**), $_\infty$[Ag(tetz)] (**4-1-2-C7**), $_\infty$[Ag(1,2,3-triz)-(PPh$_3$)$_2$] (**4-1-2-C5**), $_\infty$[Ag(1,2,4-triz)(PPh$_3$)$_2$] (**4-1-2-C6**), $_\infty$[Ag(tetz)(PPh$_3$)$_2$] (**4-1-2-C8**).

Chen and coworkers have described supramolecular isomers of silver(I) imidazolates $_\infty$[Ag(imd)] (**4-1-2-C1**) (**4-1-2-L1** = imidazole ; Himd) which were hydro(solvo)thermally prepared *via* variations of the reaction conditions such as solvent and additive.[38] Two new supramolecular isomers (**4-1-2-C1a**, **4-1-2-C1b**) of $_\infty$[Ag(imd)] [**4-1-2-C1a**, P21/c, a 11.460(1), b 16.882(2), c 9.303(1) Å, β 106.61(1); **4-1-2-C1b**, Pbca, a 10.175(2), b 6.8415(9), c 23.881(2) Å] exhibiting two-dimensional networks through short interchain silver(I)···silver(I) contacts [3.1595(5) Å for **4-1-2-C1a** and 3.4445(5) Å for **4-1-2-C1b**] were structurally established to be different from two known superstructures.

4-1-2-L1   4-1-2-C1

Before a report on X-ray crystallography,[38] Nomiya and coworkers had described that a neutral, silver(I)-N bonding, polymeric compound $_\infty$[Ag(imd)] (**4-1-2-C1**), showed a wide spectrum of effective antimicrobial activities against bacteria, yeast and mold.[39] The activities against a wide range of mold are of particular note. This polymeric solid is sparingly soluble in all solvents. The monomeric, cationic, water-soluble silver(I)-N bonding complex with a neutral Himd ligand, [Ag(Himd)$_2$](NO$_3$) (**4-1-2-C2**), has also shown a wide spectrum of effective antimicrobial activities. These activities observed here were significantly different from those of the oligo-meric silver(I)-S bonding complexes; the latter have shown narrow spectra. It has been proposed that the silver(I)-N bonding is one of the key factors for the wide spectra of antimicrobial activities, and the potential targets for the inhibition of bacteria and yeast by these silver(I) complexes are proteins, but not nucleic acids. The physicochemical properties of (**4-1-2-C1**), by comparison with those of (**4-1-2-C2**), with various measurements (FTIR, laser Raman scattering spectroscopy, electron spectroscopy for chemical analysis (ESCA or also known as X-ray photoelectron spectroscopy XPS), and solid ^{13}C CP-MAS NMR spectroscopy) are described.

The polymeric complexes, $_\infty$[Ag(1,2,3-triz)] (**4-1-2-C3**) [**4-1-2-L2** = 1,2,3-triazole; 1,2,3-Htriz] and $_\infty$[Ag(**1,2,4-triz**)] (**4-1-2-C4**) [**4-1-2-L3** =1,2,4-triazole; 1,2,4-Htriz], were obtained as noncrystalline, colorless powder-solids.[39] **4-1-2-C3** and **4-1-2-C4** were fully characterized by elemental analyses, TG/DTA and FTIR in the solid state and by various solution NMR (^{31}P, ^{109}Ag, ^{1}H and ^{13}C) spectroscopies and molecular-weight measurements in solution. Furthermore, two polymeric silver(I)-PPh$_3$ complexes with N-containing heterocycles, $_\infty$[Ag(**1,2,3-triz**)(PPh$_3$)$_2$] (**4-1-2-C5**) and $_\infty$[Ag(**1,2,4-triz**)(PPh$_3$)$_2$] (**4-1-2-C6**), were synthesized from reactions of polymeric precursors **4-1-2-C3** and **4-1-2-C4** with 3 equivalents of PPh$_3$ in CH$_2$Cl$_2$, respectively.

4-1-2-L2   4-1-2-L3

The antimicrobial activities of complexes **4-1-2-C3** to **4-1-2-C6** were com-pared and the key factors affecting them discussed. **4-1-2-C4** showed a wide spectrum of excellent antibacterial activities, but **4-1-2-C3** showed poor activ-ities. On the other hand, polymeric PPh$_3$ derivatives, **4-1-2-C5** and **4-1-2-C6**, did not show any activity. Nomiya and coworkers have also described at least one of the key factors determining antimicrobial effects, and their spectra are the types of atoms coordinated to silver(I) and their bonding properties, rather than the polymerization degree of the silver(I) complexes, their solubility in water and whether or not the ligands are neutral.[40,41]

The crystal structures of **4-1-2-C5** and **4-1-2-C6** were determined by single-crystal X-ray diffraction. **4-1-2-C5** and **4-1-2-C6** in the solid state were helical polymers consisting of AgN$_2$P$_2$ cores formed by bridging triazolate anions and two PPh$_3$ ligands.

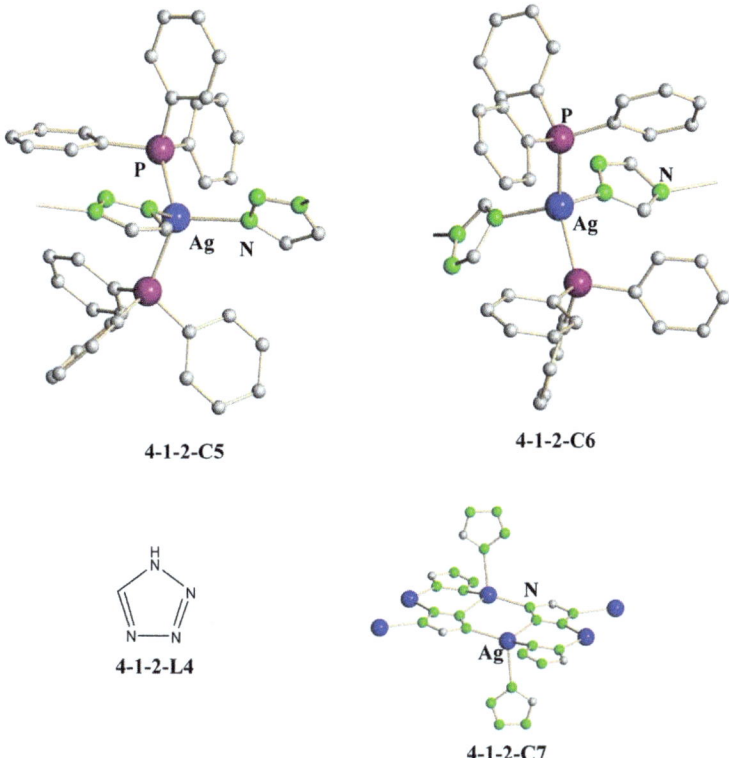

4-1-2-C5

4-1-2-C6

4-1-2-L4

4-1-2-C7

Ciani and coworkers have studied the reactions of tetrazole (**4-1-2-L4** ; Htetz) with different silver(I) salts, such as $AgNO_3$, $AgBF_4$ and $Ag(CF_3SO_3)$, producing single crystals of $_{\infty}[Ag(tetz)]$ (**4-1-2-C7**) suitable for X-ray structure analysis.[42] It consists of a three-dimensional neutral network based on triconnected and tetraconnected silver centers in the ratio 1 : 1, which are joined through tetrazolate anions, one half of which act as exo-tridentate and one half as exo-tetradentate ligands. Compound **4-1-2-C7** has shown effective activities against Gram-positive and Gram-negative bacteria and yeast, but no activities against mold.[43]

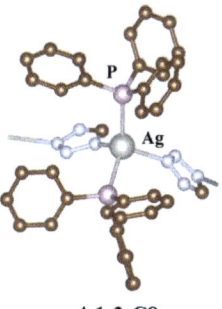

4-1-2-C8

A silver(I)-triphenylphosphine complex with a heterocyclic ligand, $_{\infty}[Ag-(tetz)(PPh_3)_2]$ (**4-1-2-C8**), was synthesized from a reaction in dichloromethane

of the precursor complex **4-1-2-C7** with PPh₃ and isolated as colorless
needle crystals for **4-1-2-C8**.⁴³ The crystal structure of **4-1-2-C8** was determined by single-crystal X-ray diffraction. **4-1-2-C8** was also fully characterized
by elemental analyses, TG/DTA and FTIR in the solid state, solution NMR
(^{31}P, ^{1}H and ^{13}C, as well as ^{109}Ag) spectroscopies and solution molecular-
weight measurements. **4-1-2-C8** is composed of a helical polymer, a 2_1 helix
with a pitch of 9.471 Å, formed by a bridged tetrazolate of an AgNP₂ core in
the solid state, but it was present as a monomer in solution. The molecular
structure of **4-1-2-C8** was compared with those of the corresponding triazole
complexes, *i.e.* silver(I) complexes $_\infty$[Ag(1,2,3-triz)(PPh₃)₂] (**4-1-2-C5**) and
$_\infty$[Ag(1,2,4-triz)(PPh₃)₂] (**4-1-2-C6**), both as helical polymers. The antimicrobial activities of **4-1-2-C8** and its precursor **4-1-2-C7** were compared and
the key factors affecting them discussed. The MIC values of **4-1-2-C8** were
estimated as > 1000 μg/mL for bacteria, yeast and mold showing no activity.

7.4.1.3   Polymeric and Nonpolymeric Silver(I)-Sulfur/Nitrogen Bonding Complexes and their Phosphine Derivatives

{Na[Ag(mba)]·H₂O}$_n$ (n = 21–27) (yellow) (**4-1-3-C1**), $_\infty$[Ag(Hmba)] (orange)
(**4-1-3-C2**), K₁₂[Ag₈(2-mba)₁₀]·12H₂O (**4-1-3-C3**), [Ag(2-Hmna)]₆·4H₂O (**4-1-
3-C5**), {Na[Ag(2-mna)]·H₂O}₆ (**4-1-3-C6**), $_\infty$[Ag(pfbt)] (**4-1-3-C7**), $_\infty$[Ag(2-
Hmpa)] (**4-1-3-C9**), [Ag(2-Hmba)(PPh₃)]₄·0.5CHCl₃·0.5EtOH (**4-1-3-C4**),
[Ag(pfbt)(PPh₃)]₆ (**4-1-3-C8**), H₄[Ag₈(2-mpa)₆(PPh₃)₆]·EtOH (**4-1-3-C10**).

Nomiya and coworkers have stated that thiosalicylic acid or 2-mercapto-
benzoic acid (**4-1-3-L1**; **H₂tsa** or **H₂mba**) forms 2 types of anionic, light-, and
thermally stable silver(I) complexes with marked antimicrobial activities
for bacteria, yeast, and molds; one is yellow and water soluble
$_\infty${Na[Ag(mba)]·H₂O} (**4-1-3-C1**) and the other is orange, and water insoluble
$_\infty$[Ag(Hmba)] (**4-1-3-C2**), both of which can be converted into the other and
neither of which crystallizes.⁴⁴ Both are complexes with a compound of
metal : H₂mba (**4-1-3-L1**) = 1 : 1. **4-1-3-C2** is probably polymeric, whereas **4-1-
3-C1** is an oligomer with a formula of {Na[Ag(mba)·H₂O]}$_n$ (n = 12–14;
M_W = 3600–4200). Both complexes were characterized by complete elemental
analyses, TG/DTA, FTIR, and ESCA spectra in the solid state, and **4-1-3-C1**
also was characterized by molecular-weight measurement in aqueous solution
and by ^{109}Ag and 2D-NMR (^{1}H-H COSY, ^{1}H-^{13}C COSY, and HMBC)
measurements in aqueous solution.

A water-soluble silver(I) complex showing effective antibacterial activities,
K₁₂[Ag₈(mba)₁₀]·12H₂O (**4-1-3-C3**) was prepared from an Ag₂O: **4-1-3-L1**:
KOH = 1:4:8 molar ratio reaction in aqueous media and its crystal structure
was determined.⁴⁵ X-ray crystallography revealed that [Ag₈(mba)₁₀]$^{12-}$ was a
discrete, anionic octanuclear silver(I) cluster consisting of two butterfly-type
Ag₄S₄ subunits bridged by two μ₃-S atoms.

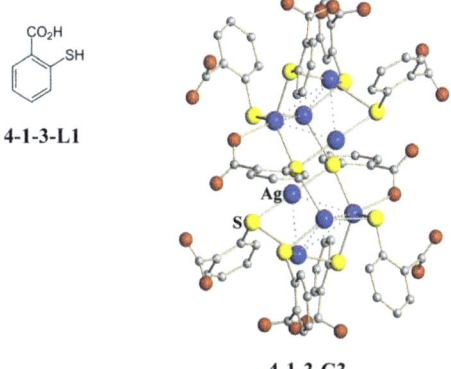

4-1-3-L1

4-1-3-C3

A tetranuclear silver(I) cluster, [Ag(2-Hmba)(PPh$_3$)]$_4$ · 0.5CHCl$_3$ · 0.5EtOH (**4-1-3-C4**) was prepared from the silver(I)-O bonding precursor $_\infty$[Ag$_2$(*R*-pyrrld)(*S*-pyrrld)] (Hpyrrld = 2-pyrrolidone-5-carboxylic acid), PPh$_3$ and H$_2$mba (**4-1-3-L1**) in an organic solvent.[46] The crystal structure of **4-1-3-C4a** was composed of a racemic mixture of a propeller cluster having three leaves constructed with three Ag(μ_3-S)(μ_2-S)P units and one Ag(μ_2-S)$_3$P unit (**4-1-3-C4b, 4-1-3-C4c**). The pyrrld-ligand in the precursor plays an important role in the formation of **4-1-3-C4**, although this ligand disappeared from the product. **4-1-3-C4** showed no antimicrobial activities for bacteria, yeast and molds.

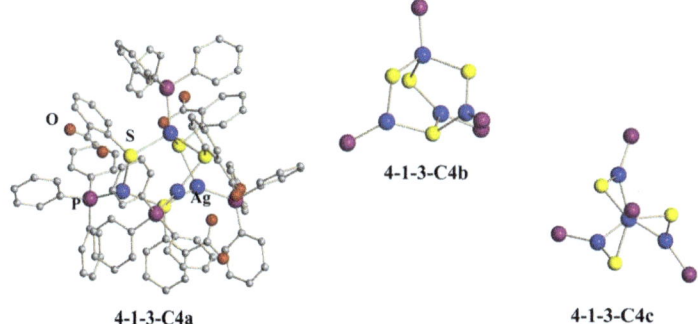

4-1-3-C4a

4-1-3-C4b

4-1-3-C4c

Two types of silver(I) 2-mercaptonicotinate (**4-1-3-L2** ; H$_2$mna) were synthesized by a reaction with Ag$_2$O suspended in aqueous media; one was a water-insoluble, DMSO-soluble yellow powder [Ag(Hmna)]$_6$ · 4H$_2$O (**4-1-3-C5**) formed under acidic conditions, and the other was a water-soluble, yellow powder {Na[Ag-(mna)] · H$_2$O}$_6$ (**4-1-3-C6**) formed under alkaline conditions.[47] These two complexes were interconverted on changing the acidity of the solution. Crystallization by vapor diffusion of the DMSO and aqueous solutions with external acetone gave pale yellow cubic crystals for **4-1-3-C5** and pale yellow plate crystals for **4-1-3-C6**. The molecular structure of **4-1-3-C5** consists of a discrete, hexanuclear silver(I) cluster with two silver(I) triangles linked by mercaptonicotinate anions, the geometry around each silver(I) atom being constructed by one aromatic N atom, two μ-S atoms and two weak silver(I)–silver(I) interactions. The structure of

4-1-3-C6 was determined by Bau and coworkers,[48] in which an $[Ag(mna)]_6^{6-}$ cluster has an Ag_6S_6 core and an overall shape of a twisted hexagonal cylinder with six sulfur atoms and six silver atoms alternating on a puckered drum-like surface. Each Ag atom is trigonally coordinated by one N and two S ligands.

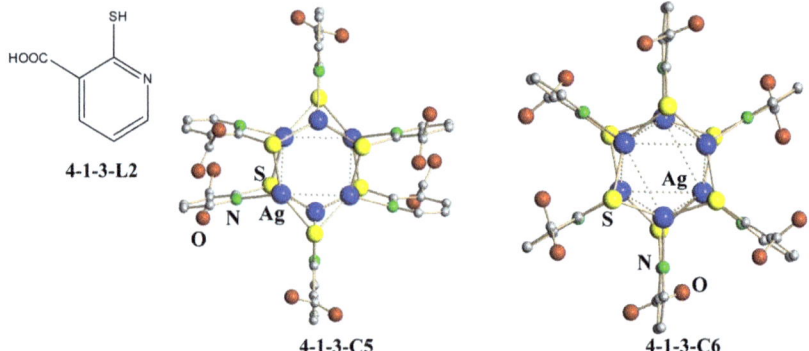

4-1-3-L2   **4-1-3-C5**   **4-1-3-C6**

These complexes were also characterized by elemental analysis, TG/DTA, FTIR, ^{1}H- and ^{13}C-NMR, as well as ^{109}Ag-NMR for **4-1-3-C6**. Noteworthy antimicrobial activities for them were observed. **4-1-3-C5** showed effective activities against four bacteria, one yeast and six molds. On the other hand, **4-1-3-C6** showed effective activities against two Gram-negative bacteria and four molds, but no activity against two Gram-positive bacteria, two yeasts and two molds. These antimicrobial activities are different from those of the previous Ag–S bonded complexes, $_\infty$[Ag(Hmba)] (**4-1-3-C2**) and $\{Na[Ag(mba)] \cdot H_2O\}_n$ ($n = 21$–27) (**4-1-3-C1**), which have shown effective activities against the four bacteria and one yeast, but no activity against the six molds.

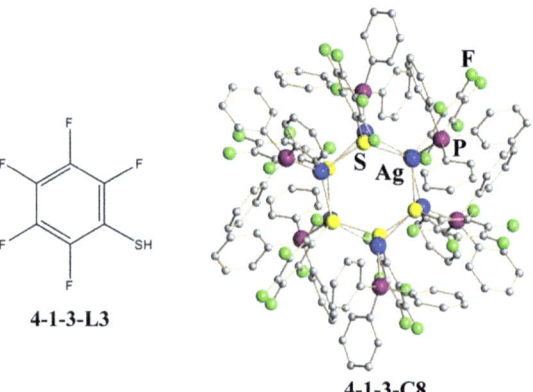

4-1-3-L3

4-1-3-C8

A polymeric compound, $_\infty$[Ag(pfbt)] (**4-1-3-C7**; Hpfbt = pentafluorobenzenethiol; **4-1-3-L3**), with higher purity was prepared in good yield using light-stable and water-soluble silver(I) carboxylate, $_\infty$[Ag(pyrrld)]$_2$ (Hpyrrld = 2-pyrrolidone-5-carboxylic acid), as the silver(I) source.[49] Furthermore, a nanoscale drum-like hexanuclear silver(I) cluster [Ag(pfbt)(PPh$_3$)]$_6$ (**4-1-3-C8**), which showed arrays of channels based on its self-assembly in the solid, was obtained by a 1 : 1 molar-ratio reaction of the insoluble polymeric precursor $_\infty$[Ag(pfbt)] (**4-1-3-C7**) and PPh$_3$ in chloroform. The synthetic yield and purity of **4-1-3-C8** were strongly dependent on the purity of **4-1-3-C7**. **4-1-3-C7** and **4-1-3-C8** have shown no antimicrobial activities.

The molecular structure of [H$_4$Ag$_8$(2-mpa)$_6$(PPh$_3$)$_6$]·EtOH (**4-1-3-C10a**; 2-H$_2$mpa = 2-mercaptopropionic acid; **4-1-3-L4**), prepared by a 1 : 5 molar-ratio reaction of polymeric $_\infty$[Ag(2-Hmpa)] (**4-1-3-C9**) suspended in EtOH with PPh$_3$ in CHCl$_3$, was a wheel-type octanuclear silver(I) cluster; the rim of the wheel was constructed with six outside Ag(μ_3-S)$_2$P units and the axle was formed by two central Ag(μ_3-S)$_3$ units with Ag–Ag separation, 3.122 Å (**4-1-3-C10b**).[50] **4-1-3-C9** showed effective antimicrobial activity for selected bacteria, but **4-1-3-C10** showed no antimicrobial activity.

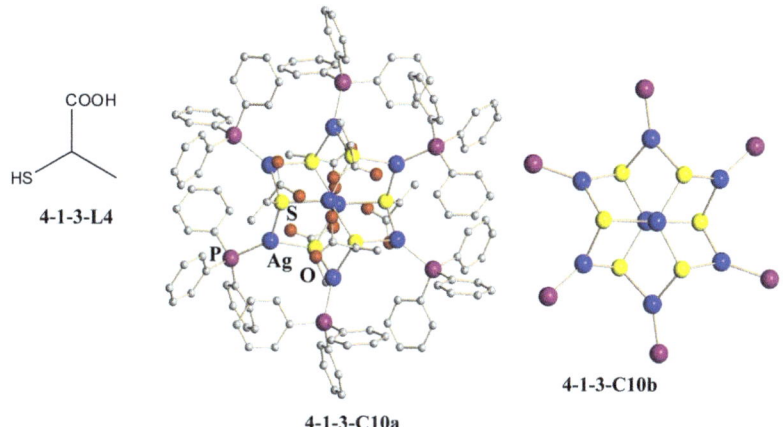

4-1-3-C10b

4-1-3-C10a

7.4.1.4 Polymeric Silver(I)–Oxygen/Nitrogen Bonding Complexes and their Phosphine Derivatives

$_\infty${[Ag(*R*-pyrrld)]$_2$} (**4-1-4-C1**), $_\infty${[Ag(*R,S*-pyrrld)]$_2$} (**4-1-4-C2**), $_\infty${[Ag(*S*-pyrrld)]$_2$} (**4-1-4-C3**) $_\infty${[Ag(*S*-othf)]$_2$} (**4-1-4-C8**), $_\infty${[Ag(*R*-othf)]$_2$} (**4-1-4-C9**), $_\infty${[Ag(*R,S*-othf)]$_2$} (**4-1-4-C10**), $_\infty${[Ag$_2$(*S*-ca)$_2$]} (**4-1-4-C11**), $_\infty${[Ag$_2$(*S*-ca)$_2$(*S*-Hca)$_2$]} (**4-1-4-C12**), $_\infty${[Ag$_2$(*R*-ca)$_2$(*R*-Hca)$_2$]} (**4-1-4-C13**), $_\infty${[Ag$_2$(ca)$_2$(Hca)$_2$]} (**4-1-4-C14**), [Ag$_2$(*R*-pyrrld)$_2$(H$_2$O)(PPh$_3$)$_2$]·H$_2$O (**4-1-4-C4**), [Ag$_2$(*S*-pyrrld)$_2$-(H$_2$O)(PPh$_3$)$_2$]·H$_2$O (**4-1-4-C5**), [Ag(*S*-pyrrld)(PPh$_3$)$_2$] (**4-1-4-C6**), $_\infty${[Ag(*R,S*-pyrrld)(PPh$_3$)]$_2$} (**4-1-4-C7**) and [Ag$_2$(Himdc)(PPh$_3$)$_2$]$_2$ (**4-1-4-C15**).

Nomiya and coworkers have described the synthesis and characterization of two water-soluble silver(I) complexes $_\infty${[Ag(*R*-pyrrld)]$_2$} **4-1-4-C1** (*R*-Hpyrrld; *R*-(+)-2-pyrrolidone-5-carboxylic acid (**4-1-4-L1**)) and $_\infty${[Ag$_2$(*R*-pyrrld)(*S*-pyrrld)]} (**4-1-4-C2**). Their wide-ranging effective antimicrobial activities were also described and their crystal structures determined.[1] Single-crystal X-ray analysis reveals that **4-1-4-C1** and **4-1-4-C2** in the solid state are a right-handed chiral helical polymer and an achiral polymer sheet, respectively, formed by self-assembly of the [Ag$_2$(carboxylato-*O,O'*)$_2$] unit [Ag–Ag 2.899(2) Å for **4-1-4-C1** and 2.875(2) Å for **4-1-4-C2**; O–Ag–O 160.3(5), 163.2(5)° for **4-1-4-C1** and 163.1(2)° for **4-1-4-C2**]. The helicity of **4-1-4-C1** in the solid state was accomplished by a connection of only one oxo group in one dimeric core to one of the silver(I) centers of the adjacent dimeric unit. It is elucidated that the chiral helical polymer **4-1-4-C1** is a mirror image of the enantiomer $_\infty${[Ag(S-pyrrld)]$_2$} (**4-1-4-C3**). The bonding modes of **4-1-4-C1** and **4-1-4-C3** are quite different from those of the achiral polymer sheet **4-1-4-C2**. Complexes **4-1-4-C1** and **4-1-4-C3** have also been characterized by elemental analysis, TG/DTA, FTIR, and solution (^{1}H and ^{13}C) and solid-state ^{13}C NMR spectroscopy. The wide-ranging effective antimicrobial activities observed in the complexes suggest that weak silver(I)–oxygen bonding properties play a key role.

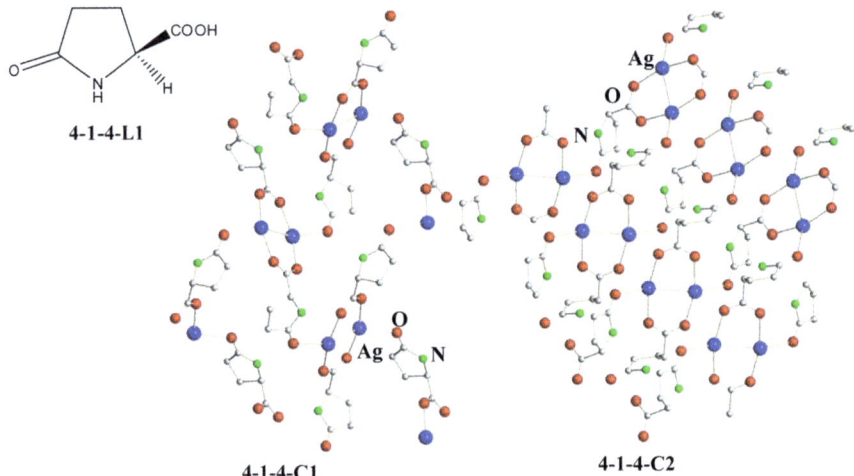

4-1-4-L1

4-1-4-C1   **4-1-4-C2**

Using light-stable dimeric silver(I) carboxylate precursors $_\infty${[Ag(pyrrld)]$_2$} formed with chiral and racemic forms of a Hpyrrld (**4-1-4-L1**) ligand, four light-stable, triphenylphosphinesilver(I) complexes consisting of both a hard Lewis base (O atom) and a soft Lewis base (P atom) were prepared, *i.e.* [Ag$_2$(R-pyrrld)$_2$(H$_2$O)(PPh$_3$)$_2$]·H$_2$O (**4-1-4-C4**), [Ag$_2$(S-pyrrld)$_2$(H$_2$O)(PPh$_3$)$_2$]·H$_2$O (**4-1-4-C5**), [Ag(S-pyrrld)(PPh$_3$)$_2$]$_2$ (**4-1-4-C6**) and $_\infty${[Ag(R,S-pyrrld)(PPh$_3$)]$_2$} (**4-1-4-C7**).[4] Their solid-state and solution structures were unequivocally characterized by elemental analysis, TG/DTA, FTIR, X-ray structure analysis, molecular weight measurements in EtOH by the vaporimetric method, solution (^{1}H, ^{13}C, and ^{31}P) NMR, and solid-state ^{31}P CPMAS NMR spectroscopy. Two sets of the enantiomeric complexes were isolated as (**4-1-4-C4** and **4-1-4-C5**) and **4-1-4-C7**. X-ray crystallography revealed that these complexes possessed different Ag–O bonding modes, depending on the number of PPh$_3$ ligands and the chirality of the pyrrld–ligand. Complexes **4-1-4-C4** to **4-1-4-C7** behaved as monomeric species in EtOH and CD$_2$Cl$_2$. The antimicrobial activites of the silver(I) complexes in a water–suspension system against selected bacteria, yeast, and molds, were significantly correlated with the number of coordinating PPh$_3$ ligands per silver(I) atom in the complexes.

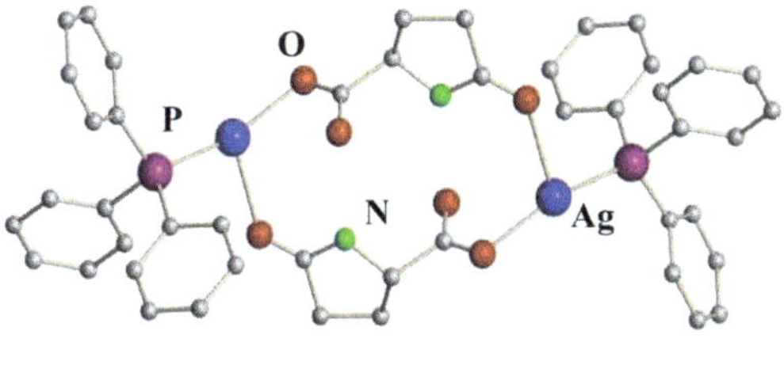

4-1-4-C7

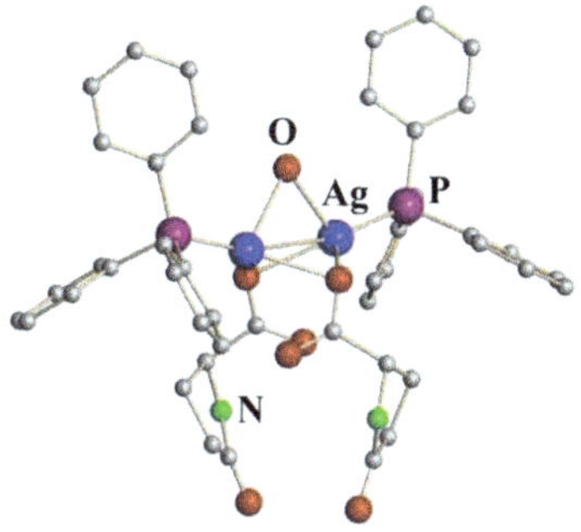

4-1-4-C4

Three water-soluble silver(I) complexes, $_\infty${[Ag(S-othf)]$_2$} (**4-1-4-L2** = S-5-oxo-2-tetrahydrofuran carboxylic acid; S-Hothf, **4-1-4-C8**), $_\infty${[Ag(R-othf)]$_2$} (**4-1-4-L3** = R-5-oxo-2-tetrahydrofuran carboxylic acid; R-Hothf, **4-1-4-C9**) and $_\infty${[Ag$_2$(R-othf)(S-othf)]} (**4-1-4-C10**), showed effective antibacterial and antifungal activities.[51] They were synthesized and their crystal structures determined. Single-crystal X-ray analysis revealed that **4-1-4-C8** and **4-1-4-C9** in the solid state are a left- and a right-handed chiral helical polymer, respectively, formed by self-assembly of noncentrosymmetric, bis-carboxylato-bridged bis(carboxylato-O,O′)disilver dimers (Ag-Ag distances 2.822(1) Å for **4-1-4-C8** and 2.823(2) Å for **4-1-4-C9**; O–Ag–O angles 164.0(2), 155.2(2)° for **4-1-4-C8** and 163.8(3), 154.5(3)° for **4-1-4-C9**). The helicity of **4-1-4-C8** and **4-1-4-C9** in the solid state is accomplished with a connection of one oxo group in one dimeric core to one of the silver(I) centers of the adjacent dimeric unit and also with a simultaneous connection of one of the carboxylato oxygens to the silver(I) center of a different dimer.

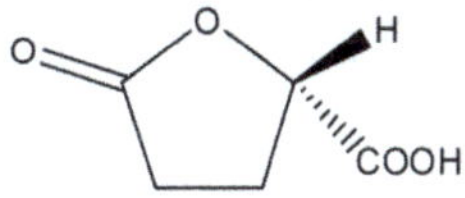

4-1-4-L2

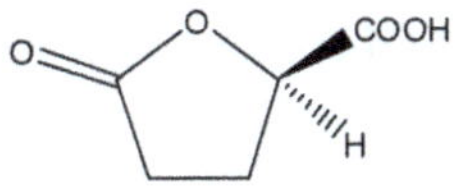

4-1-4-L3

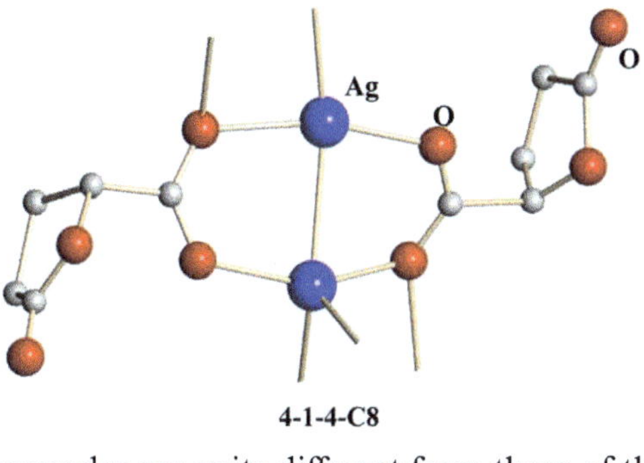

4-1-4-C8

These bonding modes are quite different from those of the stair-like polymer **4-1-4-C10** formed by self-assembly of the dimeric core (Ag–Ag 2.781(1) Å; O–Ag–O angle 164.8(1)°). The crystals of **4-1-4-C10** with an achiral polymer structure were identical to those of **4-1-4-C10** obtained from an aqueous solution containing equal amounts of **4-1-4-C8** and **4-1-4-C9**, evidencing the presence of a ligand replacement between **4-1-4-C8** and **4-1-4-C9** in aqueous solution.

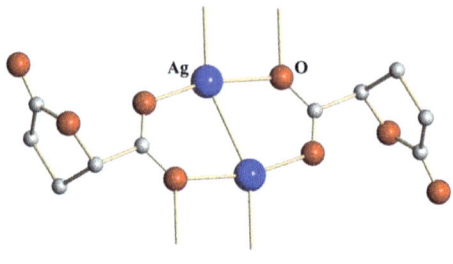

4-1-4-C10

Complexes **4-1-4-C8** to **4-1-4-C10** were also characterized by elemental analysis, TG/DTA, FTIR, and ^{1}H and ^{13}C NMR spectroscopy. The wide spectra of effective antibacterial and antifungal activities observed in **4-1-4-C8** to **4-1-4-C10** suggested that the weaker silver(I)–O bonding properties play a key role in the antimicrobial activities.

Water-soluble, relatively light-stable, chiral and achiral silver(I) complexes $_\infty\{[Ag_2(ca)_2]\}$ and $_\infty\{[Ag_2(ca)_2(Hca)_2]\}$ (*R*- and *S*-Hca $=(1R,4S)$- and $(1S,4R)$-4,7,7-trimethyl-3-oxo-2-oxabicyclo[2.2.1]heptane-1-carboxylic acid, respectively; **4-1-4-L4**) prepared from the reaction of Ag$_2$O with chiral and racemic Hca in 1 : 2 and 1 : 4 molar ratios were characterized by elemental analysis, TG/DTA, FTIR, and solution (^{1}H, ^{13}C and ^{109}Ag) and solid-state (^{13}C) NMR spectroscopy.[52] Crystallography revealed that unique 2$_1$ helical polymer and zigzag structures were formed on self-assembly of the dimeric units in the crystals of $_\infty\{[Ag_2(S\text{-}ca)_2]\}$ (**4-1-4-C11**) and three others, ($_\infty\{[Ag_2(S\text{-}ca)_2(S\text{-}Hca)_2]\}$ (**4-1-4-C12**), $_\infty\{[Ag_2(R\text{-}ca)_2(R\text{-}Hca)_2]\}$ (**4-1-4-C13**), $_\infty\{[Ag_2(ca)_2(Hca)_2]\}$ (**4-1-4-C14**)). In the crystal of **4-1-4-C11,** two 2$_1$ helices and a loop were observed in the stair-like polymer structure, whereas a zigzag and a loop were seen in the crystals of the three $_\infty\{[Ag_2(ca)_2(Hca)_2]\}$.

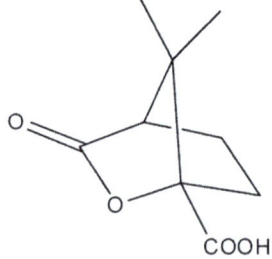

4-1-4-L4

Carbon NMR spectra in the solid state and in D$_2$O indicated that these polymeric structures were loosely bound and fast ligand-exchange reactions took place in aqueous solution. Complexes $_\infty\{[Ag_2(ca)_2]\}$ and $_\infty\{[Ag_2(ca)_2(Hca)_2]\}$ showed a wide spectrum of effective antimicrobial activity as anticipated for weak silver(I)–O bonding complexes. Similar antimicrobial activity of $_\infty${[Ag$_2$(ca)$_2$]} and $_\infty${[Ag$_2$(ca)$_2$(Hca)$_2$]} against selected micro-organisms suggested that ligand exchangeability played an important role, as well as the coordination geometry of the silver(I) ion.

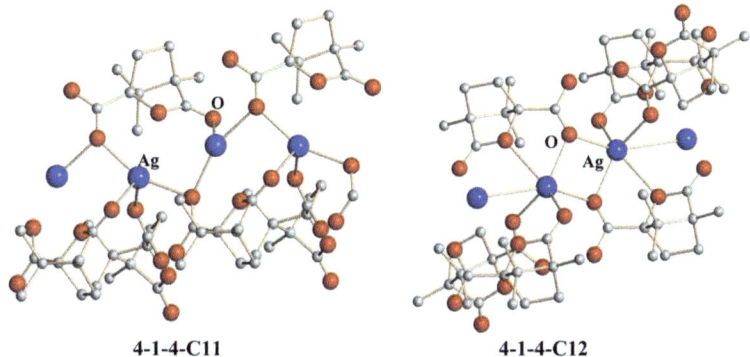

| 4-1-4-C11 | 4-1-4-C12 |

A tetranuclear silver(I) cluster, [Ag$_2$(Himdc)(PPh$_3$)$_2$]$_2$ (**4-1-4-C15**; H$_3$imdc = imidazole-4,5-dicarboxylic acid; **4-1-4-L5**) was prepared by a reaction of the Ag–O bonding precursor [Ag(R-pyrrld)]$_2$ (Hpyrrld = 2-pyrrolidone-5-carboxylic acid), with PPh$_3$ and H$_3$imdc. **4-1-4-C15** was a "bivalve"-like Ag$_4$ cluster, which has shown no antimicrobial activities.[53]

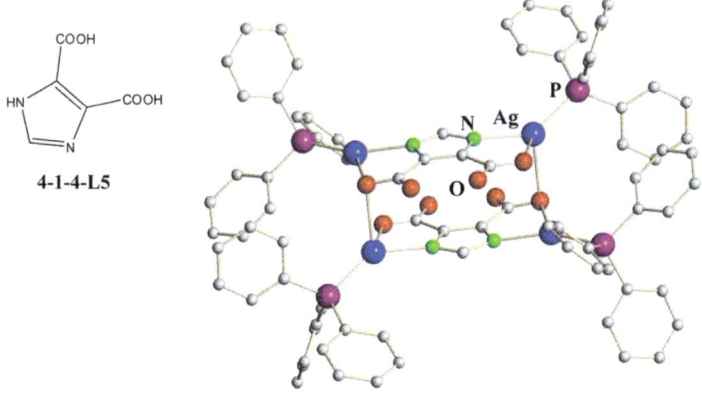

4-1-4-L5

4-1-4-C15

7.4.1.5 Polymeric Silver(I) Complexes with Amino Acids

$_\infty${{[Ag(L-Hhis)]·0.2EtOH}$_2$} (**4-1-5-C1**), $_\infty$[Ag(L-Hhis)] (**4-1-5-C2**), $_\infty$[Ag(D-Hhis)] (**4-1-5-C3**), $_\infty$[Ag$_2$(D-Hhis)(L-Hhis)] (**4-1-5-C4**), $_\infty${[Ag$_2$(D-asp)(L-asp)]·1.5H$_2$O} (**4-1-5-C5**), $_\infty${[Ag(gly)]$_2$·H$_2$O} (**4-1-5-C6**), $_\infty$[Ag(L-asn)] (**4-1-5-C7**), $_\infty$[Ag(D-asn)] (**4-1-5-C8**), $_\infty${{[Ag(L-Harg)]·0.5H$_2$O}$_2$} (**4-1-5-C9**), $_\infty${[Ag(L-acmet)]$_2$} (**4-1-5-C10**), $_\infty${[Ag$_2$(D-acmet)(L-acmet)]} (**4-1-5-C11**), $_\infty${[Ag$_2$(D-met)(L-met)]} (**4-1-5-C12**), $_\infty${[Ag(L-mecys)]$_2$} (**4-1-5-C13**), $_\infty${[Ag(L-Hcys)]} (**4-1-5-C14**), $_\infty${[Ag(Hacgly)]$_2$} (**4-1-5-C15**), $_\infty${[Ag(L-Hachis)]} (**4-1-5-C16**).

Nomiya and coworkers have described the synthesis of a water-soluble, silver(I)–N bonding complex $_\infty\{[Ag(L\text{-Hhis})]\cdot 0.2\text{EtOH}\}_2$ (**4-1-5-C1**) (**4-1-5-L1** = L-histidine; L-H$_2$his), showing a wide spectrum of effective antibacterial and antifungal activities. In aqueous solution **4-1-5-C1** was present as a dimer, whereas in the solid state it was present as a hydrogen-bonding polymer.[1] Crystallization of **4-1-5-C1** by slow evaporation and/or vapor diffusion gave water-insoluble crystals, $_\infty[Ag(L\text{-Hhis})]$ (**4-1-5-C2**), showing modest anti-microbial activities. Of particular note is the fact that in **4-1-5-C1** activities against many molds are observed. These activities mostly correspond to those of the silver(I)–N bonding complexes $_\infty[Ag(\text{imd})]$ (**4-1-2-C1**) and $[Ag(\text{Himd})_2]NO_3$ (**4-1-2-C2**). Thus, the two difficulties in the practical application, namely, the water insolubility of $_\infty[Ag(\text{imd})]$ and the light instability of $[Ag(\text{Himd})_2]NO_3$, were improved by the finding of **4-1-5-C1**. The modest activities of **4-1-5-C2** cannot be simply attributed to its low solubility in water, because both $_\infty[Ag(\text{imd})]$ and $[Ag(\text{Himd})_2]NO_3$ show effective activities despite quite different solubilities in water. The water-soluble powder **4-1-5-C1** is a polymer in the solid state, formed by intermolecular H-bonding interactions between dimeric $[Ag(\text{Hhis})]_2$ cores, while **4-1-5-C2** is a different polymer without a core unit. X-ray crystallography revealed that **4-1-5-C2** was a left-handed helical polymer consisting of a bent, 2-coordinate silver(I) atom bonding to the N_{amino} atom of one Hhis$^-$ ligand and the N_π atom of a different Hhis$^-$ ligand.

4-1-5-L1

4-1-5-C1

4-1-5-C2

Of particular note is the fact that $O_{carboxyl}$ atoms do not participate in the coordination. The FTIR and the solid-state ^{13}C and ^{15}N NMR spectra showed that the dimeric core of **4-1-5-C1** was formed through Ag–N bonds. The molecular ion of **4-1-5-C1** was detected by the positive-ion electrospray ionization (ESI) mass spectrometry. For **4-1-5-C1** and **4-1-5-C2**, character-ization by elemental analysis, TG/DTA, FTIR, and variable-temperature solid-state ^{13}C NMR and room-temperature ^{15}N NMR measurements was performed, and for **4-1-5-C1**, characterization by solution molecular weight measurements and solution (^{109}Ag, ^{1}H, and ^{13}C) NMR spectroscopies was also carried out.

Another helical polymer, $_\infty[Ag(D\text{-Hhis})]_2$ (**4-1-5-C3**), *i.e.* an enantiomer of **4-1-5-C2,** was also obtained. An achiral coordination polymer, $_\infty[Ag_2(D\text{-Hhis})(L\text{-Hhis})]$ (**4-1-5-C4**), was prepared by slow diffusion of two aqueous solutions of chiral complexes, $_\infty$[Ag(D-Hhis)]$_2$ (**4-1-5-C3**) and $_\infty$[Ag(L-Hhis)] (**4-1-5-C1**).[2] The crystal structure of **4-1-5-C4** consists of a linkage of meso-form dimer units through two kinds of silver(I)···silver(I) contacts. The crystals of **4-1-5-C4** exhibited different self-assembly from those of chiral helical polymers. The formation of **4-1-5-C4** from the two aqueous solutions indicated that ligand exchange around the silver(I) atoms took place in water.

4-1-5-C3   4-1-5-C4

The antimicrobial activities of **4-1-5-C4** showed a wider spectrum than those of hydrated Ag$^+$ (AgNO$_3$), which exhibited effective activities against two Gram-negative bacteria. The wider antimicrobial activities of these silver(I) histidinates support the fact that the nature of the atom that coordinates with the silver(I) center and its bonding properties (rather than the solubility, charge, chirality, or degree of polymerization of the complexes) and the ease of ligand replacement are the key factors that lead to those activities.

As model compounds for silver(I)–protein interactions, silver(I) complexes, (**4-1-5-C5** to **4-1-5-C8**) with three amino-acid ligands, *i.e.* aspartic acid (**4-1-5-L2**; H$_2$asp), glycine (**4-1-5-L3**; Hgly), L-asparagine (**4-1-5-L4**; L-Hasn) and D-asparagine (**4-1-5-L5**; D-Hasn), were prepared and characterized by elemental analysis, FTIR, TG/DTA, solution (^{1}H and ^{13}C) NMR and solid-state (^{13}C and ^{15}N) NMR spectroscopy.[54] The crystal structures of four silver(I) complexes, $_\infty$\{[Ag$_2$(D-Hasp)(L-Hasp)]·1.5H$_2$O\} (**4-1-5-C5**), $_\infty$\{[Ag(gly)]$_2$·H$_2$O\} (**4-1-5-C6**), $_\infty$[Ag(L-asn)] (**4-1-5-C7**) and $_\infty$[Ag(D-asn)] (**4-1-5-C8**), were determined. The bonding modes of the silver(I) center were different for each complex, and also different from those of the two reported silver(I) histidinates with only Ag-N bonds, *i.e.* water-soluble powder $_\infty$[\{[Ag(Hhis)]·0.2EtOH\}$_2$] (**4-1-5-C1**) and water-insoluble crystals $_\infty$[Ag(Hhis)] (**4-1-5-C2**). Silver(I) complexes formed by amino acids with N and O donor atoms and without S atom have been classified into four types (I–IV) based on the bonding modes of the silver(I) center.

4-1-5-L2   4-1-5-L3   4-1-5-L4   4-1-5-L5

4-1-5-C5   4-1-5-C7   4-1-5-C8

4-1-5-C6

4-1-5-C5 belongs to **I** that contains only Ag–O bonds, **4-1-5-C6** belongs to **II**, in which the two-coordinate O–Ag–O and N–Ag–N bonding units are alternately repeated, **4-1-5-C7** and **4-1-5-C8** belong to **III**, in which the two-coordinate N–Ag–O bonding units are repeated, and complexes that contain only Ag–N bonds, as found in the two silver(I) histidinates **4-1-5-C1** and **4-1-5-C2**, belong to **IV**.

These complexes (types **I–IV**) showed a wide spectrum of effective activities against Gram-negative (*E. coli, P. aeruginosa*) and Gram-positive (*Bacilus subtilis* and *S. aureus*) bacteria and yeasts (*C. albicans* and *Saccharomyces cerevisiae*), and even against more than 11 tested molds. In fact, the anti-microbial activities are in contrast to those of Ag–S bonding complexes and many phosphine–silver(I) complexes with $AgNP_2$, $AgNP_3$, $AgSP_2$ and $AgSP_3$ cores that have shown no antimicrobial activities. These results primarily in-dicate that, rather than solubility, charge and chirality of the complexes, the types of coordination donor atom attached to the silver(I) center, namely, the weaker Ag–O and Ag–N bonding properties, play a key role in the wide spectrum of antimicrobial activities observed.

A chiral coordination polymer of silver(I) *L*-argininate (**4-1-5-L6**; *L*-Harg) as nitrate, *i.e.* $_\infty\{\{[Ag(L\text{-Harg})](NO_3)\}_2 \cdot H_2O\}$ (**4-1-5-C9**), was prepared by vapor diffusion using a water–methanol mixed solution containing $AgNO_3$ and neutral *L*-arginine (*L*-Harg) as the inner solution and ether as the external solvent, followed by characterization by CHN analysis, FTIR, TG/DTA, sin-gle-crystal X-ray analysis and (^{1}H and ^{13}C) NMR spectroscopy.[55] In the crystal, each silver(I) ion was coordinated by N_{amino} of one *L*-Harg and $O_{carboxy}$ of the other. Linear polymer chains consisting of repeating N_{amino}–Ag–$O_{carboxy}$ bonding units were connected by Ag$\cdots$Ag (3.0440(5) Å) interaction and intermolecular hydrogen bonding to form a grid-like sheet. The sheets were connected by hydrogen bonds and electrostatic interaction between guanidium groups and NO_3^-.

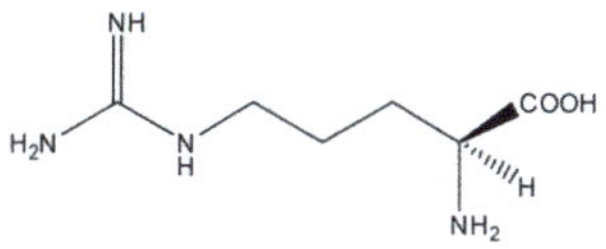

4-1-5-L6

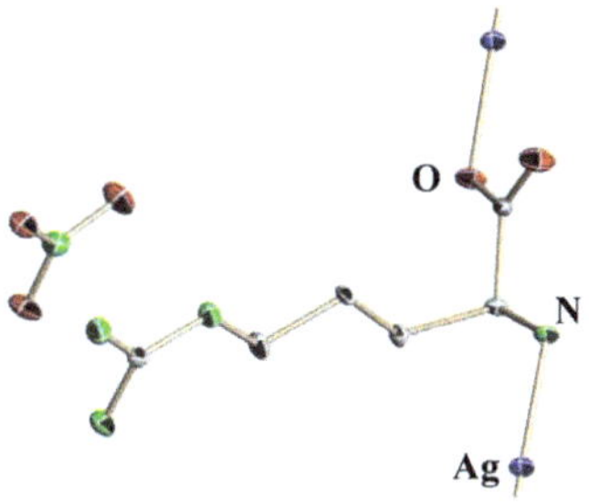

4-1-5-C9

Complex **4-1-5-C9** inhibited moderately the growth of four bacteria and one yeast, the spectrum of which is different from that of AgNO$_3$. The antimicrobial activities of **4-1-5-C9** were less effective in comparison with those of $_\infty$[Ag(*L*-asn)] (**4-1-5-C7**), although both silver(I) complexes belong to the same type (**III**) having an O–Ag–N polymer structure. This may be caused by the multiple-bonded sheet structure of **4-1-5-C9**.

The reaction of *L*- and *DL*-N-acetylmethionine (**4-1-5-L7**: Hacmet) and Ag$_2$O in water at ambient temperature afforded the remarkably light-stable silver complexes $_\infty${[Ag(*L*-acmet)]} (**4-1-5-C10**) and $_\infty${[Ag$_2$(*D*-acmet)(*L*-acmet)]} (**4-1-5-C11**), respectively.[56] The color of the solids and aqueous solutions of **4-1-5-C10** and **4-1-5-C11** did not change for more than one month under air without any shields. The light stability of these two silver(I) complexes is much higher than that of silver(I) methioninate $_\infty${[Ag$_2$(*D*-met)(*L*-met)]} (**4-1-5-C12**) (Hmet = methionine; **4-1-5-L8**), silver(I) S-methyl-*L*-cysteinate $_\infty${[Ag(*L*-mecys)]} (**4-1-5-C13**) (*L*-Hmecys = S-methyl-*L*-cysteine; **4-1-5-L9**), and silver(I) *L*-cysteinate $_\infty${[Ag(*L*-Hcys)]} (**4-1-5-C14**) (*L*-H$_2$cys = *L*-cysteine; **4-1-5-L10**). X-ray crystallography of **4-1-5-C10** obtained by vapor diffusion revealed that ladder-like coordination polymers with two O- and two S donor atoms were formed. The acetyl group of acmet$^-$ prevents chelate formation of the ligand to the metal center, which is frequently observed in amino-acid–metal complexes, but allows formation of hydrogen bonds between the ligands in the crystals of **4-1-5-C10**. **4-1-5-C10** to **4-1-5-C13** showed a wide spectrum of effective antimicrobial activities

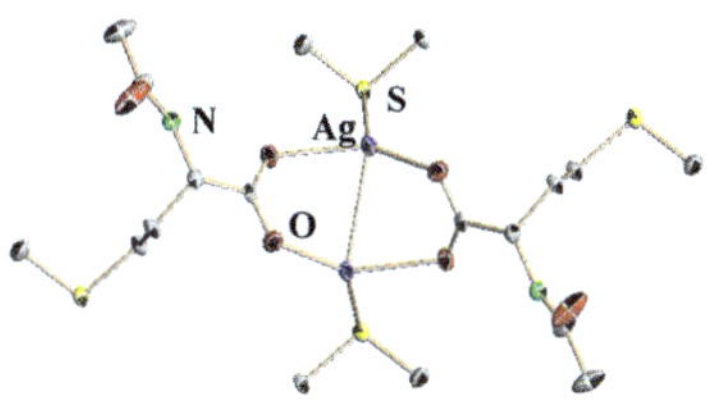

4-1-5-C10

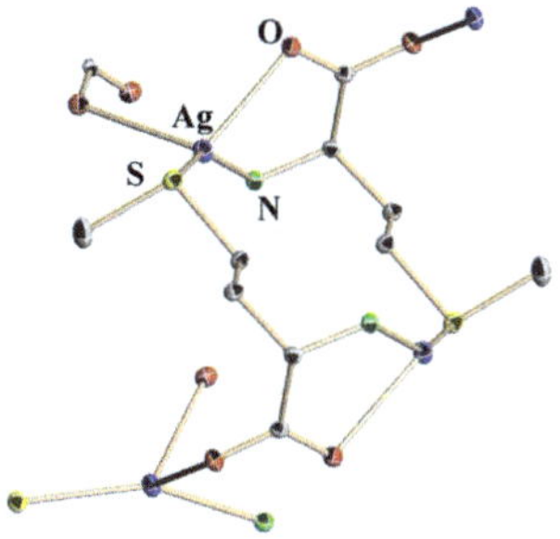

4-1-5-C12

against Gram-negative bacteria and yeasts, which was comparable to that of water-soluble Ag–O bonding complexes. The antimicrobial activity of **4-1-5-C13** was slightly less effective than those of complexes **4-1-5-C10** to **4-1-5-C12** with a methionine backbone. Water-insoluble **4-1-5-C14** showed no activities against selected bacteria, yeasts and molds.

4-1-5-C13

| **4-1-5-L7** | **4-1-5-L8** | **4-1-5-L9** | **4-1-5-L10** |

The reaction in water of silver oxide with *N*-acetylglycine (**4-1-5-L11**; Hacgly) possessing a partial structure, *i.e.* the O=C–N–C–COOH moiety, afforded light-stable and water-soluble dinuclear silver(I) complex $_\infty${[Ag(acgly)]$_2$} (**4-1-5-C15**).[3] X-ray crystallography revealed that **4-1-5-C15** in the solid state formed a ladder polymeric structure based on a bis(carboxylato-O,O′)-bridged centrosymmetric Ag$_2$O$_4$ core, which was different from the two previously reported structures of silver(I) glycinate. **4-1-5-C15,** which only comprises labile Ag–O bonding, showed a wide range of antimicrobial activities against selected bacteria, yeasts and molds. These results support Nomiya's hypothesis that weak bonding, which is readily replaced by substrates, including biomolecules, plays the key role in the antimicrobial activities of silver(I) complexes. **4-1-5-C15** can also work as useful silver(I) precursor for novel silver(I) cluster synthesis.

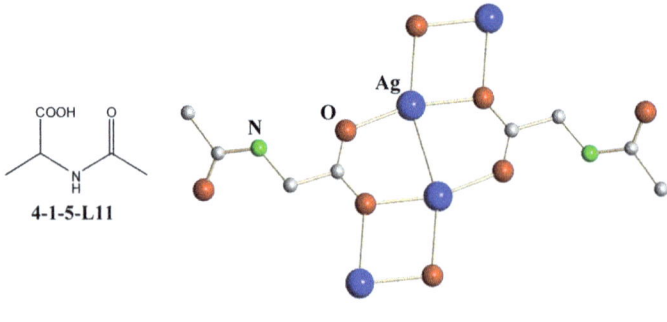

4-1-5-L11

4-1-5-C15

The 1 : 2 molar ratio reaction of Ag$_2$O with *N*-acetyl-*L*-histidine (**4-1-5-L12**; *L*-H$_2$achis) possessing the partial structure O=C–N–C–COOH, as well as imidazole units, afforded the light-stable and water-insoluble silver(I) complex, $_\infty$[Ag(*L*-Hachis)] (**4-1-5-C16**).[57] This complex showed modest antimicrobial activity against selected bacteria and yeasts in a water–suspension system. X-ray crystallography revealed that the complex in the solid state forms a helical polymeric structure with an $AgNO_3$ core based on chelating Hachis⁻ (one carboxylate O atom and one imidazole N atom) and two O atoms of the other two Hachis⁻ ligands that can be described as $_∞${[Ag(*L*-Hachis)]}; this differs from the structure of the related silver(I) complex $_∞${[Ag(*L*-Hhis)$_2$]} (**4-1-5-C2**) based on an AgN_2 core. Polymer chains of **4-1-5-C16** form three-dimensional intermolecular hydrogen-bonding networks in the crystal structure.

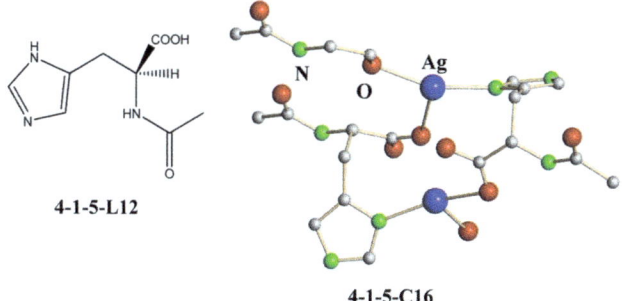

4-1-5-L12

4-1-5-C16

7.4.1.6 Monomeric and Dimeric N-Heterocyclic Carbene (NHC) Silver(I) Complexes

N-Heterocyclic carbene (NHC) studies first began with the metallation of imidazol-2-ylidenes by Wanzlick and Schonherr and Öfele in 1968.[58,59] Stemming from this research, Arduengo and coworkers discovered and isolated the first stable free carbene in 1991 and NHCs have proven since then to be versatile ligands in both transition-metal and main-group coordination chemistry.[60] The first silver carbene complex (**4-1-6-C1**) was synthesized and its structure was revealed by Arduengo and coworkers in 1993 for using in catalysis.[61] Arduengo's method required harsh conditions and often led to decomposition of the metal complex. To remedy this problem, Wang and Lin developed a synthetic method that generated the carbene *in situ* by using a silver salt as a base.[62]

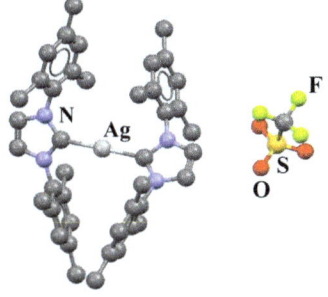

4-1-6-C1

After the Wang/Lin report, various carbene ligands were prepared and their silver complexes were extensively studied by Youngs and coworkers,[63–66,71–74]

Özdemir and coworkers[67,70] and Tacke and coworkers.[68,69] Youngs and coworkers prepared and determined the molecular structure of the silver(I) – imidazole cyclophane *gem*-diol complex, **4-1-6-C2**, which is half as effective as an antibiotic than 0.5% $AgNO_3$, with about the same amount of silver. Mats made of nanofibers of tecophilic® polymer formed by electrospinning of solution containing **4-1-6-C2** released nanosilver particles, which in turn sustained the antimicrobial activity of the mats over a long period of time. The fiber mat was found to kill *S. aureus* at the same rate as 0.5% $AgNO_3$, with zero colonies on an agar plate, and about 6 times faster than silver sulfadiazine cream. The silver mats were found to be effective against *E. coli*, *P. aeruginosa*, *S. aureus*, *C. albicans*, *Aspergillus niger*, and *S. cerevisiae*. TEM and scanning electron microscopy were used to characterize the fiber mats. The acute toxicity of the ligand (imidazolium cyclophane *gem*-diol dichloride) was assessed by intravenous administration to rats, with an LD_{50} of 100 mg/kg of rat.[73]

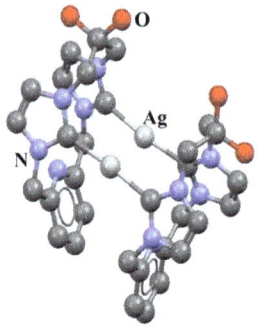

4-1-6-C2

Studies by Youngs and coworkers indicate that silver N-heterocyclic carbene complexes have been shown to have great potential as antimicrobial agents, affecting a wide spectrum of both Gram-positive and Gram-negative bacteria.[63–66,71,73]

Among a series of N-heterocyclic carbene silver complexes, **4-1-6-C2** to **4-1-6-C6** are currently being explored due to their high efficacy against antibiotic-resistant bacteria, biosafety level 3 bacteria (*Burkholderia pseudomallei*, *Burkholderia mallei*, *Bacillus anthracis*, methicillin-resistant, *S. aureus* and *Yersinia pestis*), and various types of cancer as well.[74] The molecular structure of **4-1-6-C3** is shown below.

C3   **4-1-6-C3** to **4-1-6-C6**

R=Me   **C4**
=hexyl   **C5**
=CH_2Nap **C6**

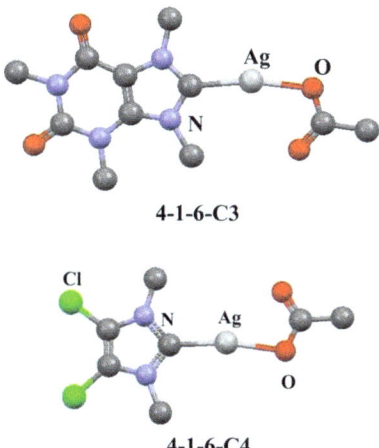

4-1-6-C3

4-1-6-C4

Ag–NHC complexes **4-1-6-C7** and **4-1-6-C8** also showed clinically relevant activity against a silver-resistant strain of *E. coli* based on MIC testing.

R=CH$_2$CH$_2$OH **C7**
=CH$_2$Naph **C8**

Youngs and coworkers think that the antimicrobial abilities of silver–NHC complexes are due to the strong binding ability of NHCs to silver that can result in more stable complexes that can slowly release silver ions, thus retaining the antimicrobial effect over a longer period of time.[63]

7.4.1.7 Monomeric Silver(I)–Coumarin Complexes

[Ag(7-OHCca)] (**4-1-7-C1**), [Ag(6-OHCca)], (**4-1-7-C2**) [Ag(phen)$_2$(hnc)] (**4-1-7-C3**).

The antifungal activity and mode of action of a range of silver(I)-coumarin complexes has been examined by Creaven *et. al.*[75] The most potent silver(I)-coumarin complexes **4-1-7-C1** to **4-1-7-C3** with three ligands **4-1-7-L1** to **4-1-7-L3** (**4-1-7-L1** = 7-hydroxycoumarin-3-carboxylic acid, **4-1-7-L2** = 6-hydroxycoumarin-3-carboxylic acid, **4-1-7-L3** = 4-oxy-3-nitrocoumarin (hnc)), respectively, namely [Ag(7-OHCca)] (**4-1-7-C1**), [Ag(6-OHCca)] (**4-1-7-C2**) and [Ag(phen)$_2$(hnc)] (**4-1-7-C3**) were synthesized and the X-ray structure of **4-1-7-C3** was determined. The MIC$_{80}$ values against *C. albicans* were reported as 69.1 μM for **4-1-7-C1**, 34.1 μM for **4-1-7-C2** and 4.6 μM for **4-1-7-C3**.

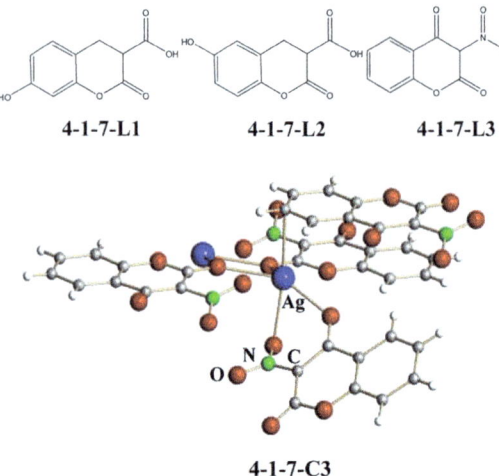

4-1-7-L1   **4-1-7-L2**   **4-1-7-L3**

4-1-7-C3

Creaven and coworkers have stated that these compounds also reduced respiration, lowered the ergosterol content of cells and increased the trans-membrane leakage of amino acids.[76] A number of the complexes disrupted cytochrome synthesis in the cell and induced the appearance of morphologic features consistent with cell death by apoptosis. These compounds appear to act by disrupting the synthesis of cytochromes that directly affects the cell's ability to respire. A reduction in respiration leads to a depletion in ergosterol biosynthesis and a consequent disruption of the integrity of the cell membrane. Disruption of cytochrome biosynthesis may induce the onset of apoptosis, which is triggered by alteration in the location of cytochrome c. Silver(I)–coumarin complexes demonstrate good antifungal activity and manifest a mode of action distinct from that of conventional azole and polyene drugs, thus raising the possibility of their use when resistance to conventional drug has emerged or in combination with such drugs. Creaven and coworkers have also stated the following: while polyene antifungals bind ergosterol in the fungal cell membrane and azoles inhibit ergosterol biosynthesis, the compounds described here demonstrated to exhibit strong antifungal activities that are mediated through the disruption of the respiratory function and the induction of apoptosis.[75]

7.4.1.8 *Hexameric and Tetrameric Silver(I)–Thiosemicarbazone Complexes*

Thiosemicarbazones have attracted considerable interest because of their chemistry and potentially beneficial biological activities, such as antitumor, antibacterial, antiviral and antimalarial activities. Thiosemicarbazones are versatile ligands because they can coordinate to the metal either as a neutral ligand or as a deprotonated ligand through the S, N, N atoms, and also structural isomers (*E*-, *E'*- and *Z*-forms) are often seen.[77] The coordination chemistry of thiosemicarbazones is also of interest because of their biological activities such as therapeutic or diagnostic PET (positron emission tomography) or SPECT (single-photon emission computed tomography) imaging agents.[78]

Although silver(I) complexes with thiosemicarbazones were reported more than 20 years ago,[79] the first structural report showing a hexameric silver(I) cluster (**4-1-8-C1**) prepared by AgCF$_3$SO$_3$ and 2-salicylaldehyde thiosemicarbazone (**4-1-8-L1**) was published only in 2004, probably due to their light-unstable properties.[80]

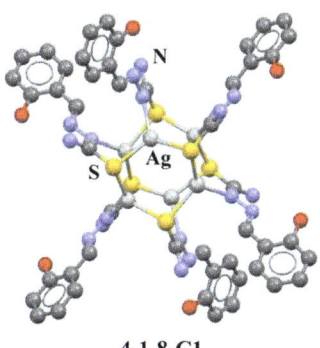

4-1-8-C1

A neutral silver(I)-thiosemicarbazone complex ([Ag(mtsc)$_4$] (**4-1-8-C2**)) was obtained from reactions of a tridentate ^4N-morpholyl 2-acetylpyridine thiosemicarbazone ligand (N'-[1-(2-pyridyl)ethylidene]morpholine- 4-carbothiohydrazide, **4-1-8-L2** or Hmtsc) and silver(I) sources containing Ag–O bonds (Ag$_2$O, Ag(OAc), silver(I) 2-pyrrolidone-5-carboxylate (**4-1-4-C1** to **4-1-4-C3**), silver(I) 5-oxo-2- tetrahydrofurancarboxylate (**4-1-4-C8** to **4-1-4-C10**), and silver(I) complexes with camphanic acid (**4-1-4-C11** to **4-1-4-C14**)). Complex **4-1-8-C2** was extremely light stable and was well characterized by elemental analysis, TG/DTA, FTIR and ESI-MS and solution (^{1}H, ^{13}C and ^{31}P) NMR spectroscopy. Single-crystal X-ray analysis revealed that **4-1-8-C2** is a tetramer.

4-1-8-L2

4-1-8-C2

Table 7.1 Antimicrobial activities of **4-1-8-C2**, free ligand **4-1-8-L2**, Ag$_2$O and CHCl$_3$ evaluated by minimum inhibitory concentration (MIC; µg mL^{-1}).

Compound solvent	**4-1-8-C2·2CHCl$_3$** H$_2$O	**4-1-8-C2·2CHCl$_3$** CHCl$_3$	**4-1-8-L2** H$_2$O	Ag$_2$O H$_2$O	Ag$_2$O CHCl$_3$
E. coli	>1000	125	250		>1000
B. subtilis	>1000	62.5	125	500	>1000
S. aureus	>1000	1000	250	1000	>1000
P. aeruginosa	>1000	500	125		>1000
C. albicans	>1000	31.3	31.3		>1000
S. cerevisiae	>1000	31.3	31.3		>1000
A. brasiliensis	>1000	>1000	125		>1000
P. citrinum	>1000	>1000	125		>1000

The antimicrobial activities of the water-insoluble complex were evaluated by minimum inhibitory concentration (MIC; µg mL^{-1}) in a water–suspension system and in CHCl$_3$ solution (Table 7.1).

The free ligand (**4-1-8-L1**) showed a wide spectrum of moderate to effective antimicrobial activities against the test organisms (*E. coli*, *Bacillus subtilis*, *S. aureus*, *P. aeruginosa*, *C. albicans*, *S. cerevisiae*, *Aspergillus brasiliensis* and *Penicillium citrinum*), particularly, against yeasts (*C. albicans* and *S. cerevisiae*), and water-soluble starting materials, *i.e.* Ag–O bonded complexes, showed superior antimicrobial activities against all the selected micro-organisms. However, the MIC values of **4-1-8-C2** for selected bacteria, yeasts and molds in water suspension were larger than 1000, which indicated no antimicrobial activity. When **4-1-8-C2** was dissolved in CHCl$_3$ and added to the test culture, it showed modest activities for bacteria and was effective against yeasts, but showed no activity against molds. A blank test adding CHCl$_3$ alone to the culture cells did not have any effect on the selected micro-organisms. The different antimicrobial spectral patterns between the ligand and the CHCl$_3$ solution of **4-1-8-C2** indicates that some of the $_\infty$[Ag(mtsc)] species yielded by the equilibrium in solution interacted with the biomolecules of the micro-organisms. **4-1-8-C2** did not inhibit the selected micro-organisms when contacted in the solid state. The lack of activity in the water–suspension system could be attributed to the extraordinary stability or the low solubility of this complex.[81]

The reaction of **4-1-8-C2** with PPh$_3$ produced two dimeric isomers, namely, [Ag(µ(S)-mtsc)(PPh$_3$)]$_2$ (**4-1-8-C3**) and [(PPh$_3$)$_2$Ag(µ(S)-mtsc)$_2$Ag] (**4-1-8-C4**), which had no antimicrobial activities against selected micro-organisms. Solution NMR spectroscopy and solution molecular weight measurement indicated that fast equilibrium occurred for the dimer, and it was dissociated in the solution.

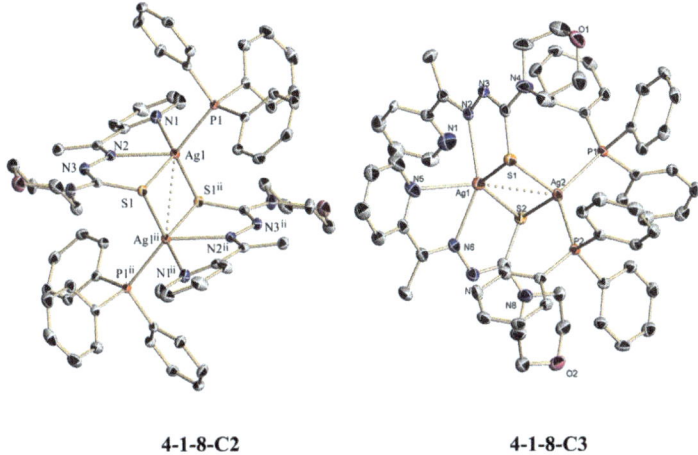

4-1-8-C2   **4-1-8-C3**

7.4.1.9   Dimeric Silver(I)–Hinokitiol Complex

2-Hydroxy-4-isopropylcyclohepta-2,4,6-trienone (Hinokitiol (**4-1-9-L1** or Hhino), also called β–thujaplicin) was found in an extract of wood (Thujapalicata and Taiwan Hinoki).[82,83] It is a 4-isopropyl-substituted compound of a 7-membered ring organic compound, tropolone and exhibits a wide range of biological activities including antimicrobial activities against Gram-positive and Gram-negative bacteria, yeasts and molds.[84,85] The monoanion of **4-1-9-L1** forms various metal–chelate complexes $[M(hino)_x]_n$ through two unequivalent oxygen donor atoms and their antimicrobial activities, in particular those of silver(I), iron(III) and copper(II) complexes, are reported to be significantly changed from that of **4-1-9-L1**.[86] Nomiya and coworkers prepared silver complex (**4-1-9-C1**) and characterized it by elemental analysis, thermogravimetric and differential thermal analysis (TG/DTA), FTIR and solution (^{1}H and ^{13}C) NMR spectroscopy.[87] The crystal structure of **4-1-9-C1** was determined by Rietveld analysis based on X-ray powder diffraction (XPD) data, which revealed **4-1-9-C1** as a dimeric, silver(I)–oxygen bonding complex ([Ag(hino)]$_2$) through two unequivalent oxygen donor atoms of an anion of **4-1-9-L1**, forming a *trans*-Ag$_2$O$_4$ core geometry.

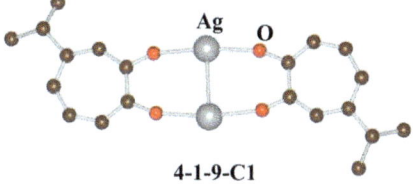

4-1-9-C1

The antimicrobial activities of **4-1-9-C1** were evaluated by minimum inhibitory concentration (MIC; μg mL^{-1}). Nomiya and coworkers found that the spectral pattern of the antimicrobial activity of **4-1-9-C1** was very similar to those reported for silver(I) complexes with an Ag$_2$O$_4$ core, the ligand itself of which showed no activity, **4-1-4-C1** to **–C3, C8–C14,** as described in Section

7.4.1.4.[87] They proposed that the antimicrobial activities of **4-1-9-C1** were due to the action of the weak Ag–O bonding property based on accumulated data of molecular structures of silver(I) complexes and antimicrobial activities. They suggested that the antimicrobial activities of the silver(I)–oxygen bonding complexes are due to a direct interaction or complexation of the silver(I) ion with biological ligands such as protein, enzyme and membrane, and that the coordinating ligands of the silver(I) complexes play the role of carrier of the silver(I) ion to the biological system.

7.4.2 Other Metal Complexes

7.4.2.1 Monomeric Gold(I)–Phosphine Complexes

[Au(dppp)$_2$]Cl(**4-2-1-C1**), [Au(dppey)$_2$]Cl (**4-2-1-C2**), [AuCl(TPPTS)] (**4-2-1-C3**), Na$_3$[Au(tsa)$_2$] · 5H$_2$O (**4-2-1-C4**), {K$_3$[Au(mba)$_2$]}$_2$ (**4-2-1-C5**), [Au(pz)(PPh$_3$)] (**4-2-1-C6**), [Au(im)(PPh$_3$)] (**4-2-1-C7**), [Au(1,2,3-triz)(PPh$_3$)] (**4-2-1-C8**), [Au(1,2,4-triz)(PPh$_3$)]$_2$ (**4-2-1-C9**), [Au(tetz)(PPh$_3$)] (**4-2-1-C10**), [Au(6-mna)(PPh$_3$)] (**4-2-1-C11**), [Au(*D*-Hpen)(PPh$_3$)] (**4-2-1-C12**), [Au(*DL*-Hpen)(PPh$_3$)] (**4-2-1-C13**), [Au(4-mba)(PPh$_3$)] (**4-2-1-C14**), [Au(2-Hmba)(PPh$_3$)] (**4-2-1-C15**), [Au(3-Hmba)(PPh$_3$)] (**4-2-1-C16**), [Au(2-Hmpa)(PPh$_3$)] (**4-2-1-C17**), [Au(*R,S*-Hpyrrld)(PPh$_3$)] · CHCl$_3$ (**4-2-1-C18**), [Au(*R,S*-othf)(PPh$_3$)] (**4-2-1-C19**).

The antibacterial and antifungal activities of diphosphine Au(I) complexes were assessed.[88] Gold(I) complexes with 1,3-bis(diphenylphosphino)propane (**4-2-1-L1**; dppp), [Au(dppp)$_2$]Cl(**4-2-1-C1**), and *cis*-bis(diphenylphosphino)ethylene (**4-2-1-L2**; dppey), [Au(dppey)$_2$]Cl (**4-2-1-C2**), showed modest activity against three of the 12 bacterial strains tested, but all complexes exhibited antifungal activity against three strains of *C. albicans* in a "defined" medium, [Ag(depe)$_2$]NO$_3$ (depe = 1,2-bis(diethylphosphino)ethane) and [Au(dppp)$_2$]Cl (**4-2-1-C1**) having comparable activity to fungizone. The antifungal activity of the complexes is reduced in Sabouraud's broth medium, and lost altogether for silver(I) complexes.

4-2-1-L1   4-2-1-L2

Contel and coworkers have described the preparation of water-soluble compounds of the type [AuCl(PR$_3$)] with alkylbis(*m*-sulfonated-phenyl) (*m*C$_6$H$_4$SO$_3$Na)$_2$ (**4-2-1-L3**) and dialkyl-(*m*-sulfonated-phenyl) (*m*C$_6$H$_4$SO$_3$Na) (R = *n*Bu, Cp) phosphanes (**4-2-1-L4**).[89] Dialkylphosphane compounds generate water-soluble nanoparticles with a radius of 10–15 nm when dissolved in water. The antimicrobial activity of these complexes against Gram-positive and Gram-negative bacteria and yeast has been evaluated. These compounds display moderate to high antibacterial activity. The more lipophilic compounds are also potent against fungi. Their cytotoxic properties have been analyzed *in vitro* by utilizing human Jurkat T-cell acute lymphoblastic leukemia cells.

Compounds with dialkyl-(*m*-sulfonated-phenyl) dialkyl-(*m*-sulfonated-phenyl) ($mC_6H_4SO_3Na$) phosphanes displayed moderate to high cytotoxicity in this cell line. The cell death mechanism involves mainly early apoptosis. The biological activity of a compound with commercial *m*-trisulfonated-triphenylphosphane (TPPTS; **4-2-1-L5**), [AuCl(TPPTS)] (**4-2-1-C3**), has also been evaluated to compare the effects of the higher basicity and lipophilicity of the alkyl- and dialkyl-(*m*-sulfonated-phenyl)phosphanes on these compounds.

Nomiya and coworkers have prepared and isolated the sodium salt of a water-soluble, anionic, and monomeric 1:2 complex of gold(I) with a dianion of thiosalicylic acid or 2-mercaptobenzoic acid (**4-1-3-L1**; H_2tsa or H_2mba), $Na_3[Au(tsa)_2] \cdot 5H_2O$ (**4-2-1-C4**), as colorless needle crystals through a stoichiometric reaction of $NaAuCl_4 : H_2$tsa $: NaOH = 1{:}4{:}8$ molar ratio in aqueous/EtOH solution. In this reaction, the tsa^{2-} ligand played the role of a reducing agent for the starting gold(III) ion and also of donor ligands coordinating to the reduced gold(I).[90] This compound was characterized by complete elemental analyses, TG/DTA, FTIR, 2-dimensional-NMR (^{1}H-^{1}H COSY, ^{1}H-^{13}C HMBC, and ^{1}H-^{13}C HMQC) spectroscopy, and molecular mass measurement based on the cryoscopic method. This complex was a monomeric species of gold(I) of $Na_3[Au(tsa)_2] \cdot 5H_2O$ in the solid state, but not a polymeric species even in aqueous solution. A full assignment of seven C and four proton resonances in the coordinated tsa^{2-} ligand was achieved by the 2-dimensional ^{1}H-^{13}C HMBC NMR technique. **4-2-1-C4** showed modest antimicrobial activities against two Gram-positive bacteria and two Gram-negative bacteria.

4-2-1-L3   **4-2-1-L4**   **4-2-1-L5**

Medicinally active, water-soluble and anionic gold(I) complex {$K_3[Au(mba)_2]$}$_2$ (**4-2-1-C5**) formed by 2-mercaptobenzoic acid (**4-1-3-L1**; H_2mba) was isolated and its crystal structure has been determined.[91] The molecular structure in the solid state comprised a dimeric $(AuS_2)_2$ core through a short gold(I)–gold(I) contact (3.1555(7) Å), *i.e.* aurophilic interaction. **4-2-1-C5** has also shown selective and modest antibacterial activities against two Gram-positive bacteria and two Gram-negative bacteria.

4-2-1-C5

Two isomeric gold(I)–triphenylphosphine complexes with nitrogen-containing heterocycles, [Au(L)(PPh₃)] (HL = pyrazole ; Hpz (**4-2-1-L6**; **4-2-1-C6**), imidazole; Himd (**4-1-2-L1**; **4-2-1-C7**)), were isolated as colorless cubic crystals for **4-2-1-C6** and colorless plate crystals for **4-2-1-C7**, respectively.[92] The crystal structures of **4-2-1-C6** and **4-2-1-C7** were determined by single-crystal X-ray diffraction. These complexes were also fully characterized by complete elemental analyses, TG/DTA and FTIR in the solid state and by solution NMR (^{31}P, ^{1}H and ^{13}C) spectroscopy and molecular weight measurements in acetone solution. These complexes consisted of a monomeric 2-coordinate AuNP core both in the solid state and in solution. The molecular structures of **4-2-1-C6** and **4-2-1-C7** were compared with those of related gold(I) complexes, [Au(1,2,3-triz)(PPh₃)] (**4-2-1-C8**, 1,2,3-Htriz = **4-1-2-L2**), [Au(1,2,4-triz)(PPh₃)]₂ (**4-2-1-C9**, 1,2,4-Htriz = **4-1-2-L3**) as a dimer through a gold(I)–gold(I) bond in the solid state, and [Au(tetz)(PPh₃)] (**4-2-1-C10**, Htetz = **4-1-2-L4**). Selective and effective antimicrobial activities against two Gram-positive bacteria (*B. subtilis, S. aureus*) and modest activities against one yeast (*C. albicans*) found in these gold(I) complexes **4-1-2-C6** to **4-2-1-C10** are noteworthy, in contrast to poor activities observed in the corresponding silver(I) complexes.

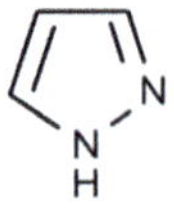

4-2-1-L6

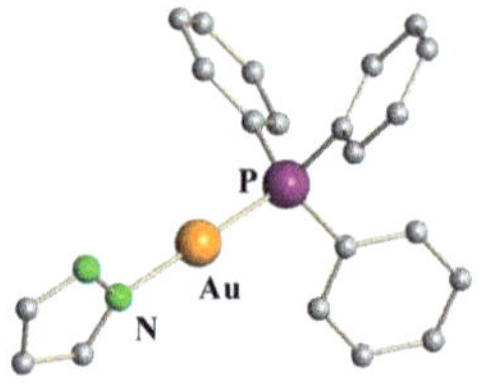

4-2-1-C6

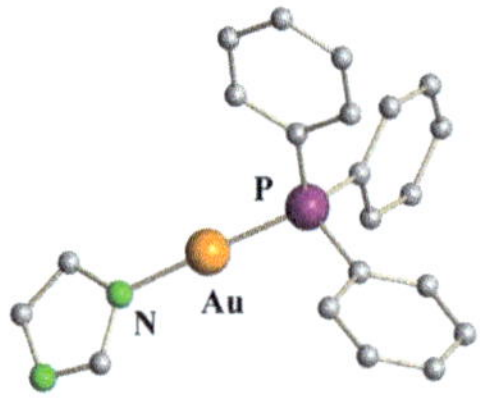

4-2-1-C7

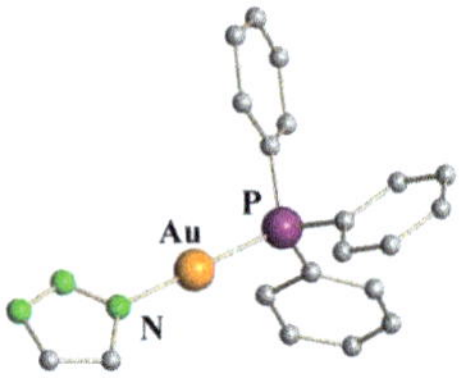

4-2-1-C8

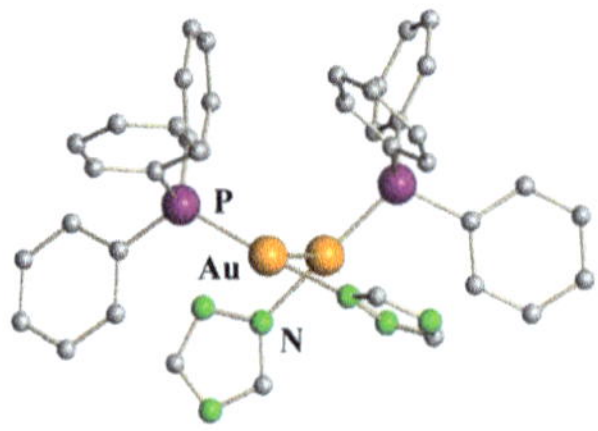

4-2-1-C9

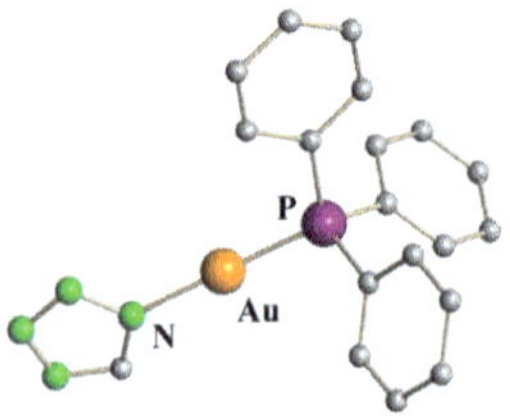

4-2-1-C10

Selective and effective antimicrobial activities against Gram-positive bacteria (*B. subtilis* and/or *S. aureus*) were found in 2-coordinate gold(I)–PPh₃ complexes with AuSP and AuNP cores, *i.e.* [Au(6-mna)(PPh₃)] (**4-2-1-C11**,

6-H$_2$mna = 6-mercapto-nicotinic acid; **4-2-1-L7**), [Au(*D*-Hpen)(PPh$_3$)] (**4-2-1-C12**, *D*-H$_2$pen = penicillamine; **4-2-1-L8**), [Au(*DL*-Hpen)(PPh$_3$)] (**4-2-1-C13**), [Au(4-mba)(PPh$_3$)] (**4-2-1-C14**, 4-H$_2$mba = 4-mercaptobenzoic acid; **4-2-1-L9**), [Au(pz)(PPh$_3$)] (**4-2-1-C6**, Hpz = pyrazole; **4-2-1-L6**), [Au(im)(PPh$_3$)] (**4-2-1-C7**, Him = imidazole), [Au(1,2,3-triz)(PPh$_3$)] (**4-2-1-C8**, 1,2,3-Htriz = 1,2,3-triazole; **4-1-2-L2**), [Au(1,2,4-triz)(PPh$_3$)] (**4-2-1-C9**, 1,2,4-Htriz = 1,2,4-triazole; **4-1-2-L3**), whereas no activity was observed in 2-coordinate AuSP core complexes [Au(2-Hmba)(PPh$_3$)] (**4-2-1-C15**) and [Au(3-Hmba)(PPh$_3$)] (**4-2-1-C16**).[93] The two AuSP core complexes, [Au(2-Hmpa)(PPh$_3$)] [H$_2$mpa = mercaptopropionic acid] (**4-2-1-C17**) and [Au(6-Hmna)(PPh$_3$)] (**4-2-1-C11**), were prepared and characterized by elemental analysis, FTIR, TG/DTA, and (^{31}P, ^{1}H and ^{13}C) NMR spectroscopy. The crystal structures of **4-2-1-C17** and **4-2-1-C11** were determined as a supramolecular arrangement of the 2-coordinate AuSP core. Both **4-2-1-C17** and **4-2-1-C11** significantly showed antibacterial activities. As a model reaction of phosphinegold (I) complexes with the cysteine residue in the biological ligands, Nomiya and coworkers examined if the ligand exchange reactions of the aromatic anions L$_1^-$ in [Au(L$_1$)(PPh$_3$)] (HL$_1$ = 6-H$_2$mna, 2-H$_2$mna, 2-H$_2$mba, Hpz, Him, 1,2,3-Htriz, 1,2,4-Htriz) with aliphatic thiols HL$_2$ (HL$_2$ = 2-H$_2$mpa, *D*-H$_2$pen) occurred under mild conditions and, also, if the "reverse" reactions, namely, the ligand exchange reactions of the thiolate anions in [Au(2-Hmpa)(PPh$_3$)] **4-2-1-C17**, [Au(*D*-Hpen)(PPh$_3$)] **4-2-1-C12** and [Au(2-Hmba)(PPh$_3$)] **4-2-1-C15** with the free ligands HL$_1$ took place under similar conditions. The relationships of the ligand exchangeability among the 2-coordinate gold(I) complexes (**4-2-1-C17**, **4-2-1-C15**, **4-2-1-C12**) were revealed. **4-2-1-C15** was substitution inert, whereas complexes **4-2-1-C17** and **4-2-1-C12** were substitution labile. The ligand exchangeability of gold(I)–S and gold(I)–N bonds in the 2-coordinate phosphinegold(I) complexes with AuSP and AuNP cores for forming new AuSP cores, with retention of the gold(I)–P bond, was closely related to the observed activities against Gram-positive bacteria, and the ease of the ligand-exchange reaction was strongly related to the intensity of the activities.

4-2-1-L7

4-2-1-L8

4-2-1-C11

4-2-1-C17

Two gold(I) complexes with hard (O donor) and soft (P donor) Lewis bases, [Au(R,S-pyrrld)(PPh$_3$)] · CHCl$_3$ (**4-2-1-C18**) (Hpyrrld = 2-pyrrolidone-5-carboxylic acid; **4-1-4-L1**) and [Au(R,S-othf)(PPh$_3$)] (**4-2-1-C19**) (Hothf = 5-oxo-2-tetrahydro-furancarboxylic acid), were prepared by an AgCl elimination reaction in CHCl$_3$ between [AuCl(PPh$_3$)] and silver(I)–O bonding precursors, such as [Ag$_2$(R-pyrrld)(S-pyrrld)] and [Ag$_2$(R-othf)(S-othf)].[5] The molecular structure of **4-2-1-C18** was determined as a discrete monomer of the 2-coordinate AuOP core. In the preparation of **4-2-1-C18** and **4-2-1-C19**, the use of the Ag–O bonding precursors is crucial. Both complexes **4-2-1-C18** and **4-2-1-C19** showed selective antimicrobial activities against Gram-positive bacteria and yeasts.

4-2-1-C18

7.4.2.2   Monomeric Zinc–Pyrithione (ZPT) Complex

Zinc pyrithione, bis(1-hydroxy-2(1H)-pyridinethionato-O, S)zinc, (**4-2-2-C1** or ZPT) is a zinc(II) complex of pyrithione or 1-hydroxy-2(1H)pyridinethione, (**4-2-2-L1**). Despite decades of commercial use, there is little understanding of its antimicrobial mechanism of action.

4-2-2-L1

Saunders and coworkers used a combination of genome-wide approaches (yeast deletion mutants and microarrays) and traditional methods (gene constructs and atomic emission) to characterize the activity of **4-2-2-C1** against model yeast, *S. cerevisiae*.[94]

4-2-2-C1 was similarly potent under anaerobic conditions in which superoxide dismutase is not required (Figure 7.1a). The finding that **4-2-2-C1** is as effective anaerobically as aerobically suggests that the mechanism of **4-2-2-C1** toxicity is not due to increased oxidative damage. Atomic emission spectroscopy of **4-2-2-C1**-treated cells to test for iron starvation and search for other metal imbalances indicated that the iron levels showed little or no change upon **4-2-2-C1** treatment (Figure 7.1b). Atomic absorption analysis showed that **4-2-2-C1** treatment resulted in a significant increase in cellular copper levels and, at most, only a small increase in zinc. From these results, Saunders

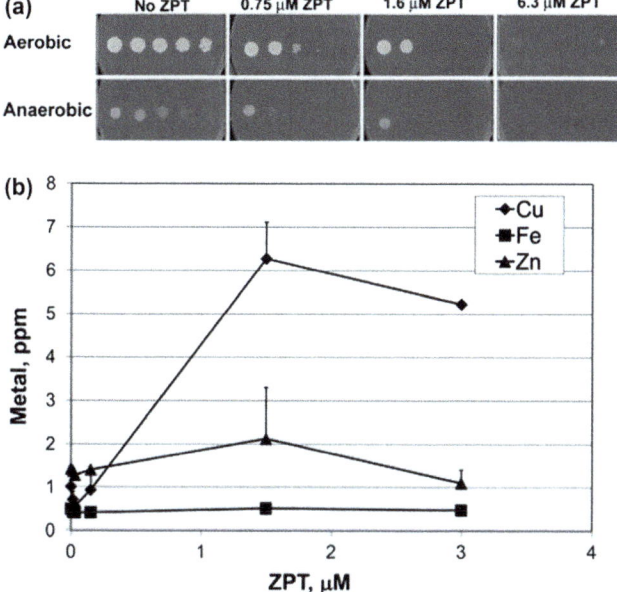

Figure 7.1 ZPT is similarly potent under anaerobic and aerobic conditions and causes a metal imbalance. (a) *S. cerevisiae* strain DY1457 was grown in a YPD (yeast extract peptone dextrose) medium and spotted onto YPD plates with the indicated amount of ZPT. Cultures were incubated for 69.5 h under either aerobic or anaerobic conditions. (b) Atomic emission of *S. cerevisiae* W303 in response to ZPT. Cultures were grown aerobically for 10 h in a YPD medium.
(Reprinted with permission from ref. 94; copyright 2011, American Society for Microbiology).

and coworkers came to the conclusion that **4-2-2-C1** acts through an increase in cellular copper levels that leads to loss of activity of iron–sulfur cluster-containing proteins. A model in Figure 7.2 is presented in which pyrithione acts as a copper ionophore, enabling copper to enter cells and distribute across intracellular membranes. This is the first report of a metalligand complex that inhibits fungal growth by increasing the cellular level of a different metal. **4-2-2-C1** was also found to mediate growth inhibition through an increase in copper in the scalp fungus *Malassezia globosa*.

Among the metal complexes of **4-2-2-L1**, only the crystal structure of copper pyrithione (**4-2-2-C2**) was reported by Bond and coworkers as shown below.[95]

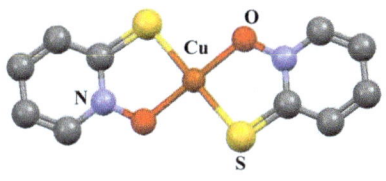

4-2-2-C2

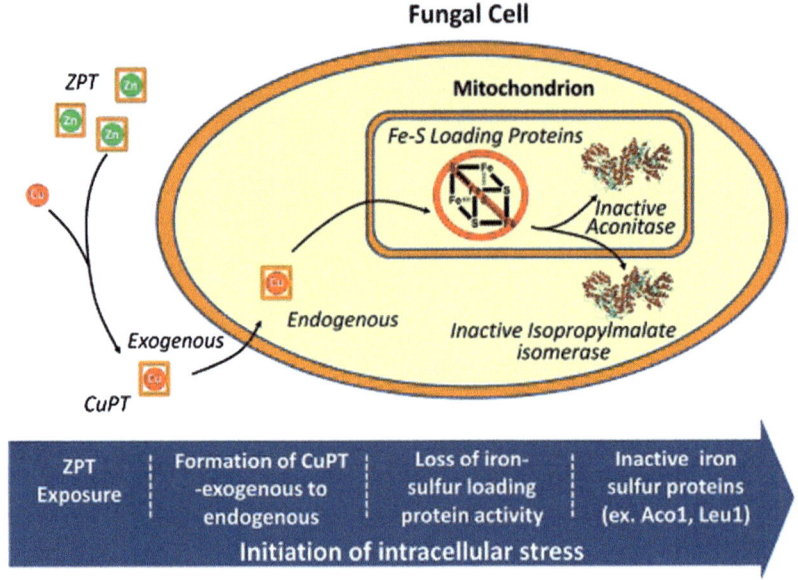

Figure 7.2 A model for the inactivation of the representative Fe–S proteins aconitase (Aco1) and isopropylmalate isomerase (Leu1). Some of the pyrithione exchanges zinc for copper and transports copper across the plasma membrane and intracellular membranes. The Fe–S protein assembly is damaged, leading to loss of Aco1 and Leu1 activity.
(Reprinted with permission from ref. 94; copyright 2011, American Society for Microbiology).

7.4.2.3 *Polymeric Chitosan–Metal Complexes*

Du and coworkers have prepared chitosan (**4-2-3-L1**) metal complexes with bivalent metal ions, including copper(II), zinc(II), and iron(II), and characterized by FTIR, XRD, atomic absorption spectroscopy (AAS) and elemental analysis.[96]

4-2-1-L1

The crystalline and structural properties of chitosan-metal complexes are different from those of chitosan, and the –NH₂, –OH groups in the chitosan molecule are considered to be the dominating reactive sites. The *in vitro* antimicrobial activities of the obtained chitosan–metal complexes, which were found to be much better than free chitosan and metal salts, were examined against two Gram-positive bacteria (*S. aureus* and *S. epidermidis*), two Gram-negative bacteria (*E. coli* and *P. aeruginosa*) and two fungi (*C. albicans* and

Candida parapsilosis). The results indicated that the inhibitory effects of chitosan–metal complexes were dependent on the properties of the metal ions, the molecular weight and degree of deacetylation of chitosan and environmental pH values. Electron microscopy confirmed that the exposure of *S. aureus* to the chitosan–copper(II) complex resulted in the disruption of the cell envelope.

To clarify the antibacterial mechanism action of chitosan–metal complexes with different M_w, they investigated the *S. aureus* cells treated with chitosan–copper(II) and chiooligosaccharide–copper(II) complexes using TEM. Through TEM observation, the integrity of most *S. aureus* cells was disrupted after exposure to a chitosan–copper(II) complex for a short time. Figures 7.3d–g are the representative morphologies of *S. aureus* cells after treatment of the chitosan–copper(II) complex. They presented the procedure of morphological changes of incubated cells (g), partial disruption of the outer membrane (f), disruption of the cytoplasmic membrane (e) and release of the cytoplasmic constituents (d). In particular, the cell in fission in (e) showed total disruption of the outer membrane, while the intramembrane was kept integral, which indicates that the target site of the chitosan–copper(II) complex is the cell envelope. At the same time, no obvious disruption of the cell membrane was observed after treatment of chitooligosaccharide–copper(II). As for chitooligosaccharide–copper(II), there may be other antimicrobial mechanisms. It was reported by Liu and coworkers that the FITC-labeled chitosan oligomers were observed inside the *E. coli* cell, which were thought to block transcription from DNA.[97] Probably, chitooligosaccharide–copper(II) has a similar mechanism as chitosan oligomers, but that cannot be confirmed here through TEM. The effect of the deacetylation degree on the antimicrobial activity of the chitosan-metal complex was studied. As shown in Figure 7.3a, under identical experimental conditions, the antimicrobial activities of chitosan–metal complexes were enhanced by increasing the deacetylation degree. This may be due to the extra amine groups in the complexes, the greater

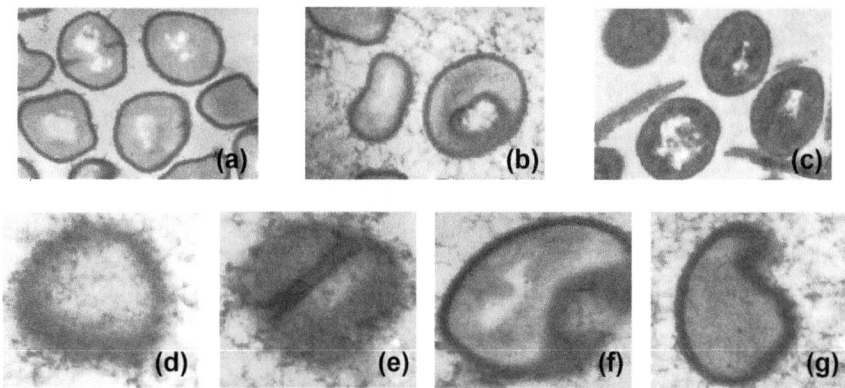

Figure 7.3 Transmission electron photograph of intact *S. Aureus* cells (a) and *S. Aureus* cells after treatment with chitosan–copper(II) complex (b), chitooligosaccharide-copper(II) complex (c). (d)–(g) are the representative morphologies of *S. Aureus* cells exposed to chitosan–copper(II). (Reprinted with permission from ref. 96; copyright 2005, Springer).

the cationic charge density obtained. For the same reason, the chitosan–metal complexes with a greater deacetylation degree could more easily adsorb on the outer membrane of a micro-organism and inhibit its growth.

All the results show that chitosan–metal complexes are promising candidates for novel antimicrobial agents that could be used in cosmetics, food, textiles, *etc.*

7.4.2.4 Monomeric and Dimeric Thiosemicarbazone and Schiff Base Metal Complexes

Copper(II) complex (**4-2-4-C1**), nickel(II) complexes (**4-2-4-C2 to 4-2-4-C4**), zinc(II) complexes (**4-2-4-C5 to 4-2-4-C7**), antimony(III) complexes (**4-2-4-C8 to 4-2-4-C10**) and bismuth(III) complexes (**4-2-4-C11, 4-2-4-C12**).

Schiff bases are organic compounds containing C=N moieties. If the compound contains other heteroatoms near an azomethine moiety, it can coordinate metal ions in a chelate manner. Some Schiff-base compounds are known to date with a variety of substitutent groups and heteroatoms.[98] According to a report by Rosu *et al.*,[99] the metal complexes of Schiff-base ligands, especially copper(II) (**4-2-4-C1**) and vanadium(IV) complexes, exhibit higher antibacterial activity as compared to the free ligand (N-(4-amino-2,3-dimethyl-1-phenyl-3-pyrazolin-5-one)pyridoxaldimine, **4-2-4-L1**) and metal salts, according to Searl's concept and Tweedy's chelation theory.[100,101]

4-2-4-L1   4-2-4-C1

Among the Schiff bases, derivatives containing C=N-N-C(S)- moieties are called thiosemicarbazones, which are of considerable interest because of their potentially beneficial biological activities, such as antitumor, antibacterial, antiviral, and antimalarial activities, and several of them are already used in medical practice.[78,102] Their biological activities are considered to be due to their ability to form chelates with metal ions. The biological activities of the metal complexes differ from those of either the ligand or the metal ion itself, and increased or decreased biological activities are reported for various metal complexes as well as the well-studied metal complexes, such as copper(II), zinc(II) and iron(III) complexes.[77,103,104]

Nomiya and coworkers prepared several transition and nontransition-metal complexes with thiosemicarbazones (**4-1-8-L2, 4-2-4-L2 to 4-2-4-L5**) to investigate the relationship between the structure and antimicrobial activity of various metal complexes with multidentate 2-acetylpyridine thiosemicarbazone ligands.[105–110]

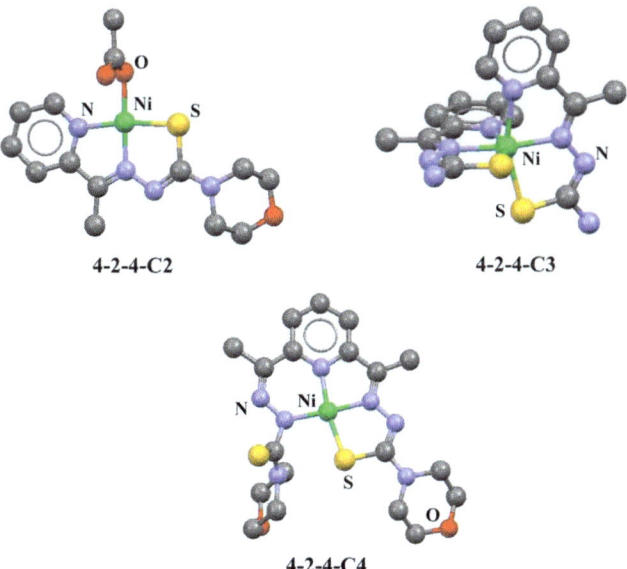

The observed results so far lead us to conclude that the structure factors that govern antimicrobial activities are strongly dependent on the central metal ion. The structure–activity correlation of nickel(II) complexes with thiosemicarbazone ligands has been attributed to their ligand-replacement ability because they show selective and effective activities against two bacteria, only when they have a labile 4-coordinate structure having one tridentate ligand and one replaceable monodentate ligand.[105,106]

In contrast, the antimicrobial activities of the zinc(II) complex (**4-2-4-C5** to **4-2-4-C7**) were influenced by ^4N-moieties more than by the geometry around the central metal ion.[107,108]

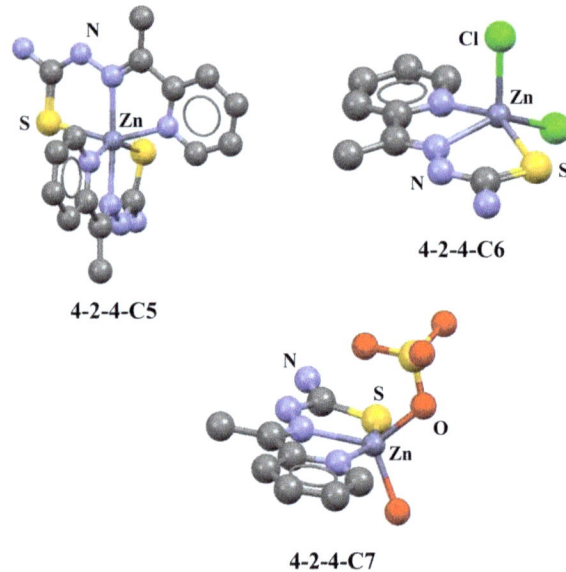

4-2-4-C5

4-2-4-C6

4-2-4-C7

Although the importance of sulfur chelation was commonly observed in antimony(III) (**4-2-4-C8** to **4-2-4-C10**) and bismuth(III) (**4-2-4-C11** to **4-2-4-C12**) complexes, the solubilities, molecular structures, selectivities and effectiveness in antimicrobial activities were different, suggesting that the action mechanism of two metal complexes in their antimicrobial activities are not the same.[109,110]

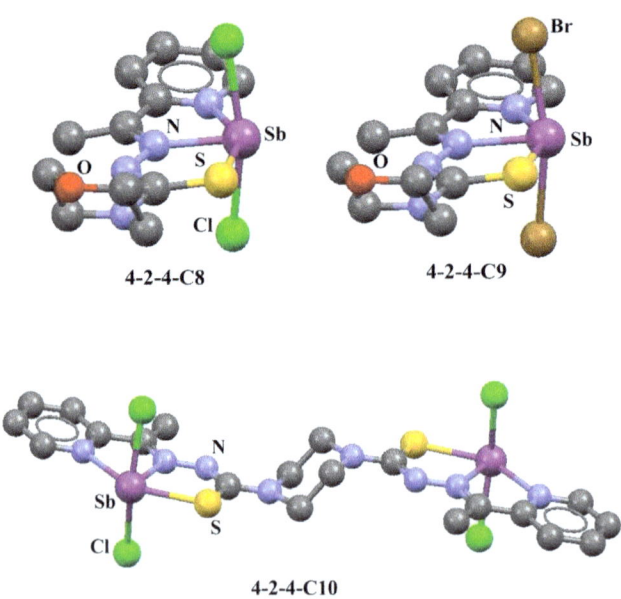

4-2-4-C8

4-2-4-C9

4-2-4-C10

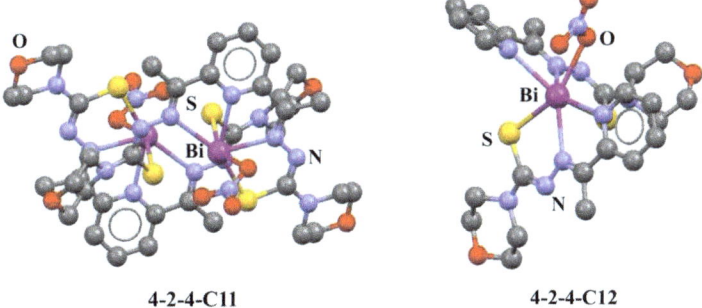

4-2-4-C11   4-2-4-C12

7.4.2.5 Monomeric and Dimeric Metal Hinokitiol Complexes

Alkali-metal salts (**4-2-5-C1** to **4-2-5-C4**), [Mg(hino)$_2$(EtOH)]$_2$ (**4-2-5-C5**), [Mn(hino)$_2$(MeOH)]$_2$ (**4-2-5-C6**), [Ni(hino)$_2$(EtOH)]$_2$ (**4-2-5-C7**), [Co(hino)$_2$(EtOH)]$_2$ (**4-2-5-C8**), [Zn(hino)$_2$(EtOH)]$_2$ (**4-2-5-C9**), *cis*-[MoO$_2$(hino)$_2$] (**4-2-5-C10**), *cis*-[WO$_2$(hino)$_2$] (**4-2-5-C11**), [Pd(hino)$_2$] (**4-2-5-C12**), [Cu(hino)$_2$] (**4-2-5-C13**), [Fe(hino)$_3$] (**4-2-5-C14**), [In(hino)$_3$] (**4-2-5-C15**), [Ti(hino)$_2$Cl$_2$] (**4-2-5-C16**) and [Hf(hino)$_4$]·H$_2$O (**4-2-5-C17**).

Hinokitiol (**4-1-9-L1** or Hhino), also called β-thujaplicin, is a 4-isopropyl-substitution compound of a 7-membered ring organic compound, tropolone, as described in Section 7.4.1.9.[82,83] The monoanion of **4-1-9-L1** forms various metal–chelate complexes through two unequivalent oxygen donor atoms. Various metal complexes with **4-1-9-L1** were prepared and characterized by elemental analysis, TG/DTA, FTIR and solution (^{1}H and ^{13}C) NMR spectroscopy.

Tropolone

Single-crystal X-ray structure analysis revealed that hinokitiol-metal complexes can be classified into several families,[111] *i.e.* Type A [M^{II}(hino)$_2$(L)]$_2$ (M = Mg^{II}, Mn^{II}, Ni^{II}, Zn^{II},[112] Co^{II},[87]; L = EtOH or MeOH), with a dimeric structure consisting of one bridging hino$^-$ anion, one chelating hino$^-$ anion and one alcohol or water molecule; Type B, with octahedral, *cis*-dioxo, bis-chelate complexes *cis*-[M^{VI}O$_2$(hino)$_2$] (M = Mo^{VI}, W^{VI}); Type C, with a square planar complex [M^{II}(hino)$_2$] (M = Cu^{II} and Pd^{II});[113–115] Type D, with tris-chelate, 7-coordinate complexes having one inert electron pair [M^{III}(hino)$_3$] (M = Sb^{III}, Bi^{III}); Type D', with bis-chelate, pseudo-6-coordinate complexes and one inert electron pair $[M^{III}(hino)_2X]$ ($M = Sb^{III}$, $X = Br$); Type E, with tris-chelate, 6-coordinate complexes having Δ and Λ isomers $[M^{III}(hino)_3]$ ($M = In^{III},[116] Al^{III}, Fe^{III}$); Type E' with a *bis*-chelate, 6-coordinate complex $[M^{IV}(hino)_2X_2]$ ($M = Sn^{IV},[117] Ti^{IV}$, $X = F$, Cl); Type F, with water-soluble alkali-metal salts $[M^{I}(hino)]$ (M = alkali metal ion), Type G $[Ag(hino)_2]$ (see Section 7.4.1.9), and Type H, with *tetrakis*-chelate, 8-coordinate complexes $[M^{IV}(hino)_4]$($M = Zr^{IV}$, Hf^{IV}). These structural features of hinokitiol-metal complexes were more complicated because of a 4-substituted structure compared to those of metal complexes with a related ligand, tropolone.

The antimicrobial activities of these complexes, evaluated in terms of minimum inhibitory concentration (MIC; $\mu g\,mL^{-1}$) in two systems, were compared to elucidate the relationship between structure and antimicrobial activity. Investigation on structure–activity relations in the antimicrobial activities of hinokitiolato–metal complexes, activity tests for water-insoluble complexes, were performed by Nomiya and coworkers in a water–suspension system, except that water-soluble Type F complexes and their activities were compared with each other within the same class.

4-1-9-L1 showed a wide range of effective activities against bacteria, yeasts and molds, except for one Gram-negative bacterium (*P. aeruginosa*). Nomiya and coworkers proposed that the antimicrobial activities of **4-1-9-C1** $[Ag(hino)]_2$, a Type G complex with a *trans* Ag_2O_4 core geometry, was due to the action of the weak Ag–O bonding property based on accumulated data of molecular structures of silver(I) complexes and antimicrobial activities. The silver complex of hinokitiol may be accounted for by ligand exchangeability. The antimicrobial activities of **4-1-9-C1** were significantly enhanced, whereas those of other metal complexes were suppressed, compared with those of the neutral **4-1-9-L1** (Hhino) and anionic hino⁻ molecules.

Both the neutral and monoanions of **4-1-9-L1** (type F, water-soluble alkali-metal complexes) show noteworthy antimicrobial activities, *i.e.* they show a wide spectrum of activities against bacteria, yeasts and molds. Water-soluble alkali-metal salts (**4-2-5-C1** to **4-2-5-C4**) showed enhanced activities against two bacteria (*E. coli* and *B. subtilis*), two yeasts (*C. albicans* and *S. cerevisiae*) and two molds (*A. brasiliensis* and *P. citrinum*), but depressed activities against two bacteria (*S. aureus* and *P. aeruginosa*), which show the activities of the monoanion of **4-1-9-L1**.

4-2-5-C1   4-2-5-C2   4-2-5-C3   4-2-5-C4

Hinokitiolato-metal complexes belong to type F

When the MIC test of water-insoluble complexes was conducted as a water–suspension system, modest antimicrobial activities were observed against selected micro-organisms. The spectral patterns of the antimicrobial activities of the metal complexes were different from each other and their MIC values also varied, indicating that the antimicrobial activities of hinokitiolato–metal complexes depend on the metal center rather than on the geometry around the metal ions in a water–suspension system.

Type A complexes: the magnesium complex (**4-2-5-C5**) showed moderate activity against a Gram-negative bacterium (*E. coli*), two yeasts (*C. albicans* and *S. cerevisiae*) and a mold (*P. citrinum*), and modest activity against two Gram-positive bacteria (*B. subtilis* and *S. aureus*) and a mold (*A. brasiliensis* (*niger*)). The manganese(II) complex (**4-2-5-C6**) showed effective activity against a yeast (*S. cerevisiae*) and moderate activity against a Gram-positive (*B. subtilis*) and a Gram-negative bacterium (*E. coli*), a yeast (*C. albicans*) and two molds (*A. brasiliensis* and *P. citrinum*). The nickel(II) complex (**4-2-5-C7**) showed modest activity against two Gram-positive bacteria (*B. subtilis* and *S. aureus*) and two molds (*A. niger* and *P. citrinum*). In contrast, cobalt(II) (**4-2-5-C8**) and zinc(II) complexes (**4-2-5-C9**) showed almost no activity against the same microbes, except that the cobalt complex showed modest activity against a Gram-positive bacterium (*S. aureus*) and a yeast (*S. cerevisiae*).

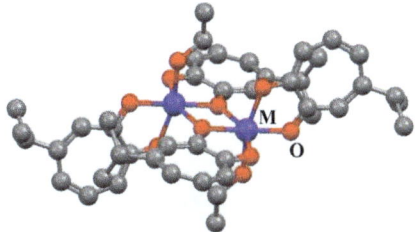

Type A $[M^{II}(hino)_2(L)]_2$

Among type B complexes, the molybdenum(VI) complex (**4-2-5-C10**) showed no activity against the selected microbes, while the tungsten(VI) complex (**4-2-5-C11**) showed a wide spectrum of modest activity against a Gram-positive bacterium (*S. aureus*), two yeasts and two molds. The tungsten(VI) complex showed weaker activity, but the spectral pattern was similar to that of the monoanion of **4-1-9-L1**.

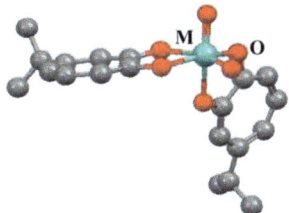

Type B *cis*-$[M^{VI}O_2(hino)_2]$

Among type C complexes, the palladium(II) complex (**4-2-5-C12**) and copper(II) complexes (**4-2-5-C13**) of rod and plate crystals showed almost no activity against the selected microbes. Only the copper complex with needle crystals showed modest activity against bacteria, albeit no activities against selected yeasts and molds.

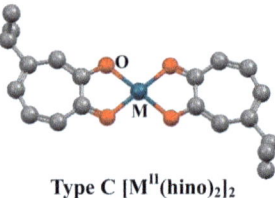

Type C [M^{II}(hino)$_2$]$_2$

Complexes of types D, D′, E, and H showed almost no activity against the same microbes in a water–suspension system. The barium complex showed modest activity against selected bacteria, and effective activities against yeasts and molds, with a spectral pattern similar to that of the monoanion of the ligand.

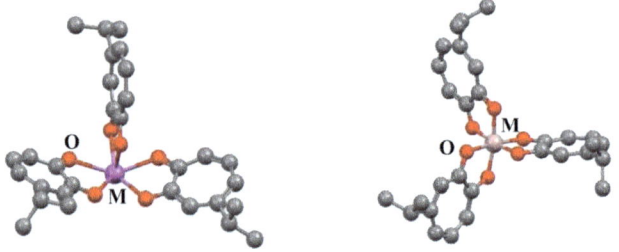

Type D [M^{III}(hino)$_3$] (left) and **Type E** (right) **[M^{III}(hino)$_3$]**

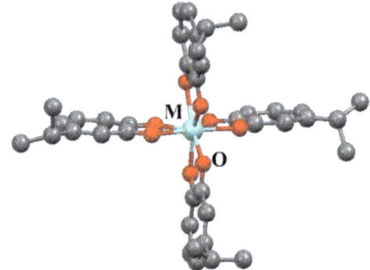

Type H [M^{IV}(hino)$_4$]

Interestingly, the CHCl$_3$-soluble metal complexes ([Fe(hino)$_3$] (**4-2-5-C14**), [In(hino)$_3$] (**4-2-5-C15**), [Ti(hino)$_2$Cl$_2$] (**4-2-5-C16**) and [Hf(hino)$_4$]·H$_2$O(**4-2-5-C17**)), which showed no activity in solid-phase contact in a water–suspension system were effective against the selected micro-organisms. An iron(III) complex (**4-2-5-C14**) was remarkably effective against all of the selected bacteria, yeasts and molds (MIC < 2). An indium(III) complex (**4-2-5-C15**) of the same type (Type A) was also effective against all of the selected bacteria, yeasts and molds, but the MIC values were larger than those of the iron(III) complexes. The titanium(IV) complex ((**4-2-5-C16**), Type E′) and hafnium(IV) complex (**4-2-5-C17**, Type H) were also effective against bacteria, but showed no activity against yeasts and only modest activity against molds in $CHCl_3$ solutions, and the spectral pattern was different from that of hino$^-$ in a homogeneous system. These results indicated that the hinokitiolato–metal complexes inhibited the growth of the selected micro-organisms much more effectively by interfacial biphasic contact than by solid-phase contact. Single molecules could interact with micro-organisms more easily than in solid-state stacking, and a hydrophobic solvent should also aid in interaction. The activities of metal complexes in organic solution also depended on the metal center rather than the geometry around the metal ions, as in MIC tests in a water–suspension system. In both heterogeneous antimicrobial test systems, the activities of metal complexes depended on the metal center itself more than on the geometry around the metal ion.

7.5 Summary

1) It has been considered that the antimicrobial behavior of the aqueous silver(I) ion itself attacks the targets, mainly proteins existing either in the outer membrane or inside the cell, and other materials such as nucleic acids. On the other hand, the antimicrobial behavior of silver nano-particles is attributable to light-driven formation of active oxygen species by silver(I) ions such as the hydroxyl radical and the interaction of silver nanoparticles themselves with micro-organism membranes.

2) The antimicrobial mechanism is different among aqueous silver(I) ions and most the silver(I)–oxygen bonding complexes, although the aqueous silver(I) ion is one of silver(I)–oxygen bonding complexes or a silver(I) complex with aqua ligands, $[Ag(H_2O)_n]^+$. The reasons for this are that the geometrical structure around the silver center, coordination number of the silver atom, ligand-exchange rate, and stability constant (as a complex) are significantly different among them, resulting in different inter-actions with organisms. Thus, the antimicrobial mechanism of the aqueous silver(I) ion is essentially different from those of silver(I)–sulfur, –nitrogen, and –phosphine bonding complexes.

3) A study of silver-resistant organisms would be useful for understanding the antimicrobial mechanism of silver materials.

4) Antimicrobial behaviors are essentially the same among the polymeric and nonpolymeric materials of silver(I) complexes. The antimicrobial behavior of the silver(I) complexes and other metal complexes is essen-tially different, *i.e.* the antimicrobial mechanism is dependent on the metal centers of the complexes.

5) Although the mechanism and mode of action of the silver(I) complexes are still unclear, there are several points to be made. The antimicrobial activities stem from: (1) slow release of silver(I) ions from the complexes, (2) ligand exchange of silver(I) complexes with "bioligands" of the or-ganisms form a "silver(I)–bacterium complex", "silver(I)–yeast com-plex", "silver(I)–mold complex" and (3) other active species formation such as hydroxyl radicals and so forth. Point (1) indicates that the anti-microbial activities of silver complexes are essentially the same as that of

the Ag^+ ion itself, while point (2) suggests that the antimicrobial activities of silver complexes are due to the direct interaction of the silver(I) ion with the cell membrane, *i.e.* outside of the cell. Electron microscope measurements would be useful for understanding the difference in the antimicrobial mechanisms of the silver(I) ion itself and silver(I) complexes.

6) Although this is the case for only silver(I) complexes, we can say that the coordinating donor atoms significantly determine whether the antimicrobial spectrum of the silver(I) complexes is wide or narrow. This is ascribed to the ease of ligand exchange. Thus, by choosing the appropriate ligands, it is possible to design silver(I) complexes that show only antibacterial activities, and both antibacterial and antifungal activities.

7) In addition to (6), what are the factors determining the strengths of the antimicrobial activities, *i.e.* how can we understand the different MIC values within the silver(I) complexes with the same donor atoms? In fact, we know the different MICs within the silver(I)–nitrogen bonding complexes and within the silver(I)–oxygen bonding complexes. Solubility is one of such factors, because generally a homogeneous system shows smaller MICs than a heterogeneous system, or a water–suspension system. Other factors, if present, that determine different MIC values within silver(I) complexes with the same donor atoms, should also be studied.

8) At the proof-reading stage, we found the two related and important references with regard to the interaction of silver(I) ions with peptides and bacteria[118] and silver(I) complex formation with amino acids (cysteine, penicillamine and glutathione),[119] which are cited in the References.

References

1. K. Nomiya, S. Takahashi, R. Noguchi, S. Nemoto, T. Takayama and M. Oda, *Inorg. Chem.*, 2000, **39**, 3301.
2. N. C. Kasuga, Y. Takagi, S. Tsuruta, W. Kuwana, R. Yoshikawa and K. Nomiya, *Inorg. Chim. Acta*, 2011, **368**, 44.
3. N. C. Kasuga, R. Yamamoto, A. Hara, A. Amano and K. Nomiya, *Inorg. Chim. Acta*, 2006, **359**, 4412.
4. R. Noguchi, A. Sugie, Y. Okamoto, A. Hara and K. Nomiya, *Bull. Chem. Soc. Jpn.*, 2005, **78**, 1953.
5. R. Noguchi, A. Hara, A. Sugie and K. Nomiya, *Inorg. Chem. Commun.*, 2006, **9**, 355.
6. J. L. Clement and P. S. Jarrett, *Metal Based Drug*, 1994, **1**, 467.
7. N. P. Farrell, *Uses of Inorganic Chemistry Complexes in Medicine*, RSC, Cambridge, 1999, p 7.
8. M. C. Gimeno and A. Laguna, *Comprehensive Coordination Chemistry II*, Elsevier, Oxford, 2004, Vol. 6, p 911.
9. S. Ahmad, A. A. Isab, S. Ali and A. R. Al-Arfaj, *Polyhedron*, 2006, **25**, 1633.
10. C. F. Shaw II, *Chem. Rev.*, 1999, **99**, 2589.

11. A. B. G. Lansdown, *J. Wound Care*, 2002, **11**, 125.
12. A. D. Russell and W. B. Hugo, *Prog. Med. Chem.*, 1994, **31**, 351.
13. A. B. G. Lansdown, *Crit. Rev. Toxicol.*, 2007, **37**, 237.
14. Q. L. Feng, J. Wu, G. Q. Chen, F. Z. Cui, T. N. Kim and J. O. Kim, *J. Biomed. Mater. Res.*, 2000, **52**, 662.
15. K. B. Holt and A. J. Bard, *Biochemistry*, 2005, **44**, 13214.
16. W. K. Jung, H. C. Koo, K. W. Kim, S. Shin, S. H. Ki and Y. H. Park, *Appl. Environ. Microbiol.*, 2008, **74**, 2171.
17. X. Yang, W. Yang, Q. Wang, H. Li, K. Wang, L. Yang and W. Liu, *Talanta*, 2010, **81**, 1508.
18. Y. Matsumura, K. Yoshikata, S. Kunisaki and T. Tsuchido, *Appl. Environ. Microbiol.*, 2003, **69**, 4278.
19. L. S. Nair and C. T. Laurencin, *J. Biomed. Nanotechnol.*, 2007, **3**, 301.
20. J. Jain, S. Arora, J. M. Rajwade, P. Omray, S. Khandelwal and K. M. Panknikar, *Mol. Pharmaceut.*, 2009, **6**, 1388.
21. T. N. V. K. V. Prasad, V. S. R. Kambala and R. Naidu, *Curr. Nanosci.*, 2011, **7**, 531.
22. R. R. Arvizo, S. Bhattacharyya, R. A. Kudgus, K. Giri, R. Bhattacharya and P. Mukherjee, *Chem. Soc. Rev.*, 2012, **41**, 2943.
23. A. Taglietti, Y. A. D. Fernandez, E. Amato, L. Cucca, G. Daccaro, P. Grisoli, V. Necchi, P. Pallavicini, L. Pasotti and M. Patrini, *Langmuir*, 2012, **28**, 8140.
24. J. S. Kim, E. Kuk, K. N. Yu, J.-H. Kim, S. J. Park, H. J. Lee, S. H. Kim, Y. K. Park, Y. H. Park, C.-Y. Hwang, Y.-K. Kim, Y.-S. Lee, D. H. Jeong and M.-H. Cho, *Nanomed. Nanotech. Biol. Med.*, 2007, **3**, 95.
25. H. Lara, A. V. Nilida, L. del C. I. Turrent and C. R. Padilla, *World J. Microbiol. Biotechnol.*, 2010, **26**, 615.
26. I. Sondi and B. Salopek-Sondi, *J. Colloid Interface Sci.*, 2004, **275**, 177.
27. J. R. Morones, J. L. Elechiguerral, A. Camacho, K. Holt, J. B Kouri, J. T. Ramirez and M. J. Yacaman, *Nanotechnology*, 2005, **16**, 2346.
28. N. Duran, P. D. Marcato, R. De Conti, O. L. Alves, F. T. M. Costa and M. Brocchi, *J. Brazil. Chem. Soc.*, 2010, **21**, 949.
29. S. P. Fricker, *Toxicol. in Vitro*, 1994, **8**, 879.
30. Z. Guo and P. J. Sadler, *Angew. Chem. Int. Ed.*, 1999, **38**, 1512.
31. C. L. Fox, S. Modak, J. W. Stanford and P. L. Fox, *Scand. J. Plast. Reconst. Surg.*, 1979, **13**, 89.
32. D. S. Cook and M. F. Turner, *J. Chem. Soc., Perkin Trans. 2*, 1975, 1021.
33. N. C. Baenziger and A. W. Struss, *Inorg. Chem.*, 1976, **15**, 1807.
34. T. N. C. Wells, P. Scully, G. Paravicini, A. E. I. Proudfoot and M. A. Payton, *Biochemistry*, 1995, **34**, 7896.
35. A. Mastrolorenzo, A. Scozzafava and C. T. Supuran, *Eur. J. Pharmt. Sci.*, 2000, **11**, 99.
36. A. Mastrolorenzo, A. Scozzafava and C. T. Supuran, *J. Enzym. Inhib.*, 2000, **15**, 517.

37. J. P. Pirnay, D. D. Vos, C. Cochez, F. Bilocq, J. Pirson, M. Struelens, L. Duinslaeger, P. Cornelis, M. Zizi and A. Vanderkelen, *J. Clin. Microbiol.*, 2003, **41**, 1192.

38. X. C. Huang, J. P. Zhang and X. M. Chen, *Cryst. Growth Des.*, 2006, **6**, 1194.

39. K. Nomiya, K. Tsuda and N. C. Kasuga, *J. Chem. Soc., Dalton Trans.*, 1998, 1653.

40. K. Nomiya, K. Tsuda, T. Sudoh and M. Oda, *J. Inorg. Biochem.*, 1997, **68**, 39.

41. K. Nomiya, K. Tsuda, Y. Tanabe and H. Nagano, *J. Inorg. Biochem.*, 1998, **69**, 9.

42. L. Carlucci, G. Ciani and D. M. Proserpio, *Angew. Chem., Int. Ed.*, 1999, **38(23)**, 3488.

43. K. Nomiya, R. Noguchi and M. Oda, *Inorg. Chim. Acta*, 2000, **298**, 24.

44. K. Nomiya, Y. Kondoh, K. Onoue, N. C. Kasuga, H. Nagano, T. Sudoh and S. Sakuma, *J. Inorg. Biochem.*, 1995, **58**, 255.

45. K. Nomiya, R. Noguchi and C. N. Kato, *Chem. Lett.*, 2000, 162.

46. R. Noguchi, A. Hara, A. Sugie, S. Tanabe and K. Nomiya, *Chem. Lett.*, 2005, **34**, 578.

47. K. Nomiya, S. Takahashi and R. Noguchi, *J. Chem. Soc., Dalton Trans.*, 2000, 2091.

48. I. Tsyba, B. B-K. Mui, R. Bau, R. Noguchi and K. Nomiya, *Inorg. Chem.*, 2003, **42**, 8028.

49. R. Noguchi, A. Hara, A. Sugie and K. Nomiya, *Inorg. Chem. Commun.*, 2006, **9**, 60.

50. K. Nomiya, H. Yokoyama, R. Noguchi and K. Machida, *Chem. Lett.*, 2002, 922.

51. K. Nomiya, S. Takahashi and R. Noguchi, *Dalton Trans.*, 2000, **8**, 1343.

52. N. C. Kasuga, A. Sugie and K. Nomiya, *Dalton Trans.*, 2004, 3732.

53. R. Noguchi, A. Sugie, A. Hara and K. Nomiya, *Inorg. Chem. Commun.*, 2006, **9**, 107.

54. K. Nomiya and H. Yokoyama, *J. Chem. Soc., Dalton Trans.*, 2012, 2483.

55. A. Takayama, R. Yoshikawa, S. Iyoku, N. C. Kasuga and K. Nomiya, *Polyhedron*, 2013, **52**, 844.

56. N. C. Kasuga, R. Yoshikawa, Y. Sakai and K. Nomiya, *Inorg. Chem.*, 2012, **51**, 1640.

57. N. C. Kasuga, S. Tsuruta, A. Amano and K. Nomiya, *Acta Crystallogr., Sect. E: Struct. Rep. Online*, 2007, **E63**, m2440.

58. H. W. Wanzlick and H. J. Schonherr, *Angew. Chem. Int. Ed. Engl.*, 1968, **7**, 141.

59. K. Öfele, *J. Organomet. Chem.*, 1968, **12**, P42.

60. A. J. Arduengo, R. L. Harlow and M. Kline, *J. Am. Chem. Soc.*, 1991, **113**, 361.

61. A. J. Arduengo III, H. V. R. Dias, J. C. Calabrese and F. Davidson, *Organometallics*, 1993, **12**, 3405.

62. H. M. J. Wang and I. J. B. Lin, *Organometallics*, 1998, **17**, 972.

63. A. Kascatan-Nebioglu, M. J. Panzner, C. A. Tessier, C. L. Cannon and W. J. Youngs, *Coord. Chem. Rev.*, 2007, **251**, 884.
64. K. M. Hindi, M. J. Panzner, C. A. Tessier, C. L. Cannon and W. J. Youngs, *Chem. Rev.*, 2009, **109**, 3859.
65. A. Kascatan-Nebioglu, A. Melaiye, K. Hindi, S. Durmus, M. J. Panzner, L. A. Hogue, R. J. Mallet, C. E. Hovis, M. Coughenour, S. D. Crosby, A. Milsted, D. L. Ely, C. A. Tessier, C. L. Cannon and W. J. Youngs, *J. Med. Chem.*, 2006, **49**, 6811.
66. G. Roymahapatr, S. M. Mandal, W. F. Porto, T. Samanta, S. Giri, J. Dinda, O. L. Franco and P. K. Chattaraj, *Curr. Med. Chem.*, 2012, **19**, 4184.
67. B. Yiğit, Y. Gök, I. Özdemir and S. Günal, *J. Coord. Chem.*, 2012, **65**, 371.
68. S. Patil, A. Deally, B. Gleeson, F. Hackenberg, H. Müller-Bunz, F. Paradisi and M. Tacke, *Z. Anorg. Allg. Chem.*, 2011, **637**, 386.
69. S. Patil, K. Dietrich, A. Deally, B. Glesson, H. Müller-Bunz, F. Paradisi and M. Tacke, *Helv. Chim. Acta*, 2010, **93**, 2347.
70. I. Özdemir, S. Demir, S. Guenal, C. Arici and D. Uelkue, *Inorg. Chim. Acta*, 2010, **363**, 3803.
71. K. M. Hindi, T. J. Siciliano, S. Durmus, M. J. Panzner, D. A. Medvetz, D. V. Reddy, L. A. Hogue, C. E Hovis, J. K. Hiliard, R. J. Mallet, C. A. Tessier, C. L. Cannon and W. J. Youngs, *J. Med. Chem.*, 2008, **51**, 1577.
72. B. D. Wright, P. N. Shah, L. J. McDonald, M. L. Shaeffer, P. O. Wagers, M. J. Panzner, J. Smolen, J. Tagaev, C. A. Tessier, C. L. Cannon and W. J. Youngs, *Dalton Trans.*, 2012, **41**, 65006.
73. A. Melaiye, Z. Sun, K. Hindi, A. Misted, D. Ely, D. H. Reneker, C. A. Tessier and W. J. Youngs, *J. Am. Chem. Soc.*, 2005, **127**, 2285.
74. M. J. Panzner, A. Deeraksa, A. Smith, B. D. Wright, K. M. Hindi, A. Kascatan-Nebioglu, A. G. Torres, B. M. Judy, C. E. Hovis, J. K. Hilliard, R. J. Mallett, E. Cope, D. M. Estes, C. L. Cannon, J. G. Leid and W. J. Youngs, *Eur. J. Inorg. Chem.*, 2009, **41**, 1739.
75. B. S. Creaven, D. A. Egan, K. Kavanagh, M. McCann, M. Mahon, A. Noble, B. Thati and M. Walsh, *Polyhedron*, 2005, **24**, 949.
76. B. Thati, A. Noble, R. Rowan, B. S. Creaven, M. Walsh, M. McCann, D. Egan and K. Kavanagh, *Toxicol. in Vitro*, 2007, **21**, 801.
77. J. S. Casas, M. S. Garcia-Tasende and J. Sordo, *Coord. Chem. Rev.*, 2000, **209**, 197.
78. J. R. Dilworth and R. Hueting, *Inorg. Chim. Acta*, 2012, **389**, 3.
79. K. K. Aravindakshan and C. G. R. Nair, *Indian J. Chem., Sect. A*, 1981, **20**, 684.
80. L. J. Ashfield, A. R. Cowley, J. R. Dilworth and P. S. Donnelly, *Inorg. Chem.*, 2004, **43**, 4121.
81. K. Onodera, N. C. Kasuga, T. Takashima, A. Hara, A. Amano, H. Murakami and K. Nomiya, *Dalton Trans.*, 2007, 3646.
82. T. Nozoe, *Bull. Chem. Soc. Jpn.*, 1936, **11**, 295.

83. H. Erdtman and J. Gripenberg, *Nature*, 1948, **161**, 179.
84. M. Saniewski, A. Saniewska and S. Kanlayanarat, *S. Acta Hort.*, 2007, **755**, 133.
85. N. Uchide and H. Toyoda, *Molecules*, 2011, **16**, 2032.
86. H. Iinuma, *Nippon Kagaku Kaishi*, 1943, **64**, 742 (in Japanese).
87. K. Nomiya, A. Yoshizawa, K. Tsukagoshi, N. C. Kasuga, S. Hirakawa and J. Watanabe, *J. Inorg. Biochem.*, 2004, **98**, 46.
88. S. J. Berners-Price, R. K. Johnson, A. J. Giovenella, L. F. Faucette, C. K. Mirabelli and P. J. Sadler, *J. Inorg. Biochem.*, 1988, **33**, 285.
89. B. T. Elie, C. Levine, I. Ubarretxena-Belandia, A. Varela-Ramirez, R. J. Aguilera, R. Ovalle and M. Contel, *Eur. J. Inorg. Chem.*, 2009, 3421.
90. K. Nomiya, H. Yokoyama, H. Nagano, M. Oda and S. Sakuma, *J. Inorg. Biochem.*, 1995, **60**, 289.
91. K. Nomiya, R. Noguchi and T. Sakurai, *Chem. Lett.*, 2000, 274.
92. K. Nomiya, R. Noguchi, K. Ohsawa, K. Tsuda and M. Oda, *J. Inorg. Biochem.*, 2000, **78**, 363.
93. K. Nomiya, S. Yamamoto, H. Yokoyama, N. C. Kasuga, K. Ohyama and C. Kato, *J. Inorg. Biochem.*, 2003, **95**, 208.
94. N. L. Reeder, J. Kaplan, J. Xu, R. S. Youngquist, J. Wallace, P. Hu, K. D. Juhlin, J. R. Schwartz, R. A. Grant, A. Fieno, S. Nemeth, T. Reichiling, J. P. Tiesman, T. Mills, M. Steinke, S. L. Wang and C. W. Saunders, *Antimicrob. Agents Chemother.*, 2011, **55**, 5753.
95. A. D. Bond, N. Feeder, S. J. Teat and W. Jones, *Acta Crystallogr.*, 2001, **C57**, 1157.
96. X. Wang, Y. Du, L. Fan, H. Liu and Y. Fu, *Polym. Bull.*, 2005, **55**, 105.
97. X. F. Liu, Y. L. Guan, D. Z. Yang, Z. Li and K. D. Yao, *J. Appl. Polym. Sci*, 2001, **79**, 1324.
98. P. A. Vigato, V. Peruzzo and S. Tamburini, *Coord. Chem. Rev.*, 2012, **256**, 953.
99. T. Rosu, E. Pahontu, M. Reka-Stefana, D.-C. Ilies, R. Georgescu, S. Shova and A. Gulea, *Polyhedron*, 2012, **31**, 352.
100. J. W. Searl, R. C. Smith and S. J. Wyard, *Proc. Phys. Soc.*, 1961, **78**, 1174.
101. B. G. Tweedy, *Phytopathology*, 1964, **55**, 910.
102. P. J. Crouch and K. J. Barnham, *Acc. Chem. Res.*, 2012, **45**, 1604.
103. D. X. West, A. E. Liberta, S. B. Padhye, R. C. Chikate, P. B. Sonawane, A. S. Kumbhar and R. G. Yerande, *Coord. Chem. Rev.*, 1993, **123**, 49.
104. T. S. Lobana, R. Sharma, G. Bawa and S. Khana, *Coord. Chem. Rev.*, 2009, **253**, 977.
105. N. C. Kasuga, A. Ohashi, C. Koumo, J. Uesugi, M. Oda and K. Nomiya, *Chem. Lett.*, 1997, 609.
106. N. C. Kasuga, K. Sekino, C. Koumo, N. Shimada, M. Ishikawa and K. Nomiya, *J. Inorg. Biochem.*, 2001, **84**, 55.
107. N. C. Kasuga, K. Sekino, M. Ishikawa, A. Honda, M. Yokoyama, S. Nakano, N. Shimada, C. Koumo and K. Nomiya, *J. Inorg. Biochem.*, 2003, **96**, 298.

108. N. C. Kasuga, Y. Hara, C. Koumo, K. Sekino and K. Nomiya, *Acta Crystallogr.*, 1999, **C55**, 1264.
109. N. C. Kasuga, K. Onodera, S. Nakano, K. Hayashi and K. Nomiya, *J. Inorg. Biochem.*, 2006, **100**, 1176.
110. K. Nomiya, K. Sekino, M. Ishikawa, A. Honda, M. Yokoyama, N. C. Kasuga, H. Yokoyama, S. Nakano and K. Onodera, *J. Inorg. Biochem.*, 2004, **98**, 601.
111. K. Nomiya, K. Onodera, K. Tsukagoshi, K. Shimada, A. Yoshizawa, T. Itoyanagi, A. Sugie, S. Tsuruta, R. Sato and N. C. Kasuga, *Inorg. Chim. Acta*, 2009, **362**, 43.
112. M. C. Barret, M. F. Mahon, K. C. Molloy, W. J. Steed and P. Wright, *Inorg. Chem.*, 2001, **40**, 4384.
113. M. C. Barret, M. F. Mahon, K. C. Molloy, P. Wright and J. E. Creeth, *Polyhedron*, 2002, **21**, 1761.
114. G. M. Arvanitis, M. E. Berardini and D. M. Ho, *Acta Crystallogr.*, 2004, **C60**, m126.
115. K. Nomiya, A. Yoshizawa, N. C. Kasuga, H. Yokoyama and S. Hirakawa, *Inorg. Chim. Acta*, 2004, **357**, 1168.
116. I. Abrahams, N. Choi, K. Henrick, H. Joyce, R. W. Matthews, M. McPartlin, F. Brady and S. L. Waters, *Polyhedron*, 1994, **13**, 513.
117. M. C. Barret, M. F. Mahon, K. C. Molloy and P. Wright, *Main Group Metal Chem.*, 2000, **23**, 663.
118. S. Eckhardt, P. S. Brunetto, J. Gagnon, M. Priebe, B. Giese and K. M. Fromm, *Chem. Rev.*, 2013, **113**, 4708.
119. B. O. Leung, F. Jalilehvand, V. Mah, M. Parvez and Q. Wu, *Inorg. Chem.*, 2013, **52**, 4593.

CHAPTER 8

Electrospun Polymer Nanofibers with Antimicrobial Activities

EL-REFAIE KENAWY

Department of Chemistry, Polymer Research Group, Faculty of Science, University of Tanta, Tanta 31527, Egypt
Email: ekenawy@yahoo.com

8.1 Introduction

Over the last few decades, microbial infections have remained one of the most serious complications associated with the use of indwelling medical devices, drug-delivery systems, dentistry, water treatment, and various consumer products. Antimicrobial materials and biocides are compounds that kill or prevent the growth of pathogenic and other unwanted micro-organisms.[1] However, many low molecular weight antimicrobial materials may be toxic to the environment and their application can be short-lived.

In order to solve the problems of microbial contamination, researchers have intensively investigated antimicrobial materials containing various bioderived and synthetic compounds.

8.1.1 Overview of Fibrous Materials with Antimicrobial Activity

There have been recently numerous research efforts dedicated toward electrospun polymer nanofibers with antimicrobial activities. The introduction of antimicrobial activity to polymer fibers has attracted much attention for applications including protection of hospital personnel and first responders, reduction of odors, and water and air filters. Various techniques have been used

RSC Polymer Chemistry Series No. 10
Polymeric Materials with Antimicrobial Activity: From Synthesis to Applications
Edited by Alexandra Muñoz-Bonilla, María L. Cerrada and Marta Fernández-García
© The Royal Society of Chemistry 2014
Published by the Royal Society of Chemistry, www.rsc.org

to introduce an antimicrobial function to nanofibers, such as physical mixing or blending of the antimicrobial agent with the polymer before electrospinning, or chemical modification of polymers to introduce the intended biocidal functions onto the polymer followed by electrospinning. In the first method, the prevention of microbial colonization can be achieved by the release of antimicrobial agent physically incorporated into the polymers.[2] In spite of the importance of the ease in manufacturing of these systems, their practical applications are limited due to the short-term killing efficacy, and inability to attack airborne bacteria. In addition, the release of biocides to the environment has the potential to increase bacterial resistance to biocides. We note that, using this strategy, it is possible to prepare silver nanoparticles containing nanofibers *via* electrospinning.

An alternative strategy is the chemical modification of polymer surfaces to achieve a more permanent effect through covalent bonding of the antibacterial agent. This leads to what is called self-sterilized materials that can protect themselves from microbes and helps in killing the pathogenic micro-organisms. There is significant literature on the preparation of permanent antibacterial materials *via* the covalent coupling of quaternary ammonium and phosphonium compounds, metals, and *N*-halamine compounds to a variety of polymer surfaces.[3]

One important application for antimicrobial nanofibers is in the field of wound dressings, where nanofibrous mats can deliver antibacterial agents and other wound healing improvements. Moreover, nanofibrous mats may promote wound cleanliness by restricting bacterial invasion *via* the sieve effect.[4]

8.2 Polymer Electrospinning

Electrospinning is an old but widely used technology for nanofiber formation that employs electrical forces to produce polymer fibers with diameters ranging from a few nanometers to several micrometers. Solutions of both natural and synthetic polymers, alone or in combination, have been employed. This technology has seen a tremendous increase in research and commercial attention over the past decade. This is due to the fact that electrospinning is a simple and easy way to obtain ultrafine fibers for a variety of medical applications. A schematic description of a typical electrospinning set-up is shown in Figure 8.1.

8.2.1 Why Polymer Nanofibers as Antimicrobial Agents?

Because of their unique properties, such as small diameters of fibers and the large surface area to volume ratio, nanosized fibers from electrospinning have been of great interest in a variety of applications including filtration, tissue-engineering scaffolds, drug-release systems, enzyme stabilization, protective clothes, sensor, carbonaceous materials, and controlled-drug delivery platforms. For medical applications, the polymers must be biocompatible and of low toxicity.[3,5–9]

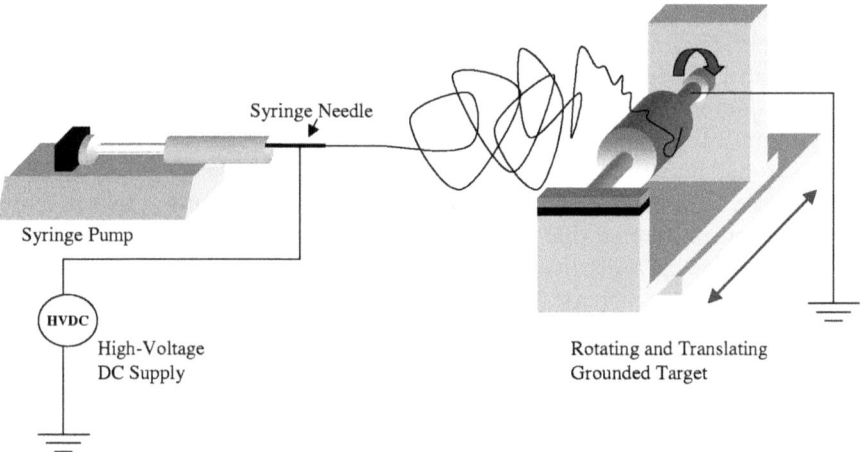

Figure 8.1   Schematic of electrospinning setup.[8,25]

In addition, the antimicrobial activity introduced into fibers and polymers has attracted increasing attention for applications including protection of hospital personnel and first responders, for reducing odor, and for antimicrobial water and air filters.[10]

In the field of wound-dressing biomaterials, nanofibrous mats perform two important functions, temporary substitute for the native extracellular matrix (ECM) and potential carrier system for the controlled delivery of antibacterial agents and other wound-healing enhancers. Because of their resemblance to the fibrillar, highly porous structure and size scale of the native ECM, plain nanofibers can inherently promote the hemostasis phase of wound healing and initiate tissue repair by facilitating cell attachment and proliferation.[11,12] They reduce wound scarring by giving cells a better roadmap for self-repair.[11] Moreover, nanofibrous mats promote wound cleanliness by restricting bacterial invasion *via* the sieve effect. The role of nanofibrous wound dressings can be further enhanced by functionalization with antimicrobial drugs and other wound-healing promoters, such as silver nanoparticles. The large surface area of the nanofibers results in efficient drug release by mass transfer, a process that can be modulated by controlling characteristics of the nanofibrous membrane and by functionalization with drug-loaded nanoparticles incorporated or adsorbed on nanofibers as well as surface graft polymerization.[4]

Also, the electrospinning technique can produce a broad range of complex architectures of nanofibers and nonwovens (Figures 8.2a–c).[13–16]

An additional advantage of these antimicrobial nanofibers polymers is that they are not yet associated with antibiotic resistance like other pharmaceuticals. Antibiotic resistance is a serious and growing phenomenon in contemporary medicine and has emerged as one of the eminent public health concerns of the 21st century, particularly as it pertains to pathogenic organisms. Nowadays, about 70 percent of the bacteria that cause infections in hospitals are resistant to at least one of the drugs most commonly used for treatment.

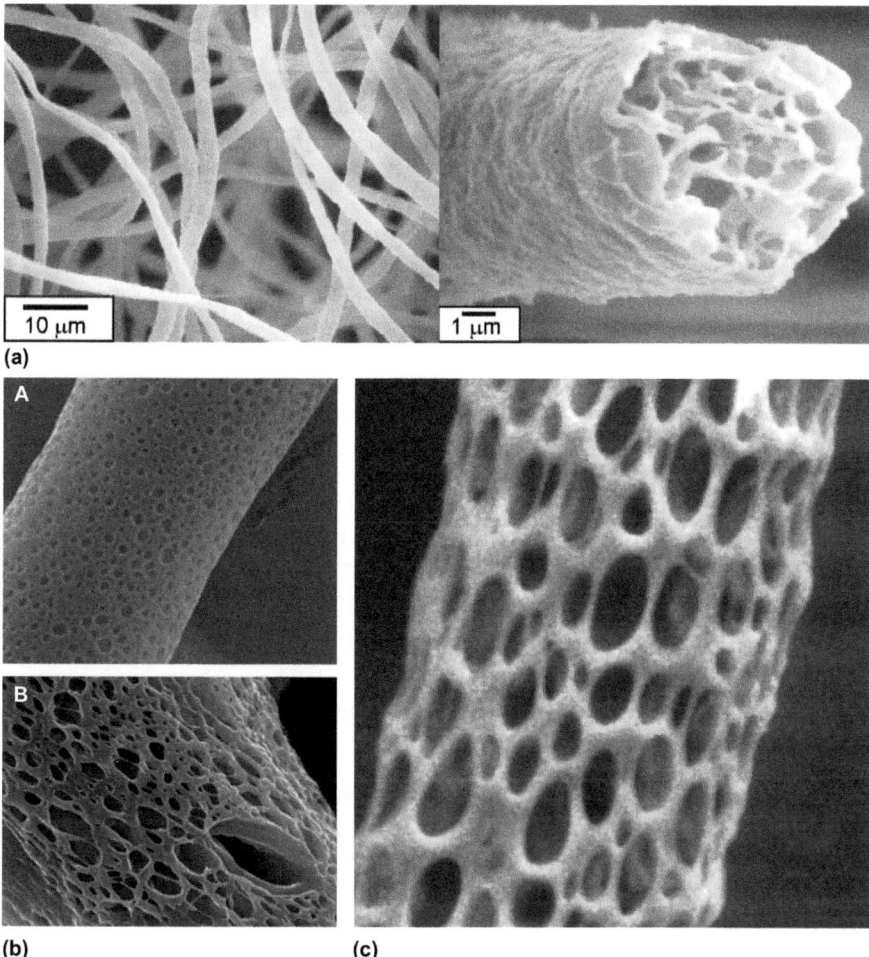

Figure 8.2 (a) PVDF fibers created by electrospinning.[14] Reprinted from *J. Supercrit. Fluids*, **53**, 142–150. Copyright (2010) with permission from Elsevier. (b) SEM images of (A) pristine poly (ε-caprolactone) (PCL) and (B) methoxy poly(ethylene glycol) (MPEG)/PCL (2 g MPEG in blend solution) mats collected into the water bath.[15] Reprinted from *Colloids Surf. B*, **88**, 587–592. Copyright (2011) with permission from Elsevier. (c) PLA fiber with porous surface spun from dichloromethane.[16] Reprinted from *Polym. Adv. Technol.*, **16**, 276–282. Copyright (2005) with permission from John Wiley and Sons.

Some organisms are resistant to all approved antibiotics and can only be treated with experimental and potentially toxic drugs. Microbial development of resistance, as well as economic incentives, has resulted in research and development in the search for new antibiotics in order to maintain a pool of effective drugs at all times.[17]

In this context, research efforts for further improvement of nanofibrous wound-healing materials have been mainly targeting wound-dressing design,

Table 8.1   Examples of polymers used to prepare fibers with antimicrobial
functionalities.

Polymer	*Biocide function*	*Type of formulation*
Chitosan	Quaternary ammonium salt	Chemical bonding[18–22]
Polyacrylonitrile	Silver nanoparticles	Physical[23]
Cellulose acetate	Silver nanoparticles	Physical[24]
Poly(ethylene-*co*-vinyl acetate)	Tetracycline HcL	Physical[25]
Nylon 6	Hydantoin derivatives	Physical[26]
Polycarbonate	Quaternary ammonium salt	Physical[27]
Polyurethane	Quaternary ammonium salt	Chemical bonding[28]
Poly(methyl methacrylate)	Silver nanoparticles	Physical[29]
Amidoxime of polyacrylonitrile	Amidoxime	Chemical bonding[30]
Poly(vinyl phenol)	Phenolic group	Chemical[31]

material properties, and manipulation of nanofibers characteristics, functionalization with antimicrobial agents, and growth factors as well as drug-release modulation. The developed materials have been usually evaluated using *in vitro* testing and less frequently *in vivo* wound healing in animal models.[13]

8.2.2   Polymers used in Electrospinning of Antimicrobial Nanofibers

Table 8.1 gives some examples from the literature of polymers used as antimicrobial nanofibers.

8.3   Antimicrobial Nanofibers

8.3.1   Antimicrobial Nanofibers by Physical Mixture

Most of the techniques used in the synthesis of bioactive polymers are based on the concept of combining the bioactive substance with polymeric materials either by physical mixing to yield a rate-controlling device, or by covalent attachment of bioactive substances to the polymer to act as a carrier for the agent.[32]

The approach mostly used to do physical mixture is that the bioactive system is dissolved or dispersed in the polymeric material. The release of the bioactive material is controlled by the diffusion through the polymeric matrix or by erosion. The prevention of microbial colonization of polymeric surfaces can be achieved by the release of biocides physically incorporated into the polymers.

The electrospinning technique provides large surface area to carry more antimicrobial agents. Embedded agents in the fiber matrix can serve as the reservoir to provide the antimicrobial power with wear due to the newly exposed surface and the potential migration from fiber matrix to the surface.

In this trend, Tan and Obendorf incorporated different *N*-halamine additives,[26] chlorinated 5,5-dimethylhydantoin (CDMH), chlorinated 2,2,5,5-tetramethyl-imidozalidin-4-one (CTMIO) and chlorinated 3-dodecyl-5,5-dimethylhydantoin (CDDMH) (Figure 8.3), into the electrospun nylon 6. Effects of *N*-halamine addition on the properties of electrospun nanofibrous membranes were also investigated. A total reduction of both *Escherichia coli* (Gram-negative bacteria) and *Staphylococcus aureus* (Gram-positive bacteria) was observed after a short contact period of 5–40 min based upon the type of *N*-halamine and the active chlorine contents. No significant leaching of *N*-halamine additives from electrospun nylon 6 membranes was observed.

Ampicillin-incorporated PMMA-nylon 6 core/shell fibers were fabricated utilizing the coaxial electrospinning technique by Sohrabi *et al.*[33] The interesting result in this work, is that there is a decrease in the fibers diameter by increasing the content of the encapsulated drug from 1% to 20%, which may be due to the increase in the conductivity of the electrospinning solution as a consequence of the ionic nature of the ampicillin sodium salt. The results reported by the authors stated that the designed drug-delivery system for all the concentrations of the encapsulated drug indicates a three-stage drug release over a period of 31 days with a sustained manner and suppressed burst release, which occurred only for 6 h. The antibacterial studies showed that there is a gradual decrease in the concentration of the bacteria with the increase in the concentration of the encapsulated drug from 1% to 20%. Higher concentrations of drug in the fibers showed enhanced drug release after 18 h incubation that resulted in the higher degree of the growth inhibition.

Preparation and electrospinning of blends of poly(vinyl alcohol) (PVA) with 2,5 dimethyl-4-hydroxy-3(2H)-furanone (DMHF) using the electrospinning technique were reported by Gule *et al.*[34] The nanofibers obtained had diameters between 150 and 300 nm and pore sizes of less than 100 nm^2.

The introduction of the furanone moiety in the nanofibers was confirmed by ATR-FTIR. DMHF-containing PVA nanofibers demonstrated good antimicrobial and cell-adhesion inhibition efficiency and did not show any signs of leaching into filtered solutions, which gives it a good chance for further investigation toward water filtration applications.

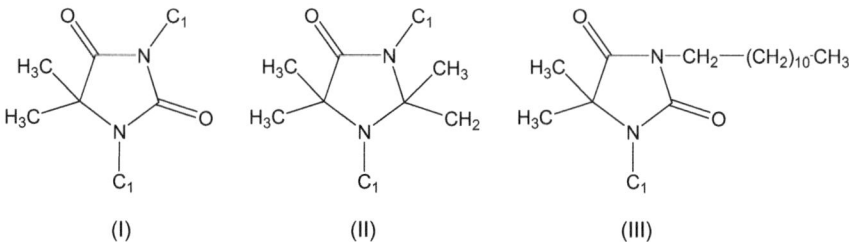

Figure 8.3 Three structurally different *N*-halamines: (I) CDMH, (II) CTMIO, and (III) CDDMH.

Lim *et al.* investigated electrospun nylon 6 nanofibers containing organic photosensitizers to demonstrate the antimicrobial properties in the application of the material to protective clothing and home appliances.[35] They used benzophenone (BP), 4,4-bis(dimethylamino)benzophenone (MK) and thiox-anthen-9-one (TX) as photosensitizers that were added during the electrospinning process. After UV (365 nm) irradiation of the photosensitizers, the intensity of peak photon excitation in the electron spin resonance spectra was increased. Antimicrobial properties of the prepared photosensitizers containing nanofibers were tested against *S. aureus* and *E. coli*. It was found that the antimicrobial properties of nylon 6 nanofibers containing MK and TX were superior to those of nylon 6 nanofibers containing conventionally used BP. The antimicrobial effects of the nanofibers for *S. aureus* were superior to those for *E. coli*. The antimicrobial activity gradually increased with the UV irradiation time.

Similar work by the same authors found that poly(vinyl alcohol) films containing organic photosensitizers provided good antimicrobial properties.[36] The results of their work showed that the photoactivity and mechanical properties of films containing various photosensitizers were significantly affected by the type and the amount of photosensitizer, and by contact areas.

Incorporation of silver nanoparticles in polymer nanofibers is a very common method to introduce the antimicrobial activities on the electospun fibers. Therefore, many researchers investigated the processing of polymers nanofibers by entrapment of nanosilver on the polymer solution before electrospinning or created the nanosilver in the fiber after electrospinning with certain treatments.[37,38]

One recent example of the latter case showed that polymer nanofibers containing Ag nanoparticles on their surface could be produced by UV irradiation of electrospun polymer nanofibers with small amounts of silver nitrate (AgNO$_3$).[24] When cellulose acetate (CA) nanofibers electrospun from CA solutions with 0.5 wt.% of AgNO$_3$ were irradiated with UV light at 245 nm, Ag nanoparticles were predominantly generated on the surface of the CA nanofibers. The number and size of the Ag nanoparticles were continuously increased up to 240 min, the results reported by the authors showed that the Ag + ions and Ag clusters diffused and aggregated on the surface of the CA nanofibers during the UV irradiation. The Ag nanoparticles with an average size of 21 nm exhibited strong antimicrobial activity.

Another example was carried out by using atmospheric helium plasma treatment to reduce the AgNO$_3$ precursor in pre-electrospinning solutions into metallic silver nanoparticles, followed by electrospinning into continuous and smooth nanofibers with Ag nanoparticles embedded in the matrix.[23] Silver nanoparticles, with diameters ranging between 3 and 6 nm, were found to be uniformly dispersed in the nanofiber matrix. The poly(acrylonitrile) (PAN) nanofibers containing Ag, exhibited slow and long-lasting silver ion release, which provided robust antibacterial activity against both Gram-positive *Bacillus cereus* and Gram-negative *E. coli* micro-organisms.

8.3.2 Antimicrobial Nanofibers by Chemical Modification of Polymers

The antimicrobial agent is covalently attached to the polymeric material before or after the electrospinning process in this type of fibers. Modification of polymer surfaces is an alternative attractive strategy to achieve a more permanent effect through covalent bonding of the antibacterial agent. There is a significant number of literature reports on the preparation of permanent antibacterial surfaces *via* the covalent coupling of poly(quaternary ammonium) compounds to a variety of polymers.[39] These antibacterial moieties were also attached to the surfaces of some woven textiles, such as cotton, wool, nylon, and polyester.[40]

Yao *et al.* reported surface modification of electrospun poly(vinylidene fluoride-*co*-hexafluoropropylene) (PVDF-HFP) fibrous membranes by poly(4-vinyl-*N*-alkylpyridinium bromide) to achieve antibacterial activities. The membranes were first subjected to plasma pretreatment followed by UV-induced surface graft copolymerization of 4-vinylpyridine (4VP) and quaternization of the grafted pyridine groups with hexyl bromide. The antibacterial activities of the modified electrospun PVDF-HFP fibrous membranes were assessed against Gram-positive *S. aureus* and Gram-negative *E. coli*. The results showed that the PVDF-HFP fibrous membranes modified with quaternized pyridinium groups are highly effective against both bacteria with killing efficiency as high as 99.99%. The conclusions from this work were that the surface modification of electrospun fibrous membranes with poly(4-vinyl-*N*-alkylpyridinium bromide) serves as an efficient and versatile approach for the development of polymeric fibers with excellent antibacterial activity.[40]

Ignatova *et al.* prepared nonwoven nanofibrous mats fibers from quaternized chitosan derivative (QCh) (Figure 8.4) by electrospinning of mixed aqueous solutions of QCh with poly(vinyl pyrrolidone) (PVP).[41]

The antibacterial activity of the quaternized chitosan electrospun mats against the Gram-positive bacteria *S. aureus* and Gram-negative bacteria *E. coli* has been reported. The crosslinked electrospun mats containing QCh were efficient in inhibiting growth of both Gram-positive and Gram-negative bacteria. This suggests that the crosslinked QCh/PVP electrospun mats are promising as materials for biomedical applications, such as wound-dressing applications.

Nonwoven nanofibrous mats from *N*-carboxyethyl chitosan and poly(acrylamide) have also been reported.[42] It was shown that the presence of

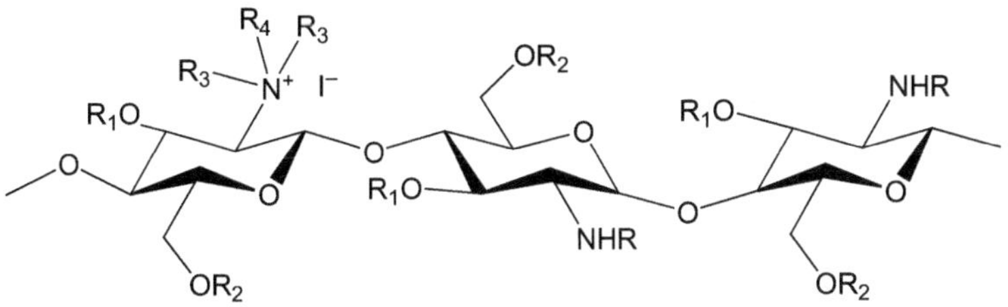

Figure 8.4 Quaternized chitosan derivative (QCh).

ionizable low molecular weight compounds (8-quinolinol derivatives) in chitosan and *N*-carboxyethylchitosan containing spinning solutions led to the decrease in the diameter of the nanofibers. The drug-containing nanofibers showed good antimicrobial and antimycotic activity. The authors also reported on the preparation of nanofibers containing quaternized chitosan (QCh) by electrospinning of mixed aqueous solutions QCh/poly(vinyl alcohol). The electrospun fibers had diameters in the range 60–200 nm and showed good antibacterial activity against Gram-positive and Gram-negative bacteria. Zhang *et al.* prepared nanofibrous membranes based on PAN with fiber diameters of 450 nm by the technique of electrospinning. The amidoxime nanofibrous membranes were prepared through treatment of PAN nanofibrous membranes with hydroxylamine (NH_2OH) aqueous solution. The –CN groups on the surface of PAN nanofibers reacted with NH_2OH molecules and led to the formation of –$C(NH_2)$ N–OH groups, which were used for coordination of Ag + ions (Figure 8.5). Subsequently, the coordinated Ag + ions were converted into silver nanoparticles (AgNP) with sizes of tens of nanometers. The antimicrobial efficacies against *S. aureus* and *E. coli* of the membranes of electrospun PAN (ESPAN) nanofibers, ESPAN surface functionalized with amidoxime groups (ASFPAN), ASFPAN coordinated with silver ions (ASF-PAN–Ag +), and ASFPAN attached with silver nanoparticles (ASFPAN–AgNP) were investigated. The study revealed that, with treatment of ESPAN membranes in 1M NH_2OH aqueous solution for 5 min, the resulting ASF-PANm membranes became antimicrobial without distinguishable morphological variations.[30]

Antimicrobial polymeric systems were prepared from poly(vinyl phenol) (PVPh) (Figure 8.6). Two systems were prepared by the electrospinning of poly(vinyl phenol) with molecular weight 20–10^3 (PVPh 20k spun) and 100×10^3 (PVPh 100k spun). An SEM photo of electrospun PVPh 100k is shown in Figure 8.7.

The antimicrobial activity of the polymer was examined against different test micro-organisms. Generally, it was found that polymer morphology and molecular weight affect the activities against test micro-organisms. For example, PVPh 20k and PVPh 100k in their powder form showed no antimicrobial activity. However, the results showed that PVPh 20k spun has antibacterial activity against *Bacillus subtilis*, and there is no growth of the tested micro-organisms.[31]

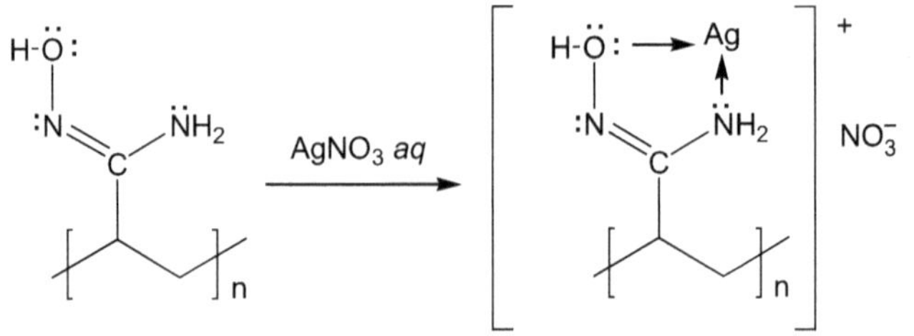

Figure 8.5 The formation of coordination bonds between a silver ion and an amidoxime group.

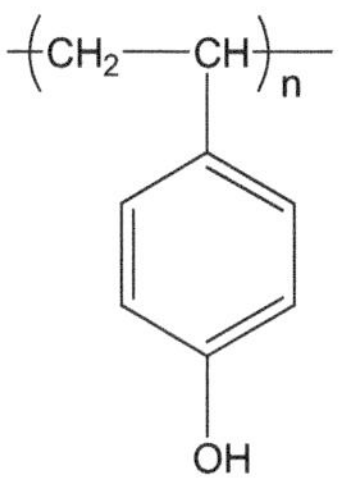

Figure 8.6 Structure of poly(vinyl phenol) (PVPh).

Figure 8.7 SEM micrograph of electrospun PVPh 100k.[31]

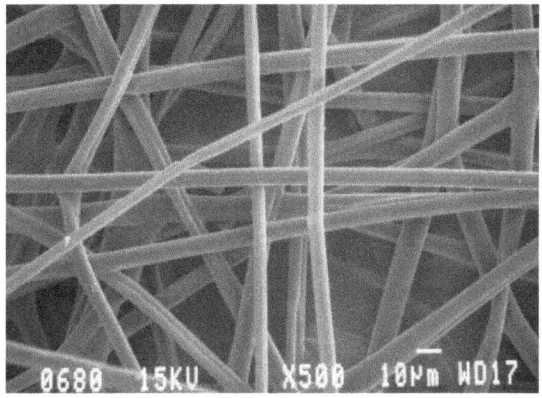

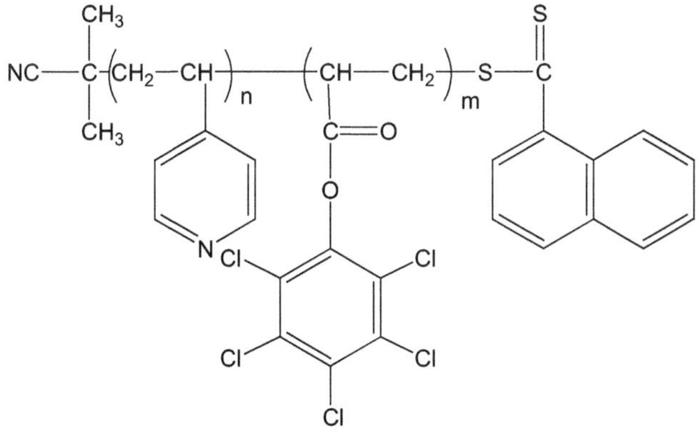

Figure 8.8 Copolymers of 4-vinylpyridine (4VP) and pentachlorophenyl acrylate (PCPA) (P4VP-*b*-PPCPA).

Antibacterial nanofibers based on block copolymers of pentachlorophenyl acrylate (PCPA) and 4VP (P4VP-*b*-PPCPA) were prepared by Qun *et al.* (Figure 8.8).[43] The quaternary ammonium salts (QASs) were generated by

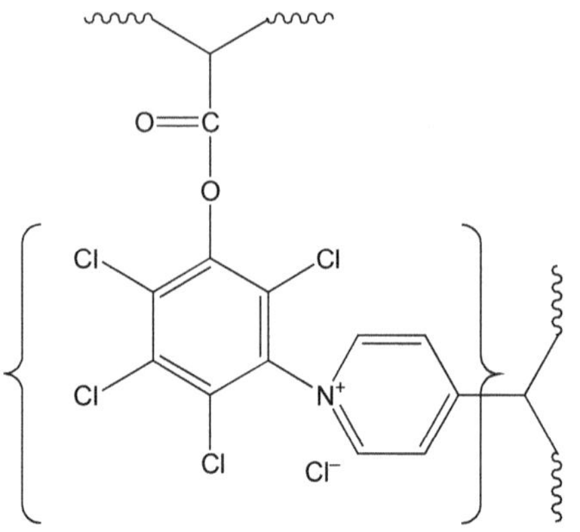

Figure 8.9 Self-quaternization of P4VP-*b*-PPCPA.

N-alkylation of pyridine groups of P4VP block and chloroaromatic compounds of PPCPA block. The polymers were electrospun from mixed tetrahydrofuran/ dimethylformamide (THF/DMF) solution and fibers with diameters in the range of 500 nm to 4 μm were obtained. The nanofibers exhibit good anti-bacterial activity to *E. coli* and *S. aureus*, 96% of *E. coli* and 99% of *S. aureus* were killed after being in contact with 50 mg nanofibers in 10 min. A high antibacterial activity of the nanofibers is attributable to the hydrophobic interaction of PPCPA block and the electrostatic interaction of QASs generated from the self-quaternization of P4VP-*b*-PPCPA (Figure 8.9). The leaching of active groups of nanofibers was largely prevented.

8.4 Examples of Applications of Antimicrobial Nanofibers

8.4.1 Drug Delivery

Electrospun fiber mats are explored as drug-delivery vehicles using tetracycline hydrochloride antibiotic as a model drug in 2002 by Kenawy *et al.*[25] The mats were made either from poly(lactic acid) (PLA), poly(ethylene-*co*-vinyl acetate) (PEVA), or from a 50:50 blend of the two (Figure 8.10). The fibers were electrospun from chloroform solutions containing a small amount of methanol to solubilize the drug. The release of the tetracycline hydrochloride from these new drug-delivery systems was followed by UV-Vis spectroscopy. Release profiles from the electrospun mats were also compared with a commercially available drug-delivery system, Actisite, as well as to cast films of the various formulations. Electrospun PEVA shows a higher release rate for tetracycline hydrochloride than the mats derived from 50/50 PLA/ PEVA or pure PLA. Electrospun PEVA released 65% of its drug content within 120 h, whereas the 50/50 material released about 50% over the same time.

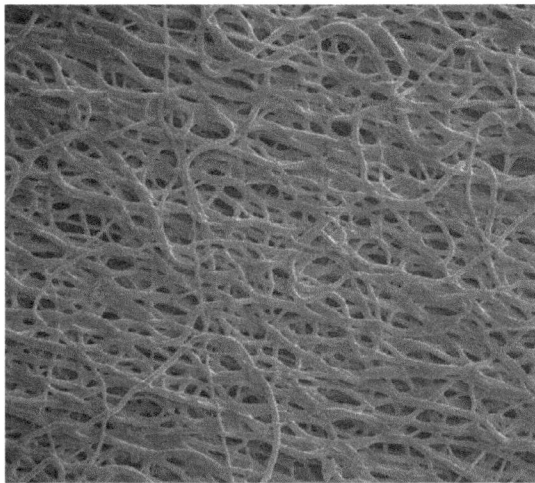

Figure 8.10 SEM micrograph of fibers from PEVA solution in chloroform containing 5% tetracycline hydrochloride.[25]

8.4.2 Wound Healing

An ideal dressing for wound healing should have certain characteristics, such as haemostatic ability, efficiency as bacterial barrier, absorption ability of excess exudates (wound fluid/pus), appropriate water-vapor transmission rate, adequate gaseous exchange ability, capacity to be shaped to the contour of the wound area, functional adhesion (*i.e.* adherent to healthy tissue but non-adherent to wound tissue), painless to patient and ease of removal, and finally low cost.[44]

Quaternized chitosan (QCh) derivatives have shown higher activity against bacteria, broader spectrum of activity and higher killing rates when compared with those based on chitosan[18] and, thus, they are potential candidates for wound-dressing applications.[19] Electrospinning in common solvent has been used for the preparation of bicomponent hybrid nanofibrous materials based on natural chitosan (Ch) or its quaternized derivative (*N,N,N*-trimethyl chitosan iodide), shown in Figure 8.11, and synthetic aliphatic polyester poly[(L-lactide)-*co*-(D,L-lactide)] (PLA).[20] The spinning solution Ch/PLA was prepared by mixing 5 wt.% chitosan solution in trifluoroacetic acid/dichloromethane (TFA/DCM) (70/30 vol.%) with 5 wt.% PLA solution in TFA/DCM (70/30 vol.%) at weight ratio of Ch:PLA = 50:50. On the other hand, the spinning solution QCh/PLA was prepared by mixing 5 wt.% QCh solution in dry DMF/DMSO (60/40, vol.%) with 5 wt.% PLA solution in dry DMF/DMSO (60/40 vol.%) at weight ratio of QCh:PLA = 30:70.

The antibacterial activities of crosslinked bicomponent Ch/PLA and QCh/PLA nanofibers against Gram-positive bacteria *S. aureus* and Gram-negative bacteria *E. coli* were evaluated by counting the viable bacteria cells that rested in a bacteria suspension after the contact of electrospun mats with the suspension. These hybrid nanofibers showed the capability to kill all the *S. aureus* and *E. coli* cells within 60 min of contact.[20]

Nanofibers containing quaternized chitosan have been prepared by electrospinning of QCh solutions mixed with poly (vinyl alcohol).[21] The quaternized chitosan used was *N*-butyl-*N*,*N*-dimethyl chitosan iodide prepared by Kim *et al.* method.[22] The spinning solutions QCh/PVA were prepared by mixing PVA solution with aqueous solutions of QCh. The total polymer concentration in the mixed solutions was 8 wt.%. The solutions for the production of crosslinked nanofibers and of solution-cast films were prepared using triethylene glycol diacrylate (TEGDA) crosslinking agent, 2,2-dimethoxy-2-phenylacetophenone (DMPA) (1%) in dimethyl sulfoxide (DMSO) and ammonium peroxydisulfate (1%) acting as photoinitiator (all in weight percent to total polymer content). Antimicrobial activity screening has demonstrated the antibacterial activity of the QCh/PVA as well as the photocrosslinked electrospun nanofibers against *S. aureus* and *E. coli*. The reduction of bacteria *E. coli* for photocrosslinked QCh/PVA nanofibers containing QCh was 98% after 120 min contact time. The obtained nanofibrous electrospun mats are promising for wound-healing applications because they could contribute to the prevention of secondary infections in wounds by *S. aureus*, resulting in limited scar formation. Another chitosan quaternary salt *N*-[(2-hydroxy-3-trimethylammonium) propyl] chitosan chloride (HTCC) (Figure 8.12), a water-soluble derivative of chitosan, was synthesized *via* the reaction between glycidyl trimethyl ammonium chloride and chitosan.[45] Solutions of PVA-HTCC blends

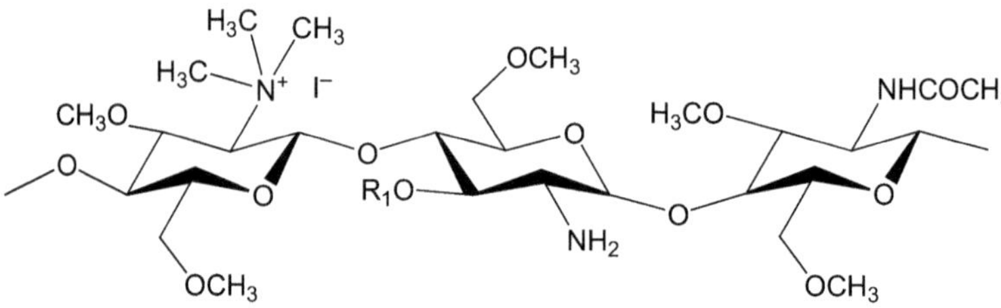

Figure 8.11   Structure of *N*,*N*,*N*-trimethyl chitosan iodide.

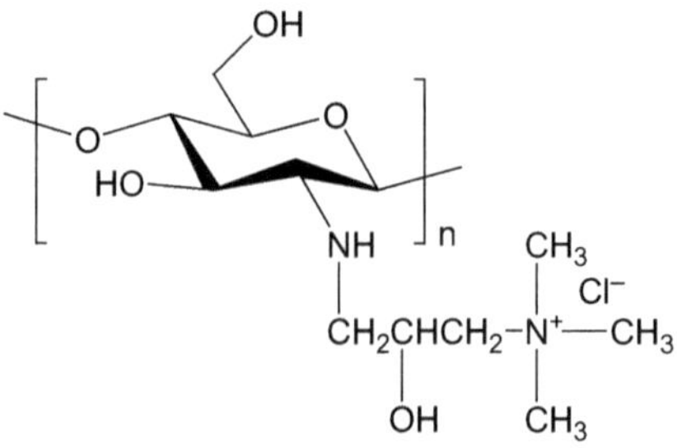

Figure 8.12   Structure of *N*-[(2-hydroxy-3-trimethylammonium) propyl] chitosan chloride (HTCC).

were electrospun. The average fiber diameter was in the range of 200–600 nm. SEM images revealed that increasing HTCC content in the blends decreases the average fiber diameter. The shear viscosity of the PVA-HTCC blend decreases as HTCC content increases in the blend, whereas the electrical conductivity of the blend increases. Electrospun nanofibrous PVA-HTCC mats showed a good antibacterial activity against the Gram-negative bacteria *E. coli* and Gram-positive bacteria *S. aureus*. This feature suggests that the PVA-HTCC electrospun mats are potentially good candidates for biomedical applications.

8.4.3 Filtration

The filters of nanofibrous membranes with antimicrobial functionality have attracted growing attention due to the concerns about qualities of purified water and/or filtered air as well as the processing costs.[46] Water and air filters (particularly those operating in the dark and damp conditions) are constantly subject to attacks from environmental micro-organisms. The micro-organisms (such as bacteria) that can be readily captured by the filters grow rapidly, resulting in the formation of biofilms. Consequently, the buildups of micro-organisms on the filter surfaces deteriorate the quality of purified water and/or filtered air. Additionally, they also have unfavorable effects on the flow of water and/or air.[30]

Antimicrobial agents, such as quaternary ammonium salts, metals, phosphonium compounds and *N*-halamine compounds, have been used to impart biocidal functions onto fibers and polymers for protection against infectious disease pathogens by using techniques such as grafting, coating, and blending.

Functional nanofibers with antimicrobial properties were prepared by electrospinning from poly(acrylonitrile)/DMSO solution containing a bioactive agent based on quaternary ammonium salts (*N,N*-didecyl-*N,N*-dimethylammonium chloride, bis-(3-aminopropyl)-dodecylamine) and 2-propanol. Addition of bioactive compound to the polymer solution (concentration of active substance in solution about 1.5 wt.%) makes it possible to obtain fibers showing good bactericidal properties. 99.84% of *E. coli* were killed and 99.99% of *S. aureus* after 6 h of contact with the fibers. As reported by the authors, the spectrophotometric investigations of the nanofibers do not indicate a residue of solvent in the bioactive nanofibers and show an increase in content of CH and CH_2 groups in relation to the pure nanofibers, which is connected to the presence of the biocide, suggesting that these nanofibers can be designed for medical and filtration applications.[47]

8.5 Future Prospects

There is ongoing research for the improvement of nanofiber properties and the scale up of the electrospinning process. Significant opportunities exist for nanofibers with antimicrobial activity to meet a variety of important needs in hygiene and human health. In the future, electrospun nanofibers will likely prove to be promising candidates for an even wider range of applications.

References

1. E.-R. Kenawy, S. D. Worley and R. Broughton, *Biomacromolecules*, 2007, **8**, 1359–1384.
2. C. Yao, X. Li, K. G. Neoh, Z. Shi and E. T. Kang, *Appl. Surf. Sci.*, 2009, **255**, 3854–3858.
3. X. Ren, H. B. Kocer, S. D. Worley, R. M. Broughton and T. S. Huang, *J. Appl. Polym. Sci.*, 2013, **127**, 3192–3197.
4. S. S. Said, A. K. Aloufy, O. M. El-Halfawy, N. A. Boraei and L. K. El-Khordagui, *Eur. J. Pharm. Biopharm.*, 2011, **79**, 108–118.
5. E.-R. Kenawy, F. I. Abdel-Hay, M. H. El-Newehy and G. E. Wnek, *Mater. Sci. Eng., A*, 2007, **459**, 390–396.
6. A. Elder, I. Lynch, K. Grieger and E.-R. Kenawy, Critical Knowledge Gaps in Nanomaterials Risk Assessment, NATO Advanced Research Workshop on Nanomaterials – Environmental Risk and Benefits, APR 27–30, 2008 Faro Portugal nanomaterials in risks and benefits, 2009, pp. 3–29.
7. E.-R. Kenawy and Y. R. Abdel-Fattah, *Macromol. Biosci.*, 2002, **2**, 261–266.
8. E.-R. Kenawy, J. M. Layman, J. R. Watkins, G. L. Bowlin, J. A. Matthews, D. Simpson and G. E. Wnek, *Biomaterials*, 2003, **24**, 907–913.
9. N. Charernsriwilaiwata, P. Opanasopita, T. Rojanarataa, T. Ngawhirunpata and P. Supaphol, *Carbohydr. Polym.*, 2010, **81**, 675–680.
10. R. Gopal, S. Kaur, Z. Ma, C. Chan, S. Ramakrishna and T. Matsuura, *J. Membr. Sci.*, 2006, **281**, 581–586.
11. P. Zahedi, I. Rezaeian, S.-O. Ranaei-Siadat, S.-H. Jafari and P. Supaphol, *Polym. Adv. Technol.*, 2010, **21**, 77–95.
12. S. Sell, C. Barnes, M. Smith, M. McClure, P. Madurantakam, J. Grant, M. McManus and G. Bowlin, *Polym. Int.*, 2007, **56**, 1349–1360.
13. S. S. Said, O. M. El-Halfawy, H. M. El-Gowelli, A. K. Aloufy, N. A. Boraei, L. K. El-Khordagui and J. Liu, *Eur. J. Pharm. Biopharm.*, 2012, **80**, 85–94.
14. Z. Shen, S.-H. Lee, M. Marquez and M. A. McHugh, *J. Supercrit. Fluids*, 2010, **53**, 142–150.
15. H. R. Pant, M. P. Neupane, B. Pant, G. Panthi, H. J. Oh, M. H. Lee and H. Y. Kim, *Colloids Surf. B*, 2011, **88**, 587–592.
16. R. Dersch, M. Steinhart, U. Boudriot, A. Greiner and J. H. Wendorff, *Polym. Adv. Technol.*, 2005, **16**, 276–282.
17. R. J. B. Pinto, S. C. M. Fernandes, C. S. R. Freire, P. Sadocco, J. Causi, C. P. Neto and T. Trindade, *Carbohydr. Res.*, 2012, **348**, 77–83.
18. Z. Jia, D. Shen and W. Xu, *Carbohydr. Res.*, 2001, **333**, 1–6.
19. M. Z. Elsabee, H. F. Naguib and R. E. Morsi, *Mater. Sci. Eng. C*, 2012, **32**, 1711–1726.
20. M. Ignatova, N. Manolova, N. Markova and I. Rashkov, *Macromol. Biosci.*, 2009, **9**, 102–111.
21. M. Ignatova, K. Starbova, N. Markova, N. Manolova and I. Rashkov, *Carbohydr. Res.*, 2006, **341**, 2098–2107.
22. C. H. Kim, J. W. Choi, H. J. Chun and K. S. Choi, *Polym. Bull.*, 1997, **38**, 387–393.

23. Q. Shi, N. Vitchuli, J. Nowak, J. M. Caldwell, F. Breidt, M. Bourham, X. Zhan and M. McCord, *Eur. Polym. J.*, 2011, **47**, 1402–1409.
24. W. K. So, J. H. Youk and W. H. Park, *Carbohydr. Polym.*, 2006, **65**, 430–434.
25. E.-R. Kenawy, G. L. Bowlin, K. Mansfield, J. Layman, D. G. Simpson, E. H. Sanders and G. E. Wnek, *J. Control. Release*, 2002, **81**, 57–64.
26. K. Tan and S. K. Obendorf, *J. Membr. Sci.*, 2007, **305**, 287–298.
27. S. J. Kim, Y. S. Nam, D. M. Rhee, H.-S. Park and W. H. Park, *Eur. Polym. J.*, 2007, **43**, 3146–3152.
28. C. Yao, X. Li, K. G. Neoh, Z. Shi and E. T. Kang, *J. Membr. Sci.*, 2008, **320**, 259–267.
29. Y. Bao, C. Lai, Z. Zhu, H. Fong and C. Jiang, *RSC Adv.*, 2013, **3**, 8998–9004.
30. L. Zhang, J. Luo, T. J. Menkhaus, H. Varadaraju, Y. Sun and H. Fong, *J. Membr. Sci.*, 2011, **369**, 499–505.
31. E.-R. Kenawy and Y. R. Abdel-Fattah, *Macromol. Biosci.*, 2002, **2**, 261–266.
32. E.-R. Kenawy, D. C. Sherrington and A. Akelah, *Eur. Polym. J.*, 1992, **28**, 841–862.
33. A. Sohrabi, P. M. Shaibani, H. Etayash, K. Kaur and T. Thundat, *Polymer*, 2013, **54**, 2699–2705.
34. N. P. Gule, M. de Kwaadsteniet, T. E. Cloete and B. Klumperman, *Water Res.*, 2013, **47**, 1049–1059.
35. K. S. Lim, K. W. Oh and S. H. Kim, *Polym. Int.*, 2012, **61**, 1519–1524.
36. K. S. Lim and K. W. Oh, *Textile Sci. Eng.*, 2011, **48**, 42–50.
37. A. M. Abdelgawad, S. M. Hudson and O. J. Rojas, *Carbohydr. Polym.*, 2013, http://dx.doi.org/10.1016/j.carbpol.2012.12.043.
38. X. Zhuang, B. Cheng, W. Kang and X. Xu, *Carbohydr. Polym.*, 2010, **82**, 524–527.
39. E.-R. Kenawy, F. I. Abdel-Hay, L. Shahada, A. E.-R. R. El-Shanshoury and M. H. El-Newehy, *J. Appl. Polym. Sci.*, 2006, **102**, 4780–4790.
40. C. Yao, X. Li, K. G. Neoh, Z. Shi and E. T. Kang, *Appl. Surf. Sci.*, 2009, **255**, 3854–3858.
41. M. Ignatova, N. Manolova and I. Rashkov, *Eur. Polym. J.*, 2007, **43**, 1112–1122.
42. R. Mincheva, N. Manolova, D. Paneva and I. Rashkov, *J. Bioact. Compat. Polym.*, 2005, **20**, 419–435.
43. X. L. Qun, Y. Fang, Y. Shan, G.-D. Fu, S. Liang, N. Shengzhe and Z. Meifang, *High Perform. Polym.*, 2010, **22**, 359–376.
44. N. Bhardwaj and S. C. Kundu, *Biotechnol. Adv.*, 2010, **28**, 325–347.
45. S. M. Alipour, M. Nouri, J. Mokhtari and S. H. Bahrami, *Carbohydr. Res.*, 2009, **344**, 2496–2501.
46. N. Daels, S. Vrieze, I. De Sampers, B. Decostere, P. Westbroek, A. Dumoulin, P. Dejans, K. De Clerck and S. W. H. Van Hulle, *Desalination*, 2011, **275**, 285–290.
47. E. Gliscinska, B. Gutarowska, B. Brycki and I. Krucinska, *J. Appl. Polym. Sci.*, 2013, **128**, 767–775.

CHAPTER 9

Biomimetic Polyurethanes

ANTONELLA PIOZZI* AND IOLANDA FRANCOLINI

Department of Chemistry, Sapienza University of Rome, P. le Aldo Moro 5, 00185 Rome
*Email: antonella.piozzi@uniroma1.it

9.1 Biomimetic Materials: Definition and Potential Applications

A biomaterial can be defined as a substance (with the exception of drugs), or a combination of substances (both synthetic and natural), employed for the treatment, the improvement or substitution of human tissues, organs or function. Since interaction with the biological system is involved, biocompatibility implies the capability of the material of exhibiting, in the host, an appropriate functional and "biomimetic" response. In recent years, the interest towards natural and synthetic polymers has steadily grown owing to their physicochemical properties that can be easily modified to fulfil different tasks. Their use has contributed to significant improvement of the quality and duration of human life.

The applications of polymers in the biomedical field are numerous and can be either of intra- or extracorporeal type. Intracorporeal applications include short-term devices, like sutures[1,2] and adhesives,[3–5] or long-term implants such as cardiac valves,[6] artificial tendons[7–9] and contact lenses.[10–12] Also, complex systems aimed at substituting the function of an organ in the emerging tissue-engineering field can be included in this class.[13–16]

Among the extracorporeal applications, intravascular and urinary catheters, filters for hemodialysis, oxygenators and detoxification circuits are the most used devices in healthcare settings.

RSC Polymer Chemistry Series No. 10
Polymeric Materials with Antimicrobial Activity: From Synthesis to Applications
Edited by Alexandra Muñoz-Bonilla, María L. Cerrada and Marta Fernández-García
© The Royal Society of Chemistry 2014
Published by the Royal Society of Chemistry, www.rsc.org

Since biocompatibility is strictly related to the specific application and location of the implanted device, initially the research aimed to obtain substantially inert materials with respect to the surrounding physiological environment thus to be tolerated. However, although a number of materials with enhanced features have been developed, the complexity and multitude of parameters involved in the regulation of tissue/material interaction made difficult the attainment of absolutely inert materials.[17–20]

The modern research approaches are therefore focused on the development of materials able to specifically interact with the biological systems by exploiting the principles of cellular recognition. These so-called "biomimetic materials" can result from the polymerization of monomers possessing intrinsic biomimetic properties[21–23] or from the surface modification of suitable commercial polymers with biologically active molecules.[24–28]

In general, any foreign material contacting body tissues or fluids generates a specific reaction depending on the particular site of implantation. Generally, a complex integrated defensive system is activated. As for blood-contacting medical devices, upon exposure to the biological environment, plasma proteins, including fibrin, albumin and fibrinogen, adsorb on the device's surface (conditioning film), allowing the adhesion and activation of platelets and leukocytes. This process, known as biofouling, is usually the first stage of a cascade of biological events, which leads to blood-clot formation, and promotes bacterial adhesion and biofilm formation on device surfaces.[29,30] The resulting infectious and thrombotic complications can impair the function of the device and lead to implant failure and life-threatening consequences, as in the case of vascular grafts.

A reaction of inflammatory type can also occur and cause the progressive intervention of macrophages and leukocytes, with a subsequent secretion of mucopolysaccharides and procollagen giving rise to the foreign body coating with a fibrous tissue. A further reaction can be of the immune type, due to the antigenic activity exerted by the material or its release products. Other adverse reactions can be the activation of the complement system and the arising of carcinogenesis or mutagenesis phenomena.[31,32]

In order to minimize or inhibit these negative responses, the ideal biomaterial should possess the following peculiar properties:

- constancy of the bulk properties, in order to maintain the functional properties of the device, and to avoid the release of potentially toxic substances, like monomers, additives, or degradation products;
- good hemocompatibility, to avoid the activation of the clotting mechanisms and formation of thrombi;
- antifouling properties, to control the type of the adsorbed proteins on the device surface and thus prevent bacterial adhesion.

In addition to biostable polymers that are used for long-term prostheses (such as hip, knee, dental, ophthalmic and cardiovascular devices) and for extracorporeal devices for purification treatments (hemodialysis, hemofiltration, plasmapheresis), also biodegradable and bioerodible polymers play an important role in several therapeutic applications, such as in drug delivery and tissue engineering.

Therefore, the chemical, physical and technological properties of biomedical polymers can be very different. Indeed, water-soluble or biodegradable polymers, like poly(ethylene glycol) or poly(lactic acid), or hydrophobic polymers and resistant to the mechanical stress and hydrolytic enzyme action, like poly(ethylene terephthalate), poly(alkyl siloxanes), fluorinated polymers and polyurethanes, are currently employed as biomaterials.

Polyurethanes (PUs) are commonly used in the biomedical field since they provide good biocompatibility, flexural endurance, high tensile strength, high tear and abrasion resistance. These properties, which can avoid the device weakening or failure, together with their processing versatility have led to the manufacturing of a wide range of high-performance medical devices including vascular prostheses, catheters, feeding tubes, surgical drains, intra-aortic balloon pumps, dialysis devices, medical garments, wound dressings and more.[31]

In addition, plasma protein absorption onto polyurethanes was found to be slower or less than onto other materials.[33–35] This finding makes polyurethanes ideal candidates for a variety of medical applications requiring unique antithrombotic and biomimetic properties.

Finally, the possibility to obtain biodegradable polyurethanes has extended the range of properties exhibited by these materials and their application in drug delivery and tissue engineering fields.[36–41]

This chapter will deal with the use of biomimetic polyurethanes in the medical field especially for the manufacturing of medical prostheses or regeneration of damaged tissues or organs. Polyurethane chemistry will be first introduced to show the importance of monomer selection and polymerization conditions on the chemical and mechanical properties of polyurethanes. The main applications of polyurethanes in biomedical field will be reviewed. Next, the experimental approaches to obtain the biomimetic polyurethanes will be described paying particular attention to polymer bulk and surface modification with proper functional groups or antifouling small molecules and macromolecules. Then, a survey of the potential applications of these materials in preventing microbial biofilm formation on medical devices or in promoting cell adhesion and proliferation in tissue engineering will be carried out. Finally, a discussion of the future direction of biomimetic polyurethanes as biomaterials will complete the chapter.

9.2 Polyurethanes: Chemistry and Main Applications of Polyurethanes in the Biomedical Field

The successful use of polyurethanes in biomedical applications is mainly related to their unique molecular architecture from which interesting and various mechanical properties stem.

Elastomeric polyurethanes are indeed synthesized as segmented block copolymers, their chemical structure consisting of alternating *soft* and *hard* segments. The *soft* segments are typically macrodiols (polyalkyldiols, polyesters or polyethers) while the *hard* segments are usually high glass transition temperature aromatic or aliphatic diisocyanates, linked with a low molecular weight chain extender.

The broad range of physical properties exhibited by polyurethanes is due to the type and molecular weight of the monomers as well as the molar ratio in which the monomers are reacted. Therefore, either very hard materials or soft and viscous ones can be obtained. In fact, owing to the partial chemical incompatibility between the *hard* and *soft* segments, a certain degree of immiscibility exists. This means that although polyurethanes are macroscopically isotropic, they are not structurally homogeneous at the microscopic level. Indeed, PUs show a two-phase structure in which *hard* segment domains are dispersed in a matrix of *soft* segments (Figure 9.1).

The physical and mechanical properties of polyurethanes depend strongly on the degree of *hard/soft* phase segregation together with the cohesion of the *hard* domains. The degree of phase segregation depends on the *hard/soft* phase ratio, the molecular weight of the *soft* segments and the different polarity of the two segments. The use of crystallizable *hard* segments as well as of *hard* and *soft*

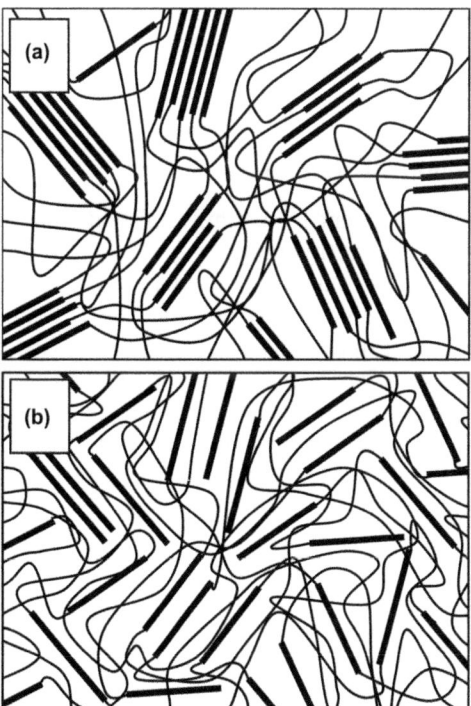

Figure 9.1 Schematic representation of the two phase structure in polyurethanes: high (a) and low (b) degree of phase segregation (——————— = *hard* segment, ⌐⌐⌐⌐ = *soft* segment).

segments with different polarity is known to contribute to the improvement of PU phase segregation and *hard* domain cohesion.[42–45]

Phase segregation in PUs was investigated by using different experimental techniques including electron microscopy,[46] small-angle X-ray scattering (SAXS),[47] infrared spectroscopy,[48] dynamo-mechanical analysis and differential scanning calorimetry (DSC).[49]

DSC permits to evaluate the glass-transition temperature (T_g) that is characteristic of the *soft* phase and can be affected by the establishment of interactions between the *hard* and *soft* segments. In particular, a large number of interactions between the two segments (phase mixing) reduces the mobility of the *soft* phase with a consequent T_g increase.[50,51] A stiffening of the *soft* phase can be brought about by the formation of hydrogen bonds established between ether or ester oxygen of the *soft* phase and -NH groups of the *hard* phase.

The introduction of functional groups in the polymer chain can be responsible for two effects: the first is a stiffening of the polymer chain, with T_g increase, owing to the introduction of groups able to form hydrogen bonds; the second is the steric-hindrance effect due to the presence of substituents that obstacle the formation of the above mentioned intrachain bonds.[52]

The domain structure of PUs can be designed and controlled at various stages during synthesis. In the following section, the chemistry of the urethane group, the main raw materials employed in the PU polymerization as well as the basic principles of the synthesis of segmented polyurethanes are presented.

9.2.1 Basic Chemical Reactions

The term polyurethane refers to a family of synthetic polymers containing urethane moieties in the repetitive unit.

The main reaction involved in polyurethane synthesis is between the isocyanate group and a hydroxy-terminated molecule (Figure 9.2).

The reaction implies a nucleophilic addition at the isocyanate N=C double bond and is catalyzed by basic (*e.g.* tertiary amines) or metal compounds (*e.g.* organotin). Aromatic isocyanates are generally more reactive than aliphatic ones, while the presence of electron-donating substituents near the isocyanate group reduces the reaction rate. Primary alcohols react faster than secondary and tertiary alcohols since the substituent groups sterically hinder the reaction.

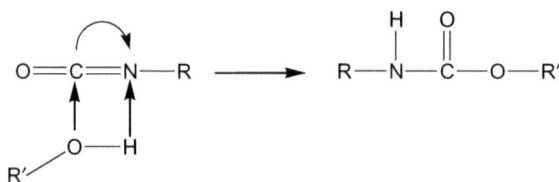

Figure 9.2   Reaction between the isocyanate and the hydroxyl group.

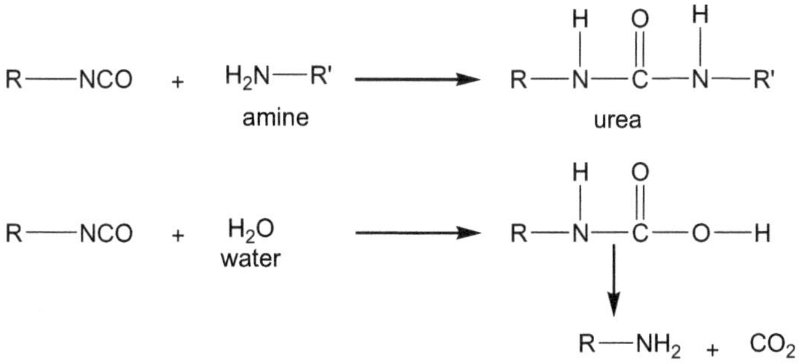

Figure 9.3 Reaction of the isocyanate group with amine or water.

If a diamine reacts with the isocyanate the formation of a urea linkage occurs (Figure 9.3). The isocyanate group can also react with water, if present in the reaction environment, leading to the formation of unstable carbamic acids that quickly decompose to amines and carbon dioxide (Figure 9.3). The liberated carbon dioxide can be used to obtain PU foams. If this is not desired, an anhydrous environment must be kept during the synthesis.

When an excess of isocyanate is present, thermally labile allophanate bonds can be formed by reaction of two isocyanate groups. The biuret formation is instead possible for reaction between a urea and isocyanate group.

9.2.2 Raw Materials

To obtain a linear polyurethane, three components, each having two reactive functional groups, are needed: i) a diisocyanate, ii) a macrodiol consisting of a flexible chain ending with hydroxyl groups and iii) a low molecular weight diol or diamine (chain extender).

As already described, the choice of the monomers strongly influences the physicochemical properties of the resulting polymer.

-Isocyanates

Both aliphatic and aromatic isocyanates can be used to synthesize polyurethanes. Toluene diisocyanate (TDI) and methylene bis-*p*-phenyl isocyanate (MDI) are the most commonly used isocyanates (Figure 9.4). MDI has superior reactivity and gives polymers possessing better physical properties. Also, aliphatic isocyanates as 1,6-hexane diisocyanate (HDI), isophorone diisocyanate (IPDI), and methylene bis(*p*-cyclohexyl isocyanate) ($H_{12}MDI$) are employed in the polyurethane synthesis. The resulting polymers show a higher hydrolysis resistance but lower chain stiffening than those having aromatic isocyanates in the *hard* segment.

-Polyols

Polyols commonly employed in polyurethane synthesis are polyethers or polyesters with a molecular weight ranging from 400–5000 $g \cdot mol^{-1}$.

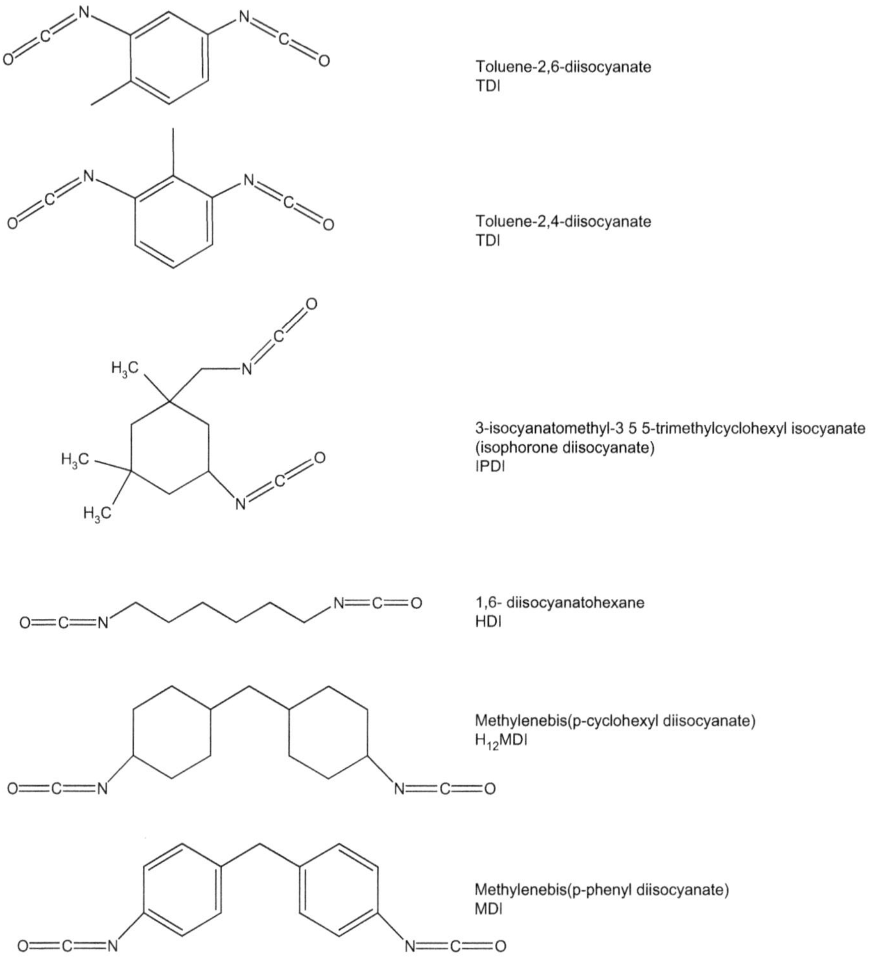

Figure 9.4   Chemical structure of commonly used diisocyanates.

Representative polyethers are poly(tetramethylene oxide) (PTMO), poly(propylene oxide) (PPO) and poly(ethylene oxide) (PEO). Polyurethanes synthesized from PTMO and PPO show resistance to oxidation and hydrolysis.[53] PTMO-based polyurethanes are more stable in a water environment and possess better mechanical properties than the PPO-based ones. Therefore, they are suitable for the manufacturing of medical devices where hydrolytic and mechanical stability is required. When PEO is used as the *soft* segment, the material can lose dimensional stability due to the hydrophilic nature of the polyether.

Polyols more recently employed to fit specific medical requirements are polyalkyl,[54] polydimethylsiloxanes[55] and polycarbonate. Among these, siloxane-based polymers have greater *in vivo* stability than polyetherurethanes.[56]

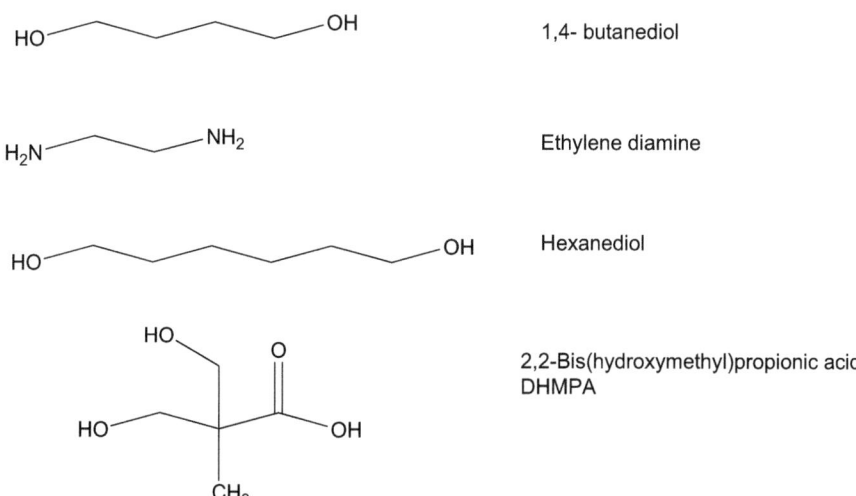

Figure 9.5 Chemical structure of commonly used chain extenders.

To produce biodegradable PUs, polyesters like poly(caprolactone diol), poly(lactide diol), poly(glycolide diol) and their copolymers can be employed as *soft* segments. The resulting poly(ester urethanes) possess relatively good physical properties together with susceptibility to hydrolytic cleavage.

-*Chain extenders*

Low molecular weight diols or diamines are used as chain extenders to increase the length of the *hard* segment and the polymer molecular weight. The role of chain extender is also to physically crosslink the *hard* and *soft* segments by H-bonds giving rise to the typical two-phase structure of polyurethanes. Commonly used diols are 1,4-butandiol, hexanediol and dihydroxymethylpropionic acid (Figure 9.5). In this latter case, the presence of a carboxylic group affects polymer *hard/soft* phase segregation and permits binding to biological active molecules, thus providing the polymer with different properties such as hemocompatibility, antimicrobial or antifouling activity.

As for diamines, ethylene diamine is the most used. In this case, poly(urethane ureas) are obtained possessing fairly good mechanical properties in terms of tensile strength but poor flexibility and processability.

9.2.3 Polyurethane Synthesis

Polyurethane synthesis has the features of both addition and condensation polymerization. In fact, although as in polyadditions no small molecule is eliminated during the synthesis, the reaction kinetics are similar to those of polycondensations.

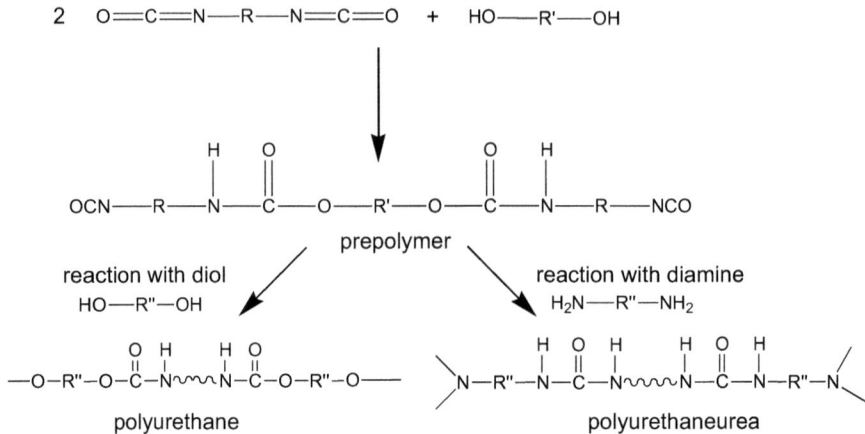

Figure 9.6 Two-step polymerization scheme for polyurethane synthesis.

Commercially, polyurethanes can be synthesized *via* either one-shot or two-step procedure. The one-shot process is easier at the industrial level but the chemistry of the reaction and, therefore, the properties of the final product are difficult to control.

The two-step procedure is instead the most employed to obtain segmented polyurethanes. The first step involves the reaction between the macrodiol and the diisocyanate to obtain a prepolymer that is a highly viscous liquid. In the second step this prepolymer reacts with a diol or diamine chain extender to produce a block copolymer. In Figure 9.6, a scheme of the two-step polymerization is reported.

Dibutyltin dilaurate and stannous octoate are widely used as catalysts in polyurethane synthesis since they are more efficient than tertiary amines in promoting the reaction between isocyanates and hydroxyl groups.

9.2.4 Applications of Polyurethanes in the Biomedical Field

Polyurethanes are considered among the most bio- and hemocompatible materials. They are, indeed, superior to other polymers such as silicone rubber, polyethylene, poly(vinyl chloride) (PVC) and poly(tetrafluoroethylene) (PTFE) widely used in medical applications. Their unique segmented structure allows the change of the polymer chemistry to achieve both flexibility and mechanical strength in the same material. In addition, the presence of hydrogen-containing groups in the polyurethane backbone permits either good interaction with biological active molecules, such as anticoagulants and antibiotics or chemical modification to introduce biorecognizable groups. For these reasons numerous medical devices ranging from catheters to artificial heart have been developed and are currently employed in medicine (Table 9.1).

Among different developed devices, the cardiovascular ones are subjected to repeated loading and deformation cycles. Therefore, in these applications a

Table 9.1   Main applications of polyurethanes in medicine.

> ➤ *Cardiovascular Applications*
>   ○ **Catheters**
>   • Intravenous (IV)
>   • Central Venous (CV)
>   • Balloon, Angiography, Angioplasty
>   • Urological
>   ○ **Pacemaker Leads Insulation**
>   ○ **Vascular Prostheses**
>   ○ **Heart Valves**
>   ○ **Cardiac Assist Device**
>   • Left Ventricular Assist Device (LVAD)
>   • Intra-Aortic Balloon Pump (IABP)
> ➤ *Artificial Organs*
>   • Total Artificial Heart (TAH)
>   • Artificial Kidneys-Hemodialysis
>   • Artificial Lung-Blood Oxygenators
> ➤ *Tissue Replacement and Augmentation*
>   • Breast Implants
>   • Wound Dressing
>   • Surgical Adhesives
> ➤ *Others*
>   • Controlled Drug Delivery
>   • Contraceptives (s)
>   • Miscellaneous Applications

high flexural endurance and fatigue resistance of the material is essential to avoid a failure of the device with life-threatening consequence for patients. Thanks to their mechanical properties, polyurethanes are materials of choice for the development of catheters, pacemaker lead insulation, vascular prostheses, heart valves and cardiac-assist devices. Here, the different types of polyurethanes used for manufacturing of these devices are listed and compared in terms of their performance. In Table 9.2, information concerning the composition of commercial biomedical polyurethanes, in terms of *soft* and *hard* segment components, is reported.

Catheters

The most used medical devices for administering and removal fluids, monitoring central venous pressure or diagnostic functions are catheters. They are manufactured by several polymers, including polyethylene, poly(vinyl chloride), silicone rubber and polyurethanes.

The important requirements that polymer materials must have for the development of catheters are suitable flexibility to make for easier insertion, a rather smooth surface to minimize tissue trauma and good column strength to avoid catheter kinking during its advancing into the body. Other factors influencing the performance of catheters are the antithrombogenicity and susceptibility to colonization by bacteria.

Table 9.2 Commercial biomedical polyurethanes.

Polyurethane	Composition	Uses
Vialon	PTMO/MDI/BD	Catheters
Biomer	Family of poly(urethanes ureas) 2000 MW PTMO/MDI/Diamines	Bladders, chamber coatings, catheters
Cardiothane 51 (Avcothane)	Mixture of poly(ether urethane) and poly(dimethyl siloxane)	Artificial heart, intra-aortic balloons, blood conduits
Pellethane	Series of thermoplastic elastomers based on PTMO/MDI/1,4-butandiol can be fabricated by injection molding	Pacemaker leads, blood bags
Corethane (Corvita)	Polycarbonate-based polyurethane	Cardiovascular applications
Tecoflex	Aliphatic from hydrogenated MDI/ PTMO	Wound dressings
Enka PUR	Poly(ether urethane), poly(ether urethane) with poly(dimethyl siloxane), poly(ester urethane)	Vascular prostheses

Catheter-related thrombosis is a complication often associated with the implant of intravascular catheters. Among intravascular catheters, central venous catheters (CVCs) are widely used for the treatment of cancer patients, needing hemodialysis or hospitalized in intensive care units (ICU). The incidence of CVC-related thrombosis ranges from 2% to 67% depending on the diagnostic procedure as well as patient and CVC features.[57] Polyurethanes are widely used to make over-the-needle intravascular catheters, central-venous access catheters and multilumen catheters. One of the first polyurethane employed for catheter manufacturing was Vialon™ developed in the early 1980s and today among the most sold biomaterials for vascular access devices. Thanks to its softening capability this polyurethane offers indwelling superiority, reduces catheter drag upon insertion and decreases the risk of catheter-related thrombosis.

Polyurethane-based CVCs were found to be less associated with local thrombosis than those in polyethylene.[57–59] However, thrombus formation related to CVC surface roughness has been reported for coronary[60] or angiographic[61] catheters made in polyurethane. These findings suggested a correlation between vascular catheter thrombogenicity and surface roughness.[62]

The strategies so far pursued to reduce thrombotic complications, involving the use of anticoagulant catheter lock solutions, heparin-bonded catheters or systemic anticoagulant prophylaxis, have been largely ineffective.[63]

Catheter-related infection is another major complication intimately linked to thrombosis and inflammation that increases patient morbidity, duration of hospitalization and mortality. To reduce the risk of development of local or systemic infections, particular attention should be paid both to the moment of catheter insertion in the patient and during all the time of its implantation.[64,65]

Risk factors in the development of catheter colonization and bloodstream infection include patient factors (increased risk associated with malignancy, neutropenia, and shock) and treatment-related factors (increased risk associated with total parenteral nutrition, ICU admission for any reason, and endotracheal intubation). Other risk factors are prolonged catheter indwelling time, lack of asepsis during CVC insertion and frequent manipulation of the catheter.

Factors promoting bacterial colonization are: a) the nature, chemical composition and surface features (hydrophobicity, flaws, roughness, *etc.*) of catheters; b) the formation of a protein (albumin, fibrinogen, fibronectin, *etc.*) film as the host response to the device implantation and the possible proliferation of host epithelial cells on the catheter surface;[66] c) the ability of micro-organisms to express adhesins[67–69] or to produce an exopolysaccharidic matrix (slime) able to mediate the final phases of the microbial colonization.[70]

The ability of the catheter to resist bacterial adhesion depends on the surface properties. First, surface roughness seems to play a primary role in the initial attachment of bacteria. In fact, the presence of microcavities or microfractures allows bacteria anchoring and a temporary protection for the bacteria with respect to the action of the host fluids is provided.[71] Another important surface feature is the hydrophilicity. In fact, bacteria are known to be better adsorbed onto hydrophobic surfaces. In this regard, the two-phase structure of polyurethanes consisting of hydrophilic/hydrophobic segments could reduce bacterial-adherence phenomena thanks to the formation of large hydrophilic domains and to their exposure on the surface when the material interacts with biological fluids. Notwithstanding the fact that polyurethanes are less susceptible to bacterial adhesion than other polymers, bacterial contamination is not completely prevented. Therefore, surface modifications of polyurethanes by hydrophilic coatings or loading with antimicrobial agents are needed to reduce catheter-related infections. Several catheters with hydrophilic coatings that reduce bacterial adhesion have been proposed[72–75] and are commercially available, as the Hydrocath catheter, which is a polyurethane catheter coated with a thin, hydrophilic layer of poly(N-vinyl pyrrolidone).

Another approach for the prevention of microbial colonization consists in providing the catheter surface with negative charges to establish electrostatic repulsion between the device and the negatively charged cell wall of micro-organisms. Liu *et al.*[76] reported that if a 10 μA electric current is applied to conducting catheters, they are able to generate a zone of inhibition of bacterial growth after incubation in nutrient agar seeded with bacteria. The bactericidal activity[77] of low amperage electric current is due to hydrogen peroxide and free chlorine produced by electrolysis at the catheter surface.

Pacemaker lead insulation

Pacemaker implantation is needed to correct anomalies in heart rhythm. A pacemaker consists of a pulse generator with lead connector, electrodes and a lead wire connecting the two. The first cardiac pacing lead was introduced in 1967.[78] The lead connector was made from insulating material such as polyethylene or silicone rubber. As a consequence of endocardial problems caused by these polymers, polyurethanes were also employed for the manufacturing of leads.[79] The good mechanical properties of polyurethanes, in terms of tensile strength and tear resistance, as well as their low coefficient of friction and smoothness allowed the diameter of leads to be reduced, making easier the insertion and also permitting the introduction of more than one lead into a vein.

Pellethane™ is the most used polyurethane type for these applications. However, the pacemaker leads insulated with this material have shown poor *in vivo* stability, degrading after the implantation.[80] This degradation was attributed to environmental stress surface cracking (ESC) and metal-induced oxidation (MIO). To overcome the surface cracking drawback a Pellethane type possessing a higher *hard* segment content was used for lead manufacturing. Studies have shown that these lead types were less sensitive to surface cracking than those previously utilized.[81]

The degradation of the material can also be promoted by the presence of metal ions derived from corrosion processes of pacemaker metal wires. The complexation of these ions with the polyurethane can occur causing polymer stiffening and surface cracking.[82] The use of novel alloys containing nickel and chromium enhanced the corrosion resistance, increasing the lead lifetime.

Although silicone insulating leads do not show the ESC and MIO drawbacks, they have to be manufactured with a slightly larger diameter and thickness than equivalent polyurethane leads.

The more recent approaches to overcome lead failure regard the use of antioxidants as additives for polyurethanes,[83] the coextrusion of silicone and polyurethane[84] or the synthesis of silicone–polyurethane copolymers.[85] However, a material that combines the biostability and flexibility of silicone rubber and the strength and abrasion resistance of polyurethanes has not yet been identified.

Indeed, a recent study analyzing the defects of 990 right ventricular defibrillation leads that were implanted between 1992 and 2005 showed that lead failure is an overall problem encountered in all models.[86] This study also showed that the annual rate of lead defects increased with time and reached 20% in 10-year-old leads.

Vascular prostheses

Cardiovascular diseases are the main cause of mortality in the world. Vascular grafts are used to replace, bypass or maintain the function of damaged, occluded or diseased blood vessels. They can be obtained through the use of either blood vessels of the patient's own body (autogenous grafts) or synthetic materials (artificial grafts).

The most used materials to manufacture synthetic grafts are poly(ethylene terephthalate) (PET, Dacron®), PTFE and PUs.

Venous autografts are more satisfactory with respect to synthetic vascular ones (longer patency time and minor complications). A systematic review of the available literature studies comparing the patency of saphenous vein and PTFE bypass demonstrated that the patency of venous bypasses was superior to that of PTFE bypasses at all time intervals studied.[87] The authors concluded that, if a saphenous vein is available, a venous bypass should be chosen at all times, even if patients have an anticipated short life expectancy (< 2 years). If the saphenous vein is absent or not suitable for bypass grafting, PTFE is a good alternative as bypass material. Indeed, some patients need prosthetic grafts due to issues related to autogenous veins including phlebitis or varicosities.

An ideal vascular graft should possess several features: biocompatibility; mechanical strength and compliance for long-term devices; thromboresistance; availability in various sizes; ability to withstand infection; satisfactory graft healing;[88] and reasonable manufacturing costs.

The prosthetic grafts based on Dacron® or ePTFE (expanded poly(ethylene terephthalate), Gore-Tex®) materials showed a long-term patency rate and a good compliance match between graft and native vessel both in large (> 8 mm) and medium-caliber (6–8 mm) replacements.[89] Probably, in the case of large and medium-caliber grafts thrombus formation can be avoided by massive blood flow.[90] As for small caliber vessels (< 6 mm), the used polymers have not proved to be satisfactory since a high incidence of occlusion, caused by thrombosis, or an excessive neointimal hyperplasia, located at distal anastomosis of prosthetic grafts, occurs.

PUs are known to be one of the most blood-compatible biomaterials. In addition, their structural versatility allows materials with mechanical properties resembling those of natural blood vessels. For these features, polyurethanes played and still play an essential role in the development of small vascular grafts.

The better thromboresistance of polyurethanes with respect to PET and PTFE can indeed allow the development of a thinner fibrin layer. The formation of a stable neointima and the lining of the graft lumen with endothelial cells contribute to the long-term patency of the polyurethane prosthesis.[91]

Besides the blood compatibility, the surface texture and porosity are important characteristics for development of long-term healing small synthetic vascular prostheses.[92,93] Indeed, although rough surfaces can favor thrombus formation, a microporous structure, particularly for small diameter grafts, is highly desirable as the pore type and size affect either the speed and extent of tissue ingrowth or the development and stability of endothelial layer.[94,95]

The first PU vascular graft has been developed by using polyester as soft segment (Vascugraft®). This prosthesis type showed good biocompatibility but a poor chemical stability *in vivo*.[96,97] The biodegradation, due to hydrolytic instability of ester bonds, can, in addition, promote the release of toxic substances for the body.[98] As a consequence, polyether-based PUs have been manufactured (Pulse-Tec®). However, it must be remembered that the poly(ether urethane)-based devices are also susceptible to degradation, particularly to the oxidative type one.[99] In addition, the poor surface porosity of these grafts delays tissue ingrowth.

Thoratec® is a prosthesis made of poly(ether urethane urea) (PUU). Clinical studies showed that PUU graft allows an earlier access with respect to the ePTFE-based one without compromising long-term performance.[100] The PUU-based grafts can be manufactured with or without mechanical reinforcement and possessing a more or less microporous structure.

To overcome the degradation drawback, a new generation of PU grafts based on poly(carbonate urethanes) (PCUs) has also been introduced (Corvita®).[101,102] The superior biostability of these prostheses is due either to the lack of ester and ether linkages or to the structure externally reinforced with Dacron®.

Nowadays, however, thrombi formation and unfavorable healing process are still the most common causes that contribute to the mid- or long-term vascular graft failure.

The two strategies pursued to optimize tissue–biomaterial interactions improved graft performance but have not yet led to a solution of the failure causes of vascular grafts. The first approach has been addressed to modification of device surface properties (hydrophilicity improvement, chemical group or biologically active substance introduction) in order to minimize thrombi and emboli generation and to increase the graft lifetime. In particular, the immobilization onto device surface of natural materials such as gelatin or collagen,[103,104] anticoagulants like heparin,[105–107] polymers like poly(ethylene glycol)[108] and growth factors[109,110] is the most used strategy.

The second approach consists in miming the biological function of the native vessel by endothelial cell (EC) seeding. Vascular sealants, particularly gelatin and collagen, fibronectin or RGD tripeptide (Arg–Gly–Asp sequence) have been used as substrates to promote seeding of ECs on the luminal surface of the graft.[110–112]

In a recent study, the performance of five polyurethane vascular grafts in terms of device function, healing features and material biostability in a canine model has been investigated.[113] In the studied period after implantation, ranging from 1 to 6 months, all the grafts showed thrombogenicity and intimal hyperplasia at various degrees. These findings showed that, among different parameters influencing the performance of vascular devices, the structural design and material selection are the most important ones.

Certainly, the most promising approach to enhance the patency of small-diameter vascular grafts concerns the development of completely tissue engineered vascular grafts. In fact, the use of tissue-engineering techniques allows promotion of the complete growth of new vascular conduits from autogenous material.[114]

Heart valves

The heart is a pump made of muscle tissue. The heart has four pumping chambers: two upper chambers, called atria, and two lower chambers, called ventricles. To keep the blood flowing through the heart, there are valves between each of the pumping chambers.

Valve replacement is required when heart valves become damaged or diseased. Artificial valves, in use since 1960, can be of two types: mechanical and biological (xenograft). While the implant of a mechanical valve requires long-term anticoagulant administration, problems related to the use of biological ones, made of porcine valves or bovine pericardium, concern durability and calcification of the tissue.

Polyurethanes, thanks to their excellent mechanical properties and blood compatibility, have been used to manufacture prosthetic heart valves. A number of polyurethane heart valves, mostly trileaflet type, have been prepared and tested.[115,116] These studies showed that the fabrication procedure is the most important factor influencing device thrombogenicity. Indeed, a smoothness surface and a suitable thickness of the valve minimize thrombus formation allowing the normal functioning of heart.

Valve durability can also be compromised by material stiffening owing to calcification problems. Calcium phosphate deposition occurs on the surface of prosthetic aortic and mitral valves due to the presence of cellular material and thrombotic debris or to the complexation of calcium ions with the soft segment.[117,118] Also, high mechanical stresses provoked by manufacturing can contribute to calcification of polyurethane valves.[119]

To enhance hemocompatibility and to reduce calcification, polyurethanes have been chemically modified by introducing on their surface negative charge groups like sulfonate or phosphonate groups.[120] The modified valves displayed both lower calcification and poor platelet adhesion and thrombus formation with respect to unmodified devices.

Cardiac-assist devices

When a heart does not work effectively there is a need to use mechanical methods to assist and support the failure circulation. The main methods for pumping or preventing regurgitation of the blood exploit the use of left ventricular assist devices (LVAD) and intra-aortic balloon pumps (IABPs). Both LVAD and IABP are considered to be transient devices.

To date, indeed, all polymer materials have not shown satisfactory results in terms of blood compatibility and mechanical properties to be used for long-term application (totally implantable artificial heart).

Because of their good flexure strength and wear resistance, polyurethanes are used in the manufacturing of blood-contacting diaphragm within the pump. Polyurethane is also textured to provide blood cell adherence.

Different types of polyurethanes have been used to develop ventricular assist devices. As for the poly(ether urethane)-based LVADs, they showed systemic inflammatory response and insufficient blood compatibility. These drawbacks were ascribed to continuous cycling of the chamber.[121]

Poly(ether urethane)s (PEUs), in addition, are water-permeable and degradable materials. Moisture, particularly, going through a diaphragm can influence the motor working and then the long-term performance of LVAD. To decrease moisture transmission either a composite material (PEU/butyl rubber/ PEU)[122] or a segmented polycarbonate polyurethane[123] have been used.

The use of LVADs based on polyurethane with a poly(dimethyl siloxane) component (Avcothane-51) and constructed in different configurations has been displayed with opposing results. On the one hand, good hemocompatibility in short-term tests[124] and, on the other hand, thrombus formation due to mechanical faults.[125]

To increase blood compatibility of VDAs without affecting their mechanical properties, the device surface has been modified by adding low amounts of additives,[126] collagen lining[127] or by using textured surfaces.[128]

Nowadays, the major drawback concerning the use of polyurethanes in ventricular assist devices is related to moisture transmission in device regions that should be totally hermetic. This affects the reliability of the device and provokes its failure.

Intra-aortic balloon pumps (IABPs) are the most commonly used forms of mechanical circulatory assistance. Their physiological objectives are to reduce the workload of the heart and to increase coronary artery perfusion. The device, generally employed for a short term (few days), consists of a cylindrical polyurethane balloon placed in the descending thoracic aorta by way of a femoral artery and connected by a catheter to an external cyclical pump.

Polymer materials and technology suitable for the development and manufacture of IABPs have been described by Brash *et al.*[129] Both polyurethanes selected in this work and the procedures for the fabrication of high-quality IABPs are still adopted today.

Complications following IABP insertion are mainly caused by surgical accidents related to the insertion procedure rather than to device failure or biocompatibility of materials used. However, polyurethanes used for the development of these devices have to assure abrasion and fatigue resistance of the balloon membrane as well as facility in the insertion to the catheter of the IABP and kink resistance.

9.3 Biomimetic Polyurethanes in Medical Prostheses

In modern medicine, medical devices are used for a wide range of applications. Some of these devices, including heart valves, artificial joints and pacemakers, are permanently implanted in the patients to repair or replace damaged parts of the body. Other devices, such as intravascular and urinary catheters, are temporary implanted and allow clinicians delivering drugs or nutrition fluids into the body, to expel fluids out of the body and to monitor the status of critically ill patients.

Among the temporary implanted devices, intravascular catheters are extensively used. They can be divided into two broad categories: (i) those used for short-time vascular access, such as peripheral venous catheters, or the non-tunneled central venous catheters (CVCs); (ii) those used for long-term (indwelling) vascular access, such as the tunneled CVCs. Indwelling catheters require surgical insertion, while short-time devices can be usually inserted percutaneously.

In all these applications, the control of phenomena occurring at the biological tissue/material interface is of paramount importance for the success of the implanted medical device. In particular, limiting the nonspecific adsorption of proteins, cells and bacteria onto the biomaterial surfaces is important for the development of medical devices with specific levels of performance.

Soon after the exposure of the device to physiologic fluids and tissues, the device surfaces are coated by a film of biological material mainly consisting of proteins, polysaccharides and cells. This biofouling process occurs spontaneously and, in many cases, represents an adverse event that can lead to implant failure. Important clinical issues associated with the biofouling process include the occlusion of cardiovascular implants by thrombus formation or the development of medical device-related infections supported by microbial biofilms growing on the device surface.

To understand how these two complications can be prevented, a brief overview of phenomena occurring at the biological fluids/device interface is needed. In particular, the attention will be mainly focused on the description of the mechanism of formation of microbial biofilms since they are considered the causative agents of the device-related infections. As reported in the following sections, in the case of blood-contacting medical devices, biofilm development and the occurrence of device-related infections increase the risk of thrombosis, evidencing a relationship between these two complications. The infectious risks as well as the severity of the infections, in terms of morbidity and mortality, are strongly dependent on the type of implanted device. In the following sections, data concerning the incidence of medical device-related infections on the healthcare system will be first reported for the widest clinical use devices. Then, the main strategies to prevent such infections will be discussed paying particular attention to those based on biomimetic polyurethanes that are considered the most promising approaches.

9.3.1 Mechanism of Biofilm Formation

Medical device-related infections contribute to about 60% of all healthcare-associated infections.[130] It is well known that micro-organisms can colonize a wide variety of medical devices leading to the formation of a heterogeneous mushroom-like structure known as biofilm (Figure 9.7).

Biofilm development can put patients at risk for local and systemic infectious complications, including local site infections, catheter-related bloodstream infections (CR-BSIs) and endocarditis.[131]

Microbial biofilms are known to be microbial sessile communities, often multispecies, irreversibly attached to a surface and encased in a matrix of polysaccharides, proteins and nucleic acids produced by the bacteria themselves. The mechanism of biofilm formation is a multistage process (Figure 9.8). In the first stage, free-floating micro-organisms approaching the surface reversibly adhere to it *via* van der Waals or electrostatic interactions (*reversible adhesion*). The strength of this initial adhesion depends on both bacterial surface properties (hydrophobicity, net charge, *etc.*) and, of most interest, the physicochemical features of material surface.

In the second stage, micro-organisms irreversibly adhere to the surface by using cell-adhesion structures, such as pili, or by the secretion of specific

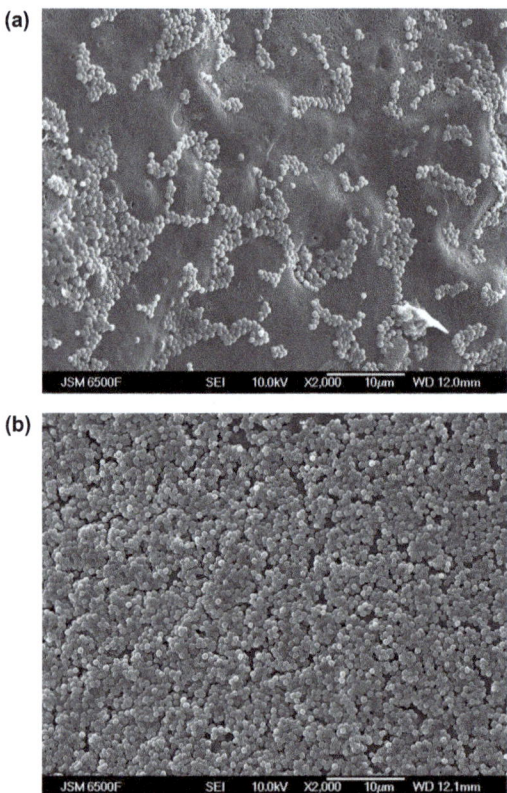

Figure 9.7 Initial adhesion (a) and mature biofilm (b) of *S. epidermidis* onto a polyurethane surface.

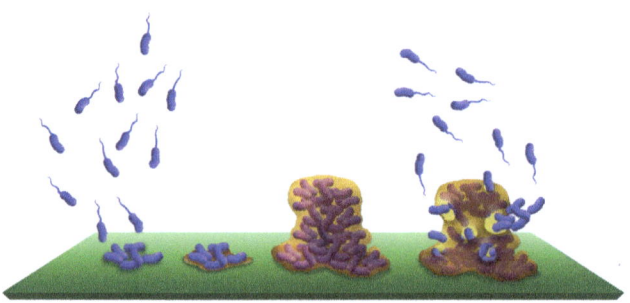

Figure 9.8 Stages involving biofilm formation on a surface.

adhesins. For example, after adhesion to a substratum *Staphylococcus epidermidis* is able to produce a polysaccharide intercellular adhesin (PIA) that binds the cells together and promotes the formation of microcolonies.[132] The ability of bacteria to produce these sticky extracellular components is a prerequisite for biofilm formation. The third stage concerns the maturation of biofilm architecture. In this stage, the regulation of gene transcription is possible by the accumulation of signal molecules that allows bacteria communicating with each other. This cell-density-dependent bacterial intercellular signaling mechanism is known as quorum sensing and enables bacteria to express some genes able to coordinate the behavior of the whole bacterial community. Finally, the fourth stage of biofilm formation is represented by the detachment of single cells or cell aggregates from the biofilm. This is an essential stage of the biofilm life cycle since allows micro-organisms spreading and colonizing of other surfaces.

Microbial biofilms are recognized as a significant virulence factor.[133,134] In fact, once they are established, it becomes extremely difficult to eradicate them since micro-organisms growing in biofilms are, in general, much more resistant to antibiotics than planktonic cells. The resulting infections are more resistant to both the host immune defenses and antimicrobial agents.[135] Therefore, often, patients must undergo revision surgery for the debridement of infected tissue or, in many cases, implant removal and replacement.[136]

The mechanism involved in the increased drug resistance of biofilms[137] is rather complex and seems to include several factors. First, the penetration of antimicrobial agents through the biofilm matrix is usually slow or incomplete. Secondly, a physiological adaptation of bacteria to the heterogeneous chemical environment (pH, oxygen concentration) existing in biofilms seems to occur.[137,138] As a consequence, subpopulations of persisters have been shown to be present in the biofilm.[139] Finally, there is a good plasmid exchange among bacteria in the biofilm, this contributes to amplify the antibiotic resistance phenomena.

Biofilm formation has been well documented on several types of medical devices, including intravascular and urinary catheters, prosthetic hips, knees, contact lenses and dental implants.[140–144] The most common pathogens involved in medical device-related infections are *Staphylococcus aureus*, *Staphylococcus epidermidis*, *Pseudomonas aeruginosa* and *Escherichia coli*. Lately, also *Candida* spp. have been demonstrated to be able to form *in vivo* biofilms usually in association with *Staphylococcus aureus*.[145,146]

9.3.2 Incidence of Medical Device-Related Infections

Implantation of a medical device always provokes a host response that intrinsically increases the risk of infection. The type of implanted device, the material used for its manufacturing, the body site of the implant as well as patient features all contribute to the overall risk of device-related infection. Here, the data concerning the infectious risk related to the widest clinical use devices are reported and discussed.

Central venous catheters

Among medical devices, intravascular medical devices have been estimated to be associated with an infection rate of about 15% and an attributable mortality that can exceed 25%.[147] Among intravascular devices, central venous catheters are responsible for the highest proportion of bloodstream infections. In particular, nontunneled-CVCs are the most commonly used CVCs and account for an estimated 90% of all catheter-related bloodstream infections.[148] CVC-related infections are also the most common cause of nosocomial endocarditis.[131]

The pathogenesis of the nontunneled CVC-related infection is often related to extra-luminal colonization of the catheter, which originates from the skin and, less commonly, from hematogenous seeding of the catheter tip. The intraluminal colonization of the hub and lumen of the CVC is instead the most common route of infection for tunneled-CVCs or implantable intravascular devices. Coagulase-negative staphylococci, *S. aureus* and *C. albicans* are the most commonly isolated pathogens in CVC-related infections.[148]

Thrombosis is another complication often associated with the implant of intravascular catheters.[149] A prospective multicenter study carried out on ICU patients requiring internal jugular or subclavian catheterization[150] evidenced a dependence of the thrombosis risk on the site of catheter insertion. In particular, a risk of thrombosis with internal jugular vein site of insertion higher than the subclavian site was found. This study also found a relationship between CVC-related thrombosis and infection. In fact, the risk of catheter-related sepsis was 2.62-fold higher when thrombosis occurred. This finding was confirmed by a more recent study performed on 105 patients undergoing intensive chemotherapy.[151] In particular, a temporal association of CVC-related infection and subsequent thrombosis was found. Besides, the absolute risk of developing a thrombotic event increased with the severity of CVC-related infection. In particular, a 57% thrombosis risk was observed in patients with CVC-associated septicemia versus 27% in patients with a local CVC infection.

The correlation between catheter-related infection and thrombosis seems to be due to several factors. First, within hours from the implant, the device surfaces are coated with plasma proteins and blood cells, starting the body's coagulation cascade. The result is the formation of a fibrin sheath covering the device surface. This fibrin sheath cannot only occlude the catheter lumen but also become a physical trap for micro-organisms coming from the patient skin through the catheter insertion site. Once near the device surface, micro-organisms can irreversibly adhere by the production of specific adhesins able to recognize the host proteins layered on the device surface. Finally, the systemic inflammation occurring as a consequence of the infection developed at the catheter implant site results in activation of the body's coagulation response, further increasing the risk of a catheter-related thrombotic event.[152]

Urinary catheters

Urinary tract infection (UTI) is the most common hospital acquired infection, the major one being associated with the use of indwelling urinary catheters. About half of all hospitalized adults who have urinary catheters for longer than a week will get urinary tract infection.[153] Catheter-associated urinary tract infections affects 449 334 patients per year in United States hospitals.[154,155]

Escherichia coli is the most common pathogen causing these infections. Other major involved species are *Pseudomonas aeruginosa, Enterococcus spp, Klebsiella spp, or Enterobacter spp.*[156,157]

There are two the main routes of infection: i) the intraluminal route, in which bacteria or fungi are introduced *via* the inside lumen of the catheter; ii) the transurethral route, where micro-organisms access along the thin layer of mucus present between the catheter and the urethra.

Orthopedic implants

The implant of orthopedic devices is of great relevance for public health due to the increasing number of aged and disabled people. Thanks to the improvement in the surgical procedures as well as in the biocompatibility of the materials employed to manufacture orthopedic prostheses, the risk for orthopedic device-related infections is nowadays low and less than 1–2%. However, the total number of patients suffering these infections is high due to the numerous devices annually implanted. In fact, more than 200 000 and more than 400 000 hip replacements and primary total knee arthroplasties are performed each year in the United States, respectively.[158] On the other hand, the number of total hip and knee arthropasty in the United States has been estimated to grow by 137% and 601% from 2005 to 2030, respectively, as shown by a recent projection by Kurtz and colleagues.[159]

The early prosthetic joint infection, defined as infection of the implant site within the first 3 months after surgery, is usually caused by *S. aureus* or *S. epidermidis*. Delayed infection (within 2 years from surgery) is instead caused by micro-organisms of low virulence such as coagulase-negative staphylococci. The manifestation of infection after 2 years from surgery is usually related to hematogenous seeding from other infected body sites.

When an implant-related infection occurs, simple debridement procedures with retention of prosthesis and chemotherapy with antimicrobial agents are not always effective to eradicate the infection. Therefore, a revision surgery for prosthesis removal and replacement is usually needed with a consequent increase of the hospitalization time of the involved patients and significant associated health costs.[160]

9.3.3 Biomimetic Polyurethanes to Prevent Medical Device-Related Infections

As described in the first part of this chapter, polyurethanes are considered the materials of choice for the manufacturing of intravascular devices. The clinical complications associated with the implantation of these devices are thrombosis and local or systemic infections. As already underlined, these two drawbacks are closely interrelated. The device coating with plasma proteins and other blood components, as host reaction to the foreign body, generates a condition of immunosuppression at the implant site[133] and an altered device surface to which bacteria can more readily adhere. In particular, reduced levels of

Table 9.3 Possible antibiofilm strategies.

Step in biofilm formation	*Possible antibiofilm strategy*
Microbial adhesion and colonization of the device surface	Antifouling polymers Antimicrobial devices
Biofilm formation	Interference with the microbial communication mechanism by quorum sensing interfering molecules
Biofilm dispersal	Biofilm matrix disaggregating enzymes Electric current or ultrasound

bactericidal enzymes and of superoxide production have been found in neutrophils recovered from the surface of medical implants.[161] Moreover, studies carried out on *S. aureus* showed the presence of specific microbial surface proteins able to bind fibronectin[162,163] or fibrinogen.[164] Proteins with broader specificity able to bind more than one type of plasma protein were also found in *S. aureus*.[165,166]

Once adhered, bacteria colonize the entire device surface forming a thick and heterogeneous biofilm. The resulting local inflammation further activates the body's coagulation response with an increased related thrombosis risk. Since a temporal association of infection and subsequent thrombosis was found, the strategies aimed at the prevention of medical device-related infections may also contribute to significantly decrease the occurrence of device-related thrombosis.

On the basis of the mechanism of biofilm formation, possible antibiofilm strategies[167] can address one of the main steps of the biofilm life cycle (Figure 9.8).

In Table 9.3 the main strategies pursued to control medical device-related infections are listed. The two principal approaches to prevent medical device-related infections are (i) the development of polymers or polymer surfaces with antifouling properties, (ii) development of polymers with antimicrobial properties.

9.3.3.1 Antifouling Polyurethanes

Microbial adhesion is an essential step in the pathogenesis of medical device-related infections. Given the hydrophobic nature of microbial surfaces,[168] prevention of bacterial adhesion can be achieved by coating the device surface with hydrophilic polymers. In fact, device surface modification by the immobilization of antifouling small molecules or macromolecules is considered a promising antibiofilm approach. This strategy is particularly interesting since it avoids the use of drugs (either by systemic administration or by local release from medicated devices) that, when administered over long periods of time, may be associated with undesired side effects.

However, as will be described in the following sections, the main limit of this strategy is that the efficacy in preventing biofilm formation is strongly dependent on the bacterial species involved in device colonization. In fact, the physicochemical properties of micro-organisms can vary widely among species and generalizations at the species or even strain level are often impossible.

PEG-containing polyurethanes

On the basis of the empirical criteria recently proposed by Ostuni and colleagues,[169] antifouling polymers should be hydrophilic, electrically neutral and possess hydrogen-bond acceptors. Accordingly, several polymer classes have been explored,[170] including polyacrylates,[171] polyzwitterions[172] and poly(ethylene glycol) (PEG) derivates.[173,174] Of these, PEG is the most widely studied antifouling polymer since it offers resistance to protein adsorption, nonimmunogenicity and antithrombogenicity. Its ability to impart protein resistance is believed to be related to both hydration and steric effects.[175] PEG has been grafted onto the surface of a series of materials, including glass,[176] gold,[177] poly(ethylene terephthalate)[178] and polyurethanes.[179–184] Although PEG possesses unique properties of nontoxicity and biocompatibility, a number of limitations have been associated with PEG grafting, including stability (autoxidation) and poor functionality.[185]

As for polyurethanes, PEG was either introduced in the polymer backbone[186,187] or grafted to the polymer side chain by urethane[182] or allophanate linkages,[188] Michael addition onto main chain double bonds[189] and click chemistry.[179] In general, a decrease in platelet adsorption on PEG-modified polyurethanes was observed in comparison with pristine polymers.[182,190] Recently, a polyurethane with improved hemocompatibility was obtained by first surface grafting PEG and then conjugating lysine.[191] This functionalized polyurethane was able to reduce both platelet and plasma protein adhesion while promoted the adsorption of plasminogen that when suitably activated lyses fibrin. More recently a PEG-containing polyurethane hydrogel was employed as a coating for poly(dimethyl siloxane)-based neural electrodes to improve their biocompatibility.[187]

The behavior of PEG-functionalized polyurethanes towards bacterial adhesion has also been investigated. Corneillie and colleagues[186] synthesized segmented polyurethanes at different PEG content employing in the polymer synthesis PEG, poly(propylene oxide) (PPO) and poly(tetramethylene oxide) (PTMO) in different ratios. Samples containing PEG/PPO or PEG/PTMO in 1/1 molar ratio were able to significantly reduce the adhesion of *S. aureus* and *Enterococcus faecalis* with respect to glass, but they allowed the adhesion of *P. aeruginosa*. A reduction of *P. aeruginosa* adhesion on the polymers was instead obtained if samples were incubated in urine instead of phosphate buffer, indicating an influence of the condition film.

The influence of culture media as well as of PEG molecular weight and end-group type on bacterial adhesion was investigated in a study by Park and colleagues.[181] These authors grafted Pellethane® with PEG, either 1000 or 3400 molecular weight, possessing different chain end groups (-OH, -NH$_2$ and SO$_3$, respectively). The adhesion of *S. epidermidis* and *E. coli* was studied in tryptic soy broth (TSB), brain heart infusion (BHI) and human plasma. In the case of PEG1000-modified surfaces, no reduction of *S. epidermidis* adhesion was observed in TSB media regardless of the PEG chain end group. On the contrary, in plasma the adhesion was reduced to different degrees depending on PEG terminal groups. In particular, due to a different protein adsorption on the polymer surface, the sulfonated PEG1000-containing polyurethane was the most repellent one. If longer PEG chains were used in the grafting, *S. epidermidis* adhesion was reduced both in TSB and plasma. As for *E. coli*, a significant reduction in adherent bacteria was found in all the adopted experimental conditions. The same authors also covalently bonded heparin on the PEG3000-functionalized polyurethane, obtaining a reduction of *S. epidermidis* and *E. coli* adhesion comparable to that obtained with the sulfonated PEG1000-modified polyurethane. This finding suggests that the sulfonate groups are responsible for the antifouling features of heparin.

In conclusion, the ability of PEG-containing polyurethanes to reduce bacterial adhesion has had variable degrees of success, depending on the PEG's molecular weight, degree of branching and surface packing density. In addition, the activity of PEG-modified surfaces seems to be strongly dependent on both the bacterial species and the culture media used in the experiments.

Heparin-coated and heparin-like polyurethanes

Heparin, a highly sulfated glycosaminoglycan, has attracted particular attention in the field of hemocompatible biomaterials since it possesses a number of biological functions, such as anticoagulant activity, cell-growth stimulation and antivirus ability. It is, in fact, able to reversibly bind to many biofunctional proteins, such as antithrombin III (ATIII) and platelet factor 4.[192] Moreover, it contains distinct recognition sites for growth factors, including basic fibroblast growth factor and vascular endothelial growth factor.[192]

The physical adsorption or chemical grafting of heparin to artificial surfaces has been shown to be a successful strategy to improve device hemocompatibility.[193–195] Our group has developed different hydrophilic polyurethanes able to bind significant amounts of heparin ionically[196] or covalently.[4]

Heparin coatings or bindings have also been shown to prevent microbial adhesion and colonization *in vitro* and *in vivo*.[197] This activity is presumably related to heparin's ability to favor albumin adsorption while reducing fibrinogen adsorption. In fact, many micro-organisms, including coagulase negative staphylococci (CNS), have specific receptors for fibrinogen.[198]

In a recent pilot study,[199] the encrustation and biofilm formation was evaluated onto heparin-coated ureteral stents (Heparius) and compared with uncoated polyurethane stents. This prospective randomized study considered 20 heparin-coated and uncoated stents inserted into obstructed ureters and left implanted for periods ranging from 2 to 6 weeks. Scanning electron microscopy observations of the surfaces of the removed stents showed that heparin-coated stents were unaffected by encrustation phenomena. On the contrary, after

6 weeks of indwelling time, all the uncoated stents had different degrees and forms of deposits on their surfaces. In addition, the surfaces of 33% of uncoated stents resulted colonized by bacteria, while no biofilms were detectable on heparin-coated stents.

Also, the heparinization of CVCs and dialysis catheters resulted in a significant reduction of catheter-related infections, as recently confirmed by a randomized-controlled clinical trial of heparin-coated and uncoated non-tunneled CVCs inserted in 246 patients[200] and a retrospective comparative analysis of coated and uncoated tunneled dialysis catheters.[201]

An alternative strategy to the heparin coating is represented by the development of heparin-like polymers that is polymers able to mime the biological behavior of heparin.

The first heparin-like polyurethanes were sulfonated PTMO-based polymers obtained by Cooper's group.[202] Different levels of sulfonated groups in the *hard* segments were introduced by nucleophilic substitution of urethane hydrogens with propyl sulfonate groups. Platelet deposition onto these materials was reduced as the degree of sulfonation increased. The anticoagulant effect of these polymers seems to be related to their ability to complex thrombin and to interfere with fibrin polymerization as shown by Silver and colleagues[203] who synthesized similar water-soluble sulfonated polyurethanes.

A series of sulfonated polyurethanes containing the anionic groups in the soft segment were also developed.[204] These polymers showed good film-forming ability, typical elastomeric behavior and high compatibility between the *hard* and *soft* segments.

Very recently, our group synthesized biomimetic segmented polyurethanes containing sulfate or sulfamate groups in the *hard* segment by employing in the polymer synthesis a carboxylated diol, dihydroxymethylpropionic acid, as chain extender.[28] The resulting carboxylated polyurethane was first amidated with different functional amines and then reacted with pyridine–SO_3 or DMF–SO_3 adducts. The obtained –SO_3H-containing polymers, possessing a degree of sulfonation ranging from 30 to 40%, showed good hemocompatibility and inhibited *S. epidermidis* adhesion. These features seemed to be related to both polymer phase segregation and the density of –SO_3H groups.

Other studies confirmed that the concentration of sulfate groups in the polyurethane seems to be a key parameter to impart antifouling properties to heparin-like polyurethanes. In fact, a pellethane sulfonated at a theoretical degree of 20% of the NH groups of the urethane linkages failed in reducing *S. aureus* adhesion.[205] On the contrary, the coating of a polyurethane with a highly sulfonated hyaluronan resulted in the inhibition of *S. epidermidis* adhesion with respect to the pristine polymer.[206]

In conclusion, although heparin-coated devices are in general more expensive than the uncoated counterparts,[199] their activity justifies the cost. In addition, for long indwelling times, if we consider that the use of heparin-coated devices would result in a reduction of the number of device exchange procedures, the total costs for heparin-coated devices should be reduced compared with uncoated ones. As for permanent devices, the decrease in patient morbidity and in invasive procedures due to the reduced number of infected devices should further justify the use of the more expensive heparin-coated devices.

Hydrophilic coatings for polyurethanes

Several hydrophilic coatings for medical devices have been proposed to reduce bacterial adhesion.[72,75,207] Morra and Cassineli reported an *in vitro* study in which polyurethane catheters coated with hyaluronan were exposed to 10^9 CFUs/mL *S. epidermidis* concentration. While bacteria were able to colonize the surface of the uncoated catheter, they did not adhere to the soft, hydrated surface of the modified ones.[75]

More recently, Choong *et al.*[208] reported encrustation, a problem especially related to the use of urinary catheters, onto polyurethane coated with hyaluronic acid catheters lower than onto uncoated ones in both *in vitro* and *in vivo* models.

Other polysaccharides than hyaluronic acid were investigated as antifouling macromolecules including dextran,[209] dermatan sulfate[210] and chitosan.[211,212] In particular, chitosan has been shown to possess antibacterial activity against different micro-organisms including *S. aureus* and *E. coli*.[213] The coating of a segmented polyurethane urethral catheters with a chitosan/poly(vinyl alcohol) hydrogel decreased protein absorption and provided the catheter with antibacterial properties against *S. aureus*, *P. aeruginosa* and *E. coli*.[214]

Another hydrophilic polymer studied as antifouling coating is poly(N-vinyl pyrrolidone) (Hydromer®). Polyurethane catheters coated with Hydromer® significantly reduced the adhesion of five strains of *S. epidermidis* and one strain of *S. aureus*.[207] Similarly, a marked 1 to 3 log reduction in adherence to Hydromer-coated polyurethane biliary stents of both Gram-negative and Gram-positive bacteria was found with respect to the uncoated catheters.[72] A reduction of fibrinogen and fibronectin adsorption to poly(N-vinyl pyrrolidone) (PVP)-coated polyurethane catheters was also reported and related to the reduction in the adhesion of *S. aureus* and *S. epidermidis*. More recently, a PVP-coating was applied to polyurethane and evaluated in terms of ability to control bacterial adhesion and encrustation for application as urinary tract biomaterial.[215] While encrustation was on the PVP-coated polyurethane less than on uncoated polyurethane and silicone catheters, bacterial adhesion depended on the surface hydrophilicity of the employed bacteria.

9.3.3.2 Antimicrobial Polyurethanes

Medical devices coated with antimicrobial or antiseptic agents possessing different antimicrobial spectra and duration of activity have been shown to reduce bacterial colonization and biofilm formation on biomaterial surfaces. The advantage of this strategy is that the drug can be released at high dose locally at the site of infection even for long times.

The use of antimicrobial-coated medical devices has had variable clinical success, with differences in the degree of protection depending on: a) the type and location of the device, b) the type of antimicrobial agent used for the device coating, c) the concentration of the drug on the surface of the coated device and d) the kinetics of drug release from the device surface to the surrounding environment.

Different experimental approaches have been described for the development of medicated devices, including i) the entrapping of the antimicrobial agent into the polymer during the manufacturing process; ii) the adsorption of one or two antimicrobial substances on the device surface after its manufacturing; iii) the preparation of intrinsically antimicrobial polymers possessing either groups with antimicrobial activity, such as quaternary ammonium functionalities, or covalently bound antibiotics or antimicrobial agents.

Antimicrobial entrapped-polyurethanes

The entrapping of antimicrobial agents into the polymeric matrix of the medical device during polymer synthesis or device manufacturing has the advantage to allow the loading of significant amounts of drug, thus prolonging the duration of antimicrobial activity of the device. However, this approach is limited to the antimicrobial agents that do not degrade at the temperature of polymer processing. Other downsides of this strategy to be considered are i) possible drug diffusion limitations from the polymer bulk to the surface where the drug can exert its activity[216] and ii) possible reduction in the polymer mechanical properties.[217]

Ciprofloxacin-HCl salt and lipophilic ciprofloxacin-betaine were incorporated into polyurethanes by Schierholz and colleagues to develop anti-infectious properties of this biomaterial.[218] The hydrophilic ciprofloxacin-HCl salt showed a fast initial release rate, whereas ciprofloxacin-betaine was characterized by a more continuous release behavior. This finding suggested a higher diffusion ability of the lipophilic ciprofloxacin-betaine in the polymer than its salt. The authors stated that a good affinity between the drug and the polymer is required for a sustained and prolonged release.

Also, the release of rifampin loaded within the matrix of biodegradable poly(ester urethanes)[219] was controlled by a diffusion-dependent release mechanism. In fact, after an initial burst release of antibiotic in a time interval of 0–30 min, the drug release depended on polymer swelling ability.

To control the release of the drug entrapped in the polymer matrix, the use of pore-former agents, such as PEG or albumin, to be entrapped in polyurethanes together with antibiotics has been investigated.[52,220] In particular, cefadroxil was entrapped in a polyurethane together with either PEG, D-mannitol, or bovine serum albumin (BSA).[220] An increase of release rate of cefadroxil was obtained by increasing the pore-former loading concentration. In particular, the release rate of cefadroxil was the highest when BSA was used as a pore former and the lowest with PEG1450. The authors stated that this finding was related to an incomplete formation of channels or pores due to PEG1450 high solubility in tetrahydrofuran that is the solvent used to prepare polymer/pore former mixtures.

A sustained drug release was instead obtained by entrapping two antibiotics cefamandole nafate and rifampin in a polyurethane matrix together with PEG at different molecular weights.[216] In this case, a long-lasting antimicrobial activity (up to 23 days) against *S. aureus* was obtained with the PEG10000-containing polymer.

More recently, polymer nanocomposites were investigated in their ability to control drug release and prolong device antimicrobial activity.[216,221] In particular, our group developed a nanostructured polymer consisting of cefamandole-loaded albumin or poly(allyl amine) nanoparticles entrapped in a polyurethane matrix.[216] The experimental findings showed how the hydrophilicity and porosity of the nanoparticle/polymer systems as well as drug/polymer interactions play important roles in the control of both antibiotic release and duration of antimicrobial activity.

Antibacterial polyurethane–montmorillonite nanocomposites at sustained drug release were also prepared by using chlorhexidine diacetate as nanofiller dispersant and antibacterial agent.[221] Results showed that higher nanofiller content allowed a better antimicrobial activity against *S. epidermidis*.

Antimicrobial or antiseptic-coated polyurethanes

The device coating with antimicrobial or antiseptic agents is undoubtedly the most investigated approach to prevent microbial colonization and biofilm formation on the device surface.

This approach has the advantage of being relatively simple and applicable to all the types of antimicrobial agents. However, important limitations of this strategy can be an insufficient drug loading for a prolonged antimicrobial effect and the uncontrolled, rapid release of the adsorbed drug in the very first hours following implantation. In fact, most of the available antimicrobial-coated devices release the majority of the adsorbed drug in the first hour from the implantation. This burst release is followed by a long-lasting phase of slow release at very low drug concentrations. This behavior causes a short-term antimicrobial activity and can contribute to the development of antimicrobial resistance.

The most experimented antimicrobial coatings are those based on silver, chlorhexidine and antibiotic/antifungal molecules. Different methods have been investigated to incorporate silver in polymeric substrates. A conventional approach is the direct deposition of metallic silver onto the device surface[222,223] by vapor coating, sputter coating, or ion beam coating. Another method to coat a substrate with silver involves deposition or electrochemical deposition of silver from solution.[224] An approach based on silver impregnation rather than coating has been recently developed by impregnation of nanoparticulate silver metal into silicone devices.[225,226] Recent innovative methods to incorporate silver in polymers consisted in the coordination of silver ions by anionic polyurethanes[227] or in the synthesis of silver nanoparticles on the surface of polyurethane hemodialysis catheters by photoreduction of a silver salt in alcohol.[228] Different types of silver-coated devices were experimented *in vitro* and *in vivo* studies,[229–231] including urinary catheters[232,233] and central venous catheters.

Silver coatings were shown to be effective in reducing catheter-related infections particularly when applied to urinary catheters.[234] Conflicting results were instead obtained with silver-coated CVCs. In fact, although some clinical studies demonstrated the efficacy of silver-coated CVCs in reducing bacterial colonization compared with the control ones,[235,236] the majority of the performed clinical trials failed in showing a statistically significant reduction either in catheter colonization or in catheter-related bloodstream infection.[230,237–239] The poor protection of silver in preventing CVC-related infections may be due to either the lower susceptibility of staphylococci to silver ions than Gram-negative bacteria[240] or deactivation of silver due to its interaction with blood components such as albumin.[241] A recent *in vitro* study[242] showed that silver nanoparticle coatings applied to CVCs or the release of silver ions from the catheter may affect the coagulation of contacting blood. In particular, both an acceleration of thrombin formation and stronger platelet activation upon contact of the silver nanoparticle-containing coatings with plasma was found compared to other coatings. These results further discourage the application of silver coatings to blood-contacting medical devices.

Device coating with chlorhexidine/silver sulfadiazine (CH/SS) has also been extensively studied in the early 1990s[243–245] and was anti-infective for short (approximately 1 week) insertion times.[246,247] In the early 2000s, a new generation of CH/SS-treated CVCs were marked containing silver sulfadiazine and chlorhexidine on the external surface and chlorhexidine in the catheter lumen. Three clinical trials[248–250] performed on this second-generation CH/SS-coated catheters showed a significant decrease in microbial colonization on the treated catheters in comparison with the untreated ones. However, the incidence of catheter-related bloodstream infection was similar in both groups of catheters.[248,250]

A series of antibiotic-coated devices with different spectrum of activity have been developed in the last decades and investigated both *in vitro* and *in vivo*. One of the first investigations was performed by Sherertz and colleagues[251] on intravascular catheters coated with one of the following antimicrobial agents: dicloxacillin, clindamycin, ciprofloxacin, chlorhexidine, cefotaxime, cefuroxime or fusidic acid. The results obtained employing a rabbit model indicated that only catheter coating with dicloxacillin, clindamycin, chlorhexidine or fusidic acid decreased the risk of infection compared with uncoated control catheters.

Three studies performed by Raad and Darouiche's group in the early 1990s allowed selecting the couple of antibiotics minocycline/rifampicin (M/R) as the best synergistic combination of being adsorbed on catheter surfaces to prevent related infections. In fact, minocycline and rifampin demonstrated activity against all the methicillin-sensitive *S. aureus* and methicillin-resistant *S. epidermidis* and *S. aureus* isolates recovered from patients with vascular catheter-related bacteremia over a 10-year period.[252] In addition, M/R combination was able to prevent catheter colonization by slime-producing strains of *S. epidermidis* and *S. aureus in vitro*[253] and in a rabbit model.[254]

Clinical trials performed on M/R-coated CVCs showed the ability of these catheters to reduce bacterial colonization and catheter-related infections both for short (approximately 7 days)[255] and long-term implantations.[256,257]

In particular, two large randomized clinical trials on either long-term M/R coated nontunneled[256] or tunneled[257] CVCs have shown their efficacy in reducing catheter-related bloodstream infections up to a 60-day catheterization time. The long-term activity of the M/R-coated catheters, also confirmed by a recent systematic review,[258] seems to be associated with the good affinity of the drugs for the catheter realized by precoating the catheter surface with the cationic surfactant tri-iododecylmethyl ammonium chloride (TDMAC). In fact, a multicenter randomized clinical trial performed on M/R-coated indwelling catheters showed that rifampin and minocycline continued to be detected on the surfaces of the catheters for at least 2 weeks after placement but no antibiotic was detected in the serum samples obtained from patients during catheterization.[259]

This finding highlighted that the concentration of the drug on the catheter surface as well as the kinetics of drug release from the device play key roles to obtain antimicrobial-coated catheters with long-lasting antimicrobial activity.

Polyurethanes able to absorb high amounts of antibiotics on their surfaces and to control drug release have been obtained by our group[260,261] by introducing in the polymer side chain specific functional groups able to interact with the selected drugs. The obtained results showed that either the adsorbed antibiotic amount or the kinetics of drug release were closely linked to the strength of polymer–drug interactions. Our approach is similar to the use of TDMAC, but in our case the antimicrobial chelant groups are covalently bonded in the polymer chain.

Although M/R-coated CVCs showed a good activity against a significant number of Gram-positive and Gram-negative bacteria, they were not active against *P. aeruginosa* and *Candida* spp.[262] In addition, a prospective randomized double-blind controlled multicenter trial on catheters coated with M/R on both the internal and external lumen showed an increase in *Candida* spp. colonization when the catheter remained implanted for 43 days in ICU patients.[263] Therefore, to fight fungal colonization of catheter surfaces, a new rifampicin-miconazole impregnated polyurethane catheter was developed.[264]

A prospective controlled nonblind randomized clinical trial performed in two German university hospitals on 223 CVC-bearing adult patients[265] showed the efficacy of rifampicin–miconazole-impregnated CVCs in reducing catheter colonization and related infection. A further analysis of the data obtained in this study performed in 2010 by the same authors[266] showed a pronounced reduction of catheter colonization and related infection when rifampicin–miconazole-impregnated CVCs were implanted in male, overweight and oncology patients. In particular, CR-BSIs were reduced from 26% to 2.3% in the cancer subgroup, suggesting possible benefits of these CVCs for high-risk patients.

Intrinsically antimicrobial polyurethanes

In the last decade, the development of intrinsically antimicrobial polymers has emerged as a promising alternative approach to prevent biofilm formation on medical devices. The advantage of this strategy is that the polymer matrix will not release antimicrobial substances but micro-organisms will be killed on contact.

In this regard, cationic polymers have gained much attention due to their ability to bind to the negatively charged bacterial cell surface.[267,268] Some examples include polyacrylates with side-chain biguanide groups, polycationic biocides with pendant phosphonium salts and quaternary ammonium functionalized poly(propylene imine) dendrimers.[267] Different factors affect their activity such as polymer molecular weight, the type of counterion, hydrophilic–hydrophobic balance and the alkyl chain length attached to charged group. These and other aspects will be discussed in another chapter of this book. Herein, the investigation concerning the development of cationic polyurethanes only will be discussed.

Cationic polyurethanes possessing phosphonate or quaternized amine groups were synthesized by Flemming and colleagues.[205] The quaternized amine polyurethanes reduced the number of adherent bacteria to only a few hundred after only 1 h. A reduction in bacterial adhesion was also found on the surface of the zwitterionic phosphonated polyurethane but to less extent. In fact, the phosphonated polyurethane showed a decreasing factor of 10 in bacteria after 24 h.

More recently, a series of quaternized polyurethanes were prepared by using N,N-bis(2-hydroxyethyl)isonicotinamide (BIN) as the chain extender in poly(tetramethylene oxide)-based segmented polyurethanes.[269] Three materials containing increasing hard segment fractions and therefore increasing concentrations of BIN were synthesized. The pyridine ring in BIN was quaternized with different alkyl halides. All the quaternized polyurethanes possessed good mechanical properties as well as bactericidal activity against *S. aureus* but only bacteriostatic activity against *E. coli*.

Polyurethanes having random copolymer 1,3-propylene oxide soft blocks with alkylammonium (C6 or C12 alkyl ammonium chain lengths) and either trifluoroethoxy or PEGylated side chains were recently prepared as polymer-surface modifiers.[270] In this work, a biomimetic amphiphilic polycation approach was followed since it takes compositional guidance from the ratio of charged to uncharged groups in naturally occurring antimicrobials, such as magainins and defensins. The coatings were highly effective against the Gram-negative *P. aeruginosa* and *E. coli*. The antimicrobial effectiveness of these materials was more strongly influenced by alkyl ammonium chain length.[271]

Another strategy to obtain intrinsically antimicrobial polymers concerns the covalent binding of antimicrobial agents, either antiseptic or antibiotic molecules, in the polymer main or side chain. To exert antimicrobial activity, the drug must be bonded to the polymer by hydrolyzable linkages to allow its release in the surrounding environment.

Woo *et al.* first covalently bonded ciprofloxacin to biodegradable polyurethane based on polycaprolactone diol and hexane diisocyanate.[272] This polymer was shown to be able to release the drug in the presence of cholesterol esterase up to 30 days, while its activity against *P. aeruginosa* lasted for 10 days.

In another approach, a drug precursor was grafted to polyurethanes and further activated to release the active molecule.[273] In particular, polyurethanes with soft blocks containing semifluorinated and 5,5-dimethylhydantoin pendant groups were prepared and employed as coatings for a conventional poly(tetramethylene oxide)-based polyurethane. Then, the polyurethane was activated by converting near-surface amide groups to chloramide with hypochlorite. After activation, the modified PU was effective against both Gram-positive (*S. aureus*) and Gram-negative (*P. aeruginosa* and *E. coli*) bacteria.

9.4 Biomimetic Polyurethanes in Tissue Engineering

Tissue engineering is an emerging field aiming to replace compromised tissues or organs by implanting biological substitutes.[274–276] The process generally involves the presence of reparative cells, a structural template (scaffold), transport of nutrients and metabolites, and molecular and mechanical regulatory factors.[277] The scaffold should be made of highly biocompatible and biodegradable material that does not elicit an immunological or foreign-body reaction. Moreover, in the case of load-bearing tissues, the scaffold must provide sufficient temporary mechanical support and degrade at a rate matching the rate of the new tissue formation.

Essential properties of scaffold are porosity, suitable pore size, permeability (for the diffusion of nutrient, transport of proteins and waste removal), and optimal surface properties for cell adhesion. Indeed, surface characteristics, such as hydrophilicity, surface charge density, surface micromorphology, free energy, and specific chemical groups deeply affect the adhesion, spreading and signaling of cells.

Several materials have been investigated to develop scaffolds for tissue engineering, including hybrid materials,[278,279] natural and synthetic polymers.[280–284] Natural polymers have the potential advantage of biological recognition, while the advantage of synthetic polymers is reproducible large-scale production with controlled properties of strength, degradation, and microstructure. Among synthetic polymers, the most commonly used in tissue engineering are polyesters such as polylactide (PLA), polyglycolide (PGA), polycaprolactone (PCL), poly(3-hydroxybutyrate) (PHB).[285–287]

To tune the degradation rate and mechanical properties of aliphatic polyesters according to tissue-engineering application, modifications of their chemical composition, crystallinity, molecular weight, molecular-weight distribution, glass-transition temperature and hydrophobicity have been performed. In particular, a series of block copolyesters based on PCL, PLA, PGA and polyethylene oxide (PEO) were prepared.[288–291] Some of these systems displayed a good cell growth and tissue formation. In order to enhance material cytocompatibility and then its biomimetic features, active groups on polymer surface by the synthesis of copolymers containing L-lactic acid and L-lysine units[287] or by plasma treatments[292] were introduced.

Although polyesters are well tolerated by cells in culture and in tissues, the mechanical properties of these polymers are poorly compatible with elastic and strong natural tissues such as heart, blood vessels, skeletal muscle, tendon, and so forth. Indeed, polyesters can be too stiff and brittle with low elongation or very soft with relatively low strength.

Elastomeric biodegradable polyurethanes are promising candidates for regeneration of different soft tissues, including cardiac muscle,[293] blood vessels,[294,295] skeletal muscle,[296] cartilage[297] and for tendon, ligament, and skin repair. This class of materials is also under investigation for hard-tissue regeneration such as bone-tissue repair.[298–300] In the past, the research was mainly focused on the development of nondegradable polyurethanes for long-term implants including catheters, pacemaker-lead insulation, vascular prostheses and cardiac-assistance devices, all being already described in Section 9.2.5. On the contrary, in the last 20 years biodegradable polyurethanes have also gained interest for tissue-engineering applications since a wide range of chemical, mechanical and biological properties can be easily obtained.

To be successful, a scaffolding material should mimic one or more characteristics of the natural extracellular matrix (ECM). Any biomaterial that matches the biodegradability or mechanical properties of natural tissue components can then be considered biomimetic.

As for biodegradability, a biomimetic approach consists in synthesizing polymers that similarly to collagen or other ECM components degrade by cell-secreted enzymes. The elastic properties typical of natural ECM can be instead mimicked by synthesizing block or segmented polymers at different composition. The choice of suitable blocks or segments as well as the introduction of enzymatic recognition sites or specific functional groups could fully attain the features of tissues that have to be regenerated or replaced.

To further favor biomolecular recognition of materials by cells, polymers can be either adsorbed with bioactive molecules, that once released from the material modulate new tissue formation, or grafted with cell-binding peptides, that specifically interact with cell receptors.

Tissue healing and regeneration can be compromised by infection development. Therefore, a sustained release of an antimicrobial agent from the scaffold is recently under investigation to prevent microbial adhesion and biofilm formation onto the implant. In this way, a local antibiotic concentration higher than that obtained with systemic parenteral administration can be achieved.

Since also scaffold morphology and structure affect tissue formation, different porosities and surface to volume ratios can be obtained by using specific fabrication methods such as thermally induced phase separation, electrospinning and water foaming. In this way, the physical properties of scaffolds can be widely tuned to obtain substitutes ranging from blood vessels to bone.

9.4.1 Biodegradable Polyurethanes for Soft-Tissue Regeneration

Blood-vessel tissue engineering

Many patients with vascular diseases need blood-vessel substitutes to restore vascular continuity. Surgical operations for small-diameter diseased vessels involve bypass grafting with autologous arteries or veins. As already mentioned in Section 9.2.5, synthetic materials can provoke neointimal hyperplasia and thrombosis when used to bypass arteries less than 6 mm in diameter. Thus, the development of tissue-engineered grafts can represent a winning approach to enhance the patency of such vascular devices.

The elastic properties displayed by polyurethanes make them serious candidate as scaffolding materials for vascular tissue. Initially, bioresorbable grafts were directly implanted in the body with the problem of tissue ingrowth *in vivo*. Next, it was seen that *in vitro* cell seeding on biodegradable scaffolds could better promote the regeneration of functional tissues. Indeed, Yue *et al.*[301] seeded smooth muscle cells (SMC) on a microporous biodegradable scaffold made of polyurethane (95%) and poly-L-lactide (5%). Seeded and nonseeded vascular grafts were then implanted into the abdominal aorta of rats. The patency of seeded grafts was superior to that of nonseeded devices.

More recently, PCL-PU fibrous composite scaffolds possessing a luminal PCL surface and a highly porous antiluminal polyurethane surface forming the vessel-wall substitute were developed.[302] These novel composites showed strong attachment and proliferation of human endothelial cells with formation of a cell monolayer. Polyurethanes with nontoxic degradation products were developed by using PCL macrodiol, 1,4-diisocyanante butane (BDI) and lysine ethyl ester or putrescine as chain extenders.[303] Putrescine was chosen because it is a polyamine essential for cell growth and proliferation. Surface modification of both scaffold types by coupling of RGDS growth factor (Arg-Gly-Asp-Ser) improved the adhesion of human endothelial cells (> 160% than polystyrene control) showing the potential application of these materials for soft-tissue engineering.

To achieve high cellular density and then favor cell infiltration, Stankus *et al.* have developed a tissue-engineered construct made mainly of cells and reinforced with an elastomeric fiber matrix.[304] In particular, SMCs were electrosprayed during the electrospinning of poly(ester urethane urea) fiber. The SMC microintegrated scaffold was strong, flexible and anisotropic with tensile strength ranging from 2 to 6.5 MPa and strain at break from 850 to 1700%, depending on material axis.

Woodhouse and colleagues[305] synthesized a series of biodegradable polyurethanes containing the amino acid L-phenylalanine as chain extender. PCL diol or polyethylene oxides with different molecular weights were employed as the *soft* segment obtaining polymers with different bulk hydrophilicity. These polyurethanes were proposed for making cardiac tissue.

To further encourage cell adhesion, in a more recent study, embryonic stem-cell-derived cardiomyocytes were seeded on these same polyurethane films precoated with gelatin, laminin or collagen IV.[306] After 30 days, cells cultured on laminin and collagen IV exhibited preferential adhesion. The authors suggested that these systems may hold potential for the repair of damaged heart tissue.

In order to develop polyurethane-based scaffolds for longer-term vascular tissue engineering, degradable polar hydrophobic ionic polyurethanes (D-PHI) containing a lysine-based polycarbonate divinyl oligomer were synthesized and then reacted with a lysine-based crosslinker.[295] An increase of the elastic modulus and a reduction of scaffold elongation at yield and swelling were found by increasing polymer crosslinker concentration. A vascular smooth muscle cells adhered onto the scaffolds showed a gradual increase in cell density within 2 weeks of culture.

Cartilage tissue engineering

Articular cartilage is a flexible connective tissue covering the ends of bones where they come together to form joints. The main function of cartilage is to transmit loads between bones allowing the motion of the joints. Cartilage can be damaged by injury or normal wear and tear. Since such lesions do not spontaneously heal, cartilage repair and regeneration by tissue-engineering approaches are major challenges in orthopedic field.

Poly(ester urethanes) have been investigated as scaffolding materials for chondrocytes. The first study was performed by using DegraPol®, a degradable block poly(ester urethane), consisting of crystallizable blocks of α-ω-dihydroxy-oligo[((R)-3-hydroxybutyrate-*co*-(R)-3-hydroxyvalerate)-*block*-ethylene glycol and blocks of poly(ε-caprolactone) diol linked with a lysine methyl ester diisocyanoate.[307] Chondrocytes and osteoblasts, isolated from adult male rats, were cultured on this poly(ester urethane). Both types of cells exhibited relatively high cell adhesion on DegraPol®-foam.

Electronspun DegraPol® membranes were also investigated as fibrous scaffolds for skeletal muscle tissue engineering by Riboldi *et al.*[296] Thanks to the elasticity of the fibrous material and to the reticulated structure of the membrane, electronspun DegraPol® membranes, showed a tensile modulus value (E) around 10 MPa matching the range of skeletal muscle elasticity. The good cellular response obtained in adhesion and differentiation experiments evidenced the suitability of such poly(ester urethane) as scaffold for skeletal muscle-tissue engineering.

To support attachment and proliferation of primary chondrocytes for cartilage regeneration, Grad *et al.* evaluated as scaffolding material a linear biodegradable porous polyurethane synthesized from hexamethylene diisocyanate, polycaprolactone diol (molecular weight = 530 Da) and isosorbide diol as chain extender.[308] Although the polyurethane scaffold showed good mechanical properties, dynamic cell seeding allowed the only incorporation of 60–70% of cells into the scaffold 2 days after seeding. This study demonstrated that the physical and geometrical properties of the scaffold can markedly influence cell–polymer interactions.

Meniscus tissue engineering

Porous polymer scaffolds can be also used for meniscal lesion repair. Meniscus damage can develop as a result of an injury or from degenerative changes. The fibro-cartilage-like tissue of the meniscus has a limited regenerative capacity.

Tissue engineering could offer solutions for treatment of meniscus lesions or for replacement of the whole meniscus. Poly(ester urethane)s based on ε-capro-lactone, L-lactide or both have been demonstrated to be particularly suitable for tissue engineering of the meniscus. Klompmaker *et al.* implanted polyurethanes at different porosities (ranging from 50 to 500 μm) in the meniscus of 29 rabbits demonstrating that for optimal ingrowth and incorporation of partial or total meniscus prosthesis pore size must be in the range 150–500 μm.[309]

Porous biodegradable polyurethane foams have been prepared by using three different polymers: a methylene diphenyl diisocyanate-based polyurethane (Estane 5701-F1), a lysine diisocyanate-based polyurethane, and a poly(ε-caprolactone)-based polyurethane.[310] Two different combinations of techni-ques, freeze drying/salt leaching or *in situ* polymerization/salt leaching, have been used to prepare low-density polymer materials with reproducible structure and good mechanical properties. Foams prepared by salt leaching in combin-ation with a freeze-drying method showed excellent ingrowth of fibrocartila-ginous tissue when implanted in dogs.

Similar results were obtained by using a poly(ester urethane)s based on the aliphatic diisocyanate 1,4-trans-cyclohexanediisocyanate, poly(ε-caprolactone) and cyclohexane dimethanol as chain extender.[311] Van Tienen *et al.* developed polymer foams by using two series of polyurethanes based on 50/50 ε-capro-lactone/L-lactide, 1,4-butanediisocynate and 1,4-butanediol as chain ex-tender.[312] It was evidenced that, although scaffold micropores of at least 30 μm enabled ingrowth of well-vascularized cellular connective tissue before the polymer degradation occurred, the low scaffold compression modulus (74 kPa) was not suitable to develop implants for meniscal repair. On the contrary, scaffolds with lower micropores (pore size of 10–15 μm) even though they possessed better mechanical properties (compression modulus of 460 kPa) degraded before tissue ingrowth was completed.

9.4.2 Biodegradable Polyurethanes for Hard-Tissue Regeneration

Bone-tissue engineering

Bone grafts are often needed to promote the healing of bone defects. Besides autograft, a standard procedure involving bone harvested from the patient's own body (often from the iliac crest), it is possible to use a biodegradable or bioresorbable synthetic material with similar mechanical properties to bone. In particular, a tissue-engineering approach to repair or regenerate bone-tissue function involves the combined use of osteoconductive biodegradable scaffolds, osteogenic cells and osteoinductive bioactive factors. Signaling molecules, such as growth factors, can be incorporated into the scaffold in order to enhance their osteoconductive properties. As for mechanical properties, scaffolding materials should optimally match those of the tissue to be replaced. Among bioresorbable and biodegradable polymers for bone-tissue engineering, polyhydroxyacids occupy a prominent position. However, porous scaffolds manufactured from these crystalline polymers can undergo irreversible deformation under load with significant reduction of their porosity.

Biodegradable polyurethanes thanks to their tuneable elastomeric properties are instead excellent materials for bone-tissue engineering. Different polyurethanes based on nontoxic diisocyanates have been developed for this application. Zhang *et al.* synthesized a sponge-like polyurethane scaffold from lysine diisocyante (LDI) and glycerol.[299] To promote osteoblast differentiation, ascorbic acid (AA) was copolymerized with LDI-glycerol. LDI-glycerol-AA scaffolds supported the *in vitro* growth of mouse osteoblastic precursor cells (OPCs). Gorna and Gogolewski synthesized many crosslinked 3D biodegradable polyurethane scaffolds (foams) with different hydrophilicity.[298] Foams were prepared by using hexamethylene diisocyanate, PEO as hydrophilic component, and PCL, amine-based polyol or sucrose-based polyol as hydrophobic component. Water was used as both chain extender and foaming agent. In this study, citric acid as a calcium-complexing agent and several inorganic fillers were used for *in vitro* calcification experiments. The authors demonstrated that pore size and geometry, water adsorption and degradation *in vitro* as well as the mechanical properties of the scaffolds were influenced by the material's chemical composition. All polyurethane scaffolds induced the deposition of calcium phosphate.

The same authors implanted these scaffolds into monocortical defects in the iliac crest of healthy sheep.[313] After 6 months, the defects were healed with formation of new cancellous bone with denser structure than the native one. In addition, it was shown that a high hydrophilic component in the polymer allowed deposition of more calcium phosphate.

Kavlock *et al.* synthesized a series of poly(ester urethane urea)s (PEUURs) with different content of *hard* segments ranging from 20 to 40 wt.%.[314] These polymers have been prepared from 1,4-diisocyanatobutane, a poly(caprolactone) diol *soft* segment and a tyramine-1,4-diisocyanatobutane-tyramine chain extender. By varying the PCL molecular weight, PEUURs with different storage modulus have been obtained. Differential scanning calorimetry and wide-angle X-ray scattering measurements showed that the storage modulus was correlated with PCL crystallinity. All films made of these PEUURs allowed the adhesion of bone marrow stromal cells.

To obtain porous scaffolds with enhanced affinity toward cells and tissues, Gogolewski and Gorna synthesized another series of poly(ester urethane)s using aliphatic diisocyanate, PCL and a osteoconductive molecule isosorbide diol (1,4,3,6-dianhydro-D-sorbitol) as chain extender.[315] The scaffolds were prepared by using a combined salt-leaching–phase-inverse technique. The influence of polymer dissolution solvents, the type of nonsolvent, the porogen size and concentration on the scaffold structure was studied. To further promote the osteoconductive properties of these scaffolds, hydroxyapatite (HA) was loaded at different concentrations. These scaffolds, seeded with osteoblasts and implanted into subcutaneous tissue of nude mice, promoted bone formation in their structure within 5 weeks of implantation.

To prevent scaffold-related infection complications, a biodegradable HA/ polyurethane composite scaffold containing ceftazidime-loaded ethyl cellulose (EC) microspheres was recently developed by Liu et al.[316] The antibiotic-loaded EC microspheres were uniformly distributed in the HA/PU scaffold matrix and did not affect the scaffold pore structure. The incorporation of EC microspheres into the composite scaffold significantly reduced the initial burst release of drug and controlled its release up to 60 days.

Other methods to improve scaffold bioactivity include the use of bioceramic fillers that can be incorporated into polymer matrix or deposited on the scaffold surface. Bioactive glass is a material that can promote the formation of strong tissue–material interactions. Bil et al. coated polyurethane (PUR) and poly-urethane/poly(D,L-lactide) acid (PUR/PDLLA)-based scaffolds with Bio-glass® particles by a slurry-dipping method.[317] The scaffolds showed a homogenous coating of the bioactive agent indicating a good adhesion of the glass particles to two polymers. In vitro studies performed in simulated body fluid showed that the Bioglass® coating makes polyurethane scaffolds highly bioactive to promote the formation of a carbonate hydroxyapatite layer on their surfaces.

To investigate the effects of the release of osteoinductive molecules on new bone tissue formation, Li et al. encapsulated recombinant human bone mor-phogenic protein (rhBMP-2) into biodegradable polyurethane/microsphere composite scaffolds.[318] Polyurethanes were made from hexamethylene diiso-cyanate trimer plus polyester triol (prepared from 60% caprolactone, 30% glycolide and 10% D,L-lactide) alone or in a 50/50 mixture with PEG600. Two different incorporation methods for rhBMP-2 were used. In the first method the osteoinductive agent dry powder was added prior to the foaming reaction. In the second one, rhBMP-2 was encapsulated in PLGA microspheres with dif-ferent sizes prior to the foaming reaction. The scaffolds were implanted in rat femoral plug. The results indicated that scaffolds incorporating rhBMP-2 powder displayed a more extensive formation of new bone tissue. This behavior was associated with the burst release of the osteoinductive agent (35%) from these scaffolds followed by a sustained release for 21 days. The burst release of rhBMP-2 was reduced by its encapsulation in PLGA microspheres suggesting that a burst release within the first days is important to enhance new bone-tissue formation.

Acrylic bone cements are currently used to treat bone defects such as frac-tures. Since the biodegradable sponge-like polyurethanes cannot be arthros-copically used and cured in situ, injectable biodegradable polyurethane networks have been developed for this application.

Two component injectable polyurethanes were synthesized by Bonzani et al. using a two-step process involving the reaction between two prepolymers.[300] In particular, the prepolymer A was based either on pentaerythritol (PE) and ethyl lysine diisocyanate (ELDI) or on pentaerythritol, ELDI and LLA or DLA hydroxyacid. The prepolymer B was instead based on pentaerythritol and glycolic acid (GA). Four different types of in situ cured polymers were prepared from different prepolymer mixtures. One of the developed networks contained

10% β-tricalcium phosphate granules (β-TCP). The polymers with or without β-TCP showed good compressive strengths and moduli over 2 GPa. These mechanical properties were superior to those of other injectable bone cements. Cellular cultures carried out by using primary human osteoblasts evidenced cell viability and a good proliferation that increased from day 1 to day 7. This finding was ascribed to the hydrophilic nature of the polyurethane networks. The incorporation of β-TCP enhanced surface wettability, cell viability and proliferation. The two-step process used by Bonzani *et al.*,[300] makes the processing of the materials difficult due to the high viscosity of polyester prepolymers. To reduce the viscosity, the polyester polyol can be poured out in a large excess of polyisocyanate. This procedure is called quasi-prepolymer process.

Biodegradable polyurethane networks by using a two-step quasi-prepolymer process were obtained by Guelcher *et al.*[319] In particular, PCL and PCL-*co*-PGA-*co*-PLA triols were reacted with polyisocyanates with different functionality, LDI (lysine diisocyanate) and LTI (lysine triisocyanate). The effect of the type of diisocyanate on polymer mechanical and biological properties as well as of the polyester triol composition on polyurethane degradation was studied. Viability of MC3T3 osteoblast-like cells seeded onto the PU network surface demonstrated the nontoxicity of these polymers. These polyurethane networks also supported MC3T3 cell attachment and proliferation.

Although biodegradable polyurethane scaffolds have been shown to promote bone regeneration *in vivo*, the scaffold ability to control infections has been so far poorly investigated. Antibiotic-releasing PMMA beads are generally used in clinics to treat infected fractures. Unfortunately, these beads have to be surgically removed before implanting a bone graft. To overcome this drawback, Guelcher's group incorporated antibiotics into injectable polyurethane scaffolds. In particular, tobramycin was incorporated into a polyurethane network based on the hexamethylene diisocyanate trimer (HDIt) and a trifunctional polyester polyol.[320] Tobramycin release, tuned by PEG addition, was 4–5-times greater than equivalent volumes of PMMA beads. Similar results were also observed for colistin and tigecycline.

Polyurethane scaffolds prepared employing either HDIt or lysine triisocyanate (LTI) as monomers were loaded with vancomycin free base (V-FB) or hydrochloride (V-HCl).[321] The release profiles of vancomycin from the two types of scaffolds were similar, while the drug cumulative release from PU(HDIt) scaffold was generally less than PU(LTI) one.

The PU scaffolds containing V-FB showed a minimal burst release followed by a sustained release for a period of 6 weeks due to V-FB hydrophobicity higher than V-HCl. When implanted in a contaminated critical-sized fat femoral segmental defect, the performance of PU(LTI)/V-FB was comparable to PMMA/V-HCl beads. To reduce infection and at the same time promote bone regeneration, Guelcher *et al.* loaded the recombinant human bone morphogenic protein (rhBMP-2) and vancomycin into their polyurethane scaffolds.[322] This dual delivery approach resulted in a good bone formation but only in a modest protection against infection.

9.5 Future Perspectives

In recent decades, the intense research activity focused on polyurethanes has allowed the understanding of the structure–property relationship in these polymers and the obtainment of a wide range of materials. Either rigid thermosetting or soft elastomer polyurethanes are available on the market. The tuneable mechanical properties together with the excellent biocompatibility make polyurethanes today irreplaceable materials for numerous biomedical applications. In particular, biostable polyurethanes are and will remain the materials of choice for the manufacturing of intravascular devices that, more than other devices, need high flexural endurance and fatigue resistance as well as good hemocompatibility.

In the last 20 years, the exponential growth of life sciences has expanded the knowledge of mechanisms involved in biomaterial/tissue interaction as well as of factors affecting tissue regeneration. These understandings became the inspiration for biomaterial scientists to develop novel polyurethane-based materials able to mimic the biological systems on multiple levels.

The synthesis of biomimetic polyurethanes has allowed great advances to be made in the field of implantable medical devices. In fact, a series of polyurethanes able either to prevent, at least in the short term, or delay infectious and thrombosis complications have been developed. Their use in clinics has permitted significant improvements in the quality of life of patients. However, research efforts are still needed in this field, particularly to develop medical devices resistant to microbial colonization for long-term applications. In this regard, future research must strive to better understand the molecular mechanism governing microbial adhesion to artificial surfaces in order to develop new intrinsically antifouling materials. To avoid the emergence of antibiotic resistance, strategies not using antibiotics should be mainly employed.

The recent availability of biodegradable polyurethanes has extended the application of polyurethanes in tissue engineering. The possibility to control polymer degradation rate as well as setting up new manufacturing technologies enabled scientists to design polyurethane-based scaffolds with various biomimetic features for soft- and hard-tissue regeneration.

The use of biomimetic polyurethanes as scaffolding materials is today booming and the authors envision its continuous expansion in the decades to come. This could allow replacing not only parts of tissues but also complex systems as whole organs. Therefore, we will see more of biomimetic polyurethanes in the future, not only in the scientific community but also on the market.

References

1. A. M. Reed and D. K. Gilding, *Polymer*, 1981, **22**, 494–498.
2. C. Krishna, S. Pillai and C. P. Sharma, *J. Biomater. Appl.*, 2010, **25**, 291–366.
3. T. E. Lipatova, *Adv. Polym. Sci.*, 1986, **79**, 65–93.
4. P. Ferreira, A. F. M. Silva, M. I. Pinto and M. H. Gil, *J. Mater. Sci.: Mater. Med.*, 2008, **19**, 111–120.

5. A. Serrero, S. Trombotto, Y. Bayon, P. Gravagna, S. Montanari and L. David, *Biomacromolecules*, 2001, **12**, 1556–1566.
6. Y. S. Morsi, I. E. Birchall and F. L. Rosenfeldt, *Int. J. Artif. Organs*, 2004, **27**, 445–451.
7. J. Hunter, *Am. J. Surg.*, 1965, **109**, 325–331.
8. S. Iannace, G. Sabatini, L. Ambrosio and L. Nicolais, *Biomaterials*, 1995, **16**, 675–680.
9. J. C.-H. Goh, H.-W. Ouyang, S.-H. Teoh, C. K. C. Chan and E.-H. L. Lee, *Tissue Eng.*, 2003, **9**, 31–44.
10. G. L. Grobe III, P. L. Valint Jr and D. M. Ammon Jr, *J. Biomed. Mater. Res.*, 1996, **32**, 45–54.
11. P. C. Nicolson and J. Vogt, *Biomaterials*, 2001, **22**, 3273–3283.
12. A. K. Mishra, R. Narayan, T. M. Aminabhavi, S. K. Pradhan and K. V. S. N. Raju, *J. Appl. Polym. Sci.*, 2012, **125**, E67–E75.
13. T. Shinoka, C. Breuer, R. E. Tanel, G. Zund, T. Miura, P. X. Ma, R. Langer, J. P. Vacanti and J. E. Mayer, *Ann. Thorac. Surg.*, 1995, **60**, S513–S516.
14. U. A. Stock and J. E. Jr Mayer, *J. Long Term. Eff. Med. Implants*, 2001, **11**, 249–260.
15. D. F. Stamatialis, B. J. Papenburg, M. Girones, S Saiful, S. N. M. Bettahalli, S. Schmitmeier and M. Wessling, *J. Membrane Sci.*, 2008, **308**, 1–34.
16. T. Tsuruta, *Biopolymers Liquid Crystalline Polymers Phase Emulsion*, ed. Springer Berlin Heidelberg, Series Advances in Polymer Science, Hardback Germany, 1996, Chapter 126, 1–51.
17. B. D. Ratner, *J. Biomat. Sci.-Polym. Ed.*, 1993, **4**, 3.
18. A. Alanazi, C. Nojiri, T. Kido, T. Noguchi, Y. Ohgoe, T. Matsuda, K. Hirakuri, A. Funakubo, K. Sakai and Y. Fukui, *Artif. Organs*, 2000, **24**, 624–627.
19. S. Ramakrishna, J. Mayer, E. Wintermantel and K. M. Leong, *Compos. Sci. Technol.*, 2001, **61**, 1189–1224.
20. X. Liu, P. K. Chu and C. Ding, *Mater. Sci. Eng. R*, 2004, **47**, 49–121.
21. K. A. Marx, J. S. Lee and C. Sung, *Biomacromolecules*, 2004, **5**, 1869–1876.
22. W. Feng, J. Brash and S. Zhu, *J. Polym. Sci. A: Polym. Chem.*, 2004, **42**, 2931–2942.
23. C. de las Heras Alarcón, S. Pennadam and C. Alexander, *Chem. Soc. Rev.*, 2005, **34**, 276–285.
24. W. Marconi, F. Benvenuti and A. Piozzi, *Biomaterials*, 1997, **18**, 885–891.
25. K. M. Shakesheff, S. M. Cannizzaro and R. Langer, *J. Biomat. Sci. Polym. Ed.*, 1998, **9**, 507–518.
26. W. Marconi, R. Marcone and A. Piozzi, *Macromol. Chem. Phys.*, 2000, **201**, 715–721.
27. R. A. Quirk, W. C. Chan, M. C. Davies, S. J. B. Tendler and K. M. Shakesheff, *Biomaterials*, 2001, **22**, 865–872.
28. I. Francolini, F. Crisante, A. Martinelli, L. D'Ilario and A. Piozzi, *Acta Biomater.*, 2012, **8**, 549–558.
29. M. Habash and G. Reid, *J. Clin. Pharmacol.*, 1999, **39**, 887–898.

30. B. Gottenbos, H. J. Busscher, H. C. van der Mei and P. Nieuwenhuis, *J. Mater. Sci. Mater. Med.*, 2002, **13**, 717–720.
31. N. M. K. Lamba, K. A. Woodhouse and S. L. Cooper. *Polyurethanes in Biomedical Applications.* ed. N. M. K. Lamba, K. A. Woodhouse, and S. L. Cooper, CRC Press LLC, Boca Raton FL USA, 1st edn, 1998, chapter 6, 115–142.
32. S. W. Kim, R. G. Lee, H. Oster, D. Coleman, J. D. Andrade, D. J. Lentz and D. Olsen, *Trans. Am. Soc. Artif. Intern. Organs*, 1974, **20**, 449–455.
33. J. H. Silver, C. W. Myers, F. Lim and S. L. Cooper, *Biomaterials*, 1994, **15**, 695–704.
34. Y. Wu, F. I. Simonovsky, B. D. Ratner and T. A. Horbett, *J. Biomed. Mater. Res.*, 2005, **74A**, 722–738.
35. J. M. Morais, F. Papadimitrakopoulos and D. J. Burgess, *AAPS J.*, 2010, **12**, 188–196.
36. L. Zhou, L. Yu, M. Ding, J. Li, H. Tan, Z. Wang and Q. Fu, *Macromolecules*, 2011, **44**, 857–864.
37. M. Mahkam, M. G. Assadi, R. Zahedifar, M. Ramesh and S. Daravan, *J. Bioact. Compat. Pol.*, 2004, **19**, 45–53.
38. J. D. Fromstein and K. A. Woodhouse, *J. Biomat. Sci. Polym. Ed.*, 2002, **13**, 391–406.
39. Y. D. Wang, Y. M. Kim and R. Lange, *J. Biomed. Mater. Res. A*, 2003, **66A**, 192–197.
40. M. Sokolsky-Papkov, K. Agashi, A. Olaye, K. Shakesheff and A. J. Domb, *Adv. Drug Deliver. Rev.*, 2007, **59**, 187–206.
41. S. A. Guelcher, *Tissue Eng. Part B: Rev*, 2008, **14**, 3–17.
42. M. S. Sánchez-Adsuar, M. M. Pastor-Blas and J. M. Martín-Martínez, *J. Adhes.*, 1998, **67**, 327–342.
43. C. Prisacariu, C. P. Buckley and A. A. Caraculacu, *Polymer*, 2005, **46**, 3884–3894.
44. J. D. Merline, C. P. Reghunadhan Nair, C. Gouri, G. G. Bandyopadhayay and K. N. Ninan, *J. Appl. Polym. Sci.*, 2008, **107**, 4082–4092.
45. R. Hernandez, J. Weksler, A. Padsalgikar, T. Choi, E. Angelo, J. S. Lin, L.-C. Xu, C. A. Siedlecki and J. Runt, *Macromolecules*, 2008, **41**, 9767–9776.
46. C. H. Y. Chen-Tsai, E. L. Thomas and W. J. Macknight, *Polymer*, 1986, **27**, 659–666.
47. L. Yingjie, G. Tong and C. Benjamin, *Macromolecules*, 1992, **25**, 1737–1742.
48. T. Hun-Jan, M. W. Curtis, Y. Xiaozhen, W. J. MacKnight and H. L. Shaw, *Macromolecules*, 1994, **27**, 7146–7151.
49. D. K. Lee, H. B. Tsai and J. L. Stanford, *J. Polym. Res.*, 1996, **3**, 221–225.
50. R. W. Seymour and S. L. Cooper, *Macromolecules*, 1973, **6**, 48–53.
51. C. S. Paik Sung, C. B. Hu and C. S. Wu, *Macromolecules*, 1980, **13**, 111–116.
52. V. Ruggeri, I. Francolini, G. Donelli and A. Piozzi, *J. Biomed. Mater. Res. A*, 2007, **81A**, 287–298.

53. R. E. Marchant, *Handbook of Polymer Degradation*, ed. S. H. Hamid, M. B. Amin and A. G. Maadhah, Marcel Dekker, Inc, New York, 1st edn, 1992, 617–631.

54. P. L. Kuo, J. M. Chang and T. L. Wan, *J. Appl. Polym. Sci.*, 1998, **69**, 1635–1643.

55. A. Stanciu, A. Airinei, D. Timpua, A. Ioanid, C. Ioan and V. Bulacovschi, *Eur. Polym. J.*, 1999, **35**, 1959–1965.

56. P. A. Gunatillake, G. F. Meijs and S. J. McCarthy, *Biomedical Applications of Polyurethanes*, ed. P. Vermette, H. J. Griesser, G. Laroche and R. Guidoin, Eurekah.com Landes Bioscience, Georgetown, Texas, USA, 1st edn, 2001, Chapter 6, 160–174.

57. H. O. Efsing, B. Lindblad, J. Mark and T. Wolff, *World J. Surg.*, 1983, **7**, 419–423.

58. T. Pottecher, M. Forrler, P. Picardat, D. Krause, J. P. Bellocq and J. C. Otteni, *Eur. J. Anaesth.*, 1984, **1**, 361–372.

59. C. J. Van Rooden, M. E. T. Tesselaar, S. Osanto, F. R. Rosendaal and M. V. Huisman, *J. Thromb. Haemost.*, 2005, **3**, 2409–2419.

60. M. G. Bourassa, M. Cantin, E. B. Sandborn and E. Pederson, *Circulation*, 1976, **53**, 992–996.

61. G. D. Wilner, W. J. Casarella, R. Baier and C. M. Fenoglio, *Circ. Res.*, 1978, **43**, 424–428.

62. J. F. Hecker and L. A. Scandrett, *J. Biomed. Mater. Res.*, 1985, **19**, 381–395.

63. A. Y. Lee and P. W. Kamphuisen, *J. Thromb. Haemost.*, 2012, **10**, 1491–1499.

64. N. P. O'Grady, M. Alexander, E. P. Dellinger, J. L. Gerberding, S. O. Heard, D. G. Maki, H. Masur, R. D. McCormick, L. A. Mermel, M. L. Pearson, I. I. Raad, A. Randolph and R. A. Weinstein, *Infect. Control Hosp. Epidemiol.*, 2002, **23**, 759–769.

65. G. Donelli, P. De Paoli, G. Fadda, P. Marone, G. Nicoletti and P. E. Varaldo, *J. Chemother.*, 2001, **13**, 251–262.

66. P. Francois, P. Vaudaux, T. J. Foster and D. P. Lew, *Infect. Control Hosp. Epidemiol.*, 1996, **17**, 514–519.

67. M. Nilsson, J. I. Frykberg, J. I. Flock, L. Pei, M. Lindberg and B. Guss, *Infect. Immun.*, 1998, **66**, 2666–2673.

68. M. E. Rupp, J. S. Ulphani, P. D. Fey, K. Bartscht and D. Mack, *Infect. Immun.*, 1999, **67**, 2627–2632.

69. S. Zanetti, A. Angioi, P. L. Fiori and G. Fadda, *New Microbiol.*, 1994, **17**, 297–305.

70. L. Baldassarri, G. Donelli, A. Gelosia, A. W. Simpson and G. D. Christensen, *Infect. Immun.*, 1997, **65**, 1522–1526.

71. G. Donelli, A. Gelosia, L. Baldassarri, M. Mignozzi, E. Fiscarelli and G. Rizzoni, *Electron Microscopy*, 1992, **3**, 513–514, EUREM 92, Granada, Spain.

72. B. Jansen, L. P. Goodman and D. Ruiten, *Gastrointest. Endosc.*, 1993, **39**, 670–673.

73. M. J. Bridgett, M. C. Davies, S. P. Denyer and P. R. Eldridge, *Biomaterials*, 1993, **14**, 184–188.

74. K. G. Kristinsson, *J. Med. Microbiol.*, 1989, **28**, 249–257.
75. M. Morra and C. Cassineli, *J. Biomat. Sci.: Polym. Ed.*, 1999, **10**, 1107–1124.
76. W. K. Liu, S. E. Tebbs, P. O. Byrne and T. S. J. Elliott, *J. Infection*, 1993, **27**, 261–269.
77. W. K. Liu, M. R. W. Brown and T. S. J. Elliott, *J. Antimicrob. Chemother.*, 1997, **39**, 687–695.
78. S. Furman, D. J. Escher, N. Solomon and M. Krauthamer, *J. Thorac. Cardiovasc. Surg.*, 1967, **54**, 723–727.
79. K. Stokes and K. Cobian, *Biomaterials*, 1982, **3**, 225–231.
80. A. S. Chawla, P. Blais, I. Hinberg and D. Johnson, *Biomater. Artif. Cells Artif. Organs*, 1988, **16**, 785–800.
81. G. F. Meijs, S. J. McCarthy, E. Rizzardo, Y. C. Chen, R. C. Chatelier, A. Brandwood and K. Schindhelm, *J. Biomed. Mater. Res.*, 1993, **27**, 345–356.
82. K. Stokes, P. Urbanski and J. Upton, *J. Biomater. Sci.: Polym. Ed.*, 1990, **1**, 207–230.
83. J. M. Anderson, A. Hiltner, M. J. Wiggins and M. A. Schubert, *Polym. Intern.*, 1998, **46**, 163–171.
84. K. E. Cobian, M. J. Ebert, P. B. McIntyre and D. W. Mayer US005796044, 1998.
85. C. Jenney, J. Tan, A. Karicherla, J. Burke and J. Helland, *Heart Rhythm.*, 2005, **2**, S318–S319.
86. T. Kleeman, T. Becker, K. Doenges, M. Vater, J. Senges, S. Schneider, W. Saggau, U. Weisse and K. Seidl, *Circulation*, 2007, **115**, 2474–2480.
87. P. Klinkert, P. N. Post, P. J. Breslau and J. H. van Bockel, *Eur. J. Vasc. Endovasc.*, 2004, **27**, 357–362.
88. J. D. Kakisis, C. D. Liapis, C. Breuer and B. E. Sumpio, *J. Vasc. Surg.*, 2005, **41**, 349–354.
89. M. Prager, P. Polterauer, H. J. Bohmig, O. Wagner, A. Fugl, G. Kretschmer, M. Plohner, J. Nanobashvili and I. Huk, *Surgery*, 2001, **130**, 408–415.
90. D. C. Brewster, *J. Vasc. Surg.*, 1997, **25**, 365–379.
91. S. K. Williams, T. Carter, P. K. Park, D. G. Rose, T. Schneider and B. E. Jarrell, *J. Biomed. Mater. Res.*, 1992, **26**, 103–117.
92. C. Nojiri, K. Senshu and T. Okano, *Artif. Organs*, 1995, **19**, 32–39.
93. S. H. Hsu, H. J. Tseng and M. S. Wu, *Artif. Organs*, 2000, **24**, 119–123.
94. M. A. Contreras, W. C. Quist and F. W. Logerfo, *Microsurg.*, 2000, **20**, 15–21.
95. Z. Zhang, Z. Wang, S. Liu and M. Kodama, *Biomaterials*, 2004, **25**, 177–187.
96. R. Guidon, M. Sigot, M. King and M. F. Sigot-Luizard, *Biomaterials*, 1992, **13**, 281–288.
97. Z. Zhang, Y. Marois, R. G. Guidoin, P. Bull, M. Marois, T. How, G. Laroche and M. W. King, *Biomaterials*, 1997, **18**, 113–124.

98. L. Pinchuk, *J. Biomater. Sci.; Polym. Ed.*, 1994, **6**, 225–267.
99. J. P. Santerre, R. S. Labow, D. G. Duguay, D. Erfle and G. A. Adams, *J. Biomed. Mater. Res.*, 1994, **28**, 1187–1199.
100. M. H. Glickman, G. K. Stokes, J. R. Ross, E. D. Shuman, W. C. 3ʳᵈ Sternbergh, J. S. Lindberg, S. M. Money and M. I. Lorber, *J. Vasc. Surg.*, 2001, **34**, 465–472.
101. H. J. Salacinski, N. R. Tai, R. J. Carson, A. Edwards, G. Hamilton and A. M. Seifalian, *J. Biomed. Mater. Res.*, 2002, **59**, 207–218.
102. E. M. Christenson, J. M. Anderson and A. Hiltner, *J. Biomed. Mater. Res.*, 2004, **70**, 245–255.
103. A. M. Seifalian, A. Tiwari, G. Hamilton and H. J. Salacinski, *Artif. Organs*, 2002, **26**, 307–310.
104. Y. Zhu, C. Gao, T. He and J. Shen, *Biomaterials*, 2004, **25**, 423–427.
105. B. H. Walpoth, R. Rogulenko, E. Tikhvinskaia, S. Gogolewski, T. Schaffner, O. M. Hess and U. Althaus, *Circulation*, 1998, **98**, 319–327.
106. B. S. Conklin, E. R. Richter, K. L. Kreutziger, D. S. Zhong and C. Chen, *Med. Eng. Phys.*, 2002, **24**, 173–183.
107. X. H. Wang, D. Yin and R. Zhang, *J. Bioact. Compat. Polym.*, 2007, **22**, 323–341.
108. G. K. Young, H. K. Young, D. P. Ki, H. J. Lee, W. K. Lee, H. D. Park, S. H. Kim, G. S. Lee and D. J. Ahn, *Biomaterials*, 2001, **22**, 2115–2123.
109. U. Hersel, C. Dahmen and H. Kessler, *Biomaterials*, 2003, **24**, 4385–4416.
110. B. Krijgsman, A. M. Seifalian, H. J. Salacinski, N. R. Tai, G. Punshon, B. J. Fuller and G. Hamilton, *Tissue Eng*, 2002, **8**, 673–680.
111. C. Gao, J. Guan, Y. Zhu and J. Shen, *Macromol. Biosci.*, 2003, **3**, 157–162.
112. H. J. Salacinski, A. Tiwari, G. Hamilton and A. M. Seifalian, *Med. Biol. Eng. Comput.*, 2001, **39**, 609–618.
113. X. Xie, A. Eberhart, R. Guidoin, Y. Marios, Y. Douville and Z. Zhang, *J. Biomat. Sci.*, 2010, **21**, 1239–1259.
114. A. Ratcliffe, *Matrix Biol.*, 2000, **19**, 353–357.
115. H. B. Lo, M. Herold, H. Reul, K. Taguchi, M. Surmann, K. H. Hildinger, H. Lambertz, H. de Haan and S. Handt, *ASAIO Trans.*, 1988, **34**, 839–855.
116. M. Kütting, J. Roggenkamp, U. Urban, T. Schmitz-Rode and U. Steinseifer, *Expert Rev. Med. Devic.*, 2011, **8**, 227–233.
117. S. L. Hilbert, V. J. Ferrans, Y. Tomita, E. E. Eidbo and M. Jones, *J. Thorac. Cardiovasc. Surg.*, 1987, **94**, 419–422.
118. R. J. Thoma, *J. Biomater. Appl.*, 1986, **1**, 449–486.
119. K. Imachi, T. Chinzei, Y. Abe, K. Mabuchi, H. Matsuura, T. Karita, K. Iwasaki, S. Mochizuki, Y. Son, I. Saito, A. Kouno and T. Ono, *J. Artif. Organs*, 2001, **4**, 74–82.
120. I. Alferiev, S. J. Stachelek, Z. Lu, A. L. Fu, T. L. Sellaro, J. M. Connolly, R. W. Bianco, M. S. Sacks and R. J. Levy, *J. Biomed. Mater. Res.*, 2003, **66A**, 385–395.

121. J. Snow, H. Harasaki, J. Kasick, R. Whalen, R. Kiraly and Y. Nosè, *Trans. Am. Soc. Artif. Intern. Organs*, 1981, **27**, 485–489.

122. M. G. McGee, M. Szycher, S. A. Turner, W. Clay, R. Trono, G. L. Davis and J. C. Norman, *Trans. Am. Soc. Artif. Intern. Organs*, 1980, **26**, 299–303.

123. E. Tatsumi, T. Masuzawa, M. Nakamura, Y. Taenaka, T. Nishimura, S. Endo, B. Zhang, Y. Kakuta, M. Nakata, T. Nakamura, T. Nishinaka, H. Takano, K. Tsukahara and K. Tsuchimoto, *Artif. Organs*, 1999, **23**, 242–248.

124. R. M. Engelman, E. Nyilas, H. Lackner and S. J. Godwin, *J. Thorac. Cardiovasc. Surg.*, 1971, **62**, 851–858.

125. E. F. Bernstein, L. C. Cosentino, S. Reich, P. Stasz, I. D. Levine, D. R. Scott, F. D. Dorman and P. L. Blackshear, *Trans. Am. Soc. Artif. Intern. Organs*, 1974, **20**, 643–652.

126. D. J. Farrar, P. Litwak, J. H. Lawson, R. S. Ward, K. A. White, A. J. Robinson, R. Rodvien and J. D. Hill, *J. Thorac. Cardiovasc. Surg.*, 1984, **95**, 191–200.

127. W. F. Bernhard, D. G. Gernes, W. C. Clay, F. J. Schoen, R. Burgeson, R. C. Valeri, A. J. Melaragno and V. L. Poirier, *J. Thorac. Cardiovasc. Surg.*, 1988, **88**, 11–21.

128. K. A. Dasse, S. D. Chipman, C. N. Sherman, A. H. Levine and O. H. Frazier, *Trans. Am. Soc. Artif. Intern. Organs*, 1987, **33**, 418–425.

129. J. L. Brash, B. K. Fritzinger and S. D. Bruck, *J. Biomed. Mater. Res.*, 1973, **7**, 313–334.

130. J. W. Costerton, P. S. Stewart and E. P. Greenberg, *Science*, 1999, **284**, 1318–1322.

131. G. R. Corey and T. Lalani, *Int. J. Antimicrob. Agents*, 2008, **32 Suppl 1**, S26–S29.

132. S. E. Cramton, C. Gerke, N. F. Schnell, W. W. Nichols and F. Gotz, *Infect. Immun.*, 1999, **67**, 5427–5433.

133. J. M. Schierholz and J. Beuth, *J. Hosp. Infect.*, 2001, **49**, 87–154.

134. R. M. Donlan, *Emerg. Infect. Dis.*, 2001, **7**, 277–281.

135. P. S. Stewart and J. W. Costerton, *Lancet*, 2001, **358**, 135–187.

136. D. Neut, J. R. van Horn, T. G. van Kooten, H. C. van der Mei and H. J. Busscher, *Clin. Orthop. Relat. Res.*, 2003, 261–268.

137. P. S. Stewart and M. J. Franklin, *Nature Rev. Microbiol.*, 2008, **6**, 199–210.

138. G. Borriello, E. Werner, F. Roe, A. M. Kim, G. D. Ehrlich and P. S. Stewart, *Antimicrob. Agents Chemother.*, 2004, **48**, 2659–2664.

139. K. Lewis, *Nature Rev. Microbiol.*, 2007, **5**, 48–56.

140. A. G. Gristina, G. Giridhar, B. L. Gabriel, P. T. Naylor and Q. N. Myrvik, *Int. J. Artif. Organs*, 1993, **16**, 755–763.

141. G. Donelli, *Surg. Infect. (Larchmt)*, 2006, **7 Suppl 2**, S25–S27.

142. D. J. Stickler, *Nature Clin. Pract. Urol.*, 2008, **5**, 598–608.

143. A. Trampuz and A. F. Widmer, *Curr. Opin. Infect. Dis.*, 2006, **19**, 349–356.

144. K. Subramani, R. E. Jung, A. Molenberg and C. H. Hammerle, *Int. J. Oral Maxillofac. Implants*, 2009, **24**, 616–619.
145. M. E. Shirtliff, B. M. Peters and M. A. Jabra-Rizk, *FEMS Microbiol. Lett.*, 2009, **299**, 1–8.
146. M. M. Harriott and M. C. Noverr, *Antimicrob. Agents Chemother.*, 2009, **53**, 3914–3922.
147. A. S. Lynch and G. T. Robertson, *Annu. Rev. Med.*, 2008, **59**, 415–428.
148. L. A. Mermel, B. M. Farr, R. J. Sherertz, I. I. Raad, N. O'Grady, J. S. Harris and D. E. Craven, *Clin. Infect. Dis.*, 2001, **32**, 1249–1272.
149. N. Nakazawa, *Semin. Oncol. Nurs.*, 2010, **26**, 121–131.
150. J. F. Timsit, J. C. Farkas, J. M. Boyer, J. B. Martin, B. Misset, B. Renaud and J. Carlet, *Chest*, 1998, **114**, 207–213.
151. C. J. van Rooden, E. F. Schippers, R. M. Barge, F. R. Rosendaal, H. F. Guiot, F. J. van der Meer, A. E. Meinders and M. V. Huisman, *J. Clin. Oncol.*, 2005, **23**, 2655–2660.
152. M. Levi, T. van der Poll and M. Schultz, *Neth. J. Med.*, 2012, **70**, 114–120.
153. K. Schumm and T. B. Lam, *Neurourol. Urodyn.*, 2008, **27**, 738–746.
154. R. M. Klevens, J. R. Edwards, C. L. Richards Jr., T. C. Horan, R. P. Gaynes, D. A. Pollock and D. M. Cardo, *Public Health Rep.*, 2007, **122**, 160–166.
155. K. W. Lodbell, S. Stamou and J. A. Sanchez, *Surg. Clin. N. Am.*, 2012, **92**, 65–77.
156. J. A. Lohr, L. G. Donowitz and J. E. Sadler III, *Pediatrics*, 1989, **83**, 193–199.
157. J. Kalsi, M. Arya, P. Wilson and A. Mundy, *Int. J. Clin. Pract.*, 2003, **57**, 388–391.
158. A. F. Widmer, *Clin. Infect. Dis.*, 2001, **33 Suppl 2**, S94–S106.
159. S. M. Kurtz, K. L. Ong, J. Schmier, F. Mowat, K. Saleh, E. Dybvik, J. Karrholm, G. Garellick, L. I. Havelin, O. Furnes, H. Malchau and E. Lau, *J. Bone Joint Surg. Am.*, 2007, **89 Suppl 3**, 144–151.
160. D. Campoccia, L. Montanaro and C. R. Arciola, *Biomaterials*, 2006, **27**, 2331–2339.
161. W. Zimmerli, P. D. Lew and F. A. Waldvogel, *J. Clin. Invest.*, 1984, **73**, 1191–1200.
162. P. Vaudaux, R. Suzuki, F. A. Waldvogel, J. J. Morgenthaler and U. E. Nydegger, *J. Infect. Dis.*, 1984, **150**, 546–553.
163. P. E. Vaudaux, F. A. Waldvogel, J. J. Morgenthaler and U. E. Nydegger, *Infect. Immun.*, 1984, **45**, 768–774.
164. P. Vaudaux, D. Pittet, A. Haeberli, P. G. Lerch, J. J. Morgenthaler, R. A. Proctor, F. A. Waldvogel and D. P. Lew, *J. Infect. Dis.*, 1993, **167**, 633–636.
165. M. H. McGavin, D. Krajewska-Pietrasik, C. Ryden and M. Hook, *Infect. Immun.*, 1993, **61**, 2479–2485.
166. K. Jonsson, D. McDevitt, M. H. McGavin, J. M. Patti and M. Hook, *J. Biol. Chem.*, 1995, **270**, 21457–21460.

167. I. Francolini and G. Donelli, *FEMS Immunol. Med. Microbiol.*, 2010, **59**, 227–238.
168. H. C. van der Mei, A. J. Leonard, A. H. Weerkamp, P. G. Rouxhet and H. J. Busscher, *J. Bacteriol.*, 1988, **170**, 2462–2466.
169. E. Ostuni, R. G. Chapman, E. Holmlin, S. Takayama and G. M. Whitesides, *Langmuir*, 2001, **17**, 5605–5620.
170. S. Chen, L. Li, C. Zhao and J. Zheng, *Polymer*, 2010, **51**, 5283–293.
171. M. Tanaka, A. Mochizuki, N. Ishii, T. Motomura and T. Hatakeyama, *Biomacromolecules*, 2002, **3**, 36–41.
172. S. L. West, J. P. Salvage, E. J. Lobb, S. P. Armes, N. C. Billingham, A. L. Lewis, G. W. Hanlon and A. W. Lloyd, *Biomaterials*, 2004, **25**, 1195–1204.
173. M. Schuler, D. W. Hamilton, T. P. Kunzler, C. M. Sprecher, M. de Wild, D. M. Brunette, M. Textor and S. G. Tosatti, *J. Biomed. Mater. Res. B Appl. Biomater.*, 2009, **91**, 517–527.
174. V. Zoulalian, S. Zurcher, S. Tosatti, M. Textor, S. Monge and J. J. Robin, *Langmuir*, 2010, **26**, 74–82.
175. S. Chen, F. Yu, Q. Yu, Y. He and S. Jiang, *Langmuir*, 2006, **22**, 8186–8191.
176. Y. C. Tseng and K. Park, *J. Biomed. Mater. Res.*, 1992, **26**, 373–391.
177. K. L. Prime and G. M. Whitesides, *Science*, 1991, **252**, 1164–1167.
178. J. Li, D. Tan, X. Zhang, H. Tan, M. Ding, C. Wan and Q. Fu, *Colloids Surf. B Biointerfaces*, 2010, **78**, 343–350.
179. S. Rana, S. Y. Lee and J. W. Cho, *Polym. Bull.*, 2010, **64**, 401–411.
180. J. H. Lee, Y. M. Ju, W. K. Lee, K. D. Park and Y. H. Kim, *J. Biomed. Mater. Res.*, 1998, **40**, 314–323.
181. K. D. Park, Y. S. Kim, D. K. Han, Y. H. Kim, E. H. Lee, H. Suh and K. S. Choi, *Biomaterials*, 1998, **19**, 851–859.
182. J. M. Orban, T. M. Chapman, W. R. Wagner and R. Jankowski, *J. Polym. Sci. A: Polym. Chem.*, 1999, **37**, 3441–3448.
183. J. Tuominen, J. J. Lee, M. Livingstone and J. A. Halliday, *J. Control. Release*, 2005, **101**, 316–317.
184. X. Chen, W. Liu, Y. Zhao, L. Jiang, H. Xu and X. Yang, *Drug Dev. Ind. Pharm.*, 2009, **35**, 704–709.
185. M. Shen, L. Martinson, M. Wagner, D. Castner, B. Ratner and T. Horbet, *J. Biomat. Sci.: Polym. Ed.*, 2002, **13**, 367–390.
186. S. Corneillie, P. N. Lan, E. Schacht, M. Davies, A. Shard, R. Green, S. Denyer, M. Wassall, H. Whitfield and S. Choong, *Polym. Intern.*, 1998, **46**, 251–259.
187. L. Rao, H. Zhou, T. Li, C. Li and Y. Y. Duan, *Acta Biomater.*, 2012, **8**, 2233–2242.
188. J. H. Park, K. B. Lee, I. C. Kwon and Y. H. Bae, *J. Biomater. Sci. Polym. Ed.*, 2001, **12**, 629–645.
189. T. Stern, A. Penhasi and D. Cohn, *Biomaterials*, 1995, **16**, 17–23.
190. K. D. Park, K. Suzuki, W. K. Lee, J. E. Lee, Y. H. Kim, Y. Sakurai and T. Okano, *ASAIO J.*, 1996, **42**, M876–M881.

191. D. Li, H. Chen, M. W. Glenn and J. L. Brash, *Acta Biomater.*, 2009, **5**, 1864–1871.
192. I. Capila and R. J. Linhardt, *Angew. Chem. Int. Ed. Engl.*, 2002, **41**, 391–412.
193. C. Mao, Y. Qiu, H. Sang, H. Mei, A. Zhu, J. Shen and S. Lin, *Adv. Colloid Interface Sci.*, 2004, **110**, 5–17.
194. P. Y. Tseng, S. M. Rele, X. L. Sun and E. L. Chaikof, *Biomaterials*, 2006, **27**, 2627–2636.
195. A. D. Baldwin and K. L. Kiick, *Biopolymers*, 2010, **94**, 128–140.
196. W. Marconi, A. Galloppa, A. Martinelli and A. Piozzi, *Biomaterials*, 1995, **16**, 449–456.
197. P. Appelgren, U. Ransjo, L. Bindslev, F. Espersen and O. Larm, *Crit. Care Med.*, 1996, **24**, 1482–1489.
198. M. Paulsson, M. Kober, C. Freij-Larsson, M. Stollenwerk, B. Wesslen and A. Ljungh, *Biomaterials*, 1993, **14**, 845–853.
199. P. Tenke, C. R. Riedl, G. L. Jones, G. J. Williams, D. Stickler and E. Nagy, *Int. J. Antimicrob. Agents*, 2004, **23 Suppl 1**, S67–S74.
200. A. Abdelkefi, W. Achour, O. T. Ben, S. Ladeb, L. Torjman, A. Lakhal, H. A. Ben, M. Hsairi and A. A. Ben, *J. Support Oncol.*, 2007, **5**, 273–278.
201. G. Jain, M. Allon, S. Saddekni, J. Barker-Finkel and I. D. Maya, *Clin. J. Am. Soc. Nephrol.*, 2009, **4**, 1787–1790.
202. T. G. Grasel and S. L. Cooper, *J. Biomed. Mater. Res.*, 1989, **23**, 311–338.
203. J. H. Silver, A. P. Hart, E. C. Williams, S. L. Cooper, S. Charef, D. Labarre and M. Jozefowicz, *Biomaterials*, 1992, **13**, 339–343.
204. X. Wei and X. Yu, *J. Polym. Sci. Part B: Polym. Phys.*, 1997, **35**, 225–232.
205. R. G. Flemming, C. C. Capelli, S. L. Cooper and R. A. Proctor, *Biomaterials*, 2000, **21**, 273–281.
206. A. Magnani, R. Barbucci, L. Montanaro, C. Arciola and S. Lamponi, *J. Biomat. Sci.: Polym. Ed.*, 2000, **11**, 801–803.
207. M. J. Bridgett, M. C. Davies, S. P. Denyer and P. R. Eldridge, *Biomaterials*, 1993, **14**, 184–188.
208. S. K. Choong, P. Hallson, H. N. Whitfield and C. H. Fry, *BJU Int.*, 1999, **83**, 770–775.
209. C. Perrino, S. Lee, S. W. Choi, A. Maruyama and N. D. Spencer, *Langmuir*, 2008, **24**, 8850–8856.
210. F. Xu, C. E. Flanagan, A. Ruiz, W. C. Crone and K. S. Masters, *Macromol. Biosci.*, 2011, **11**, 257–266.
211. S. Sagnella and K. Mai-Ngam, *Colloids Surf. B: Biointerfaces*, 2005, **42**, 147–155.
212. Y. Wang, Q. Hong, Y. Chen, X. Lian and Y. Xiong, *Colloids Surf. B: Biointerfaces*, 2012, **100**, 77–83.
213. L. Zheng and J. Zhu, *Carbohyd. Polym.*, 2003, **54**, 527–530.
214. S. H. Yang, Y. S. Lee, F. H. Lin, J. M. Yang and K. S. Chen, *J. Biomed. Mater. Res. B: Appl. Biomater.*, 2007, **83**, 304–313.
215. M. M. Tunney and S. P. Gorman, *Biomaterials*, 2002, **23**, 4601–4608.

216. F. Crisante, I. Francolini, M. Bellusci, A. Martinelli, L. D'Ilario and A. Piozzi, *Eur. J. Pharm. Sci.*, 2009, **36**, 555–564.
217. D. S. Jones, J. G. McGovern, A. D. Woolfson, C. G. Adair and S. P. Gorman, *Pharm. Res.*, 2002, **19**, 818–824.
218. J. M. Schierholz, A. Rump and G. Pulverer, *Arzneimittelforschung*, 1997, **47**, 70–74.
219. P. Basak, B. Adhikari, I. Banerjee and T. K. Maiti, *J. Mater. Sci.: Mater. Med.*, 2009, **20 Suppl 1**, S213–S221.
220. J. E. Kim, S. R. Kim, S. H. Lee, C. H. Lee and D. D. Kim, *Int. J. Pharm.*, 2000, **201**, 29–36.
221. N. Fong, A. Simmons and L. A. Poole-Warren, *Acta Biomater.*, 2010, **6**, 2554–2561.
222. D. P. Dowling, K. Donnelly, M. L. McConnell, R. Eloy and M. N. Arnaud, *Thin Solid Films*, 2001, **398**, 602–606.
223. J. S. Groeger, A. B. Lucas, D. Coit, M. LaQuaglia, A. E. Brown, A. Turnbull and P. Exelby, *Ann. Surg.*, 1993, **218**, 206–210.
224. J. E. Gray, P. R. Norton, R. Alnouno, C. L. Marolda, M. A. Valvano and K. Griffiths, *Biomaterials*, 2003, **24**, 2759–2765.
225. F. Furno, K. S. Morley, B. Wong, B. L. Sharp, P. L. Arnold, S. M. Howdle, R. Bayston, P. D. Brown, P. D. Winship and H. J. Reid, *J. Antimicrob. Chemother.*, 2004, **54**, 1019–1024.
226. U. Samuel and J. P. Guggenbichler, *Int. J. Antimicrob. Agents*, 2004, **23 Suppl 1**, S75–S78.
227. I. Francolini, L. D'Ilario, E. Guaglianone, G. Donelli, A. Martinelli and A. Piozzi, *Acta Biomater.*, 2010, **6**, 3482–3490.
228. F. Paladini, M. Pollini, A. Talà, P. Alifano and A. Sannino, *J. Mater. Sci.: Mater. Med.*, 2012, **23**, 1983–1990.
229. R. Bambauer, P. Mestres, R. Schiel, J. M. Schneidewind, R. Goudjinou, R. Latza, R. Inniger, S. Bambauer and P. Sioshansi, *ASAIO J*, 1998, **44**, 303–308.
230. S. O. Trerotola, M. S. Johnson, H. Shah, M. A. Kraus, M. A. McKusky, W. T. Ambrosius, V. J. Harris and J. J. Snidow, *Radiology*, 1998, **207**, 491–496.
231. S. H. Hsu, H. J. Tseng and Y. C. Lin, *Biomaterials*, 2010, **31**, 6796–6808.
232. M. Beattie, *Nurs. Times*, 2011, **107**, 19–21.
233. E. J. Tobin and R. Bambauer, *Ther. Apher. Dial*, 2003, **7**, 504–509.
234. M. Beattie and J. Taylor, *J. Clin. Nurs.*, 2011, **20**, 2098–2108.
235. M. Boswald, S. Lugauer, A. Regenfus, G. G. Braun, P. Martus, C. Geis, J. Scharf, T. Bechert, J. Greil and J. P. Guggenbichler, *Infection*, 1999, **27 Suppl 1**, S56–S60.
236. R. T. Carbon, S. Lugauer, U. Geitner, A. Regenfus, M. Boswald, J. Greil, T. Bechert, S. I. Simon, H. P. Hummer and J. P. Guggenbichler, *Infection*, 1999, **27 Suppl 1**, S69–S73.
237. A. Bach, H. Eberhardt, A. Frick, H. Schmidt, B. W. Bottiger and E. Martin, *Crit. Care Med.*, 1999, **27**, 515–521.

238. R. Bambauer, R. Schiel, C. Bambauer and R. Latza, *Int. J. Nephrol.*, 2012, **2012**, 956136.
239. M. Antonelli, P. G. De, V. M. Ranieri, P. Pelaia, R. Tufano, O. Piazza, A. Zangrillo, A. Ferrario, G. A. De, E. Guaglianone and G. Donelli, *J. Hosp. Infect.*, 2012, **82**, 101–107.
240. C. Modak and C. Fox, *Biochem. Pharm.*, 1973, **22**, 2391–2404.
241. J. M. Schierholz, L. J. Lucas, A. Rump and G. Pulverer, *J. Hosp. Infect.*, 1998, **40**, 257–262.
242. K. N. Stevens, O. Crespo-Biel, E. E. van den Bosch, A. A. Dias, M. L. Knetsch, Y. B. Aldenhoff, F. H. van der Veen, J. G. Maessen, E. E. Stobberingh and L. H. Koole, *Biomaterials*, 2009, **30**, 3682–3690.
243. A. Bach, *Zentralbl Bakteriol.*, 1995, **283**, 208–214.
244. J. I. Greenfeld, L. Sampath, S. J. Popilskis, S. R. Brunnert, S. Stylianos and S. Modak, *Crit. Care Med.*, 1995, **23**, 894–900.
245. D. G. Maki, S. M. Stolz, S. Wheeler and L. A. Mermel, *Ann. Intern. Med.*, 1997, **127**, 257–266.
246. G. Donelli and I. Francolini, *J. Chemother.*, 2001, **13**, 595–606.
247. B. Walder, D. Pittet and M. R. Tramer, *Infect. Control Hosp. Epidemiol.*, 2002, **23**, 748–756.
248. C. Brun-Buisson, F. Doyon, J. P. Sollet, J. F. Cochard, Y. Cohen and G. Nitenberg, *Intensive Care Med.*, 2004, **30**, 837–843.
249. T. Ostendorf, A. Meinhold, C. Harter, H. Salwender, G. Egerer, H. K. Geiss, A. D. Ho and H. Goldschmidt, *Support Care Cancer*, 2005, **13**, 993–1000.
250. M. E. Rupp, S. J. Lisco, P. A. Lipsett, T. M. Perl, K. Keating, J. M. Civetta, L. A. Mermel, D. Lee, E. P. Dellinger, M. Donahoe, D. Giles, M. A. Pfaller, D. G. Maki and R. Sherertz, *Ann. Intern. Med.*, 2005, **143**, 570–580.
251. R. J. Sherertz, W. A. Carruth, A. A. Hampton, M. P. Byron and D. D. Solomon, *J. Infect. Dis.*, 1993, **167**, 98–106.
252. R. O. Darouiche, I. I. Raad, G. P. Bodey and D. M. Musher, *Int. J. Antimicrob. Agents*, 1995, **6**, 31–36.
253. I. Raad, R. Darouiche, R. Hachem, M. Sacilowski and G. P. Bodey, *Antimicrob. Agents Chemother.*, 1995, **39**, 2397–2400.
254. I. Raad, R. Darouiche, R. Hachem, M. Mansouri and G. P. Bodey, *J. Infect. Dis.*, 1996, **173**, 418–424.
255. R. O. Darouiche, I. I. Raad, S. O. Heard, J. I. Thornby, O. C. Wenker, A. Gabrielli, J. Berg, N. Khardori, H. Hanna, R. Hachem, R. L. Harris and G. Mayhall, *N. Engl. J. Med.*, 1999, **340**, 1–8.
256. H. Hanna, R. Benjamin, I. Chatzinikolaou, B. Alakech, D. Richardson, P. Mansfield, T. Dvorak, M. F. Munsell, R. Darouiche, H. Kantarjian and I. Raad, *J. Clin. Oncol.*, 2004, **22**, 3163–3171.
257. R. O. Darouiche, D. H. Berger, N. Khardori, C. S. Robertson, M. J. Wall Jr., M. H. Metzler, S. Shah, M. D. Mansouri, C. Cerra-Stewart, J. Versalovic, M. J. Reardon and I. I. Raad, *Ann. Surg.*, 2005, **242**, 193–200.

258. R. E. Gilbert and M. Harden, *Curr. Opin. Infect. Dis.*, 2008, **21**, 235–245.

259. I. I. Raad, R. O. Darouiche, R. Hachem, D. Abi-Said, H. Safar, T. Darnule, M. Mansouri and D. Morck, *Crit. Care Med.*, 1998, **26**, 219–224.

260. G. Donelli, I. Francolini, A. Piozzi, R. R. Di and W. Marconi, *J. Chemother.*, 2002, **14**, 501–507.

261. A. Piozzi, I. Francolini, L. Occhiaperti, R. R. Di, V. Ruggeri and G. Donelli, *J. Chemother.*, 2004, **16**, 446–452.

262. L. A. Sampath, S. M. Tambe and S. M. Modak, *Infect. Control Hosp. Epidemiol.*, 2001, **22**, 640–646.

263. C. Leon, S. Ruiz-Santana, J. Rello, M. V. de la Torre, J. Valles, F. Alvarez-Lerma, R. Sierra, P. Saavedra and F. Alvarez-Salgado, *Intensive Care Med.*, 2004, **30**, 1891–1898.

264. J. M. Schierholz, C. Fleck, J. Beuth and G. Pulverer, *J. Antimicrob. Chemother.*, 2000, **46**, 45–50.

265. N. Yucel, R. Lefering, M. Maegele, M. Max, R. Rossaint, A. Koch, R. Schwarz, M. Korenkov, J. Beuth, A. Bach, J. Schierholz, G. Pulverer and E. A. Neugebauer, *J. Antimicrob. Chemother.*, 2004, **54**, 1109–1115.

266. J. M. Schierholz, K. Nagelschmidt, M. Nagelschmidt, R. Lefering, N. Yucel and J. Beuth, *Anticancer Res.*, 2010, **30**, 1353–1358.

267. E. Kenawy, S. Worley and R. Broughton, *Biomacromolecules*, 2007, **8**, 1359–1384.

268. E. Palermo, D. Lee, A. Ramamoorthy and K. Kuroda, *J. Phys. Chem. B*, 2011, **115**, 366–375.

269. J. A. Grapski and S. L. Cooper, *Biomaterials*, 2001, **22**, 2239–2246.

270. P. Kurt, L. Wood, D. E. Ohman and K. J. Wynne, *Langmuir*, 2007, **23**, 4719–4723.

271. P. Kurt, L. J. Gamble and K. J. Wynne, *Langmuir*, 2008, **24**, 5816–5824.

272. J. L. Y. Woo, M. W. Mittelman and J. P. Santerre, *Biomaterials*, 2000, **21**, 1235–1246.

273. U. Makal, L. Wood, D. E. Ohman and K. J. Wynne, *Biomaterials*, 2006, **27**, 1316–1326.

274. R. Langer and J. P. Vacanti, *Science*, 1993, **260**, 920–926.

275. C. W. Patrick, A. G. Mikos, L. V. McIntire, *Frontiers in Tissue Engineering*, ed. C. W. Patrick, A. G. Mikos, and L. V. McIntire, Elsevier Science Ltd, Oxford, 1st edn, 1998, Chapter 1, 3–11.

276. M. S. Chapekar, *J. Biomed. Mater. Res.*, 2000, **53**, 617–620.

277. J. D. Sipe, *Ann. N. Y. Acad. Sci.*, 2002, **961**, 1–9.

278. S. Ramakarishna, J. Mayer, E. Wintermantel and K. W. Leong, *Comp. Sci. Tech.*, 2001, **61**, 1189–1224.

279. H. Niiranen and P. Tormala, *J. Mater. Sci.: Mater. Med.*, 1999, **10**, 707–718.

280. P. Buma, N. N. Ramrattan, T. G. van Tienen and R. P. H. Veth, *Biomaterials*, 2004, **25**, 1523–1532.

281. C. M. Agrawal and R. B. Ray, *J. Biomed. Mater. Res.*, 2001, **55**, 141–150.

282. L. A. Solchaga, J. E. Dennis, V. M. Goldberg and A. I. Kaplan, *J. Orthop. Res.*, 1999, **17**, 205–213.
283. P. A. Gunatillake and R. Adhikari, *Eur. Cells Mater.*, 2003, **5**, 1–16.
284. K. Tuzlakoglu, C. M. Alves, J. F. Mano and R. L. Reis, *Macromol. Biosci.*, 2004, **4**, 811–819.
285. L. Wu and J. Ding, *Biomaterials*, 2004, **25**, 5821–5830.
286. G. Q. Chen and Q. Wu, *Biomaterials*, 2005, **26**, 6565–6578.
287. D. A. Barrera, E. Zylstra, P. T. Lansbury and R. Langer, *Macromolecules*, 1995, **28**, 425–432.
288. S. Li, I. Molina, M. B. Martinez and M. Vert, *J. Mater. Sci.: Mater. Med.*, 2002, **13**, 81–89.
289. S. H. Lee, B. S. Kim, S. H. Kim, S. W. Choi, S. I. Jeong, I. K. Kwon, S. W. Kang, J. Nikolovski, D. J. Mooney, Y. K. Han and Y. H. Kim, *J. Biomed. Mater. Res.*, 2003, **66**, 29–37.
290. D. Cohn, T. Stern, M. F. Gonzales and J. Epstein, *J. Biomed. Mater. Res.*, 2002, **59**, 273–281.
291. G. Quattrociocchi, I. Francolini, A. Martinelli, L. D'Ilario and A. Piozzi, *Polym. Int.*, 2010, **59**, 1052–1057.
292. M. H. Ho, L. T. Hou, C. Y. Tu, H. J. Hsieh, J. Y. Lai, W. J. Chen and D. M. Wang, *Macromol. Biosci.*, 2006, **6**, 90–98.
293. T. C. McDevitt, K. A. Woodhouse, S. D. Hauschka, C. E. Murry and P. S. Stayton, *J. Biomed. Mater. Res. A*, 2003, **66**, 586–595.
294. J. Guan, K. L. Fujimoto, M. S. Sacks and V. R. Wagner, *Biomaterials*, 2005, **26**, 3961–3971.
295. S. Sharifpoor, R. S. Labow and J. P. Santerre, *Biomacromolecules*, 2009, **10**, 2729–2739.
296. S. A. Riboldi, M. Sampaolesi, P. Neuenschwanderc, G. Cossu and S. Mantero, *Biomaterials*, 2005, **26**, 4606–4615.
297. S. Grad, L. Kupcsik, K. Gorna, S. Gogolewski and M. Alini, *Biomaterials*, 2003, **24**, 5163–5171.
298. K. Gorna and S. Gogolewski, *J. Biomed. Mater. Res.*, 2003, **67**, 813–818.
299. J. Zhang, B. A. Doll, E. J. Beckman and J. O. Hollinger, *J. Biomed. Mater. Res.*, 2003, **67**, 389–400.
300. I. C. Bonzani, R. Adhikari, S. Houshyar, R. Mayadunne, P. Gunatillake and M. M. Stevens, *Biomaterials*, 2007, **28**, 423–433.
301. X. Yue, B. van der Lei, J. M. Schakenraad, G. H. van Oene, J. H. Kuit, J. Feijen and C. R. Wildevuur, *Surgery*, 1988, **103**, 206–212.
302. M. R. Williamson, R. Black and C. Kielty, *Biomaterials*, 2006, **27**, 3608–3616.
303. J. Guan, M. S. Sacks, E. J. Beckman and W. R. Wagner, *J. Biomed. Mater. Res.*, 2002, **61**, 493–503.
304. J. J. Stankus, J. Guan, K. Fujimoto and W. R. Wagner, *Biomaterials*, 2006, **27**, 735–744.
305. G. A. Skarja and K. A. Woodhouse, *J. Biomat. Sci.: Polym. Ed.*, 1998, **9**, 271–295.

306. C. Alperin, P. W. Zandstra and K. A. Woodhouse, *Biomaterials*, 2005, **26**, 7377–7386.
307. B. Saad, P. Neuenschwander, G. K. Uhlschmid and U. W. Suter, *Int. J. Biol. Macromol.*, 1999, **25**, 293–301.
308. S. Grad, L. Kupcsik, K. Gorna, S. Gogolewski and M. Alini, *Biomaterials*, 2003, **24**, 5163–5171.
309. J. Klompmaker, H. W. B Jansen, R. P. H. Veth, H. K. L. Nielsen, J. H. de Groot and A. J. Pennings, *Clin. Mater.*, 1993, **14**, 1–11.
310. J. H. de Groot, A. J. Nijenhuis, P. Bruin, A. J. Pennings, R. P. H. Veth, J. Klompmaker and H. W. B. Jansen, *Colloid Polym. Sci.*, 1990, **268**, 1073–1081.
311. J. H. de Groot, R. de Vrijer, A. J. Nijenhuis, J. Klompmaker, R. P. H. Veth and H. W. B. Jansen, *Biomaterials*, 1996, **17**, 163–173.
312. T. G. van Tienen, R. G. J. C. Heijkants, P. Buma, J. H. de Groot, A. J. Pennings and R. P. H. Veth, *Biomaterials*, 2002, **23**, 1731–1738.
313. S. Gogolewski and K. Gorna, *J. Biomed. Mater. Res.*, 2006, **80**, 94–101.
314. K. D. Kavlock, T. W. Pechar, J. O. Hollinger, S. A. Guelcher and A. S. Goldstein, *Acta Biomater.*, 2007, **3**, 475–484.
315. S. Gogolewski and K. Gorna, *J. Biomed. Mater. Res.*, 2006, **79**, 128–138.
316. H. Liu, L. Zhang, P. Shi, Q. Zou, Y. Zuo and Y. Li, *J. Biomed. Mater. Res. B.: Appl. Biomater.*, 2010, **95**, 36–46.
317. M. Bil, J. Ryszkowska, J. A. Roether, O. Bretcanu and A. R. Boccaccini, *Biomed. Mater.*, 2007, **2**, 93–101.
318. B. Li, T. Yoshii, A. E. Hafeman, J. S. Nyman, J. C. Wenke and S. A. Guelcher, *Biomaterials*, 2009, **30**, 6768–6779.
319. S. A. Guelcher, A. Srinivasan, J. E. Dumas, J. E. Didier, S. McBride and J. O. Hollinger, *Biomaterials*, 2008, **29**, 1762–1775.
320. A. E. Hafeman, K. J. Zienkiewicz, E. Carney, B. Litzner, C. Stratton, J. C. Wenke and S. A. Guelcher, *J. Biomater. Sci.: Polym. Ed.*, 2010, **21**, 95–112.
321. B. Li, K. V. Brown, J. C. Wenke and S. A. Guelcher, *J. Control. Release*, 2010, **145**, 221–230.
322. S. A. Guelcher, K. V. Brown, B. Li, T. Guda, B. H. Lee and J. C. Wenke, *J. Orthop. Trauma*, 2011, **25**, 477–482.

CHAPTER 10

Antimicrobial Polymeric Dental Materials

XIAOMING XU* AND STEPHEN COSTIN

Department of Comprehensive Dentistry and Biomaterials, Louisiana State University Health Sciences Center-School of Dentistry, 1100 Florida Ave, Box 137, New Orleans, LA 70119, USA
*Email: xxu@lsuhsc.edu

10.1 Introduction

Dental caries (tooth decay) remains a major worldwide health problem. The standard treatment for dental caries is to remove the carious tooth tissues and restore them with inert restorative materials such as amalgams or dental composites. Every year, billions of dollars in health care costs are accrued due to dental restorations alone.[1] In addition, greater than 50% of dental restorations are caused by the failure and replacement of previous restorations.[2] The leading cause of restoration failure is secondary (recurrent) caries.[3] Similar to primary caries, secondary caries are generated by the biofilms of acid-producing bacteria such as *Streptococcus mutans* and *Lactobacillus casei*.[4]

Traditionally, dental fillings have been done using amalgams,[5] because they have the advantages of high strength and a relatively low tendency for bacterial adhesion. However, due to the poor aesthetics and no fluoride release combined with the growing concern over the possibility of elution of highly toxic heavy metals, there has been a shift in modern dental practices away from amalgams and toward alternative restorative materials that possess better aesthetic properties while maintaining low toxicity. However, these materials have lower mechanical properties and shorter service life than dental amalgams.

RSC Polymer Chemistry Series No. 10
Polymeric Materials with Antimicrobial Activity: From Synthesis to Applications
Edited by Alexandra Muñoz-Bonilla, María L. Cerrada and Marta Fernández-García
© The Royal Society of Chemistry 2014
Published by the Royal Society of Chemistry, www.rsc.org

In order to reduce secondary caries and prolong the service life, different types of dental restorative materials have been developed, including those with fluoride-releasing and antibacterial properties. Fluoride is a well-documented anticaries element. Water fluoridation and fluoride-containing tooth paste have contributed to the reduction in dental caries over the past 40 years. Fluoride helps to prevent or reduce caries by acting as a catalyst for remineralization as well as by converting the hydroxyapatite in enamel to the much less soluble (more acid resistant) fluorapatite.[6] It has also been shown to inhibit bacteria at high concentrations.[7] The anticaries efficacy of fluoride has spurred the creation of several types of fluoride-releasing dental restorative materials, including glass ionomers (GI), resin-modified glass ionomers (RMGI), compomers and fluoride-releasing composites.[8–10]

Glass ionomers are formed by the acid–base reaction between a liquid, which contains polyalkenoic acids (*e.g.*, polymers of acrylic acid, maleic acid, *etc.*), water, and tartaric acid, and glass powders (SiO_2–Al_2O_3–CaO–CaF_2, Na_3AlF_6).[11,12] They are characterized by high levels of fluoride release and recharge but they have low mechanical properties and poor aesthetics.

In an effort to improve the mechanical properties and aesthetics of glass ionomers while maintaining a high level of fluoride release, resin-modified glass ionomers (RMGI) were developed.[13] In RMGIs, the acid polymer is modified by adding polymerizable methacrylate groups. The material is then able to set not only by the ionic interaction present in GIs but also *via* photo-polymerization. The end result is an increase in mechanical properties and aesthetics with similar fluoride release and recharge capabilities compared to glass ionomers. Unfortunately, further increasing the mechanical properties of dental materials requires a sacrifice in their fluoride-releasing ability.

Compomers (polyacid-modified composites) consist of dimethacrylate dental monomers such as bisphenol A glycerolate dimethacrylate (Bis-GMA), triethyleneglycol dimethacrylate (TEGDMA), and acid-containing dimethacrylates along with a filler consisting of unsilanized ionomer glass particles.[10] Compomers are generally light cured and thus require photoinitiators. Since compomers do not contain water, the limited acid–base reaction takes place when the light-cured material absorbs water from saliva. The physical properties and aesthetics of compomers are better than those of GIs and RMGIs, but their fluoride release and recharge capabilities are much lower. They also have high water sorption and may expand and crack the tooth structure. For restorations in high load bearing areas or for use in low caries risk patients, composites are more commonly employed.[14]

Fluoride-releasing composites are comprised of various dimethacrylate monomers, including Bis-GMA, TEGDMA, urethane dimethacrylate (UDMA), ethoxylated bisphenol A dimethacrylate (EBPADMA), and hexanediol dimethacrylate (HDDMA), along with fluoride-releasing glass filler ((Ba,Sr)$FAlSiO_x$, YbF_3) that has been treated with silane coupling agents, typically 3-methacryloylpropyltrimethoxysilane (MPTMS). Like compomers, composites are light cured and thus contain photoinitiators. Of all the modern dental restoration materials discussed here, composites have the highest strength, along with excellent aesthetics and good wear resistance. However, the fluoride-release capability of composites is generally minimal and thus the anticaries efficacy of composites is questionable. Since the level of fluoride released from the fluoride-releasing composite (and other resin-based restorative materials) is much lower than the bacteria-inhibition level (45-100 ppm),[6] the cariogenic bacterial biofilms continue to accumulate on the restorations and surrounding tooth tissues. Prevention and reduction of secondary caries in composite restorations require another mode of action.

10.2 Polymeric Dental Materials that Release Antibacterial Agents

10.2.1 Dental Materials Containing Chlorhexidine

Chlorhexidine (**1**) is a powerful antibacterial agent that is often considered to be the "gold standard" of antimicrobials (Figure 10.1).[15] It has been widely used in dental clinics and has been included in a number of dental restorative materials, including glass ionomer cements (GIC),[16,17] resin-modified glass ionomer cements (RMGIC),[18] and composites.[15] The mode of action of chlorhexidine is believed to be disruption of the bacterial cell membrane by interaction with the bisdiguanidinium moiety on the small molecule, ultimately causing cell lysis.[19] The effectiveness of chlorhexidine-releasing polymers is determined not only by the amount of drug loading but by the monomer composition and the mechanical aspects of particle size, dispersion and method of loading.[20–22]

The chlorhexidine released from polymeric dental materials varies according to the hydrophobicity/hydrophilicity of the polymer matrix.[15,23,24] More hydrophilic formulations tend to release more chlorhexidine. For example, a dental composite system based on a combination of the monomers TEGDMA and UDMA (1:1), combined with a 10 wt% solution of chlorhexidine diacetate in hydroxyethylmethacrylate (HEMA), shows significantly different relative release levels based on the ratio between the relatively hydrophobic monomers (TEGDMA/UDMA) and the chlorhexidine/HEMA solution.[15] Specifically, the composites containing only the chlorhexidine/HEMA solution release 50% of the total chlorhexidine loaded after one week, while composites containing 50% chlorhexidine/HEMA solution and 50% TEGDMA/UDMA release only

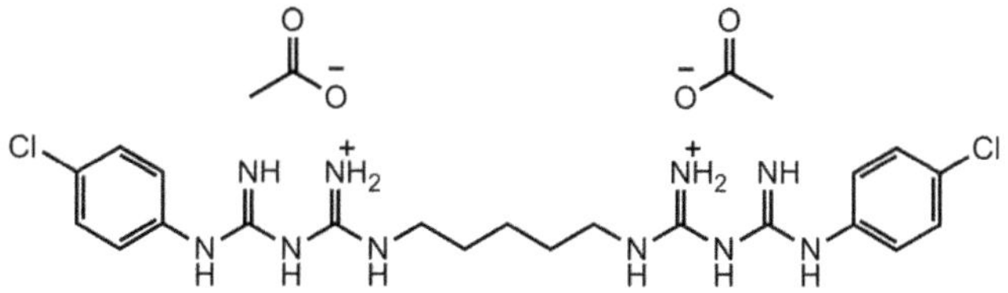

1

Figure 10.1 Structure of chlorhexidine diacetate (**1**).

Table 10.1 Effect of molecular weight (MW) on chlorhexidine release.

	Copolymer Composition (%)		MW	Chlorhexidine release
	PEMA	PHMA	(Average)	rate ($\mu g/cm^2/day$)
Copolymer I	26.1	73.9	437	1.15(0.05)
Copolymer II	25.3	74.6	276	1.81(0.10)
Copolymer III	22.4	77.6	93.3	0.44(0.06)
Copolymer IV	21.6	78.4	56.1	0.50(0.03)

5.4% of total loaded chlorhexidine in the same time period. However, the composites containing 50% chlorhexidine/HEMA solution were able to significantly inhibit bacterial growth at 1 day, although by 4 days the effects are severely diminished. Using 90% chlorhexidine/HEMA solution allows for significant bacterial reduction (10-fold) after 4 days. All composites tested show no significant difference from the control after 1 week.

In addition to the hydrophilicity of monomers, their molecular weight can affect the release of chlorhexidine from the polymer matrix.[25] Copolymers of ethyl methacrylate (EMA) and hexyl methacrylate (HMA) containing 2.5 wt% chlorhexidine showed release rates that were dependent on the molecular weight of the copolymer matrix with the general trend that the higher molecular weight copolymers release chlorhexidine at a faster rate (Table 10.1). The result is contrary to previous studies showing that release rates decrease with increasing molecular weight, presumably due to increased chain entanglements.[26–28] However, it is speculated that in this case the observed trend is due to the greater miscibility of chlorhexidine in the low molecular weight polymers.[25] As a result, the greater immiscibility observed for high molecular weight polymers leaves the drug molecules more loosely bound within the polymer matrix and thus more capable of dissolving out and releasing.

The composites discussed above incorporate chlorhexidine into the polymer matrix *via* direct mixing with the monomers. In all cases, the drug is used as-received from commercial sources. As a result, the effects of particle size on the release characteristics of the composites are unknown. By grinding chlorhexidine particles that have been frozen with liquid nitrogen, an average particle size of 0.62 μm can be obtained from the 40 μm as received particles.[22] Smaller particle sizes can potentially affect dispersion, release rates, polishability and mechanical properties of the cured composites.

Finely ground chlorhexidine acetate (10 wt%) was incorporated into composites along with nanoparticles of either amorphous calcium phosphate or calcium fluoride. In both cases, the total chlorhexidine released was around 2%, similar to previously reported results. As in the preceding studies, the composites containing finely powdered chlorhexidine show good antibacterial activity after 3 days immersion in water, the longest time studied. Additionally, the composites containing chlorhexidine maintained a pH > 6.5 over this time span, compared to pH < 5 for controls. It has been shown that at a pH < 5.5 demineralization will dominate, leading to overall dissolution of the tooth structure.[22]

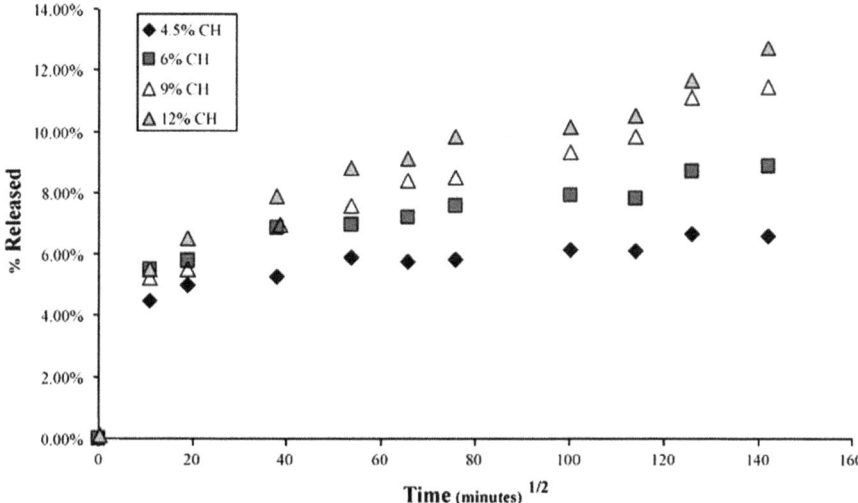

Figure 10.2 Rate of release of chlorhexidine from PEMA/THFMA with different drug loadings.
Reprinted with permission from Elsevier.[29]

With all of the factors affecting drug release and antibacterial activity, it is important not to overlook the most obvious aspect: drug loading. Various reports discuss the effect of chlorhexidine-containing materials with drug loads from 0.2 to 52 wt%.[15–21,23] As expected, drug release rates increase as drug loading increases. However, the nature of the polymer matrix appears to play a vital role in determining how the rate of release is affected by changes in drug loading. A self-curing system based on poly(ethyl methacrylate) (PEMA) and tetrahydrofurfuryl methacrylate (THFMA) reveals a significant increase in release rate over the concentration range 4.5–12 wt% (Figure 10.2).[29] This is in contrast with an ethyl methacrylate/hexyl methacrylate copolymer system, which reveals a significant increase in rate only when changing the dose from 2.5–5.0 wt%, leveling off when the dose is further increased to 7.5 wt%.[25] It appears that a minimum drug loading of 1 wt% is necessary for a significant reduction of *S. mutans* biofilms, but long-term results are dependent on several factors and must be determined on a case-by-case basis. In general, higher loadings are necessary to achieve activity for 1 week or longer.[15,22,23]

A serious drawback of directly mixing chlorhexidine with monomers in a dental composite is that dissolution of the embedded chlorhexidine leaves behind a microporous structure. As a result, the material has significantly lower mechanical properties after drug release. Although many of the above-mentioned reports do not discuss the mechanical properties of the composites, the composites containing 10 wt% chlorhexidine along with 30 wt% amorphous calcium phosphate or CaF_2 nanoparticles exhibited a 20% decrease in flexure strength over a 28-day period compared to controls without chlorhexidine (Figure 10.3).[22] Another recent study showed that a dental composite

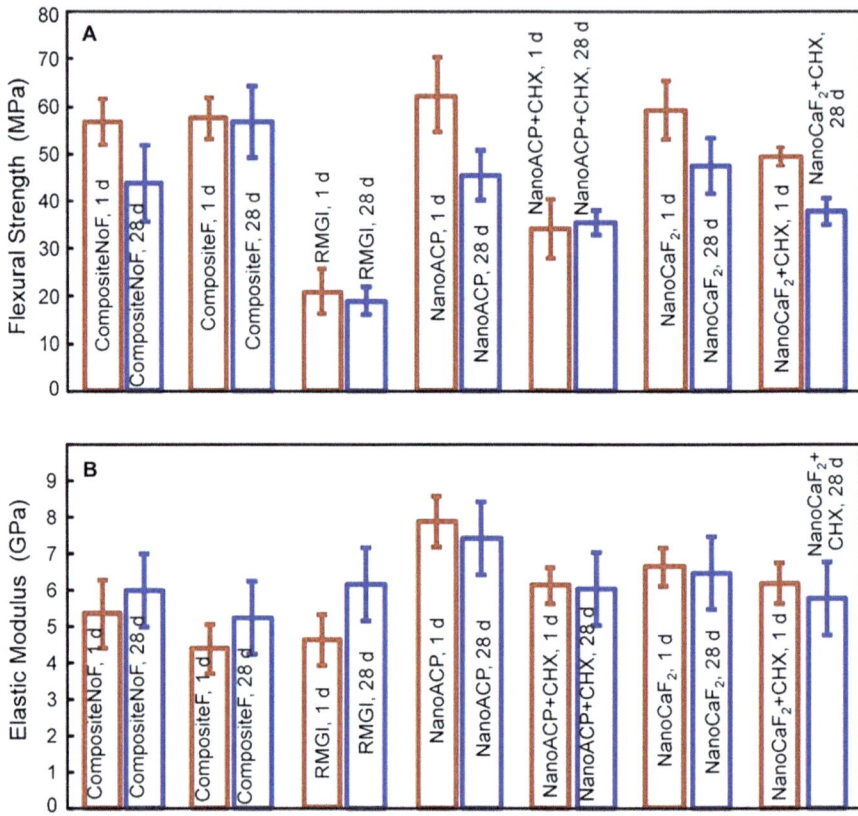

Figure 10.3   Flexure strength and modulus of composites containing chlorhexidine.
Reprinted with permission from Elsevier.[22]

containing 5% or more chlorhexidine will have significantly lower mechanical
properties than the control (containing no chlorhexidine).[30] Therefore, it remains a great challenge to achieve a sustained release of antimicrobial agents
from dental composites without adversely affecting their physical and mechanical properties.

10.2.2   Materials Containing other Antibacterial Small Molecules

Although chlorhexidine is a definitive favorite among antimicrobials used in
dental materials, there have been a number of studies utilizing other releasable
compounds including hydrocortisone,[31] triclosan[32] and octenidine.[33] All of
these compounds have been shown to be releasable when mixed in dental resins,
thus providing antibacterial activity against a number of bacterial strains
including *S. mutans*.

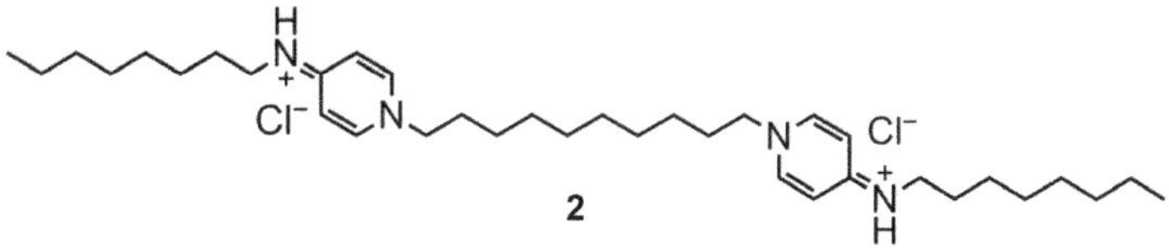

Figure 10.4 Structure of octenidine dihydrochloride (**2**).

Despite the fact that triclosan has been shown to be bactericidal and releasable from dental materials, a compomer containing 2 wt% triclosan applied directly to the tooth structure following minimally invasive removal of carious tooth structure (leaving behind some carious dentin along with caries-causing bacteria) failed to show complete inhibition of bacteria in an *in vivo* study.[32] The drug-releasing compomer is capable of reducing the number of *lactobacilli* colonies, but the total viable counts (all bacteria) was shown not to be significantly different from control.

More impressive results are found when using a composite containing octenidine. Octenidine (**2**) is a bis(aminopyridine) compound (Figure 10.4) that has long been known to possess antibacterial properties when incorporated into an antibacterial mouth rinse.[34] Until recently, however, it was not utilized as a releasable agent in dental composites.[33] In this study, dental composites based on the monomers Bis-GMA and TEGDMA were loaded with 3 wt% or 6 wt% octenidine dihydrochloride (ODH). The composites were tested for ODH release in both *in vitro* and *in situ* experiments. The *in vitro* tests were conducted after 1 week immersion in water and revealed that approximately 0.2% of the total ODH was released from the 3 wt% composites, while the composites containing 6 wt% released 0.4% of their ODH over the same time period. The *in situ* experiments were conducted by incorporating the composite specimens into acrylic splints worn by patients on the upper jaw. The patients were then asked to collect 2.0 mL saliva at set time points (2 min, 1 h, 6 h, 12 h, 18 h and 24 h) after intraoral exposure of composite specimens containing 6 wt% ODH. It was found that ODH was released at a steady state level, 0.045–0.13 µg/mL/12 resin composites.

After showing that release of ODH from composites under common oral conditions is possible, the tendency of these composites to disrupt biofilm growth was analyzed. Following both 3- and 7-day exposure to the oral environment, biofilm formation was significantly decreased relative to control. At 6 wt% loading nearly complete inhibition is observed (Figure 10.5).

Releasable drugs can also be incorporated into resorbable drug-delivery systems such as hydroxypropylcellulose films,[35] drug-carrying gels[36] and polymer-based fiber mats.[37] The advantage of this method of drug delivery is derived from the ability of the materials to release drugs with high antibacterial activity over a short period for effective periodontitis treatment before breaking down and being resorbed into the body. Because the drug is not directly incorporated into a permanent restoration, there is no detrimental effect on the mechanical properties. One example of this type of system involves metronidazole-loaded electrospun poly(L-lactide-*co*-D/L-lactide) (PLA) fiber mats.[37] In this instance, drug loadings

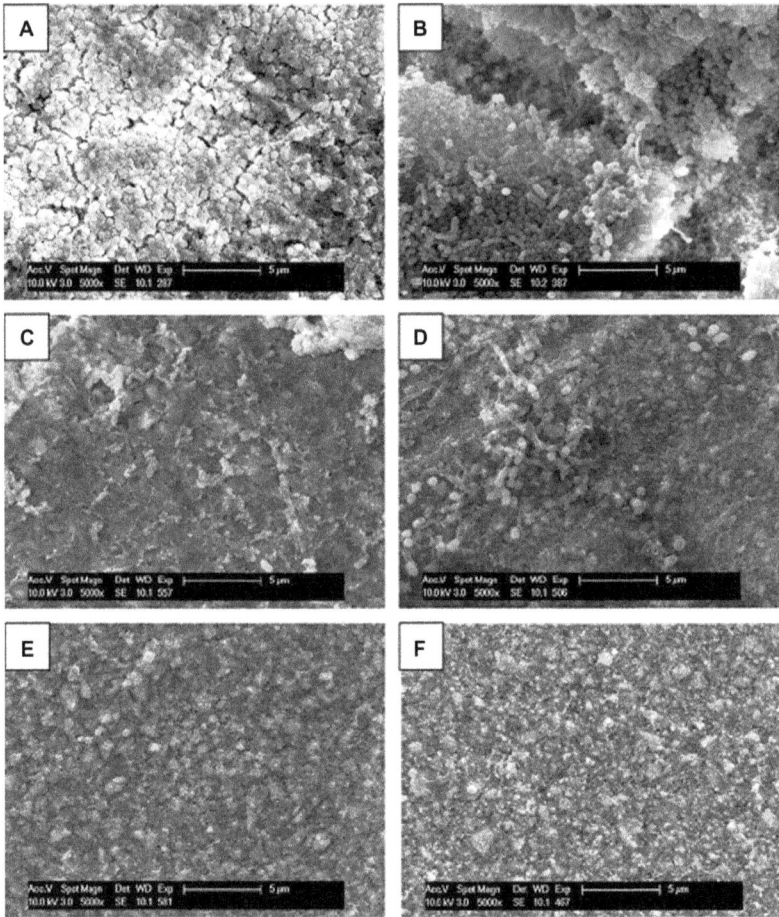

Figure 10.5 SEM micrographs of 3- (A, C, E) and 7-day (B, D, F) biofilms formed on
composites with 0 wt.% (A, B), 3 wt.% (C, D), or 6 wt.% (E, F) of
octenidine.
Reprinted with permission from Elsevier.[33]

of up to 40 wt% can be achieved. All of the fibers (incorporating between 1.0 and
40.0 wt% metronidazole) demonstrate a high initial release ($\geq$ 20% of total drug
loaded) followed by a slower, diffusion-controlled release.

   The solutions of drugs released from the fibers of varying initial content were
tested for activity against three bacterial strains; *Fusobacterium nucleatum,
Aggregatibacter actinomycetemcomitans* and *Porphyromonas gingivalis*. They
were also tested for biocompatibility with human gingival fibroblasts. In agar
diffusion tests, *F. nucleatum* and *P. gingivalis* were shown to be rather sensitive,
with inhibition being observed on the first day using solutions from the 1 wt%
drug loaded fibers and long-term (28 days) inhibition being seen for the solutions from the 40 wt% drug-loaded fibers. On the contrary, *A. actinomycetemcomitans* showed no inhibition on even the first day when tested against

solutions made from fibers containing less than 10 wt% drug. Activity on the second day was observed only with the solution from the 40 wt% drug-loaded fibers. Both the solutions resulting from drug release and the fiber mats themselves show good biocompatibility, with no more than 8% impaired cell viability being observed for the solutions and generally healthy growth observed for cells exposed directly to the fiber mats.

Not all antibacterial dental materials utilize organic molecules for bactericidal activity. TiO_2 has well-established bactericidal properties owing to its ability to generate highly reactive OH radicals under UV irradiation that can oxidize nearby organic molecules.[38] For the anatase form of TiO_2, irradiation with UV light at ≤ 385 nm allows the generation of OH radicals that oxidatively attack bacteria from the exterior to the interior causing the death of the bacteria.[39] Adhesives containing TiO_2 nanoparticles with loadings from 5–30 wt% have been investigated for on-demand antibacterial properties under UV irradiation conditions. The nanoparticles were mixed with a commercial adhesive (Adper™ Scotchbond™ 1XT) then smeared on a glass plate and light cured. *Staphylococcus epidermidis* was used as a test bacterium, being spread on glass plates with or without adhesive coating. The plates were then irradiated with UV light at either 1.2 or 7.5 mW/cm^2. Adhesives containing 10 or 20 wt% nanoparticles show nearly complete bacterial killing after 120 min irradiation at low intensity or 10 min irradiation at high intensity. Obviously, long irradiation times are impractical for clinical purposes but the authors hypothesize that by using a higher-intensity light, more acceptable irradiation times (1–2 min) can be achieved. Incorporation of nano-TiO_2 does not result in a significant decrease in the tensile bonding strength.

One obvious limitation for TiO_2 as an antimicrobial material is that the OH radicals generated by UV irradiation are short lived. After they are consumed, the antimicrobial effect will be diminished. Although repeated UV irradiation can regenerate its antimicrobial effect, it is not clinically practical for restored dental materials or implants because the strong UV irradiation will damage the surrounding tissues. Therefore, the clinical applicability of TiO_2-containing materials is questionable.

10.3 Antibacterial Dental Composites Containing Ag Compounds and Nanoparticles

The antimicrobial effects of Ag^+ ion are well known. Ag has been employed in various forms in a number of medical and dental applications and devices[40,41] because of its ability to interact with proteins, anions and receptors on cell membranes.[40] The sensitivity of bacteria to Ag^+ is related to its ability to irreversibly denature vital enzyme systems and affect DNA replication, leading to cell death.[42] Ag^+ can be introduced into dental materials in a variety of ways, including direct mixing of silver salts,[43–45] complex coordination networks[46] and addition of Ag nanoparticles.[47] The utility of Ag-containing dental materials depends on their ability to achieve antimicrobial activity without significant detrimental effects to mechanical properties, aesthetics or the ability to cure the resins.[48,49]

Ag-nanoparticle-containing resins are particularly advantageous, in part because of the high surface area to volume ratio of the nanoparticles.[50,51] The nanoparticles have been shown to have controlled release of Ag^+ ions under aqueous conditions, including those similar to the oral environment.[52] One method for creation of Ag nanoparticles for a dental acrylic resin involves first mixing $AgNO_3$ in isopropanol with an acrylic liquid for chemical reduction.[53] The synthesized nanoparticles are subsequently dispersed in an acrylic liquid before the solution is mixed with the powder portion of the system to give a doughy mixture that can thereafter be put into molds to create the desired specimens with defined shapes. In this way, complications arising from nanoparticle aggregation can be avoided.

The composites containing 0.5 wt% Ag nanoparticles show a high rate of release of Ag^+ ions in the first two days followed by a leveling off to a concentration of 0.175 mg/L.[53] Furthermore, after the incubation of *Escherichia coli* in the presence of the nanoparticle composites for 24 h, complete bacterial killing was observed with bacteria neither remaining in solution nor on the composites themselves. The mechanical properties of the composite were not adversely affected, as the flexure strength and modulus of the composite containing Ag nanoparticles were both slightly increased relative to a control composite without Ag.

Because composites in clinical settings are generally light cured, it is necessary to understand the role of Ag nanoparticles on the curing process. In some cases, extended curing times (2 min), including curing from both sides of a composite, were used to achieve a suitable degree of conversion.[54] If the polymerization of the monomers is compromised, increased elution of the monomers may result. This can result in not only reduced mechanical properties but also toxicity, as commonly employed dental monomers Bis-GMA, TEGDMA and HEMA have been shown to cause damage to DNA.[55,56] In an attempt to better define the role that Ag nanoparticles play in the light-curing process of dental monomers, composites containing from 0.0125–0.3 wt% nanoparticles were prepared.[57]

The composites were created by mixing Ag nanoparticles of diameter < 100 nm with the commercial composite Tetric Flow® (Ivoclar Vivadent). After light curing according to the manufacturer's instructions (20 s), the specimens were incubated in methanol for 24 h or 7 days, after which the solution was analyzed by GC-MS. It was shown that a number of monomers have significantly increased elution relative to control (containing no Ag nanoparticles). As a result of this study, it is clear that Ag nanoparticles can decrease the degree of conversion of composites and thus increase the elution of uncured monomers into the surrounding solution. This possibility must be taken into account when incorporating Ag nanoparticles into composites, and increased curing times along with limited nanoparticle loadings may have to be utilized.

Another major issue with Ag nanoparticle-containing resins is poor aesthetics. Both chemically cured and light-cured resins containing 0.2 wt% or more silver benzoate (AgOBz) show an amber or dark brown color instead of the desirable tooth color of composites lacking Ag.[50] This kind of color will not be acceptable in dental clinics. Interestingly, the light-cured resins are darker than the chemically cured ones with the same AgOBz loading. This effect is possibly due to photochemical reduction of the silver compound to silver metal and the larger particle sizes of Ag observed for the light-cured resins. It should also be noted that light-cured resins were only able to achieve a loading of 0.2 wt% and remain curable, whereas chemically cured resins could be cured at up to 0.6 wt% loading. The dark color of Ag absorbs light and reduces the degree of polymerization conversion and depth of cure. Inhibition of *S. mutans* by the 0.2 and 0.5 wt% chemically cured resins was analyzed, revealing partial inhibition (52%) at 0.2 wt% and nearly complete inhibition (97.5%) at the higher loading.

In recent years, silver cyanoximate complexes have been studied as light-insensitive antibacterial compounds as well as nonelectric gas sensors.[58–61] Thirty-seven cyanoxime ligands have been synthesized and the crystal structures have been determined for 25 of them by X-ray diffraction. Silver (I) complexes of 13 cyanoximates (AgL, L = cyanoxime anion) have been synthesized by mixing the alkaline aqueous solutions of cyanoximates and $AgNO_3$. These Ag complexes have various colors from bright-yellow to red and are sparingly soluble in water but soluble in polar organic solvents such as DMSO, DMF and CH_3CN. These complexes are insensitive to visible light, *i.e.* they are stable to daylight for years of continuous exposure. They have shown antibacterial activity against various bacteria and fungi such as *E. coli*, *Klebsiella pneumonia*, *Pseudomonas aeruginosa*, *S. mutans*, *Staphylococcus aureus*, *Mycobacterium fortuitum* and *Candida albicans*. However, further studies are needed to demonstrate whether the light-cured dental composites containing these silver cyanoximate complexes have acceptable color or aesthetics as well as good mechanical properties, which may vary with the concentration and the structure of the complex.

10.4 Antibacterial Dental Composites and Bonding Agents Containing Quaternary Ammonium Compounds

10.4.1 Physically Immobilized Quaternary Ammonium Compounds

Quaternary ammonium salts are widely known antibacterial agents that are commonly used to improve oral hygiene. The bactericidal activity of such compounds is thought to result from the positive charge of the ammonium group. The cationic ammonium moiety interacts with the negatively charged cell membrane of the bacteria, disrupting the electrical balance and cell membrane, causing the leakage of cell components and cell lysis.[62] In contrast to the antibacterial agents discussed previously, quaternary ammonium compounds in dental materials are generally immobilized in the resin matrix by either physical or chemical means. In this way, the antibacterial activity is expected to last for a long time. However, the consequence is that the materials can only have surface antimicrobial (contact killing) effect, giving no significant anti-bacterial activity to the surrounding solution or tooth structure, unless the uncured antimicrobial monomer (or agent) leaches out.

Cetyl pyridinium chloride (1-hexadecylpyridinium chloride, **3**) (CPC) is an aromatic ammonium salt consisting of a 16-carbon alkyl chain attached to the heteroaromatic pyridine (Figure 10.6a). It has been incorporated into a mouth-wash as well as a dental varnish due to its strong antibacterial effects.[63,64] In particular, the existence of free CPC in solution at concentrations of 3 ppm and above successfully prevents biofilm formation on a dental resin surface.[65] After mixing with an immobilizing resin matrix consisting of 10-methacryloyloxydecyl dihydrogen phosphate (MDP), HEMA, TEGDMA and a hydrophobic aromatic dimethacrylate (5:45:25:25), CPC was barely elutable, being detected at

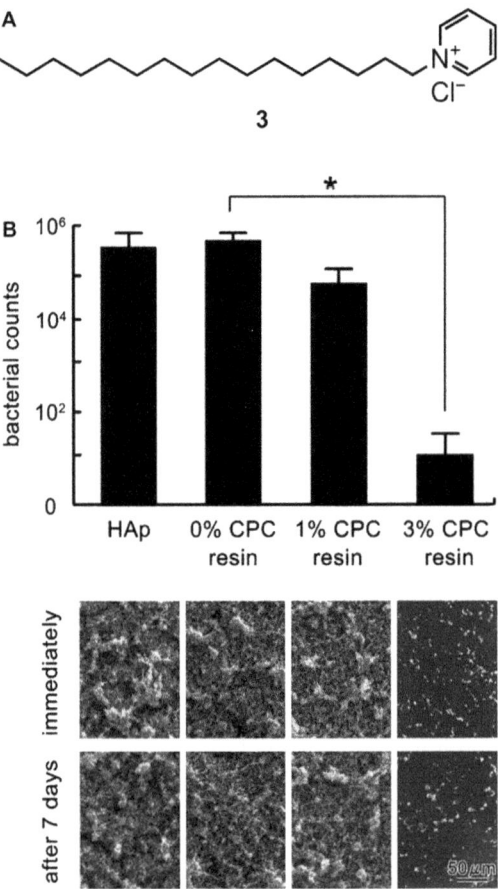

Figure 10.6   (A) Structure of CPC, (B) viable cell counts of *S. mutans* on hydro-xyapatite (HAp) with or without CPC-resin coating (upper plot) and SEM micrographs (bottom images) of *S. mutans* biofilms on HAp or CPC-resin (0 wt.%, 1 wt.%, 3 wt.%) coated HAp.
Reprinted with permission from Elsevier.[65]

concentrations < 0.11 ppm (the lowest concentration tested for the calibration). The resin containing 1 wt% CPC showed a slight (not significant) decrease in *S. mutans* biofilm formation, but the 3 wt% CPC resin shows a very significant biofilm-inhibition effect for the time period tested (up to 7 days) (Figure 10.6b).[65]

Polyethyleneimine (PEI) quaternary ammonium nanoparticles represent an alternative method for immobilizing quaternary ammonium salts in a dental composite resin.[66–68] These nanoparticles (containing octyl and methyl groups) incorporated into a dental composite resin at 1 or 2 wt% were tested for antibacterial activity against a number of strains of bacteria, including *S. aureus*, *E. coli* and *P. aeruginosa*. As anticipated, the composites showed no inhibition zone in agar diffusion tests because the antibacterial component is immobilized and thus not able to diffuse out of the resin. In the direct contact test, however, composites containing PEI nanoparticles showed significant bacterial inhibition at 1 or 2 wt% loading against nearly all bacteria tested for at least 4 weeks. In the case of *P. aeruginosa*, 2 wt% loading is necessary for significant inhibition at 4 weeks incubation time (Figure 10.7).

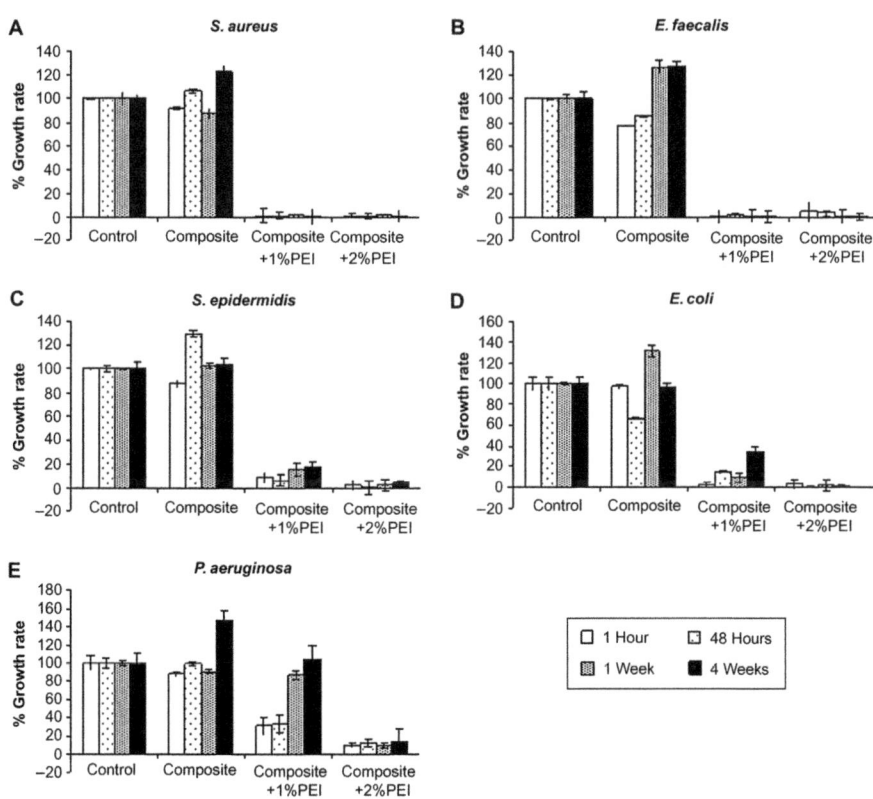

Figure 10.7 Bacterial growth rate of Gram-positive (A, B, C) and Gram-negative (D, E) bacteria after direct contact with composite resin with 0, 1, 2 wt.% polytheneimine nanoparticles (F).
Reprinted with permission from Elsevier.[67]

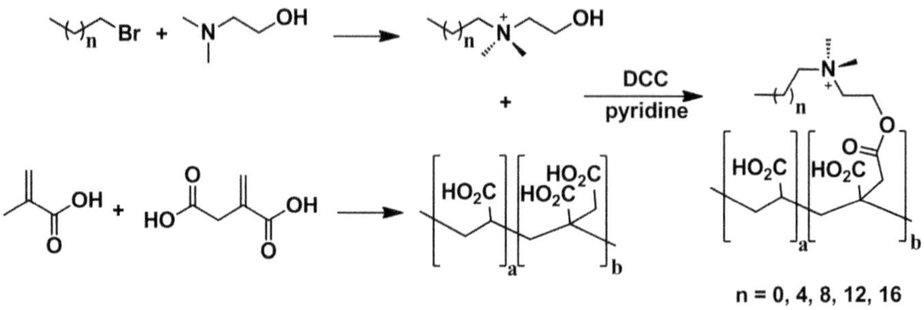

n = 0, 4, 8, 12, 16

Scheme 10.1 Synthesis of a quaternary ammonium containing polymer for use in GIC.

Other polycationic polymers have also been incorporated into dental materials.[69] A series of polyacid containing quaternary ammonium oligomers was synthesized for use in GIC (Scheme 10.1).[62,70] The combination of the new poly(quaternary ammonium salt) (PQAS) with a new poly(acrylic acid) for glass ionomer cements results in some interesting trends relating to the antibacterial activity and mechanical properties of the new materials. Specifically, the bactericidal activity of PQAS is clearly related to the length of the alkyl chain of the ammonium group. Longer chains (up to 16 C) give higher activity, and the activity steadily decreases with decreasing chain length.[62] Overall, the activity lies in the range 1.563–20 000 µg/mL for the oligomers. In the case of the RMGICs, the final *S. mutans* viability on the cements are between 18.1 and 36.4%, dependent on chain length (cements contain 5 wt% PQAS). The antibacterial activity is not diminished after 30 days.

The antibacterial activity achieved by the addition of PQAS is accompanied by a decrease in the mechanical properties. The compressive strength of the cements decreases both with increasing loading of PQAS and with increasing chain length on PQAS (*i.e.* the C_2 PQAS is stronger than the C_{16} PQAS). Comparison of the experimental cement with PQAS to a commercial one without antibacterial properties reveals a decrease in strength relative to the commercial product, but less significantly than when compared to control.

10.4.2 Chemically Immobilized Quaternary Ammonium Compounds

10.4.2.1 Dental Materials Containing MDPB

Despite some recent advances in polymeric or oligomeric quaternary ammonium salts in dental materials, the major focus of the field has long been on the development and testing of quaternary ammonium monomers. The most well known of these monomers is 12-methacryloyloxydodecyl pyridinium bromide (MDPB, **4**, Scheme 10.2).[71] MDPB can be found in dental adhesives, primers and composites.[72–74] The monomer is synthesized through a two-step procedure in which methacrylic acid is first esterified with 12-bromododecanol followed by reaction with pyridine to make the salt (Scheme 10.2).[71]

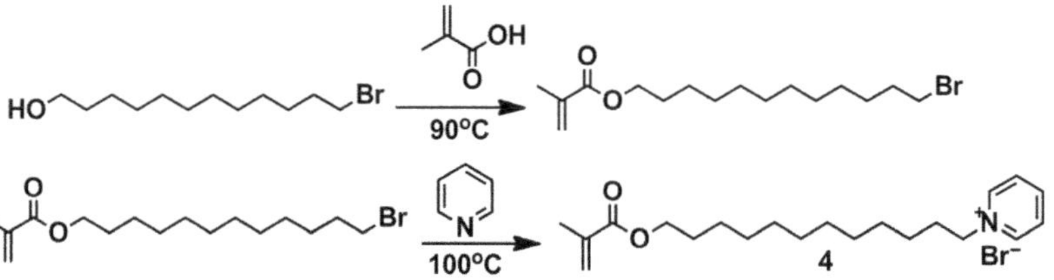

Scheme 10.2 Synthesis of antibacterial monomer MDPB.

In its monomer form, MDPB is bactericidal against a number of strains of cariogenic bacteria, including *S. mutans*, at concentrations of 62.5 μg/mL and above.[75,76] As mentioned previously, CPC partially inhibits *S. mutans* biofilm formation at a concentration of 1.0 ppm (18.8 μg/mL) and completely inhibited biofilm formation at 3.0 ppm (56.3 μg/mL) concentration, *vide supra*.[65] Thus, MDPB has less bactericidal activity than CPC. This is in accordance with the results for the PQAS, for which it was shown that bactericidal activity increases with chain length.[62] MDPB shows significant cytotoxicity at concentrations ≥40 μg/mL, lower than what is needed for antibacterial activity. However, because it is expected that the monomer will be fixed in the resin matrix, the cytotoxicity of the free monomer may not be problematic.

After polymerization in water-soluble form as either a homopolymer or a copolymer with acrylamide, the antibacterial ability of MDPB is severely reduced, with many species of *streptococci*, including *S. mutans*, showing decreased (or even minimal) cell death following contact.[77] For the water-insoluble polymer tests, 20 wt% MDPB was incorporated in unfilled dental resin disks with Bis-GMA and TEGDMA. Although a decrease in cell viability was observed for all bacteria tested, the difference is not statistically significant. It is worth noting that Ebi *et al.*[78] reported that MDPB incorporated into composite monomer compositions should be limited to 0.4 wt% to prevent agent release. This limitation causes poor reproducibility of antiplaque effects in polished composite samples. However, by prepolymerization of the monomer and dispersing the polymer into filler particles, 2.83% of polymerized MDPB can be incorporated into dental composites without deteriorating their surface roughness and other physical properties.

Besides being incorporated into composites, MDPB has been used to provide antibacterial properties to adhesive resins[79] and primers[80,81] as well. When 2.5 wt% MDPB is added to a commercial adhesive (LB bond in Liner Bond 2 system, Kuraray), a 97% inhibition of *S. mutans* growth is observed.[79] The antibacterial activity is reported to be due to a bacteriostatic effect, rather than a bactericidal effect. The bond strength of the adhesive was not adversely affected by incorporation of MDPB in either *in vitro*[79] or *in vivo*[82] tests. MDPB-containing primers have also been shown to be antibacterial both before[80] and after[83] curing, without significant cytotoxicity after curing, compared to control.[81]

10.4.2.2   Other Quaternary Ammonium Monomethacrylates

The bactericidal effects of MDPB monomer before curing and the bacterio-static effects after curing reveal the potential of QAS monomers in dental materials. As a result, many new monomers have been developed in an attempt to improve upon the antibacterial and physical/mechanical properties of QAS containing dental materials. For example, Xiao *et al.* synthesized a series of new monomers (**5–7**) and tested them for activity against several bacterial species including *S. mutans*, *S. aureus* and *L. casei* (Figure 10.8).[84] Of the three new monomers, methacryloyloxyethyl cetyl ammonium chloride (DMAE-CB) shows by far the highest activity with minimum inhibitory concentration (MIC) values in the range 1.2–4.8 µg/mL for the bacteria tested. In a separate study, the authors showed that an experimental adhesive containing 3 wt% DMAE-CB inhibited the growth of *S. mutans* in contact with the drug-containing cured resin by damaging the growth, adherence and membrane integrity of the bacteria.[85] The antibacterial activity is reported to be maintained for at least 6 months[86] without detriment to the microtensile bond strength, compared to control. The results are comparable to those using Clearfil™ Protect Bond (Kuraray), which contains the QAS monomer MDPB.[79]

A related monomer in which the alkyl chain is only 12 C long shows decreased activity relative to DMAE-CB.[87] The monomer dimethyl dodecyl (2-methacryloyloxyethyl) ammonium iodide (DDMAI) was added to a dental composite along with the monomers Bis-GMA and TEGDMA and tested for antibacterial activity against *S. mutans*. Because DMAE-CB was used as part of an adhesive and not a composite, the results are not strictly comparable. However, the decreased activity observed for DDMAI relative to DMAE-CB is in line with other reported results regarding the effect of chain length on antimicrobial activity.[62,88] When DDMAI was incorporated at 3 wt%, no antibacterial effect is observed.[87] At 5 wt% concentration, a two-fold reduction

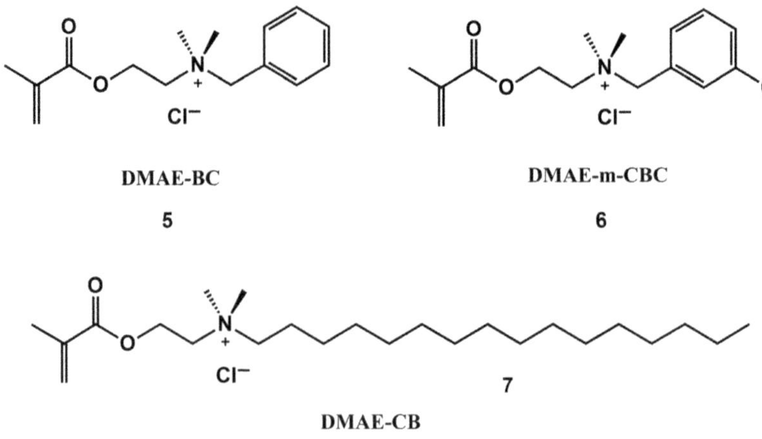

Figure 10.8   Structure of three antibacterial monomethacryate QAS monomers.

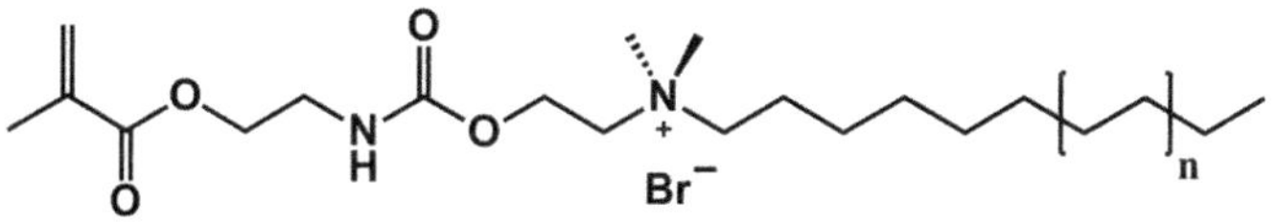

Figure 10.9 Structure of a urethane methacrylate QAS monomer.

of *S. mutans* colonies results. The composites containing 5 wt% DDMAI also show good mechanical properties with a flexure strength similar to control.

The vast majority of monomethacrylate QAS monomers are based on one of two core structures: methacryloyloxyalkyl ammonium or methacryloyloxyethyl dimethyl alkyl ammonium.[84] The resulting structures vary mostly in the length of the alkyl chain, with longer chains (up to 16 C) showing the best antibacterial activity, *vide supra*. One alternative to the monotony has been provided by Buruiana *et al.*, in the form of urethane methacrylates bearing quaternary ammonium groups (**8**, Figure 10.9).[89] These structures are similar to DMAE-CB and DDMAI, differing in that the methacrylate group has been replaced with methacryloyloxyethyl carbamate. Two new structures bearing either a C_{12} or C_{16} alkyl chain have been synthesized and incorporated into dental composite resins. The degree of conversion (DC) and agar diffusion tests reveal incomplete polymerization of the monomers, with significant inhibition zones being evident when tested against *S. mutans* (in contrast to other polymerizable, nonreleasable QAS composites).

10.4.2.3 Dental Materials Containing Quaternary Ammonium Dimethacrylate (QADM) Monomers

Until recently, the QAS monomers being employed in dental materials were all monomethacrylates. Although in many cases the use of these monomethacrylate QAS monomers does not show significant detrimental effects on the mechanical properties, the loading of the monomers must be kept at appropriately low levels in order to prevent elution of uncured monomer[78] or undesirable physical properties.[90] In addition, the inclusion of a significant portion of a monomethacrylate can decrease the degree of crosslinking, which can generate negative effects.[91] By using quaternary ammonium dimethacrylates, a potentially greater drug loading can be achieved without significant uncured monomer elution or detriment to the mechanical properties.

As aforementioned, Ag nanoparticles are able to inhibit/kill bacteria when incorporated into dental materials. However, concerns about aesthetics and mechanical properties make it difficult to load the nanoparticles at sufficiently high concentration to achieve high levels of inhibition. As a result, new materials containing both Ag nanoparticles and quaternary ammonium monomers have been developed.[92–94] In this study, a recently reported quaternary ammonium dimethacrylate (QADM) monomer (**9**, Figure 10.10)[95] was

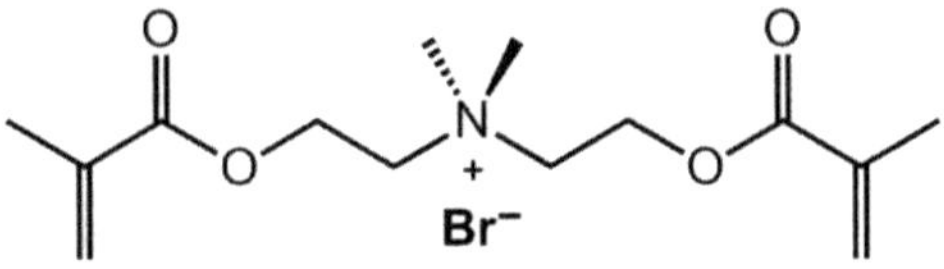

9

Figure 10.10 Structure of a QADM monomer.

included in dental composites, adhesives and primers, with or without Ag nanoparticles.

The new dental composites also include amorphous calcium phosphate (ACP) nanoparticles (in place of a portion of the glass filler) in order to enhance Ca^{2+} and PO_4^{3-} release for remineralization purposes.[93,96] The composites containing nano-ACP alone and with QADM, nano-Ag or both were all subjected to flexure strength tests to determine the effect of the various additives on the mechanical properties of the composites. The flexure strength of the various compositions are all in the range of 50–70 MPa, with the composites containing QADM being slightly lower, though not by a statistically significant margin.

For the biofilm inhibition studies, *S. mutans* was chosen because it is the primary causative agent of dental caries.[4] The bacteria were allowed to grow in the presence of composite disks for 1 or 3 days after which the cells were stained using the BacLight live/dead kit (Molecular Probes).[54] Using this method, live bacteria are stained green and dead bacteria are stained red. The composites containing either Ag nanoparticles or QADM monomer show some bacterial killing, but the best results are obtained for the composites containing both nano-Ag and QADM after 3 days. Thus, the antibacterial activity of the composites has been shown by *in vitro* tests against *S. mutans*. However, in a clinical setting, the composites are largely not in direct contact with the tooth structure. Below the composite portion of a restoration are generally a primer and an adhesive. This means that any bacteria remaining in the tooth structure following excavation of the carious material will not be in direct contact with the antibacterial composite and therefore may be less affected by the added antibacterial agents. In order to prevent the proliferation of cariogenic bacteria beneath a restoration, the antibacterial component can be incorporated into dental bonding agents (primers and adhesives).

After inclusion of 10 wt% QADM and/or 0.05 wt% nano-Ag into a dental adhesive or adhesive and primer, the biofilm inhibition ability of the system was tested using human saliva as the bacteriological source. Following 2 days incubation, the total micro-organisms as well as total *streptococci* and *S. mutans* were measured. As in the case of the composites, both QADM and nano-Ag are capable of significantly reducing biofilm growth, but the best results are obtained when the two antimicrobial agents are used in combination with one another (Figure 10.11). Furthermore, when the antimicrobials are incorporated in the primer as well as the adhesive (instead of in the adhesive alone), the

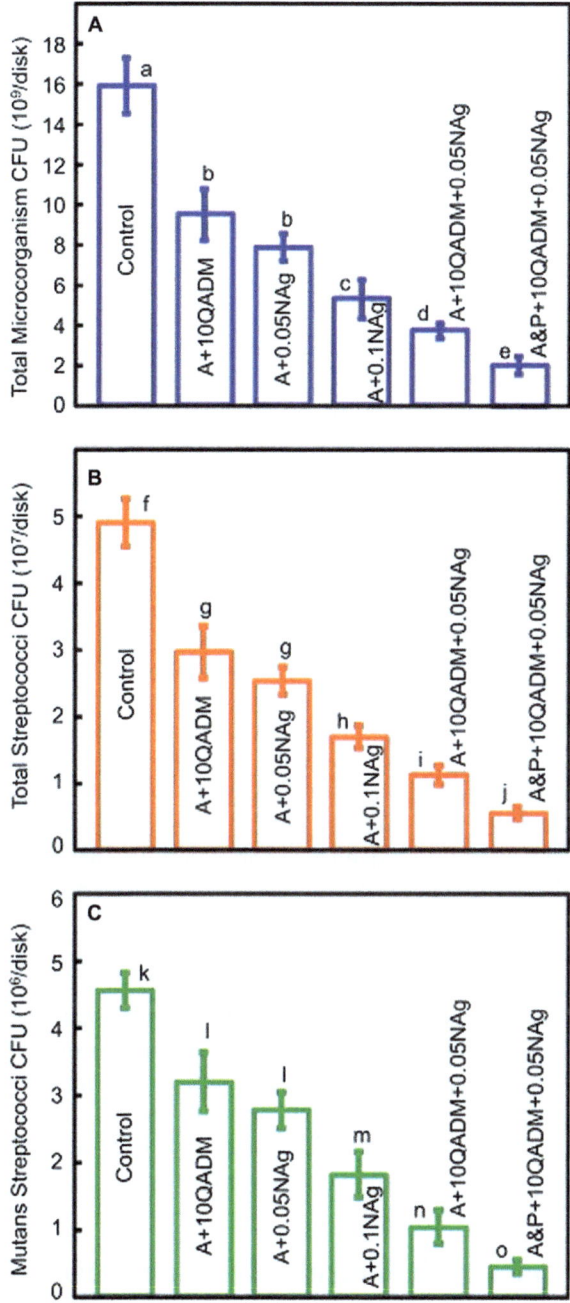

Figure 10.11 Antibacterial effects of adhesives containing Ag nanoparticles and/or QADM monomer.
Reprinted with permission from Elsevier.[92]

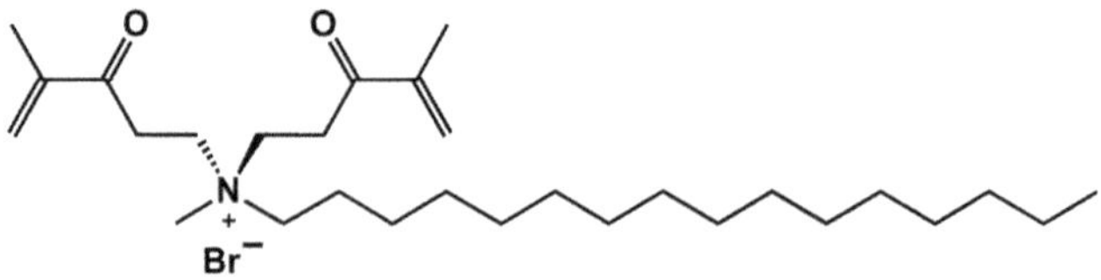

10

Figure 10.12   Structure of a QADM monomer bearing a long alkyl chain.

results again improve significantly. However, a high concentration (10%) of such a QADM is required to achieve an antibacterial effect, which indicates that its antibacterial activity is not very strong, probably because the quaternary ammonium salt (Figure 10.10) does not contain a long aliphatic chain, leading to low lipophilicity and low bactericidal effect.

In an effort to combine the mechanical properties of the dimethacrylate antibacterial monomers with the activity of their monomethacrylate counterparts, a new monomer that is a dimethacrylate version of DMAE-CB (in which a methyl group is replaced by methacryloyloxyethyl) (**10**, Figure 10.12) has been synthesized.[91] The new monomer has strong antibacterial activity comparable to DMAE-CB and is less toxic to healthy gingival fibroblasts than the commonly employed monomer Bis-GMA.

In all of the previously reported cases, the antibacterial monomers in question account for a small proportion of the total monomer content. However, one new antibacterial monomer was not used as an additive but as a base monomer.[97] When the new urethane dimethacrylate-based antibacterial monomers were used to replace Bis-GMA in the polymer matrix, a significant (albeit slight) decrease in *S. mutans* viability (when grown on composite disks) was observed. Despite a high degree of conversion (higher than the control in some cases), the composites showed decreased flexure strength and modulus, likely due to weaker intermolecular interactions caused by the long alkyl chain of the antibacterial monomer. Water sorption and solubility were also increased relative to the control.

Quaternary ammonium salts are powerful antimicrobial agents in the fight against dental caries. Because of the nature of their antibacterial effect, the problem of bacterial resistance is circumvented. Furthermore, they can be immobilized in dental resins by either noncovalent or covalent interactions. Because the compounds are generally considered to be nonleachable, a long-lasting antibacterial effect is observed in many cases.[67,71,86]

10.5   Antibacterial Fluoride-Releasing Dental Materials

The anticariogenic activity of fluoride has long been known. Fluoride can prevent caries by several mechanisms: a reduction of demineralization, enhancement of remineralization, interference of plaque formation and an inhibition of bacterial growth and metabolism.[98] The primary manner in which fluoride acts as a caries preventive is in tipping the balance between demineralization and remineralization, including the formation of acid-resistant fluorapatite (as compared to hydroxyapatite). The bactericidal effects of fluoride are generally considered unimportant due to the difference in the concentration necessary for remineralization compared to antimicrobial activity, as well as the low levels of fluoride release from composites.[98] Nonetheless, by combining fluoride-release characteristics with antimicrobial activity, a powerful anticariogenic effect can be imbued on dental restorative materials.

Significant amounts of fluoride release are possible with either GIC or RMGIC restorations.[98] Bacterial inhibition *via* these materials is generally associated with releasable reagents such as chlorhexidine (*vide supra*), but the inclusion of Zn ion (in the form of $ZnSO_4$) has also been shown to reduce bacterial growth *in vitro*[99] and caries *in vivo*,[100] despite having no significant effect or oral flora. *In vitro*, it was shown that the bacterial growth is very sensitive to Zn concentration in the early growth phase, but not in the later stages of bacterial growth.[100] The results suggest that materials that have early release of Zn ion can effectively prevent caries even without sustained release of Zn.

Addition of $ZnSO_4$ to a commercially available RMGIC (Vitremer, 3M Dental Products) results in a Zn^{2+}- and F^--releasing restorative with anticariogenic properties.[99] Also, the materials containing Zn released fluoride at a higher rate than the controls. Fluoride recharge is possible as well, but the results are short lived (1 day). Zinc release occurred only in the first day, with the amount being greatly affected by Zn loading, the 5 wt% samples releasing 0.9 ± 0.5 ppm and the 10 wt% samples releasing 7.5 ± 0.4 ppm Zn^{2+}. Growth inhibition of *S. mutans* was analyzed by agar diffusion, with the 1 h aged samples revealing inhibition zones > 2 mm, significantly higher than the control (without $ZnSO_4$). The flexure strength of the RMGIC was not affected by the addition of $ZnSO_4$ at levels up to 10 wt%.

In an effort to create a material with the mechanical properties of a composite and the F-releasing capability of a GIC or RMGIC, Xu *et al.* synthesized fluoride-releasing dimethacrylate monomers (**11–13**).[101–103] The monomers are able to form chelates with zirconium fluoride (Figure 10.13) and are miscible with common dental monomers such as Bis-GMA. Composites containing 13.7 wt% of fluoride-releasing monomer showed a dramatic increase in fluoride release relative to commercial composites.[102]

Combining fluoride release with antibacterial activity is accomplished by merging the fluoride-release monomer with one of three new antibacterial monomers (Figure 10.14).[6] The first of these monomers is 12-methacrylami-dododecyl trimethyl ammonium fluoride (**14**). It is prepared by methacrylation of 1,12-diaminododecane followed by methylation with CH_3I. The iodide is exchanged for fluoride by ion metathesis using AgF. Of the two remaining monomers, one (**15**) is a near variant of MDPB, being one C shorter and containing F^- as counterion in place of Br^-. The final monomer is methacryloyloxyundecyl benzyl dimethyl ammonium fluoride (**16**). The methacrylate monomers are prepared using standard procedures.

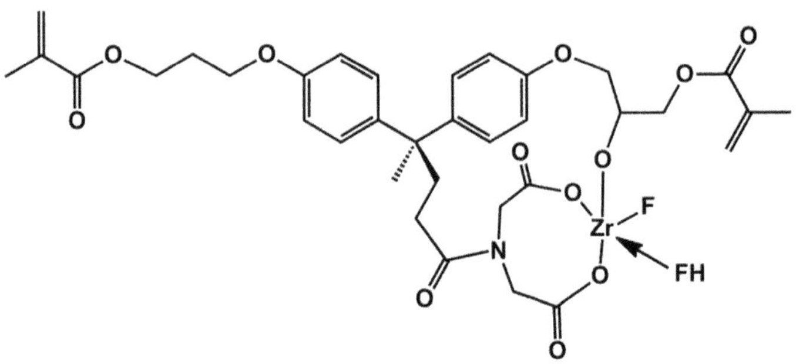

11

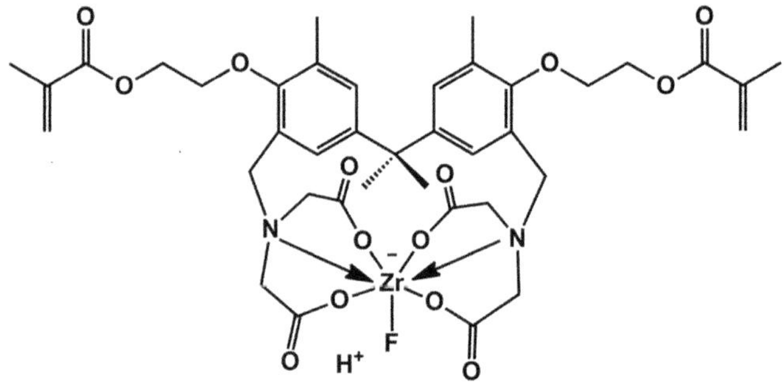

12

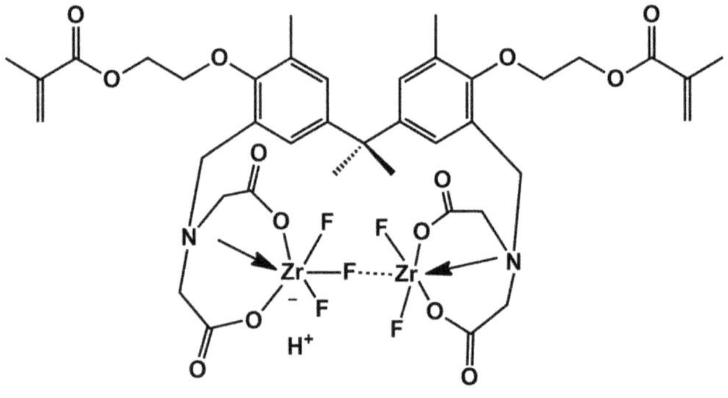

13

Figure 10.13 Structure of antibacterial monomers.

All of the new monomers were tested for antibacterial activity against *S. mutans*. The methacrylamide monomer and the pyridinium methacrylate monomer show some inhibition at 10^{-3} M concentration but not at 10^{-4} M.[6] In the case of the benzyl dimethyl ammonium monomer, however, at 10^{-4} M the monomer still shows > 98% killing of *S. mutans*. At 10^{-3} M, the bacterial killing is 99.98%. Thus, the benzyl dimethyl ammonium monomer was used for composite formulations.

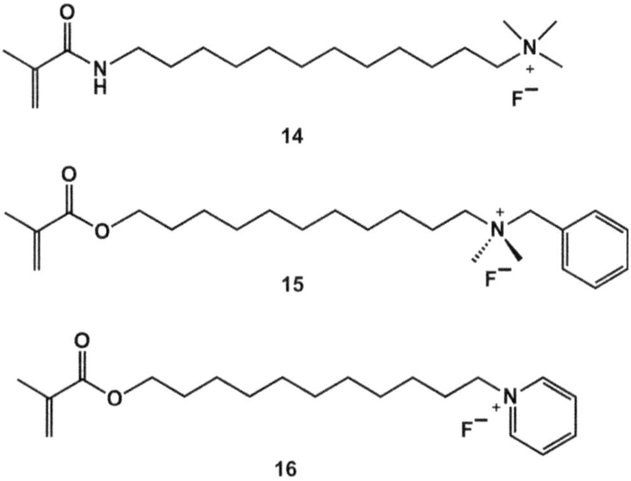

14

15

16

Figure 10.14 Structure of fluoride-releasing monomers.

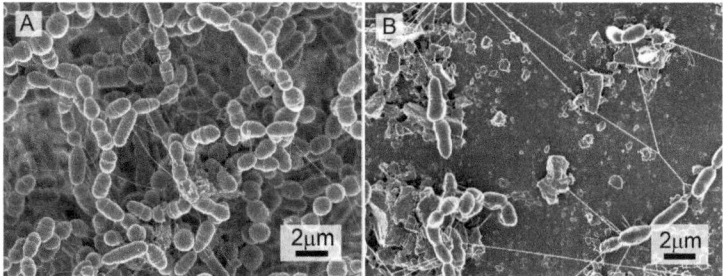

Figure 10.15 FE-SEM micrographs of *S. mutans* biofilms on composites with 0 wt.% (A) and 3 wt.% (B) antibacterial monomer.
Reprinted with permission from John Wiley and Sons.[6] © 2012 Wiley Periodicals, Inc.

The new monomer was included in dental composite resins at up to 3 wt% concentration without significant loss of mechanical properties, despite an increase in the water sorption and solubility relative to the control. In addition, the color characteristics of the composites are unchanged for all drug loadings included in this study (up to 6 wt%). When the antibacterial monomer and the fluoride-release monomer are included in the same composite, the fluoride release characteristics are considerably higher than the control (containing only a fluoride-release filler). *S. mutans* biofilms grown on composites containing at least 3 wt% antibacterial monomer were significantly reduced compared to control (Figure 10.15). The composites containing 3 wt% monomer in particular showed a > 99.9% killing rate.[6]

10.6   Antifungal Denture Materials

10.6.1   Denture Materials Containing Ag Nanoparticles

Denture stomatitis is a common disease among full denture wearers.[104,105] The principal causative agents of this affliction are fungi of *Candida* species, especially *C. albicans*. The recurrence of *Candida* infections is problematic despite attempts to permanently rid patients of the invasive species by disinfection and antifungal therapy.[106,107] As in the case of dental restorative materials, the application of materials that include releasable reagents can be used to treat/prevent infections.[108]

When planktonic cells of *C. albicans* were exposed to colloidal suspensions of Ag nanoparticles, it was shown that the fungus could be inhibited at concentrations < 1 μg/mL.[109] *Candida glabrata* is not as sensitive to Ag nanoparticles, with an MIC as high as 3.3 μg/mL (depending on nanoparticle size). Because it is well established that biofilms are far more resistant to antimicrobial agents than planktonic cells, the synthesized nanoparticles were tested for activity against mature (48 h) biofilms as well. In the case of *C. albicans*, no reduction in biofilm mass was observed. However, Ag nanoparticle suspensions were shown to be effective in reducing biofilms of *C. glabrata*. A significant reduction of biofilm mass was observed at concentrations as low as 1.6 μg/mL, with a 90% reduction possible at 108 μg/mL. In a related study, Ag nanoparticles incorporated in denture materials did not show antifungal activity, possibly due to a low level of Ag release from the resin.[110]

10.6.2   Denture Materials that Release Chlorhexidine

In Section 10.2.1, dental materials that include chlorhexidine as an antibacterial agent were discussed. In addition to being a strong antibacterial agent, chlorhexidine also shows good activity against common fungi, including *C. albicans*.[29,111] The release of chlorhexidine from acrylic-based dental materials is generally short-lived, but high levels (≥ 7.5 wt%) of drug loading can increase the time over which an effective dose can be released.[29,111] By adding 10 wt% chlorhexidine, an effective dose can be released for up to 28 days.[112]

The high initial release characteristics for chlorhexidine-loaded dental materials make preparing a long-term antifungal material difficult. After the initial burst a slow, diffusion-controlled process dominates.[111,112] However, the leachable drug quickly depletes and the material will subsequently release the drug at less than effective doses.[111] The key to permanent antifungal properties may be an effective recharge method. Cao *et al.* reported recently that denture polymers comprised of diurethane dimethacrylate (DUDMA) and methacrylic acid (MAA) are able to do just that.[104]

Disks of the polymers (with varying formulations) were loaded with chlorhexidine by soaking in aqueous solution containing the digluconate form of the drug. Using this method, chlorhexidine loading up to 45.7 μg/cm^2 is reported. Release of the drug from the samples was shown to be sensitive to pH, with a faster drug release occurring at pH 5 than pH 7. It was also shown that the resins can be quenched; that is, the drug can be nearly completely removed by treatment with ethylenediaminetetraacetic acid (EDTA) for 8 h. After quenching, the material was successfully recharged and chlorhexidine was once again loaded on the composite resins. Thus it is clear that reversible, rechargeable drug loading on denture resins is possible.

10.6.3 Denture Materials that Release Other Reagents

The release/recharge system works not only with chlorhexidine but also for a common antifungal agent, miconazole (**17**, Figure 10.16).[104] Most impressive about this is the fact that miconazole shows controlled release at antifungal concentrations for 60 days at pH 7, compared to 14 days for the chlorhexidine-containing resin. It is noteworthy that the DUDMA/MAA system allows not only drug reloading but also drug switching; *i.e.* the resins can be reloaded with a different drug than the one which was originally loaded.

Miconazole is one of a number of azole containing drugs that are fungicidal. Another azole fungicide that has been used as a component of denture materials is fluconazole (**18**, Figure 10.16). Fluconazole added to a poly(ethyl methacrylate)/tetrahydrofurfuryl methacrylate (PEM/THFM) resin at 10 wt% releases effective concentrations of drug for at least 28 days.[112] When using 25 wt% fluconazole removed from capsules (contains 60 wt% excipients), the eluted concentration remained between 4 and 240× the MIC for *Candida* species in the last week tested. Furthermore, even resistant strains were able to be inhibited for 3–21 days. Azole-resistant strains are still sensitive to chlorhexidine, and resistance to chlorhexidine has not been reported.[29]

Antifungals included in dentistry are not limited to chlorhexidine and azoles such as fluconazole. Antifungal peptides have also been used as components of coatings for dentures.[113,114] The proteins histatin 5 (Hst 5) and human β-defensin-3 (hBD-3) were used to coat Lucitone (Dentsply International) disks before *C. albicans* biofilm growth.[113] Biofilm grown on disks coated with the

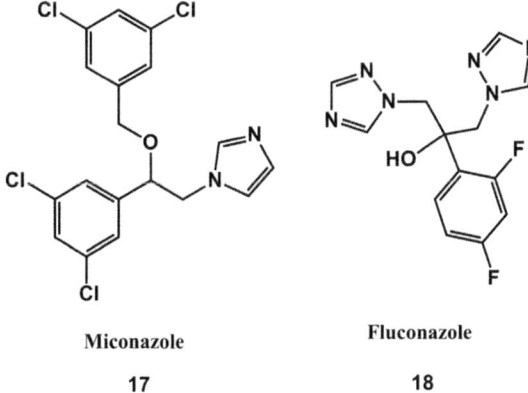

Miconazole

17

Fluconazole

18

Figure 10.16 Structure of azole-containing antifungal agents.

antifungal peptides was analyzed after 24, 48 and 72 h. In most cases, no significant differences were observed between the coated and uncoated samples. However, disks coated with Hst 5 demonstrate significant inhibition after 72 h but not before. In contrast, the peptide chromofungin, derived from chromogranin A, shows good biofilm inhibition against *C. albicans* when incorporated into multilayer polymer films.[114] Overall, the growth of *C. albicans* was reduced by 64%.

10.7   Conclusions

Oral infection by cariogenic bacteria such as *S. mutans* and fungi such as *C. albicans* is a pandemic problem worldwide. In an effort to improve overall oral health, the creation of antimicrobial dental materials has become an important target for the future of dentistry. The battle is taking place on many fronts, including the use of releasable reagents such as chlorhexidine, antibiotics and antifungals. There is also considerable effort being employed to examine the use of Ag nanoparticles and nonreleasable reagents such as quaternary ammonium salt monomers. The prevention of secondary caries is also being targeted by fluoride-releasing materials. Thus far, there is no "silver bullet" for preventive dental materials, and it is likely that one will never be found. Instead, a combination of methods is the more likely direction of future dental materials research.

References

1. A. Jokstad, S. Bayne, U. Blunck, M. Tyas and N. Wilson, *Inter. Dent. J.*, 2001, **51**, 117–158.
2. M. Bernardo, H. Luis, M. D. Martin, B. G. Leroux, T. Rue, J. Leitão and T. A. DeRouen, *J. Am. Dent. Assoc.*, 2007, **138**, 775–783.
3. R. L. Sakaguchi, *Dent. Mater.*, 2005, **21**, 3–6.
4. W. J. Loesche, *Microbiol. Rev.*, 1986, **50**, 353–380.
5. A. B. Tveit and N. R. Gjerdet, *J. Oral Rehabil.*, 1981, **8**, 237–312.
6. X. Xu, Y. Wang, S. Liao, Z. T. Wen and Y. Fan, *J. Biomed. Mater. Res. Part B*, 2012, **100B**, 1151–1162.
7. M. Maltz and C. G. Emilson, *J. Dent. Res.*, 1982, **61**, 786–790.
8. J. O. Burgess, B. Norling and J. Summitt, *J. Esthet. Dent.*, 1994, **6**, 207–215.
9. X. Xu and J. O. Burgess, *Biomaterials*, 2003, **24**, 2451–2461.
10. C. K. Chung, D. T. Millett, S. L. Creanor, W. H. Gilmour and R. H. Foye, *J. Dent. Res.*, 1998, **26**, 533–538.
11. J. W. Nicholson, *Biomaterials*, 1998, **19**, 485–494.
12. D. C. Smith, *Biomaterials*, 1998, **19**, 467–478.
13. J. F. McCabe, *Biomaterials*, 1998, **19**, 521–527.
14. D. C. Sarrett, *Dent. Mater.*, 2005, **21**, 9–20.
15. D. Leung, D. A. Spratt, J. Pratten, K. Gulabivala, N. J. Mordan and A. M. Young, *Biomaterials*, 2005, **26**, 7145–7153.

16. L. S. Türkün, M. Türkün, F. Ertuğrul, M. Ateş and S. Brugger, *J. Esthet. Restor. Dent.*, 2008, **20**, 29–44.

17. I. Lewinstein, E. Zenziper, J. Block and A. Kfir, *Int. Endodont. J.*, 2012, **45**, 1010–1017.

18. B. J. Sanders, R. L. Gregory, K. Moore and D. R. Avery, *J. Oral Rehabil.*, 2002, **29**, 553–558.

19. T. Kuyyakanond and L. B., Quesnel, *FEMS Microbiol. Lett.*, 1992, **100**, 211–215.

20. Y. Takahashi, S. Imazato, A. V. Kaneshiro, S. Ebisu, J. E. Frencken and F. R. Tay, *Dent. Mater.*, 2006, **22**, 647–652.

21. I. M. Brook, C. W. I. Douglas and R. van Noort, *Biomaterials*, 1986, **7**, 292–296.

22. L. Cheng, M. D. Weir, H. H. K. Xu, A. M. Kraigsley, N. J. Lin, S. Lin-Gibson and X. Zhou, *Dent. Mater.*, 2012, **28**, 573–583.

23. N. Hiraishi, C. K. Y. Yiu, N. M. King, F. R. Tay and D. H. Pashley, *Dent. Mater.*, 2008, **24**, 1391–1399.

24. I. M. Mehdawi, J. Pratten, D. A. Spratt, J. C. Knowles and A. M. Young, *Dent. Mater.*, 2013, **29**, 473–484.

25. P. Tallury, R. Airrabeelli, J. Li, D. Paquette and S. Kalachandra, *Dent. Mater.*, 2008, **24**, 274–279.

26. C. Ramkissoon-Ganorkar, F. Liu, M. Bandys and S. W. Kim, *J. Biomat. Sci. Polym. Ed.*, 1999, **10**, 1149–1151.

27. T. T. P. Hsu and R. Langer, *J. Biomed. Mater. Res.*, 1985, **19**, 445–460.

28. W. M. Saltzman, N. F. Sheppard, M. A. McHugh, R. B. Dause, J. A. Pratt and A. M. Dodrill, *J. Appl. Polym. Sci.*, 1993, **48**, 1493–1500.

29. M. P. Patel, A. T. Cruchley, D. C. Coleman, H. Swai, M. Braden and D. M. Williams, *Biomaterials*, 2001, **22**, 2319–2324.

30. K. J. Anusavice, N.-Z. Zhang and C. Shen, *J. Dent. Res.*, 2006, **85**, 950–954.

31. I. M. Brook and R. van Noort, *Biomaterials*, 1985, **6**, 281–285.

32. M. J. Wicht, R. Haak, S. Kneist and M. J. Noack, *Dent. Mater.*, 2005, **21**, 831–836.

33. S. Rupf, M. Balkenhol, T. O. Sahrhage, A. Baum, J. N. Chromik, K. Ruppert, D. K. Wissenbach, H. H. Maurer and M. Hannig, *Dent. Mater.*, 2012, **28**, 974–984.

34. M. R. Patters, J. Nalbandian, F. C. Nichols, C. E. Niekrash, J. E. Kennedy, R. A. Kiel and R. L. Trummel, *J. Periodont. Res.*, 1986, **21**, 154–157.

35. T. Noguchi, K. Izumizawa, M. Fukuda, S. Kitamura, Y. Suzuki and H. Ikura, *Bull. Tokyo Med. Dent. Univ.*, 1984, **31**, 145–153.

36. W. A. Soskolne, *Crit. Rev. Oral Biol. Med.*, 1997, **8**, 164–174.

37. M. Reise, R. Wyrwa, U. Müller, M. Zylinski, A. Völpel, M. Schnabelrauch, A. Berg, K. D. Jandt, D. C. Watts and B. W. Sigusch, *Dent. Mater.*, 2012, **28**, 179–188.

38. K. Welch, Y. Cai, H. Engqvist and M. Strømme, *Dent. Mater.*, 2010, **26**, 491–499.

39. M. S. Hassan, T. Amna, A. Mishra, S. I. Yun, H. Y. Kim, H. C. Kim and M. S. Khil, *J. Biomed. Nanotechnol.*, 2012, **8**, 394–404.
40. A. B. Lansdown, *J. Wound Care*, 2002, **11**, 125–130.
41. A. Balamurugan, G. Basossier, D. Laurent-Maquin, S. Pina, A. H. S. Rebelo, J. Faure and J. M. F. Ferreira, *Dent. Mater.*, 2008, **24**, 1343–1351.
42. J. R. Morones, J. L. Elechiguerra, A. Camacho, K. Holt, J. B. Kouri, J. T. Ramírez and M. J. Yacaman, *Nanotechnology*, 2005, **16**, 2346–2353.
43. K. Kawahara, K. Tsuruda, M. Morishita and M. Uchida, *Dent. Mater.*, 2000, **16**, 452–455.
44. M. Hotta, H. Nakajima, K. Yamamoto and M. Aoto, *J. Oral Rehabil.*, 1998, **25**, 485–490.
45. T. Syafiuddin, H. Hisamitsu, T. Toko, T. Igarashi, N. Goto, A. Fujishima and T. Miyazaki, *Biomaterials*, 1997, **18**, 1051–1057.
46. P. S. Brunetto, T. V. Slenters and K. M. Fromm, *Materials*, 2011, **4**, 355–367.
47. J. Husheng, H. Wensheng, W. Liqia, X. Bingshe and L. Xuguang, *Dent. Mater.*, 2008, **24**, 244–247.
48. Y. J. Cheng, D. N. Zeiger, J. A. Howarter, X. Zhang, N. J. Lin, J. M. Antonucci and S. Lin-Gibson, *J. Biomed. Mater. Res. B: Appl. Biomater.*, 2011, **97B**, 124–131.
49. K. Yoshida, M. Tanagawa and M. Atsuta, *J. Biomed. Mater. Res.*, 1999, **47**, 516–522.
50. C. Fan, L. Chu, H. R. Rawls, B. K. Norling, H. L. Cardenas and K. Whang, *Dent. Mater.*, 2011, **27**, 322–328.
51. S. J. Ahn, S. J. Lee, J. K. Kook and B. S. Lim, *Dent. Mater.*, 2009, **25**, 206–213.
52. M. Kawashita, S. Tsuneyama, F. Miyaji, T. Kokubo, H. Kozuka and K. Yamamoto, *Biomaterials*, 2000, **21**, 393–398.
53. M. Z. Kassaee, A. Akhavan, N. Sheikh and A. Sodagar, *J. Appl. Polym. Sci.*, 2008, **110**, 1699–1703.
54. R. Burgers, A. Eidt, R. Frankenberger, M. Rosentritt, H. Schweikl, G. Handel and S. Hahnel, *Archs. Oral Biol.*, 2009, **54**, 595–601.
55. E. Urcan, H. Schertan, M. Styllou, U. Haertel, R. Hickel and F. X. Reichl, *Biomaterials*, 2010, **31**, 2010–2014.
56. J. Durner, M. Debiak, A. Burkle, R. Hickel and F. X. Reichl, *Arch. Toxicol.*, 2011, **85**, 143–148.
57. J. Durner, M. Stojanovic, E. Urcan, R. Hickel and F. X. Reichl, *Dent. Mater.*, 2011, **27**, 631–636.
58. G. Glower, N. Gerasimchuk, R. Biagioni and K. V. Domasevitch, *Inorg. Chem.*, 2009, **48**, 2371–2382.
59. N. Gerasimchuk, A. N. Esaulenko, K. N. Dalley and C. Moore, *Dalton Trans.*, 2010, **39**, 749–764.
60. N. Gerasimchuk, A. Gamian, G. Glover and B. Szponar, *Inorg. Chem.*, 2010, **49**, 9863–9874.
61. N. Gerasimchuk, *Polymers*, 2011, **3**, 1475–1511.

62. D. Xie, Y. Weng, X. Guo, J. Zhao, R. L. Gregory and C. Zheng, *Dent. Mater.*, 2011, **27**, 487–496.

63. J. Moran, M. Addy, R. Jackson and R. Newcombe, *J. Clin. Periodontol.*, 2000, **27**, 37–40.

64. D. Sterinberg, M. Moldovan and D. Molukandov, *J. Antimicrob. Chemother.*, 2001, **48**, 241–243.

65. N. Namba, Y. Yoshida, N. Nagaoka, S. Takashima, K. Matsuura-Yoshimoto, H. Maeda, B. Van Meerbeek, K. Suzuki and S. Takashiba, *Dent. Mater.*, 2009, **25**, 424–430.

66. N. Beyth, I. Yudovin-Farber, R. Bahir, A. J. Domb and E. I. Weiss, *Biomaterials*, 2006, **27**, 3995–4002.

67. N. Beyth, Y. Houri-Haddad, L. Baraness-Hadar, I. Yudovin-Farber, A. J. Domb and E. I. Weiss, *Biomaterials*, 2008, **29**, 4157–4163.

68. D. K. Shvero, I. Abramovitz, N. Zaltsman, M. P. Davidi, E. I. Wiess and N. Beyth, *Int. Endodont. J.*, 2013, DOI: doi: 10.1111/iej.12054.

69. M. A. Compagnoni, A. C. Pero, S. M. M. Ramos, J. Marra, A. G. Paleari and L. S. Rodriguez, *Gerodontology*, 2013doi: 10.1111/ger.12031.

70. Y. Weng, X. Guo, R. L. Gregory and D. Xie, *J. Appl. Polym. Sci.*, 2011, **122**, 2542–2551.

71. S. Imazato, M. Torii, Y. Tsuchitani, J. F. McCabe and R. R. Russell, *J. Dent. Res.*, 1994, **73**, 1437–1443.

72. S. Imazato, *Dent. Mater.*, 2003, **19**, 449–457.

73. T. Thomé, M. P. A. Mayer, S. Imazato, V. R. Geraldo-Martins and M. M. Marques, *J. Dent.*, 2009, **37**, 705–711.

74. M. Nishida, S. Imazato, Y. Takahashi, S. Ebisu, T. Ishimoto, T. Nakano, Y. Yasuda and T. Saito, *Biomaterials*, 2010, **31**, 1518–1532.

75. S. Imazato, N. Ebi, H. Tarumi, R. R. B. Russell, T. Kaneko and S. Ebisu, *Biomaterials*, 1999, **20**, 899–903.

76. N. Izutani, S. Imazato, Y. Noiri and S. Ebisu, *Inter. Endodontic J.*, 2010, **43**, 637–645.

77. S. Imazato, R. R. B. Russell and J. F. McCabe, *J. Dent.*, 1995, **23**, 177–181.

78. N. Ebi, S. Imazato, Y. Noiri and S. Ebisu, *Dent. Mater.*, 2001, **17**, 485–491.

79. S. Imazato, Y. Kinomoto, H. Tarumi, S. Ebisu and F. R. Tay, *Dent. Mater.*, 2003, **19**, 313–319.

80. S. Imazato, A. Kuramoto, Y. Takahashi, S. Ebisu and M. C. Peters, *Dent. Mater.*, 2006, **22**, 527–532.

81. S. Imazato, H. Tarumi, N. Ebi and S. Ebisu, *J. Dent.*, 2000, **28**, 61–71.

82. S. Imazato, F. R. Tay, A. V. Kaneshiro, Y. Takahashi and S. Ebisu, *Dent. Mater.*, 2007, **23**, 170–176.

83. S. Imazato, A. Ehara, M. Torii and S. Ebisu, *J. Dent.*, 1998, **26**, 267–271.

84. Y.-H. Xiao, J.-H. Chen, M. Fang, X.-D. Xing, H. Wang, Y.-J. Wang and F. Li, *J. Oral Sci.*, 2008, **50**, 323–327.

85. F. Li, J. Chen, Z. Chai, L. Zhang, Y. Xiao, M. Fang and S. Ma, *J. Dent.*, 2009, **37**, 289–296.

86. Y.-H. Xiao, S. Ma, J.-H. Chen, Z.-G. Chai, F. Li and Y.-J. Wang, *J. Biomed. Mater. Res. Part B: Appl. Biomater.*, 2009, **90B**, 813–817.
87. J. He, E. Söderling, L. V. J. Lassila and P. K. Vallittu, *Dent. Mater.*, 2012, **28**, e110–117.
88. P. Thebault, T. E. de Givenchy, R. Levy, Y. Vandenberghe, F. Guittard and S. Geribaldi, *Eur. J. Med. Chem.*, 2009, **44**, 717–724.
89. T. Buruiana, V. Melinte, G. Costin and E. C. Buruiana, *J. Polym. Sci. Part A: Polym. Chem.*, 2011, **49**, 2615–2626.
90. S. Imazato, H. Tarumi, S. Kato and S. Ebisu, *J. Dent.*, 1999, **27**, 279–287.
91. L. Huang, Y.-H. Xiao, X.-D. Xing, F. Li, S. Ma, L.-L. Qi and J.-H. Chen, *Archs. Oral Biol.*, 2011, **56**, 367–373.
92. K. Zhang, M. A. S. Melo, L. Cheng, M. D. Weir, Y. Bai and H. H. K. Xu, *Dent. Mater.*, 2012, **28**, 842–852.
93. L. Cheng, M. D. Weir, H. H. K. Xu, J. M. Antonucci, A. M. Kraigsley, N. J. Lin, S. Lin-Gibson and X. Zhou, *Dent. Mater.*, 2012, **28**, 561–572.
94. F. Li, M. D. Weir, J. Chen and H. H. K. Xu, *Dent. Mater.*, 2013, **29**, 450–461.
95. J. M. Antonucci, D. N. Zeiger, K. Tang, S. Lin-Gibson, B. O. Fowler and N. J. Lin, *Dent. Mater.*, 2012, **28**, 219–228.
96. L. Cheng, M. D. Weir, K. Zhang, E. J. Wu, S. M. Xu, X. Zhou and H. H. K. Xu, *Dent. Mater.*, 2012, **28**, 853–862.
97. X. Liang, Q. Huang, F. Liu, J. He and Z. Lin, *J. Appl. Polym. Sci.*, 2013doi: 10.1002/app.39113.
98. A. Wiegand, W. Buchalla and T. Attin, *Dent. Mater.*, 2007, **23**, 343–362.
99. P. W. R. Osinaga, R. H. M. Grande, R. Y. Ballester, M. R. L. Simionato, C. R. M. Delgado Rodrigues and A. Muench, *Dent. Mater.*, 2003, **19**, 212–217.
100. D. G. Bates and J. M. Navia, *Archs. Oral Biol.*, 1979, **24**, 799–805.
101. X. Xu, L. Ling, R. Wang and J. O. Burgess, *Dent. Mater.*, 2006, **22**, 1014–1023.
102. X. Xu, X. Ding, L. Ling and J. O. Burgess, *J. Polym. Sci. Part A: Polym. Chem.*, 2005, **43**, 3153–3166.
103. X. Xu, L. Ling, X. Ding and J. O. Burgess, *J. Polym. Sci. Part A: Polym. Chem.*, 2004, **42**, 985–998.
104. Z. Cao, X. Sun, C.-K. Yeh and Y. Sun, *J. Dent. Res.*, 2010, **89**, 1517–1521.
105. M. H. Figueiral, A. Azul, E. Pinto, P. A. Fonseca, F. M. Branco and C. Scully, *J. Oral Rehabil.*, 2007, **34**, 448–455.
106. A. Uludamar, A. Gökhan Özyeşil and Y. K. Ozkan, *Gerodontology*, 2011, **28**, 104–110.
107. G. Geerts, M. E. Stuhlinger and N. J. Basson, *J. Oral Rehabil.*, 2008, **35**, 664–669.
108. W. M. Amin, M. H. Al-Ali, N. A. Salim and S. K. Al-Tarawneh, *Eur. J. Dent.*, 2009, **3**, 257–266.
109. D. R. Monteiro, S. Silva, M. Negri, L. F. Gorup, E. R. de Camargo, R. Oliveira, D. B. Barbosa and M. Henriques, *Lett. Appl. Microbiol.*, 2012, **54**, 383–391.

110. A. F. Wady, A. L. Machado, V. Zucolotto, C. A. Zamperini, E. Berni and C. E. Vergani, *J. Appl. Microbiol.*, 2012, **112**, 1163–1172.
111. D. J. Lamb and M. V. Martin, *Biomaterials*, 1983, **4**, 205–209.
112. N. Salim, C. Moore, N. Silikas, J. D. Satterthwaite and R. Rautemaa, *J. Dent.*, 2012, **40**, 506–512.
113. C. R. Pusateri, E. A. Monaco and M. Edgerton, *Archs. Oral Biol.*, 2009, **54**, 588–594.
114. O. Etienne, C. Gasnier, C. Taddei, J.-C. Voegel, D. Aunis, P. Schaaf, M.-H. Metz-Boutigue, A.-L. Bolcato-Bellemin and C. Egles, *Biomaterials*, 2005, **26**, 6704–6712.

CHAPTER 11

Polymeric Materials Containing Natural Compounds with Antibacterial and Virucide Properties

ANTONIO MARTÍNEZ-ABAD, GLORIA SÁNCHEZ, MARÍA JOSÉ OCIO AND JOSÉ MARÍA LAGARÓN*

Novel Materials and Nanotechnology Group, IATA-CSIC, Avda. Agustin Escardino 7, 46980 Paterna (Valencia), Spain
*Email: lagaron@iata.csic.es

11.1 Introduction

Polymers are very interesting and widely spread materials in many applications. The reason for this is related to their balanced set of physical and chemical properties as well as their ease of processing. One of the most relevant characteristics of polymers, which provides them with positive and negative aspects, is their inherent ability to allow the transport of low molecular weight components, the so-called barrier properties. Barrier properties in polymers are necessarily associated with their intrinsic capability to permit the exchange, to a higher or lower extent, of low molecular weight substances through mass-transport processes, such as sorption, permeation and migration processes. The phenomenology of permeation of low molecular weight chemical species through a polymeric matrix is generally envisaged down to the molecular level as a combination of two processes, i.e. solution of the solutes and molecular

RSC Polymer Chemistry Series No. 10
Polymeric Materials with Antimicrobial Activity: From Synthesis to Applications
Edited by Alexandra Muñoz-Bonilla, María L. Cerrada and Marta Fernández-García
© The Royal Society of Chemistry 2014
Published by the Royal Society of Chemistry, www.rsc.org

diffusion. A permeating gas is first dissolved into the upstream face of the polymer film, and then, undergoes a molecular diffusion to the downstream face of the film through typically the polymer amorphous phase, where it evaporates into the external phase again. A solution-diffusion mechanism is thus applied, which can be formally expressed in terms of permeability (P), solubility (S) and diffusion (D) coefficients as follows:

$$P = DS \qquad (11.1)$$

In many cases, the transport of low molecular weight components through polymers does not bring a benefit for applications, such as food packaging, since it results in product spoilage or losses in quality but more importantly also in safety issues. The latter effect is often the result of migration of certain low molecular weight components and additives from plastic packaging walls. However, in active packaging technologies and in antimicrobial polymer applications as a whole, controlled release of the antimicrobials is typically considered as a positive option. It is also called "intended migration" and, in this case, mass-transport properties are a positive characteristic of the polymeric materials. When natural extracts are added to polymers to render them antimicrobial, typically they are not added by melt blending because of the inherent thermal instability of many natural extracts and very often coating followed by solvent evaporation processing is used. This book chapter reviews what technologies have been devised to make polymers antimicrobial and virucidal.

11.2 Formulation of Polymer-Based Antimicrobial Films

11.2.1 Plant Extracts and their Essential Oils

11.2.1.1 Introduction

Since ancient times, mankind has served itself with spices and herbs of all kinds as additives for seasoning or in medicinal preparations imparting health or sensory benefits. Essential oils (EOs) are their aromatic oily liquids extracted from integrated plant material (flowers, buds, seeds, leaves, twigs, bark, herbs, wood, fruits or roots).[1] They are commonly extracted from plants by steam distillation. However, other methodologies, such as extractions with organic solvents or supercritical carbon dioxide, have also been explored.[2] Although they have been historically used to treat a great variety of health problems, their use as natural additives in the cosmetic, pharmaceutical and food industry has recently gained a wide interest. As they are naturally occurring substances with the "generally recognized as safe" status, they pose an alternative to other more toxic or polluting antimicrobials and comply with a current trend in many sectors of the society for more natural and environmentally friendly products. There is abundant scientific evidence in relation to the effectiveness of EOs fractions of many spices and herbs and their components as antimicrobial, antifungal, and antiviral compounds as well as other positive effects (antioxidant, anti-inflammatory, *etc.*) that are out of the scope of this chapter. The wide spectrum of antimicrobial efficacy of EOs is due to the presence of certain chemical structures, mostly phenolic compounds or terpenoids. Phenolic compounds with hydrophilic functional groups are thought to in-activate pathogens by adhesion binding, protein and cell-wall binding, enzyme inactivation and intercalation into the cell wall and/or DNA.[3] Terpenoids, due to their lipophilic character, accumulate in the lipid bilayer of the cytoplasmic membrane, leading to disruption of the membrane structure.[4] Although most of the main compounds responsible for the efficacy of EOs have been already identified, the activity of these chemical substances has been reported to be less important than that of the whole EO fraction, which suggests the activity is probably not attributable to one specific mechanism but results from a com-bination of several mechanisms involving synergistic effects with other minor components.

Many natural plant extracts have been shown to have antibacterial, anti-fungal and antiviral properties, and they are able to prevent biofilm formation. It may be expected that, due to their natural origin, EOs might not exert as much evolutionary pressure on pathogens as other synthetic biocides.[5] They are, thus, particularly interesting as natural alternatives to antibiotics, whose effectiveness has been recently waning due to increasing pathogen resistance. Therefore, natural plant extracts have lately gained much interest as potential bioactive ingredients to be incorporated into polymeric materials for drug-delivery systems in the cosmetic, pharmaceutical or textile industry or as polymer additives in active food-packaging applications.

11.2.1.2 Applications

A wide range of polymers have been studied for the incorporation of plant extracts. Either in drug-delivery devices or in food packaging applications, it is desirable that the polymeric material be nontoxic and preferably biodegrad-able. Considering the natural origin of the active compounds, much research has focused on incorporating EOs into biopolymers that also have a natural origin to fabricate natural biopolymer-based films that could be ingested or directly applied to the consumer without any toxicological concerns. Elabor-ation of edible/biodegradable polymer films and coatings has been possible thanks to the filmogenic capacity of many natural biopolymers. Among the substances used for the fabrication of these polymers we find polysaccharides, such as starch, cellulose derivatives, pectin, alginate, carrageenan, chitosan, pullulan, amaranth, and natural gums; proteins derived from plants (soy, wheat gluten, cottonseed, sorghum, kafrin, rice, bran, peanuts, corn zein and pea), and proteins derived from animal foods (meats, fish, poultry, eggs and milk) including collagen, keratin, egg white, casein, whey protein, fish myofibrillar proteins among others.[6] The biodegradable/edible nature of these biopolymers endows them with great potential for use in drug delivery or food systems. Unfortunately, they may also concurrently provide resultant films with certain undesirable physicochemical properties, such as relative weak mechanical and water-barrier properties, which may limit their applicability in active packaging.[6] In these cases, the addition of plasticizers, such as glycerol, sorbitol, polyethylene glycol (PEG) or other sugars, is a must and even when plasticizers are added to a relatively high content, physicochemical properties similar to benchmark nonbiodegradable polymers are not assured.[7] Therefore, the use of other synthetic biopolymers like poly(vinyl alcohol) (PVOH), ethylene-*co*-vinyl alcohol copolymers (EVOH), biopolyesters like polylactide (PLA) or polycaprolactone (PCL), and polyhydroxyalkanoates (PHAs), has also been explored.

One of the main drawbacks of incorporating EOs into polymers is their high chemical and thermal instability as well as a very high volatility. This problem limits the possibility of producing the films by melt compounding and compression molding, as is typically done with conventional polymers. Some authors have bypassed this problem by decreasing the time in the mixer to the minimum, for example introducing the active compound once the polymer is already in the melt state or selecting a polymer with relatively low melting temperature.[5,8–10] However, in most cases, the loss of the volatile compounds is reduced by incorporating the EOs with a casting technique. This technique involves the dispersion/dissolution of all components in a suitable solvent that is evaporated under controlled conditions. Although the most used approach in the literature is batch casting, for example letting the dissolution dry on glass petri dishes, there is also the possibility of continuously casting the dissolution onto a surface or another substrate to fabricate active coatings on surface or multilayered polymer systems.[7] Among the lesser explored possibilities of incorporating EOs into polymers, the electrospinning technique[11,12] is also found in which the EOs can be encapsulated in polymer beads or trapped inside thin polymer fibers, as well as air plasma treatments.[13] Another approach to avoid the loss of the volatile active compounds is to encapsulate the essential oils before incorporation into the selected polymer. Microencapsulation of EOs is generally achieved in two steps. First, an emulsion of the volatile compound is made in an aqueous dispersion of a wall material that also works as the emulsifier. Then, the microencapsulated emulsion must be dried under controlled conditions so as to diminish the loss of the encapsulated material by volatilization.[14] In this sense, interfacial polymerization has been successfully carried out to encapsulate galangal or thyme EO into polyurethane or poly(methyl metacrylate) (PMMA).[15,16] Freeze drying or spray drying of polymer/EO emulsions have also yielded micro- or nanocapsules with improved thermal stability.[17–19] Another interesting approach has been to encapsulate the EOs into α- and β-cyclodextrins. As an example, allyl-isothiocyanate, an active compound present in garlic or mustard, has been successfully encapsulated in α- and β-cyclodextrins to be subsequently incorporated in PLA or PCL.[20–23]

Depending on the application, the produced films should have a constant level of integrity, mechanical strength and suitable barrier properties to gases or water throughout handling and storage, especially if they are designed for food packaging. They should also have favorable sensory properties, such as appropriate color or flavor as to be accepted by the consumer. Films and coatings made from natural biopolymers can have insufficient tensile strength, high water permeability and solubility and high oxygen transmission rates. These physicochemical attributes have been modified by various strategies. The addition of the lipophilic EOs to the polymer matrix can reduce the water-vapor permeability, increase flexibility[24] and sometimes enhance mechanical strength.[25] The polymer structure can be further modified by crosslinking or photocrosslinking, gamma irradiation and complexation with polyvalent ions.[7] Additionally, inorganic fillers, like nanoclays, can enhance barrier and permeability properties of the films.[26]

11.2.1.3 Evaluation of the Antimicrobial Effectiveness

Once the polymer is fabricated, the most important thing is to confirm and assess its potential antimicrobial efficacy. Numerous studies, either in solid, liquid or vapor phase, can be performed to establish the effectiveness of anti-microbials in films and coatings. The selection of the method depends on the final application or purpose of the assay, the nature of the antimicrobial and the characteristics of target micro-organisms, among others.

The tests performed in solid or semisolid media can be used to evaluate the ability of the polymer to diffuse out of the film and inhibit growth or reduce viable counts on a contaminated surface. The film disk agar diffusion assay consists of applying a film disk containing the antimicrobial on an inoculated agar plate and, after incubation at specific conditions, the diameter of the zone where no growth occurred is measured. This test is an end point assay and gives only information of the ability of the antimicrobial incorporated in the film to diffuse and inhibit microbial growth at a prefixed time. In this assay, diffusion of antimicrobials from the film disk will depend on the size, shape, and polarity of the diffusing molecule, as well as on the chemical structure of the film and the agar.[7] The enumeration by plate count of microbial population, at selected times, from inoculated surface agar plates in contact with film disk containing the antimicrobial is a test useful to model wrapping of foods and obtained results may suggest what can happen when the film enters in contact with a contaminated surface.[27] The ASTM E2180-07 is a film surface inoculation test and consists in the enumeration by plate count of the microbial population inoculated on the surface of a film disk in contact with a semisolid media, such as an agar slurry. This assay is useful to simulate surface contamination. The results obtained may suggest what happens when microbial contamination occurs on coatings or films in contact with food and gives an idea of the barrier ability of the film to prevent an external contamination and eventually kill the bacterial population in contact with it (ASTM E2180-07). All these tests are used to model the effectiveness of the films in the presence of a contamination on a surface. In challenge tests and durability studies, the concrete surface of a food or tissue is covered or wrapped with the antimicrobial film or coating. In the challenge tests, the surface is artificially inoculated with selected bacterial strains. Durability studies measure the efficacy of the film to prevent natural indigenous bacterial contamination or proliferation as compared with a control in usually longer periods. Enumeration of the different bacteria is performed by plate count.

Evaluation methods in the liquid phase can be performed in a static or agitated environment. Broth dilution methods, such as the M-26A (Clinical and Laboratory Standards Institute), consist of incubating the films or coatings in test tubes or microtiter plates in contact with a contaminated liquid growth medium. Survival of bacteria can be ascertained by plate count or turbidity measurements. The Japanese Industrial Standard Z2801 is designed to ascertain the ability of a polymer to prevent surface contamination. In this assay, a liquid medium containing the bacterial inoculum is spread on the antimicrobial sample film and covered with another plastic to overcome surface tension. Viable bacteria are then enumerated by plate count.

As many active compounds in plant extracts are highly volatile, the evaluation of the activity of EOs in the vapor phase is particularly interesting and has been gaining in interest in recent years. Vapor phases of EOs are more effective antimicrobials than their liquid phases, because lipophilic molecules in the aqueous phase are associated in order to form micelles and thus suppress that attachment of the EOs to the organism, whereas the vapor phase allows free attachment.[28] The antimicrobial effect of EOs' vapor is usually assessed by an initial *in vitro* screening, and two main methods are used. The first is an adapted disc diffusion method, where the EO impregnated filter disc or the active film is placed on the lid of the Petri dish and then zones of inhibition are measured or bacterial colonies are enumerated.[29,30] The second methodology for this study implies placing the EO and the micro-organisms separately into a sealed environment, such as a jar or desiccator. This second approach allows several organisms to be tested at the same time and to inoculate them onto surfaces and in air, in bigger and more representative volumes.[30]

It must be remarked that the release of the preservative from a film or coating exerts a great influence on its effectiveness. For some applications, a quick release of the antimicrobial is required to control microbial growth. On the contrary, in other applications, a slow release is required so as to assure a certain level of the preservative at the surface to control the external contamination. The determination of the rate of release together with the evaluation of antimicrobial activity through the time might help to optimize the development of films and coatings as potential drug-releasing agents or active packaging materials.

11.2.2 Bacteriocins

11.2.2.1 Introduction

Bacteriocins are antibacterial peptides produced by a broad range of bacteria. The majority of Gram-positive bacteriocins identified to date are produced by lactic acid bacteria, in most cases the genus *Lactococcus*. As this type of micro-organisms is generally considered as safe, their bacteriocins have produced special interest in recent years for commercial uses.[31] Although many bacteriocins (such as pediocines, lacticin and plantaricin) have potential application in food products, the antibiotic nisin is currently the only bacteriocin approved as a GRAS (generally recognized as safe) food additive by both the Food and Drug Administration (FDA) and the World Health Organization (WHO).[32] Nisin is a nontoxic compound, heat stable and sensitive to the action of proteases. Nisin causes its biocide effects through pore formation in the bacterial membrane cell, which leads to a dissipation of transmembrane potential and pH gradient. Consequently, ATP synthesis is prevented, leading to cell death.[33] Although biocide properties of nisin have been widely demonstrated,[34] it has also been observed that it is an effective compound only against Gram-positive bacteria, such as *Listeria monocytogenes*.[35] This organism represents a high risk for food safety because of its tolerance to high levels of salt, its ability for growth at a relatively low pH and at refrigeration temperatures,[31] which could result in postprocessing contamination events. To control the presence of this micro-organism is currently the main objective of many research studies undertaken on the biocide properties of nisin and other bacteriocins (*i.e.* pediocin PA-1, bacteriocin 217) for food quality and safety purposes.[36,37] As mentioned above, bacteriocins of lactic acid bacteria are generally inactive against Gram-negative bacteria due to the resistance conferred by the outer membrane.[38] The application of other antimicrobials with bacteriocins in order to obtain synergistic effects against Gram-negative foodborne pathogens could be a means of overcoming this problem. Thus, combination of nisin with organic acids, salts or heat treatment, or high hydrostatic pressures have been studied in order to enhance bacterial inhibition.[39,40]

11.2.2.2 Applications

When directly applied to food systems, the effectiveness of nisin could be reduced due to leaching into the food matrix and/or crossreaction with food components, such as lipids or proteins.[41] Several studies have been focused on improving nisin functionality through its incorporation in packaging materials. These bioactive systems provide antimicrobial activity by releasing the bacteriocin at a continuous rate. Thus, Mauriello *et al.* obtained positive results when they studied the antimicrobial activity of this compound against *Micrococcus luteus* in milk as coated in a low-density polyethylene film. They also analyzed the release from the antimicrobial-coated film and found that pH decrease and higher temperature favored the migration. Mauriello *et al.*[42] have also shown that when bacteriocin 32Y was incorporated in polyethylene-oriented polyamide (PE-OPA) films using an industrial plant presented an antibacterial action against *L. monocytogenes* in agar plates. In this particular work, they obtained homogeneous inhibition areas, which suggested that bacteriocin was uniformly bound to the surface of the film and diffused regularly. Furthermore, they have also demonstrated that antilisterial activity was still stable after 4 months of film storage at room temperature. Guerra *et al.*[32] also demonstrated the preservation of the biocide properties of nisin when coated in a cellophane matrix to be used in the packaging of chopped meat and showed that adsorption of the antimicrobial agent to the matrix was higher at lower temperatures. McCormick *et al.*[43] studied the combination effect of nisin incorporation in wheat films and a pasteurization treatment against the growth of *L. monocytogenes* and *Salmonella typhimurium*. A significant reduction was shown in the *Listeria* population but no added inhibitory effect against *Salmonella* was evidenced as compared with pasteurization only. Release studies of nisin from various protein matrices (wheat or corn) obtained by different film-forming methods were also performed in order to determine the influence of these factors on the improvement of the antimicrobial capacity of this compound against *Lactobacillus plantarum*.[44] Basically, cast wheat gluten films preserved the highest activity after film formation. A recent study performed by Iseppi *et al.*[45] was based on the evaluation of the antilisterial activity of a PVOH based coating entrapping living bacteriocin-producer *Enterococcus casseliflavus* applied to poly(ethylene terephthalate) (PET) films. In this particular study, the authors demonstrated that the Enterocin 416K1 produced by the living bacteria was able to attack and efficiently kill *L. monocytogenes*. They demonstrated the effectiveness of this system also in food samples. The biocide results obtained with coatings entrapping *E. casseliflavus* were better than those obtained with enterocin-doped coatings. In the same line, Sánchez-González *et al.*[46] evaluated the antilisterial activity of bioactive edible films containing *L. plantarum* in protein and cellulose-based films. As a result, a significant antimicrobial activity against *Listeria innocua* was observed for polysaccharide matrices, which was not observed for protein films, where the bacteriocin production needed more time. Thus, it was demonstrated that the nature of the biopolymer and the bacteriocin concentration are key to use bioactive films in an effective way. An interesting study was carried out by Scaffaro *et al.*[47] where nisin was incorporated in ethylene-*co*-vinyl acetate copolymer (EVA) film by melt processing and its antimicrobial properties were evaluated. The authors concluded that the materials displayed antimicrobial activity even when they were prepared at very high temperatures, such as 160 °C, for a very short processing time. Marcos *et al.*[48] studied the efficacy of antimicrobial films containing nisin and high-pressure processing (HPP) as postprocessing listericidal treatments for fermented sausages. The results obtained in this work showed that PVOH films containing nisin induced a high pronounced reduction of *L. monocytogenes* counts during refrigerated storage. Surprisingly, the combination of HPP with antimicrobial film did not produce any extra protection against *L. monocytogenes* compared with the antimicrobial film alone. The authors concluded that the lack of HPP effect was attributed to a protective effect exerted by low water activity of the tested product.

In conclusion, nisin and other bacteriocins, in conjunction with other inhibitory substances, can provide a barrier for the growth of unwanted contaminating bacteria, thereby reducing the amount of chemicals added to the food, decreasing the intensity of the processing conditions, and contributing to the development of hurdle technologies.[49] In addition, the use of packaging films containing these antimicrobial compounds could be more efficient, by controlling migration of the compound into the food over time, during the transport and storage of food during distribution.[41]

11.2.3   Chlorophylls

The photodynamic process is based on the activation of nontoxic dyes by low-intensity visible light to generate singlet oxygen (type II photochemical reactions) and free radicals (type I photochemical reactions). These highly reactive species are able to interact with virtually every cell component, proteins, lipids, and nucleic acids and generate a number of reactive byproducts, such as reactive oxygen species (ROS).[50] Because of the very short lifetime of both singlet oxygen and free radicals, only those molecules that are in close proximity to the cell surface trigger extensive damage. Mechanisms of cell damage have been investigated for different photosensitizers, and it was established that they range from disruption of the cell membrane to inactivation of enzymes and DNA damage, depending on the type of cell and nature of the photosensitizer and its concentration that can result in cytotoxic effects affecting most living cells, including microbes.[51] Thus, many photosensitizers had been shown to possess antimicrobial properties.[52] As an example, Figure 11.1 shows the results of the scanning electron microscopy images performed by our group. In the nonirradiated control sample of *Staphylococcus aureus*, the bacterial surface is smooth, indicating that no damage has occurred to the cell structure. By contrast, the irradiated sample displays a very rough surface or a hollow shape, suggesting that severe damage has occurred. This micrograph supports the idea that cell dies due to cell membrane disruption and leakage of cell contents. The bacteria seem to be completely destroyed and the cell walls appear to be broken. The cell surface of *L. monocytogenes* is grainy, furrowed and crumbled.

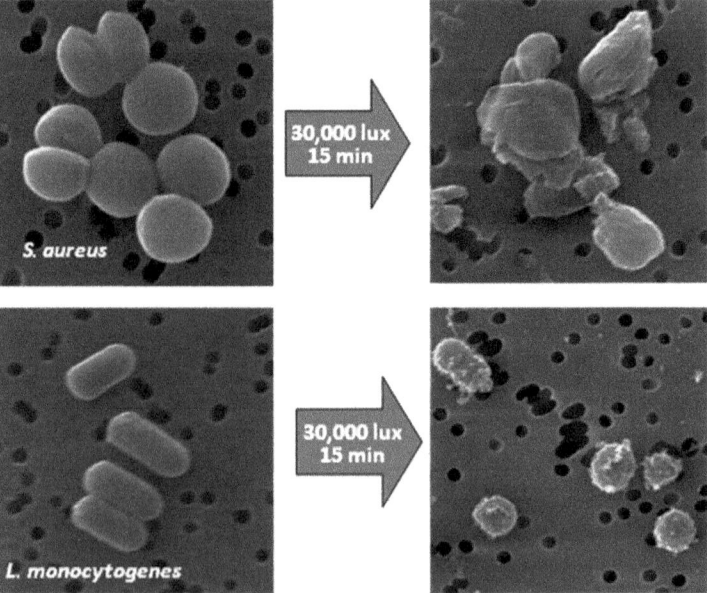

Figure 11.1   Scanning electron microscopy observations of *S. aureus* and *L. monocytogenes*.

Some bacteria are unstable and they lose their smooth, tight surface and usually look like a golf-ball. By contrast an illuminated control sample shows intact and turgid cells.

Chlorophylls are very common pigments, which give color to vegetables and several fruits. Because of their color and their physicochemical properties, they are also used as additives to food products. Their long chain of conjugated double bonds easily reacts with acid, base, oxygen and light. Upon illumination in the presence of oxygen, chlorophyllins act as photosensitizers, promoting the release of ROS including singlet oxygen.[53] As chlorophylls have a clear hydrophobic character, they can be chemically modified before being incorporated into food products, for example replacing Mg^{2+} by Cu^{2+}. The chlorophyll derivatives were found to be more stable than the original chlorophyllin, especially to acids.[54] The differences in the chemical properties (like electrical charge) of the various derivatives result in different degrees of interaction with membrane structures of the target cell.[55] Therefore, efficient formulations with different degrees of phototoxicity could potentially be fabricated to suit a wide variety of possible applications.

The natural occurrence, abundance and safe use of chlorophylls has led to their current status of generally recognized as safe (GRAS) food ingredients. Therefore, despite being extensively used as food colorants, nowadays chlorophyll precursors and chlorophyll derivatives also represent a highly interesting source of natural antimicrobials either for food preservation or photodynamic treatments in the medical field. Additionally, their ability to generate singlet oxygen even when incorporated inside polymeric matrices makes them feasible candidates for use in food packaging or other antimicrobial applications.[56]

So far, only one study has successfully developed chlorophyllin films to improve food preservation. López-Carballo *et al.*[57] incorporated chlorophyllins into gelatin film-forming solution, showing the inhibiting effect of the cast films against *S. aureus* and *L. monocytogenes*. Figure 11.2 shows the antimicrobial effect of E-140 (chlorophylls) incorporated in gelatin films over *S. aureus* after

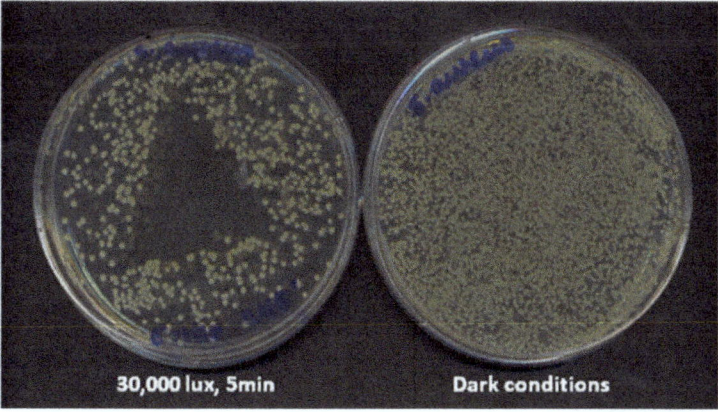

Figure 11.2   Antimicrobial film effect of E-140 incorporated in gelatin film over *S. aureus*.

30 000 lux for 5 min and also in dark conditions as a control. These films also reduced micro-organism growth in cooked frankfurters by covering them with sodium magnesium chlorophyllin-gelatin films and coating.

11.3 Formulation of Natural Extracts with Virucide Activity

The importance of foodborne diseases caused by enteric viruses is increasingly being recognized, and the World Health Organization has found that there is an upward trend in their incidence. Epidemiological evidence indicates that enteric viruses, in particular noroviruses (NV), which cause acute gastroenteritis, but also hepatitis A virus (HAV) and rotavirus (RV), are the leading cause of foodborne illness in developed countries.[58,59] In the EU, foodborne viruses (mainly human NVs) were identified to be the foremost frequently detected causative agent of foodborne outbreaks in 2011, accounting for 9.3% of the reported outbreaks.[60] Moreover, noroviruses have been listed in the top 5 highest-ranking pathogens with respect to the total cost of foodborne illness in the United States.[61]

This recognition is reflected by the attention that national and international organizations give to consider the control of foodborne viral infection in the report of the Advisory Committee on the Microbiological Safety of Food, the recent proposed guidelines for the application of food hygiene to the control of viruses for Codex Alimentarius (CX/FH/10/42/5), the scientific opinion of the European Food Safety Authority (EFSA),[62] and the expert advice on food-borne viruses for Codex Alimentarius (www.who.int/foodsafety/publications/micro/mra13/en/index.html). This latter document concluded, among other considerations, that prevention and control measures should be considered for enteric viruses in bivalve molluscan shellfish, fresh produce, or prepared foods.

The addition of antimicrobial agents into food packaging has been extensively used to control the microbiota and target specific micro-organisms to provide higher safety and enhance the quality of products. As a means of preventing recontamination with pathogens and extending the shelf-life of foods, antimicrobial packaging is one of the most promising technologies in the food industry. However, few studies have confronted the task of evaluating materials with virucidal properties in real food applications. For the first time, Martínez-Abad *et al.*[63] have evaluated active renewable packaging materials for virus control in vegetables using a norovirus surrogate. In this study, silver ions were satisfactorily incorporated into PLA films showing strong antiviral activity *in vitro*. When films were applied on food samples, antiviral activity was very much dependent on the food type and temperature. Moreover, Bright *et al.*[64] evaluated the antiviral activity of active packaging, reporting that *Feline calicivirus* (FCV) infectivity was reduced by 5 log tissue culture infection dose$_{50}$ (TCID$_{50}$)/mL when in contact with plastic coupons impregnated with 10% silver–copper zeolites.

The virucide activity of natural compounds able to be incorporated into food packaging has only been evaluated on virus suspensions (reviewed by Li *et al.*[65]); and no one has investigated their virucide activity when incorporated in food packaging. Essential oils of oregano, zataria, clove, hyssop and marjoram;[66,67] chitosan;[68,69] cranberry juice and cranberry proanthocyanidins;[70] pomegranate juice and pomegranate polyphenols;[71,72] black raspberry juice;[73] Korean red ginseng extract and ginsenosides[74] and flavonoids (myricetin, L-epicatechin, tangeretin, and naringenin)[75] have been evaluated on norovirus surrogates suspensions. Antiviral activity of *Melaleuca alternifolia* essential oil (tea tree oil) and its main components showed to be ineffective against other enteric viruses, such as polio type 1, adenovirus type 2, echovirus 9 and coxsackie virus B1.[76] Until now, only grape seed extract (GSE), and in to a lesser extent red sea weed, Iota-Lambda and Kappa carrageenans, have been evaluated on HAV with promising results, highlighting the potential of these natural compounds as an inexpensive, novel, natural alternative to reduce viral contamination and enhance food.[77,78]

Overall, the antiviral effects of the evaluated natural compounds are dependent on the virus type, titer, dose, contact time and temperature. Moreover, these studies show that foodborne viruses, typically nonenveloped viruses, are generally less sensitive to the tested natural compounds compared with bacteria and relevant medical viruses, such as herpes simplex virus or human immunodeficiency virus (Table 11.1).

Table 11.1 Efficacy of natural extract on different enteric viruses.

Essential oils (concentration)	Tested virus	Inactivation (log reduction)	Remarks	Reference
Baccharis dracunculifolia (250 µg/ml)	PV	0.7		79
Clove (1%)	FCV, MNV	3.75; 0.67		67
Hyssop (0.2%)	AdV-2, MNV	No effect		66
Marjoram (0.2%)	AdV-2, MNV	No effect		66
Mexican oregano	RV	No effect	Carvacrol alone exhibited virucide activity	80
Oregano (2%)	FCV, MNV	3.75; 1.62	Decreasing effects at lower concentrations	67
Tea tree (0.025%)	PV 1, Echo 9, Coxsackie B1, AdV-2	No effect		76
Zataria (0.1%)	FCV, MNV	4.51; 0.25		67
Other natural extracts				
Black raspberry juice (6%)	FCV, MNV	1; 1	Gallic acid or quercetin, the main phenolic compounds, did not decrease virus infectivity	73

Table 11.1 (*Continued*)

Essential oils (concentration)	Tested virus	Inactivation (log reduction)	Remarks	Reference
Carrageenans (200 µg/ml)	HAV	1		78
Chitosan (water soluble, 53 kDa)	FCV, MNV	3.4; 0.34	Increasing the MW of chitosan has no increasing antiviral effect	69
Cranberry juice (0.3 mg/ml)	FCV, MNV	5; 1.9		70
Cranberry proanthocyanidins (0.6 mg/ml)	FCV, MNV	5; 3		70
Flavonoids (1 mM)	FCV, MNV	Variable; no effect	Best inactivation rates with myricetin and l-epicatechin	75
Grape seed extract (1 mg/ml)	FCV, MNV, HAV	4.61; 1.73; 3.20		77
Korean red ginseng extract and ginsenosides (10 µg/ml)	FCV, MNV	0.83; 1.48		74
Pomegranate polyphenols (32 mg/ml)	FCV, MNV	> 5; 3.61		71

AdV: Adenovirus; Coxsackie: Coxsakiuevirus; Echo: Echovirus; FCV: Feline calicivirus; MNV: Murine norovirus; PV: Poliovirus.

11.4 Conclusions and Future Perspectives

The use of natural extracts to render polymeric materials antibacterial has significant potential applications. This is particularly the case in food-packaging applications. However, there are still a number of issues, such as low organoleptic impact, effectivity at low dosages, lack of thermal stability and volatility during processing and cost that need to be better addressed and resolved for this interesting technology to be widely used in industrial applications. Technologies are currently envisaged, such as refining the products by supercritical fluid extraction or other techniques, to enhance product efficiency and stability by, for instance, removing odor, taste and color. The understanding of the natural extracts capacity to make plastics virucidal is today an open area of research that needs to be fully addressed.

References

1. S. Burt, *Int. J. Food Microbiol.*, 2004, **94**, 223–253.
2. K. Glinel, P. Thebault, V. Humblot, C. M. Pradier and T. Jouenne, *Acta Biomater.*, 2012, **8**, 1670–1684.

3. T. Sivarooban, N. S. Hettiarachchy and M. G. Johnson, *Food Res. Int.*, 2008, **41**, 781–785.
4. H. Zhang, B. Kong, Y. L. Xiong and X. Sun, *Meat Sci.*, 2009, **81**, 686–692.
5. A. Nostro, R. Scaffaro, M. D'Arrigo, L. Botta, A. Filocamo, A. Marino and G. Bisignano, *Appl. Microbiol. Biotech.*, 2012, **96**, 1029–1038.
6. J. W. Rhim and P. K. W. Ng, *Crit. Rev. Food Sci. Nutr.*, 2007, **47**, 411–433.
7. C. A. Campos, L. N. Gerschenson and S. K. Flores, *Food Bioprocess Tech.*, 2011, **4**, 849–875.
8. I. Martínez, P. Partal, M. García-Morales, A. Guerrero and C. Gallegos, *J. Food Eng.*, 2013, **117**, 247–254.
9. M. Ahmad, S. Benjakul, P. Sumpavapol and N. P. Nirmal, *Int. J. Food Microbiol.*, 2012, **155**, 171–178.
10. D. Gómez-Martínez, P. Partal, I. Martínez, A. Guerrero and C. Gallegos, *Chem. Eng. Trans.*, 2011, **24**, 895–900.
11. S. Suganya, T. Senthil Ram, B. S. Lakshmi and V. R. Giridev, *J. Appl. Polym. Sci.*, 2011, **121**, 2893–2899.
12. C. Kriegel, K. M. Kit, D. J. McClements and J. Weiss, *J. Appl. Polym. Sci.*, 2010, **118**, 2859–2868.
13. D. Nithyakalyani, T. Ramachandran, R. Rajendran and M. Mahalakshmi, *J. Appl. Polym. Sci.*, 2013, **129**, 672–681.
14. F. Nazzaro, P. Orlando, F. Fratianni and R. Coppola, *Curr. Opin. Biotechnol.*, 2012, **23**, 182–186.
15. A. V. Podshivalov, S. Bronnikov, V. V. Zuev, T. Jiamrungraksa and S. Charuchinda, *J. Microencapsul.*, 2013, **30**, 198–203.
16. G. Q. Liu, L. L. Zhang, K. Zong, A. M. Wang and X. F. Yu, *Food Sci. Technol. Res.*, 2012, **18**, 695–704.
17. A. Esmaeili and B. Saremnia, *Ind. Crops Prod.*, 2012, **37**, 259–263.
18. C. Gomes, R. G. Moreira and E. Castell-Perez, *J. Food Sci.*, 2011, **76**, N16–N24.
19. A. Arana-Sánchez, M. Estarrón-Espinosa, E. N. Obledo-Vázquez, E. Padilla-Camberos, R. Silva-Vázquez and E. Lugo-Cervantes, *Lett. Appl. Microbiol.*, 2010, **50**, 585–590.
20. D. Plackett, A. Ghanbari-Siahkali and L. Szente, *J. Appl. Polym. Sci.*, 2007, **105**, 2850–2857.
21. M. J. Piercey, G. Mazzanti, S. M. Budge, P. J. Delaquis, A. T. Paulson and L. T. Hansen, *Food Microbiol.*, 2012, **30**, 213–218.
22. C. Gomes, R. G. Moreira and E. Castell-Perez, *J. Food Sci.*, 2011, **76**, E479–E488.
23. C. Samperio, R. Boyer, W. N. Eigel, K. W. Holland, J. S. McKinney, S. F. O'Keefe, R. Smith and J. E. Marcy, *J. Agric. Food Chem.*, 2010, **58**, 12950–12956.
24. S. Shojaee-Aliabadi, H. Hosseini, M. A. Mohammadifar, A. Mohammadi, M. Ghasemlou, S. M. Ojagh, S. M. Hosseini and R. Khaksar, *Int. J. Biol. Macromol.*, 2013, **52**, 116–124.

25. B. Giménez, M. C. Gómez-Guillén, M. E. López-Caballero, J. Gómez-Estaca and P. Montero, *Food Hydrocolloid.*, 2012, **27**, 475–486.
26. M. D. Sanchez-Garcia, A. Lopez-Rubio and J. M. Lagaron, *Trends Food Sci. Technol.*, 2010, **21**, 528–536.
27. V. Coma, *Meat Sci.*, 2008, **78**, 90–103.
28. S. Inouye, S. Abe, H. Yamaguchi and M. Asakura, *Int. J. Aromather.*, 2003, **13**, 33–41.
29. P. López, C. Sánchez, R. Batlle and C. Nerin, *J. Agric. Food Chem.*, 2007, **55**, 8814–8824.
30. K. Laird and C. Phillips, *Lett. Appl. Microbiol.*, 2012, **54**, 169–174.
31. C. M. Guinane, P. D. Cotter, C. Hill and R. P. Ross, *J. Appl. Microbiol.*, 2005, **98**, 1316–1325.
32. N. P. Guerra, C. L. Macias, A. T. Agrasar and L. P. Castro, *Lett. Appl. Microbiol.*, 2005, **40**, 106–110.
33. R. Bauer and L. M. T. Dicks, *Int. J. Food Microbiol.*, 2005, **101**, 201–216.
34. A. Rajkovic, M. Uyttendaele, T. Courtens and J. Debevere, *Food Microbiol.*, 2005, **22**, 189–197.
35. K. T. Chung, J. S. Dickson and J. D. Crouse, *Appl. Environ. Microbiol.*, 1989, **55**, 1329–1333.
36. I. Geornaras, K. E. Belk, J. A. Scanga, P. A. Kendall, G. C. Smith and J. N. Sofos, *J. Food Prot.*, 2005, **68**, 991–998.
37. E. Rodriguez, J. Calzada, J. L. Arqués, J. M. Rodriguez, M. Núñez and M. Medina, *Int. Dairy J.*, 2005, **15**, 51–57.
38. J. Lozo, Vukasinovic, I. Strahinic and L. Topisirovic, *J. Food Prot.*, 2004, **67**, 2727–2734.
39. M. Al-Holy, M. Lin and B. Rasco, *J. Food Prot.*, 2005, **68**, 512–520.
40. T. Aymerich, A. Jofré, M. Garraga and M. Hugas, *J. Food Prot.*, 2005, **68**, 173–177.
41. G. Mauriello, E. de Luca, A. la Storia, F. Villani and D. Ercolini, *Lett. Appl. Microbiol.*, 2005, **41**, 464–469.
42. G. Mauriello, D. Ercolini, A. la Storia, A. Casaburi and F. Villani, *J. Appl. Microbiol.*, 2004, **97**, 314–322.
43. K. E. McCormick, I. Y. Han, J. C. Acton, B. W. Sheldon and P. L. Dawson, *J. Food Sci.*, 2005, **70**, 52–57.
44. P. L. Dawson, D. E. Hirt, J. R. Rieck, J. C. Acton and A. Bondi, *Food Res. Int.*, 2003, **36**, 959–968.
45. R. Iseppi, S. de Niederhäusern, I. Anacarso, P. Messi, C. Sabia, F. Pilati, M. Toselli, M. D. Esposti and M. Bondi, *Soft Matter*, 2011, **7**, 8542–8548.
46. L. Sanchez-Gonzalez, J. I. Quintero and A. Chiralt, *Food Hydrocolloid.*, 2013, **33**, 92–98.
47. R. Scaffaro, L. Botta, S. Marineo and A. M. Puglia, *J. Food Prot.*, 2011, **74**, 1137–1143.
48. B. Marcos, T. Aymerich, M. Garriga and J. Arnau, *Food Control*, 2013, **30**, 325–330.
49. N. P. Guerra, A. B. Araujo, A. M. Barrera and A. Agrasar, *J. Food Prot.*, 2005, **68**, 1012–1019.

50. M. Krouit, R. Granet, P. Branland, B. Verneuil and P. Krausz, *Bioorg. Med. Chem. Lett.*, 2006, **16**, 1651–1655.
51. N. A. Romanova, L. Y. Brovko, L. Moore, E. Pometun, A. P. Savitsky, N. N. Ugarova and M. W. Griffiths, *Appl. Environ. Microbiol.*, 2003, **69**, 6393–6398.
52. H. Wang, L. Lu, S. Zhu, Y. Li and W. Cai, *Curr. Microbiol.*, 2006, **52**, 1–5.
53. B. Schoefs, *Trends Food Sci. Technol.*, 2002, **13**, 361–371.
54. B. Schoefs, *Trends Anal. Chem.*, 2003, **22**, 335–339.
55. M. Kreitner, K. H. Wagner, G. Alth, R. Ebermann, H. Foißy and I. Elmadfa, *Food Control*, 2001, **12**, 529–533.
56. J. Sherrill, S. Michielsen and I. Stojiljkovic, *J. Polym. Sci., Part A: Polym. Chem.*, 2002, **41**, 41–47.
57. G. López-Carballo, P. Hernández-Muñoz, R. Gavara and M. J. Ocio, *Int. J. Food Microbiol.*, 2008, **126**, 65–70.
58. M. Koopmans and E. Duizer, *Int. J. Food Microbiol.*, 2004, **90**, 23–41.
59. E. Scallan, R. M. Hoekstra, F. J. Angulo, R. V. Tauxe, M. A. Widdowson, S. L. Roy, J. L. Jones and P. M. Griffin, *Emerg. Infect. Dis.*, 2011, **17**, 7–22.
60. European Food Safety Authority, *EFSA J.*, 2013, **11**, 3129.
61. R. L. Scharff, *J. Food Prot.*, 2012, **75**, 123–131.
62. EFSA Panel on Biological Hazards (BIOHAZ)., *EFSA J.*, 2011, **9**, 2190.
63. A. Martínez-Abad, J. M. Lagaron, M. J. Ocio and G. Sánchez, *Int. J. Food Microbiol.*, 2013, **162**, 89–94.
64. K. Bright, E. Sicairos-Ruelas, P. Gundy and C. Gerba, *Food Environ. Virol.*, 2009, **1**, 37–41.
65. D. Li, L. Baert and M. Uyttendaele, *Food Microbiol.*, 2013, **35**, 1–9.
66. K. Kovac, M. Diez-Valcarce, P. Raspor, M. Hernández and D. Rodríguez-Lázaro, *Food Environ. Virol.*, 2012, **4**, 209–212.
67. P. Elizaquível, M. Azizkhani, R. Aznar and G. Sánchez, *Food Control*, 2013, **32**, 275–278.
68. X. Su, S. Zivanovic and D. H. D'Souza, *J. Food Prot.*, 2009, **72**, 2623–2628.
69. R. Davis, S. Zivanovic, D. H. D'Souza and P. M. Davidson, *Food Microbiol.*, 2012, **32**, 57–62.
70. X. Su, A. B. Howell and D. H. D'Souza, *Food Microbiol.*, 2010, **27**, 985–991.
71. X. Su, M. Y. Sangster and D. H. D'Souza, *Foodborne Pathog. Dis.*, 2010, **7**, 1473–1479.
72. X. Su, M. Y. Sangster and D. H. D'Souza, *Foodborne Pathog. Dis.*, 2011, **8**, 1177–1183.
73. M. Oh, S. Y. Bae, J. H. Lee, K. J. Cho and K. H. Kim., *Foodborne Pathog. Dis.*, 2012, **9**, 915–921.
74. M. H. L. Lee, B. H. Lee, J. Y. Jung, D. S. Cheon, K. T. Kim and C. Choi, *J. Ginseng Res.*, 2011, **35**, 429–435.
75. X. Su and D. H. D'Souza, *Food Environ. Virol.*, 2013, **5**, 97–102.
76. A. Garozzo, R. Timpanaro, B. Bisignano, P. M. Furneri, G. Bisignano and A. Castro, *Lett. Appl. Microbiol.*, 2009, **49**, 806–808.

77. X. Su and D. H. D'Souza, *Appl. Environ. Microbiol.*, 2011, **77**, 3982–3987.
78. S. Girond, J. M. Crance, H. Van Cuyck-Gandre, J. Renaudet and R. Deloince, *Res. Virol.*, 1991, **142**, 261–270.
79. M. C. Búfalo, A. S. Figueiredo, J. P. de Sousa, J. M. Candeias, J. K. Bastos and J. M. Sforcin, *J. Appl. Microbiol.*, 2009, **107**, 1669–1680.
80. M. R. Pilau, S. H. Alves, R. Weiblen, S. Arenhart, A. Cueto and L. T. Lovato, *Braz. J. Microbiol.*, 2011, **42**, 1616–1624.

CHAPTER 12

Carbon-Based Polymer Nanocomposites: From Material Preparation to Antimicrobial Applications

CATHERINE M. SANTOS,[a] MARIA CELESTE R. TRIA,[b] EDWARD FOSTER,[b] RIGOBERTO C. ADVINCULA[c] AND DEBORA F. RODRIGUES*[a]

[a] Department of Civil and Environmental Engineering, University of Houston, Houston, TX 77204-5003, USA; [b] Department of Chemistry, University of Houston, Houston, TX 77204-5003, USA; [c] Department of Macromolecular Science and Engineering, Case Western Reserve University, Cleveland, OH 44106, USA
*Email: dfrigiro@central.uh.edu

12.1 Introduction

Every person in their life time will, at least once, use or be in contact with a medical device. These devices range from simple catheters to life-saving implants, such as artificial hearts. Although the use of these devices has led to a better quality of life and longer patient survival, their long-term use is still an issue due to microbial contamination and colonization of their surfaces, which can cause serious infections in patients. Thus, over the past few years research efforts have been focused in the development of coating materials that can prevent microbial colonization and growth on these devices.[1–4]

RSC Polymer Chemistry Series No. 10
Polymeric Materials with Antimicrobial Activity: From Synthesis to Applications
Edited by Alexandra Muñoz-Bonilla, María L. Cerrada and Marta Fernández-García
© The Royal Society of Chemistry 2014
Published by the Royal Society of Chemistry, www.rsc.org

Fouling, or the adhesion of nonspecific biomaterials on surfaces, is a major bottleneck on the long-term use of biomedical devices. The adhesion of proteins, for example, can compromise the sensitivity of *in vitro* diagnoses[5] or it can lead to undesirable consequences including thrombus formation or fibrosis and scar tissue formation in *in vivo* applications.[6] Additionally, adhesion of bacteria onto surfaces generally leads to biofilm formation. Biofilm infections are notoriously difficult to resolve and they commonly manifest as chronic or recurrent infections. The majority of the hospital acquired infections (nosocomial), about 60–70%, are associated with bacterial contamination of implanted medical devices.[7,8] In the United States alone, approximately 2 million hospital acquired infections cost nearly $11 billion annually. Exposure to invasive medical devices is one of the most important risk factors for nosocomial infections, responsible for at least 45% of the cases. Hence, there is an urgent need to develop novel surface coatings to prevent pathogenic organisms from attaching to surfaces and causing infections.

Microbial contamination of surfaces that leads to biofilm formation occurs in several stages: The first stage involves the initial attachment of free-floating/swimming bacteria to form a single layer onto the surface by producing organic compounds that are polysaccharides or glycoproteins. If the association between the bacterium and its substrate surface persists long enough, there will be an irreversible adhesion of bacterial cells to the surfaces with the production of extracellular polymers (EPS) by the bacteria.[8,9] These EPS trap, not only more bacteria, but also diverse materials, such as organic matter, dead cells, particles, to add to the diversity and complexity of the biofilm. Finally, small parts of the biofilm can break off and release planktonic bacteria that can invade new, clean surfaces at distant sites. Biofilm-associated micro-organisms are significantly less susceptible to antibiotics and host defenses than planktonic (freely suspended) forms of the same microorganisms.

Researchers have focused on two key approaches for addressing the overarching problems in bacterial colonization: to develop biomaterials and coatings that can either prevent the adhesion of biocontaminants[3,6,10–16] or to kill pathogenic microorganisms.[17–19] The first method involves functionalization of surfaces with molecules, such as oligo(ethylene glycol) groups (OEG) and poly(ethylene glycol) (PEG), capable of inhibiting protein and bacterial adhesion into surfaces.[3,6,10–15] Although the resulting surfaces are effective in preventing microbial adhesion, they do not inactivate microbes, which is critical for preventing bacterial infections or contaminations. The second approach includes the incorporation of antimicrobial agents such as antibiotics,[20–22] silver,[23–25] cationic peptides,[26] and antimicrobial peptides.[19,27] Antibiotics have been the most commonly used antibacterial agents because of their strong antibacterial efficacy; however, problems related to the development of microbial resistance has been their major drawback.[9,28,29] Ionic silver is also considered a potent antimicrobial.[30] However, the extended exposure of tissues to silver ions have been shown to lead to a number of issues, such as argyria, which produces grayish tinges in the skin; or accumulation of silver in the liver and kidneys due to thiol-rich proteins like glutathione, making these proteins ineffective in these organs.[31,32] Polycationic peptides and photoactive species have been promising as antimicrobials, but also offer some limitations. Antimicrobial peptides (AMPs) for example can lose antimicrobial effectiveness upon surface tethering, while the application of photoactive agents is limited by the intensity of radiation.

One of the recent efforts in addressing this challenge lies in exploring carbon-based nanomaterials (CBNs) such as carbon nanotubes (CNT), graphene, and graphene oxide (GO) for potential antimicrobial applications. All of these carbonaceous nanomaterials possess: (1) unique size, shape, and structure;[33–36] (2) molecular interaction and sorption properties;[37–39] and (3) attractive optical, electrical, thermal, and mechanical properties.[40–46] More importantly, these CBNs were reported to be antibacterial.[1,47–50] Despite the many advantages that these CBNs have to offer, their practical applications in pristine forms have been very limited due to their hydrophobic nature, which reduces their dispersion and processability. In order to overcome this problem, researchers have developed new methods to incorporate CBNs, as fillers, into polymers to produce carbon-based polymer nanocomposites (CBPNs).[51,52] These CBPNs have been shown to be promising materials for antimicrobial applications with very low mammalian cell toxicity.[53,54]

This chapter is focused on the recent developments and strategies relevant to the preparation of CBPNs and its antimicrobial applications. More specifically, this chapter will describe the preparation of nanocomposites with the following carbonaceous nanomaterials: graphene, GO, and CNTs. We discuss various approaches that were done to prepare CBPNs as bulk solutions and as thin-film coatings. In addition, we present the different studies investigating the antibacterial applications of these materials.

12.2 Preparation of Carbon-Based Polymer Nanocomposites

12.2.1 Bulk-Material Preparation

The dispersion of CBNs in polymer matrices is a critical issue in the preparation of bulk polymeric nanocomposites. Better reinforcing properties are achieved if the CBNs do not form aggregates. Therefore, the CBNs must be well dispersed in polymer matrices. Currently, there are several methods used to improve the dispersion of CBNs in polymer matrices such as solution mixing, melt blending, and *in situ* polymerization method.

12.2.1.1 Solution Mixing

Bulk solution mixing methods generally involve the mixing of CBNs with a polymer, by sonication, shear mixing, or stirring. The sonication method has been shown to uniformly disperse carbon-based nanofillers in polymer matrices. The mechanical effects of ultrasonic waves come from the rapid and violent collapse of cavitation bubbles resulting in more uniformly distributed carbon-based nanofillers.[55,56] Once properly mixed, the resulting solution can be either precipitated or the solvent can be evaporated.[57]

The simplicity of solution mixing has made it a popular method for incorporating GO into a variety of polymer materials for bulk nanocomposites fabrication. For example, Zaman *et al.* have recently used solution mixing of modified GO platelets with THF (1 wt.%) followed by ultrasonication.[58] A desired amount of epoxy resin diglycidyl ether of bisphenol A (DGEBA, Araldite-F) was then dissolved in acetone (50 wt.%) using a similar method. The two batches were then mechanically mixed, followed by ultrasonication for 90 min. Acetone and THF were then evaporated, upon which a "hardener" (Jeffamine D 230 (J230)) was added and mixed for 1 h.[58] Characterization of these nanocomposites revealed well-dispersed GO platelets, with improved mechanical properties.[58] Similarly, Mejias Carpio *et al.* recently created poly(vinyl-*N*-carbazole)–GO (PVK–GO) nanocomposites.[59] As seen in Figure 12.1, the PVK–GO nanocomposites were fabricated by first sonicating GO for 1 h in deionized water.[59] Then, the GO stock solution and a diluted solution of PVK was slowly mixed to obtain a PVK/GO ratio of 97/3 wt.%, followed by further sonication for 30 min. In this study, they found that the addition of the PVK to the GO enhanced the antimicrobial property of GO due to its increased dispersion of GO into the polymer matrix.[59] Other examples of materials used to encapsulate GO using solution mixing include polystyrene (PS),[60] polycarbonate,[61] polyacrylamide,[62] polyimides,[63] poly(vinyl alcohol) (PVA),[64–67] poly(allylamine), and poly(methyl methacrylate) (PMMA).[68]

PVK has also been used in the development of carbon-nanotube polymeric nanocomposites (CNT-PVK). For instance, Ahmed and coworkers employed the solution-mixing method to prepare nanocomposites containing PVK

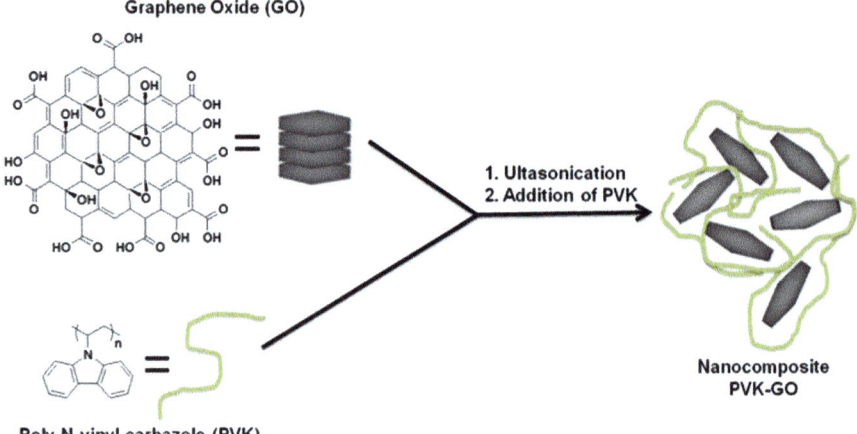

Figure 12.1   Fabrication of PVK-GO nanocomposites.
(Adapted from ref. 53).

(97 wt.%) and single-walled carbon nanotubes (SWNT) (3 wt.%). These nanocomposites were further investigated for their antimicrobial properties as suspensions in water and film coatings.[69] The toxic effects of the PVK-SWNT nanocomposite were determined for planktonic cells and biofilms of *Escherichia coli* (*E. coli*) and *Bacillus subtilis* (*B. subtilis*). The results showed that the PVK/SWNT nanocomposite had antibacterial activity on planktonic cells and biofilms. This study established that improved dispersion of SWNTs in aqueous solutions in the presence of PVK enhances the antimicrobial effects of SWNTs.[69]

Another example of CBNPs preparation was developed by Chen and coworkers.[70] In their investigation, a simple-solution mixing method to improve the dispersion and uniformity of CNTs in a polycarbonate solution using sonication was performed.[70] Similarly, Cho and coworkers prepared polyurethane (PU) and multiwalled carbon nanotubes (MWNTs) (20 wt.%) composites with better CNT dispersion.[71] In their work, the weight fractions of carboxylate-functionalized MWNTs were first dispersed in dimethylformamide solution under sonication for 1 h. PU was then added into the solution, stirred for 1 h, and then sonicated for 1 h.[71]

Using surfactants in combination with solution mixing is another route to disperse CBNs.[72,73] For example, amphiphilic palmitic acid has been shown to improve dispersion of CNTs in an epoxy matrix.[74] The hypothesis for this successful dispersion is that the hydrophobic group of the palmitic acid adheres to the carbon-based material, while the hydrophilic group induces electrostatic repulsion, reducing aggregation.[74] Similarly, poly(oxyethylene-8-lauryl) has also been shown to enhance the uniformity and interaction between epoxy-based resins and MWNTs.[75] The use of cosolvents can also tune the level of uniformity of the carbon-based material in a polymer matrix. For example, the use of trifluoroacetic acid as a cosolvent for the dispersion of MWNTs in poly(3-hexylthiophene), and PMMA, has also been reported.[76] Other examples of CNT-based nanocomposites fabricated in similar manners include PU,[77] PS,[78–80] epoxy,[81,82] PVA,[83] polyacrylonitrile,[84] and polyethylene (PE).[85]

12.2.1.2 Melt Blending

In melt mixing, a polymer melt and carbon-based nanofillers are combined and mixed under high-shear conditions, which prevents the carbon-based nanomaterials from forming aggregates. Since melt mixing does not require solvents, it is often considered more economical and compatible with industrial practices, especially when compared to solution mixing.[86] It should be mentioned, however, that research has suggested that melt mixing does not provide an equal degree of carbon-based nanofiller dispersion as solvent mixing.[87] Similarly, the thermal instability of most chemically modified graphene (*e.g.* GO), have limited the use of melt blending to a few studies using thermally reduced graphene. For example, Zhang *et al.* have used thermally expanded graphite oxide (TEGO) materials as fillers in polymeric nanocomposites.[88] TEGO is a carbon-based nanomaterial that is created by exfoliating and reducing GO by rapid heating.[89] This exfoliation method forces the GO sheets apart due to the increase in internal pressure.[90,91] After the exfoliation Zhang and coworkers, then directly add TEGO into an extruder dispersing the TEGO into a polymer matrix without the use of solvents or surfactants.[88]

Dispersion of CNTs using melt blending, on the other hand, has been more promising than GO. For example, Zhang and coworkers prepared nylon-6 CBPNs containing 1 wt.% MWNTs as filler, using melt blending methods.[92] Using SEM, it was shown that the MWNTs were well dispersed in the nylon-6 matrix, and presented enhanced mechanical properties. More recently, Xing and coworkers reported the fabrication of well-dispersed MWNTs fillers in PMMA-MWNT nanocomposites using the melt-mixing method in the presence of a room-temperature ionic liquid, *i.e.* 1-butyl-3-methylimidazolium hexa-fluorophosphate.[93] The preparation of the nanocomposites was done by first mixing the MWNTs with 1-butyl-3-methylimidazolium hexafluorophosphate using an agate mortar for 15 min at room temperature to prepare ionic liquid-modified (IL-modified) MWNTs. The nanocomposites were then prepared by direct mixing of the obtained IL-MWNTs with PMMA in a batch mixer (Toyoseiki Co. KF70V) with a twin screw using a rotation speed of 100 rpm at 170 °C for 5 min.[93] The MWNT loading content was fixed at 2 wt.% based on the polymer matrix. After melt mixing, all samples were hot pressed at 180 °C into films 500 μm thick, followed by cooling to room temperature to get the final PMMA-MWNT nanocomposite.[93] Similarly, Bocchini *et al.* using melt-blending methods, fabricated MWNTs-based nanocomposites using lin-ear low-density polyethylene (LLDPE) as polymeric material.[94] Here, they found that, when compared to pure LLDPE materials, LLDPE/MWNTs nanocomposite materials showed delayed thermal and oxidative degradation of the composite.

Similarly, Wu fabricated poly(butylene adipate-*co*-terephthalate) (PBAT) composites containing MWNTs using a melt-blending process.[95] Here, acrylic acid-grafted PBAT (PBAT-*g*-AA) and hydroxyl-functionalized MWNTs (MWNT-OH) were used to improve the compatibility and dispersibility of the MWNTs within the PBAT polymeric matrix. The functionalized PBAT-*g*-AA/ MWNT-OH composite showed enhanced antibacterial and antistatic properties due to the formation of ester bonds from the condensation of the carboxylic acid groups of PBAT-*g*-AA with the hydroxyl groups of MWNT-OH.[95] It was found that the optimal proportion of MWNT-OH in the composites was 3 wt.%, where additional loading amounts led to incompatibility between the organic and inorganic phases. In another report, Wu also evaluated the electrical conductivity properties of the nanocomposite material.[96] In this case, functionalized poly(butylene terephthalate) (PBT) PBT-*g*-AA/MWNTs-OH nanocomposites showed enhanced electrical conductivity due to the formation of ester bonds from the condensation of the carboxylic acid groups of PBAT-*g*-AA with the hydroxyl groups of MWNTs-OH.[96] Other examples of melt mixing have also been used for CNT-based nanocomposites *e.g.* polypropylene,[97,98] poly-carbonate,[99,100] PMMA,[101–103] polyoxymethylene,[104] and polyimide.[105]

12.2.1.3 In Situ *Polymerization*

Arguably one of the most used techniques to prepare CBPNs is through *in situ* polymerization methods. *In situ* polymerization methods typically involve the mixing/dispersing of fillers (*e.g.* GO, MWNT, *etc.*) in a solution of monomers, followed by polymerization in the presence of the dispersed fillers. Due to the direct polymerization of intercalated monomers with the nanofillers, the *in situ* polymerization method allows increasing the dispersion of the CBNs and formation of a strong interaction between the polymer matrices and the nanofillers.[106] For example, *in situ* polymerization has been demonstrated for several GO composite systems, including poly(vinyl acetate)[107] and poly(aniline) (PANI).[108] An *in situ* polymerization method has also been used for the preparation of polyurethane–carbon nanotubes (PU-CNT) nanocomposites. For example, PU-MWNT nanocomposites have been synthesized by *in situ* polymerization methods by first adding a calculated amount of functionalized MWNT and 1,4-butanediol to a PU solution.[109] Another method involved dispersion of desired amounts of MWNTs in poly(ε-caprolactone)diol, followed by addition of 4,4′-methylene bis(phenylisocyanate).[109] The chain extender 1,4-butanediol was added to this polymer and the final PU-MWNTs nanocomposite contained well-dispersed MWNTs.[109]

Additionally, covalent tethering of premade polymers to carbon-based nanomaterials has also been performed. Although it is difficult to covalently tether polymeric materials to pristine carbon-based nanomaterials, GO platelets have reactive functional groups that facilitate the tethering of covalent bonds between GO platelets and polymers. For instance, Hu *et al.* created 10 wt.% GO/polyimide nanocomposites *via in situ* polymerization.[110] These nanocomposites were fabricated by dispersing GO in a solution of N,N-dimethylacetamide using ultrasonication for 1 h, and subsequently adding 4,4′-diaminodiphenyl ether. The solution was then transferred and pyromellitic dianhydride was added. After which, the composite films were imidized at 350 °C for 1 h under vacuum and GO–polyimide nanocomposite materials were formed.[110] Similarly, Hu *et al.* synthesized MWNT/polyimide nanocomposites by *in situ* polymerization in the presence of acylated MWNTs.[111] Here, MWNTs were first functionalized with acyl groups, and then reacted with 3,3,4,4-biphenyltetracarboxylic dianhydride. The final MWNT-polyimide nanocomposite films were obtained by imidization of MWNT-poly(amic acid).[111]

Another *in situ* polymerization method to produce covalently bonded carbon-based nanocomposite materials is to perform surface-initiated polymerization (SIP) from functionalized GO. This method is facilitated by first covalently tethering polymer initiators to the functional groups of the GO surface. Once the GO has a polymerizable initiator covalently attached, SIP can be performed. For example, as seen in Figure 12.2, Lee *et al.* covalently tethered atom transfer radical polymerization (ATRP) initiators *via* an esterification reaction of the alcohol functional groups present on the GO surface and α-bromoisobutyryl bromide.[112] Upon performing ATRP using either, styrene,

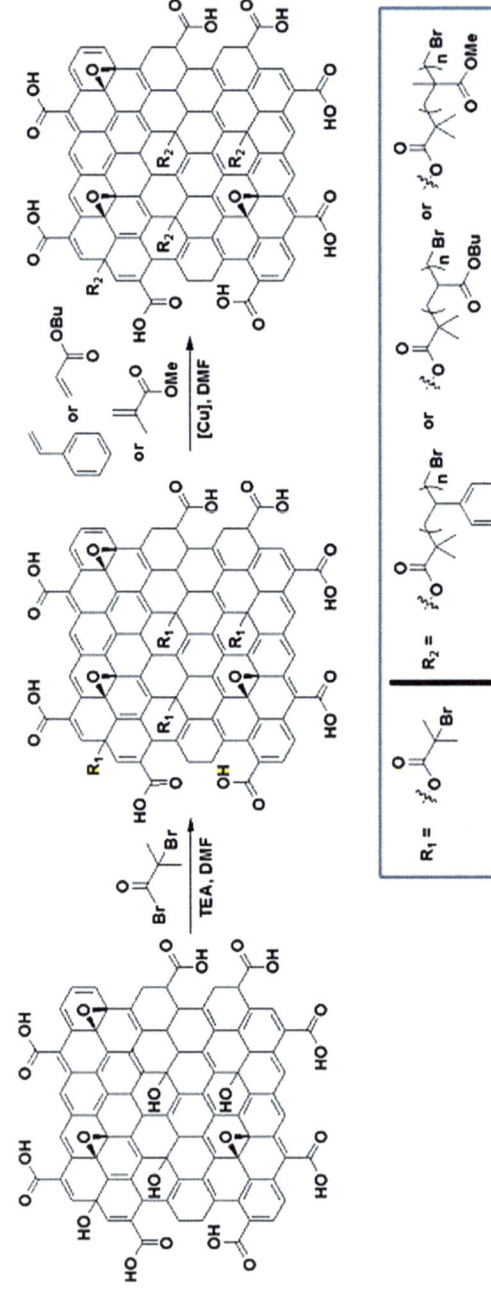

Figure 12.2 Covalent tethering of ATRP initiator to the surface of GO followed by ATRP polymerization of styrene, methyl methacrylate, or butyl acrylate. (Adapted from ref. 106).

methyl methacrylate, or butyl acrylate monomers, polymer brushes were grown in a controlled fashion from the surface of the functionalized GO, producing GO-based polymeric nanocomposites. Water-dispersible graphene with temperature-responsive polymeric surfaces have also been produced by grafting poly(N-isopropylacrylamide) (PNIPAM) from graphene *via* surface-initiated ATRP (SI-ATRP).[113] This was done by first, functionalizing the graphene surface with aminophenol groups by diazonium reaction in water. From the functionalized GO surfaces, α-bromoisobutyryl bromide groups were covalently attached to the phenol-functionalized graphene surface by esterification of α-bromoisobutyryl bromide, followed by SI-ATRP of NIPAM. Similarly SI-ATRP has also been used in conjunction with heterogeneous blending of polymer-functionalized GO in matrices composed of conducting polymers *i.e.* poly(3-hexylthiophene) (P3HT).[114] This polymer has been widely studied as a conducting material. The incorporation of carbon-based nanomaterials into conducting polymers enhances the optoelectronic properties of the device.[114] Antibacterial polyester resin as a matrix and poly[2-(dimethylamino)ethyl methacrylate] (PDMAEMA)-functionalized MWNT nanocomposites as a filler have also been prepared *via in situ* polymerization.[115] The mechanical properties of the prepared unsaturated polyester glass mat sheets were estimated by the simplified procedure designed in the laboratory.[115]

The facial method to produce carbon-based nanocomposites *via in situ* SIP has also been employed using reversible-addition fragmentation chain transfer (RAFT) radical polymerization. For example, Zhang and coworkers, covalently tethered a chain-transfer agent (CTA) to GO that served as a RAFT agent for growing PVK directly from the GO surface.[116] The PVK polymer covalently grafted onto GO had a molecular weight (M_n) of 8.05×10^3, and a polydispersity index (PDI) of 1.43.[116] Similarly, when compared to pure GO, the resulting GO-PVK nanocomposite material showed enhanced solubility in organic solvents. Likewise, combinational chemistry involving both the copper-catalyzed azide alkyne cycloaddition (click reaction) and surface-initiated RAFT (SI-RAFT) has also been performed. For example, as seen in Figure 12.3, Yang *et al.* prepared RAFT chain transfer agent (CTA)-functionalized surfaces by first synthesizing reduced graphene oxide sheets by thermal reduction and modified with a diazonium salt/propargyl *p*-aminobenzoate molecule.[117] Subsequently, an azide-containing RAFT CTA was then covalently linked to the terminal alkyne moiety *via* the click reaction and then PNIPAM was prepared *via* RAFT polymerization. Nanosized PNIPAM domains on reduced graphene oxide sheets were observed using TEM.[117] Differential scanning calorimetry (DSC) results indicated that in aqueous solution PNIPAM on reduced graphene oxide sheets presented a lower critical solution temperature (LCST) at 33.2 °C.[117]

12.2.2 Thin-Film Fabrication

Bulk composites of polymer and carbon nanomaterials have been prepared and reported to show exceptional properties. However, the application of these

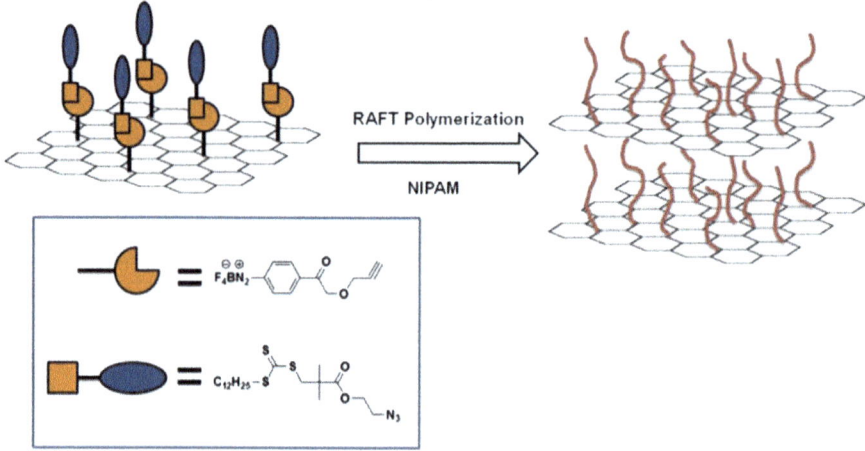

Figure 12.3   RAFT polymerization of NIPAM from click functionalized GO surfaces.
(Adapted from ref. 117).

nanocomposites into thin films is also fundamental to make them suitable for
even wider applications. The incorporation of carbon nanomaterials into a
polymer matrix facilitates better processability in making thin-film coatings.
The polymer may act as a binder, viscosifier, or template to hold, evenly spread,
or assemble the carbon nanomaterials onto interfaces. This next section discusses the reported methods in fabricating thin films of carbon-based polymer
nanocomposites on surfaces, mostly highlighting those that are used for
making antimicrobial coatings.

12.2.2.1   Casting

Among the procedures to prepare coatings of CBPNs, casting from solution
dispersions is the most common and facile route. It can be done *via* spin casting,
drop casting, dip coating, or blade coating, which can vary the thickness or
uniformity of the coating. Generally, a dispersion of the carbon material and
the desired polymer is prepared prior to its casting on a substrate
(Figure 12.4a). The cast nanocomposite is then air-dried or baked to remove
the residual solvent. Casting is the most convenient method for characterization and evaluation of the nanocomposite's properties in thin films. For
electronic applications of these nanocomposites, spin casting is conventionally
used to prepare uniform layers with controlled thickness. As an example,
functionalized graphene sheet in polystyrene as the host material can be prepared as a semiconducting film with an ambipolar field effect.[118] In this case,
the thin-film preparation can be done by spin casting a well-dispersed solution
of functionalized graphene-PS composite with dimethylhydrazine as the
reductant. The film can then be baked to remove the residual solvent and
reductant.

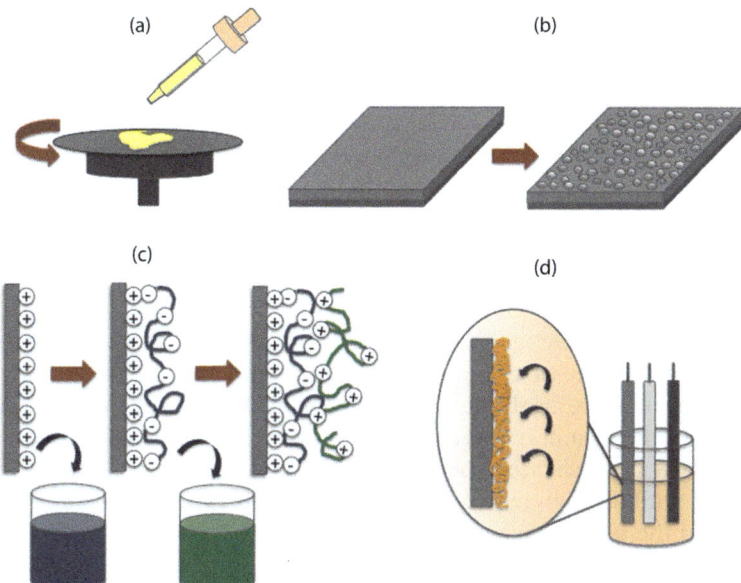

Figure 12.4 CBPNs thin-film fabrication techniques. (a) Casting (b) Wet-phase inversion, (c) Layer-by-layer assembly, and (d) Electrodeposition.

As an antimicrobial thin film, there are only a handful of reports that are based on polymer–carbon nanomaterial composite casting methods. Composites are cast either on a petri dish or a glass plate to form a film that could be assessed for antimicrobial activity. In one of the studies, carbon nanotubes have been dispersed in the common biomedical polymer poly(lactic-*co*-glycolic acid) (PLGA) to produce an antimicrobial material. This dispersion was both dip coated and spin coated on microscope slides where the hysteresis value was found to be significantly lower for spin-coated films, signifying better homogeneity throughout the surface.[119] Graphene was also mixed with the biocompatible and biodegradable chitosan polymer that produced an antibacterial film with enhanced thermal and mechanical properties.[120] The graphene–chitosan nanocomposite was cast on a petri dish for bactericidal test while free-standing films were also made from the dried cast films to test the mechanical strength of the material. GO sheets with lysozyme, a natural antibacterial protein, were also prepared *via* a solution-cast method.[121] The solution-cast mixture underwent a self-assembly process that formed a multilayered ordered structure with protein molecules inter-calated between adjacent layers of the GO sheets. In this study, it was shown that the graphene oxide sheets have the mechanical reinforcement within the film, while the lysozyme is responsible for the antimicrobial activity of the nanocomposite film.

12.2.2.2   Wet-Phase Inversion

Wet-phase inversion is very similar to the casting process except for the formation of phase-separated films that typically produces membranes (Figure 12.4b). Phase separation is induced by immersing a casted polymer in a solvent into a nonsolvent system. It could also be done by casting a dissolved polymer in a combination of solvent/nonsolvent mixture with a ratio that is just before the cloud point.

Wiesner and coworkers reported the incorporation of semidispersed MWNTs in polysulfone ultrafiltration membrane.[122] A homogeneous dispersion was prepared by adding poly(vinyl pyrrolidone) (PVP) as the porogen and polysulfone in the homogenized MWNTs in *N*-methyl-2-pyrrolidone (NMP). The mixture was then cast on a glass plate and was immersed into a water bath that served as the nonsolvent for the polymer. The morphology or the permeability of the membrane were not significantly disrupted by the addition of the MWNTs but increased the roughness. However, the membrane did not show any antibacterial activity because the MWNTs were only partially dispersed and the polymer was possibly wrapping the MWNTs, which decreased their bioavailability.

The wet-phase inversion techniques were also used by Liu and coworkers to prepare a support layer for thin-film nanocomposite membranes.[123] In the fabrication of these membranes, polysulfone served as the polymer matrix, PVP as the porogen, and MWNTs as the nanofillers.

12.2.2.3   Layer-by-Layer Assembly

The layer-by-layer (LbL) technique is another known method to prepare coatings on surfaces. This process involves the alternating adsorption of monolayers of the film constituents *via* mostly electrostatic and van der Waals interactions (Figure 12.4c). The film thickness can be easily controlled in the nanometer range by varying the number of layers. Furthermore, the adhesion of the film on the surface is improved because of the charge-to-charge interactions. Polyelectrolytes are normally used for building up the assembly on the surface. To be incorporated into the film, carbon nanomaterials are initially functionalized with groups that are capable of interacting with the other layer.

CNTs have been assembled onto surfaces with polyelectrolytes to form thin-film nanocomposites. Oxidized carbon nanotubes are widely used for this process because it contains -COOH groups on its peripheries that can turn the material into a negatively charged species at higher pH conditions. In one study, oxidized SWNTs were layered with positively charged polyelectrolytes, such as poly(ethylene imine) (PEI)[124] and poly(diallyldimethylammonium chloride) (PDDA).[125] In another study, MWNTs were also functionalized with negatively charged poly(acrylic acid) (PAA) and positively charged polyacrylamide (PAm) by direct polymerization of acrylic acid and acrylamide, respectively, on the walls of the nanotubes.[126] These oppositely charged polyelectrolyte-functionalized MWNTs were then sequentially adsorbed on the surface.

Graphene has also been incorporated into films in a LbL manner. Similar to carbon nanotubes, graphite is first oxidized prior to its deposition on the surface. The formed GO is dispersible in water and could be used for further functionalization. Kotov *et al.* reported the LbL assembly of GO with PDDA as the cationic polymer to form a nanostructured polymer–graphene composite.[127] In this study, the GO was reduced to graphene by either chemical or electrochemical means. Hydrazine hydrate was used for chemical reduction, while cyclic voltammetry was employed to electrochemically reduce the GO to graphene. Converting GO to graphene rendered a nonconductive to a conductive state condition to the surface. Similarly, Dekany and coworkers prepared a GO-PDDA nanocomposite film by the LbL technique.[128] They also converted the GO to graphene with hydrazine but added an annealing step up to 400 °C to give a 32 000-fold decrease in resistivity of the film as compared to the 27 000-fold decrease reported by Kotov and coworkers.

GO has also been functionalized with cationic and anionic polyelectrolytes on its surface for its LbL assembly on the surface. In a study presented by Park and coworkers, two different electrolytes (*i.e.* poly(styrene sulfonate) (PSS) and poly(allylamine hydrochloride) (PAH)) were used to surround the GO surface.[129] The polyelectrolyte-coated GO was then reduced to graphene and deposited on the surface. A conductive film was also produced in this manner.

Recently, LbL has also been introduced as a route for preparing antimicrobial films. Van Tassel and coworkers assembled SWNTs with biological polyelectrolytes: the cationic poly(L-lysine) (PLL) and anionic poly(L-glutamic acid) (PGA).[130] The SWNTs were dispersed in aqueous solution with the aid of the biocompatible amphiphiles, poly(oxyethylene (20) sorbitan monolaurate) (Tween 20) and phospholipid-poly(ethylene glycol) (PL-PEG). The LbL was composed of alternating layers of PLL and PGA with either SWNT-PL-PEG or SWNT-Tween 20 on the topmost surface. In another study, a different approach for preparing CBPN antimicrobial thin film by LbL was presented by Pangilinan *et al.*[131] In this study, negatively charged SWNT-COOHs were layered with anionic and cationic ATRP macroinitiators to form a film composite that exhibits a temperature-responsive antimicrobial property. This feature was made possible by growing PNIPAM brushes from the deposited macroinitiators that swell or collapse depending on the temperature of the environment. Below its LCST, PNIPAM is in its swelled configuration, exposing the underlying SWNTs that make the film antibacterial. Above its LCST, the PNIPAM switches off the antibacterial property of the film because of its collapsed state that hides the SWNTs underneath the film.

12.2.2.4 Electrodeposition

Electrodeposition is another route that has been presented as a tool to immobilize polymer–carbon nanomaterial composites on surfaces. This method uses electrochemical activation for the materials to deposit from the solution to the substrate, which is called the working electrode for this purpose. The set-up typically involves a 3-electrode system consisting of the working, reference, and counterelectrode (Figure 12.4d). Precursors are electropolymerized or electro-chemically crosslinked upon the application of a constant voltage (potentio-static) or a cyclic run of potential over a specific range (potentiodynamic). Conducting polymers are produced during the electrodeposition. In most re-ports, the conducting polymer precursor and the nanomaterial of interest are initially mixed to obtain a well-dispersed solution. This dispersion is then simultaneously electrodeposited on the surface to form the polymer-carbon material nanocomposite film.[132–135] Cheng and coworkers presented another way of making these composites where polyaniline was directly electro-deposited on a graphene paper to make high-performance flexible electrodes.[136] In their study, the graphene paper was made by first exfoliating the graphite powder through oxidation and chemically reducing it back to graphene. The suspension of the reduced graphene oxide was then filtered through a mem-brane and was vacuum dried to produce the final graphene paper. This gra-phene paper served as the working electrode for the electropolymerization of aniline on its surface. Advincula's group presented the use of a precursor polymer approach to incorporate these carbon nanomaterials into polymer composite films.[137–139] The precursor polymer is already a preformed polymer with electroactive pendants that could be crosslinked during the electro-deposition process. The use of the precursor polymer gives the advantage of providing a matrix with more aromatic groups that could yield stable and highly dispersed carbon nanomaterials through pi–pi stacking interactions.

Most of the electrodeposited carbon-based conducting polymer nano-composites have been investigated for electronic applications but only recently it has been reported to be an antimicrobial coating too. Rodrigues's group studied the bactericidal properties of different carbon-based nanomaterials that were electrodeposited with the precursor polymer, PVK. In these studies, all electrodeposited SWNTs,[69] MWNTs,[140] GO,[141] and graphene[142] nanomater-ials with PVK showed efficacy in preventing bacterial colonization.

12.2.2.5 Other Methods

There are also several papers that reported other unconventional approaches on making thin films of carbon-based polymer nanocomposites. Matrix-assisted pulsed-laser evaporation (MAPLE) was used by Fitz-Gerald's group to prepare PMMA-MWNTs thin-film composites.[143] In the MAPLE method, the polymer dissolved in the matrix of volatile absorbing solvent is called the target. The volatile solvent in the target serves as the sacrificial component that absorbs the majority of the laser energy and vaporizes while entraining the polymer on the substrate. In this particular study, MWNTs incorporated in the solution of the PMMA polymer in toluene solvent was used as the target system. The deposition of the PMMA-MWNT film through this process was studied as a function of laser fluence and polymer concentration.

Another approach described in the literature is the deposition of polymer–graphene nanocomposite from water/oil interface to a substrate.[144] In this study, aniline was chemically polymerized on the interfaces of the graphene dispersed in toluene using ammonium persulfate as oxidant. The continuous stirring of the mixture spontaneously produced a homogeneous, free-standing, and transparent film at the interface. The film was then picked up and transferred to a glass substrate.

Covalent binding of SWNTs to polyamide membranes was also reported.[145] These membranes were demonstrated to have antimicrobial properties. The method of forming the polyamide–SWNT composite involved the initial interfacial polymerization of polyamide onto commercial polysulfone ultra-filtration membranes. The polyamide surface was then made in contact with a mixture of *N*-(3-dimethylaminopropyl)-*N'*-ethylcarbodiimide hydrochloride (EDC) and *N*-hydroxysuccinimide (NHS) to activate the carboxylic groups of the surface. Ethylene diamine was then reacted with the activated group to make an amino-terminated surface that will serve as a binding site for the carbon nanomaterial. The final step is the incorporation of the oxidized SWNTs, which makes a covalent amide bond with the amino group of the surface of the membrane.

12.3   Antibacterial Properties of Carbon-Based Polymer Nanocomposites

Carbon-based nanomaterials have been recently reported to possess antimicrobial properties. However, their antimicrobial activities were limited by their known hydrophobicity, relatively high cost, and their capacity to adsorb onto bacteria, which compromises their antimicrobial applications. Several investigators have incorporated polymers onto the carbon nanomaterials, thus creating CBPNs to resolve these issues. In general, polymers serve various roles in extending the biocidal applications of CBNs, which can be categorized as follows: (a) polymers as dispersing agent; (b) polymers as linkers for antimicrobial agents; and (c) polymers as matrix for fiber formation. This section reviews the various reports on the antibacterial behavior of CBPNs where polymers were utilized differently.

12.3.1   Polymers as Dispersing Agents

Polymers were utilized to improve the dispersibility and stability of CBN and thus increase the biocidal property of the CBN. Pi–pi interaction through aromatic groups in carbon-based nanomaterials and the polymer of interest has been described to stabilize CBN onto the matrix. The use of PVK, a carbazole-containing polymer can facilitate the stable dispersion of MWNT,[140] SWNT,[69] GO,[59,137,141] and graphene[142] nanomaterials. It has been described that the presence of PVK does not impede the known antimicrobial property of the nanomaterials but rather enhances its antimicrobial behavior.[69,137] In fact, even at very low loadings of the nanomaterial ($\sim$ 3 wt.%) strong antibacterial efficacy is observed against Gram-positive and Gram-negative bacteria. This behavior was noted in both solution and coating forms. It is possible that the stable dispersion of the nanomaterial in solution due to the PVK–nanomaterial interaction resulted in increased bacterial interaction with the nanomaterial, leading to improved antimicrobial properties.

Antimicrobial investigations of carbon–polymer nanocomposites stabilized *via* ionic interactions were also evaluated. In most cases, charged polymers were used and the nanomaterial of interest was tethered using a LbL process to create a coating. Examples of this include a separate report by Aslan and coworkers in which bactericidal behavior was determined for SWNT–polypeptide films prepared by LbL of SWNT assembled with the polyelectrolytes poly(L-lysine) (PLL) and poly(L-glutamic acid) (PGA).[130] Specifically, up to 90% decrease in viabilities of *E. coli* and *S. epidermidis* were observed after 24 h exposure of the bacteria to SWNT/PLL/PGA films. They related the antimicrobial effect of this nanocomposite to the bacterial interactions with the SWNT open ends (instead of the tube walls) as the amphiphilic polymers are expected to effectively coat the walls versus the ends. Also, in another study, a combined LbL and surface-initiated polymerization approach was used to fabricate a thermoresponsive and antimicrobial CNT-PNIPAM brush films.[131] A strong dependence of bacterial inactivation was observed below and above the LCST temperature of the PNIPAM brush. This antimicrobial dependence was noted to be due to the exposure of the CNT as the brush changes its conformation from collapse to swelling above and below the LCST temperature, respectively.

12.3.2 Polymers as Linkers for Antimicrobial Agents

To improve the antimicrobial toxicity of carbonaceous materials, several investigators have tethered the carbonaceous materials in a polymer matrix containing cheaper antimicrobial agents. Zhou and Qi fabricated MWNT-epsilon-polylysine (MEP) nanocomposite by covalently grafting EP, a known antibacterial agent on MWNT using hexamethylene diisocyanate (HDI) as a coupling agent.[146] They observed that the incorporation of EP resulted in a significant increase in bacterial toxicity for *E. coli, P. aeruginosa,* and *S. aureus* as compared to pure MWNT. The higher bacterial inactivation observed with the MEP film was linked to the antiadhesive property of this film, which resulted in fewer bacterial attachments on the film as compared to the MWNT coated surface.

Similar result (*i.e.* increased biocidal activity) was also observed by Qi and coworkers upon grafting PEG-nisin on MWNTs.[147] Several studies also investigated the use of polymers such as polyamidoamine (PAMAM),[148] PAA,[149] polydopamine (PDA),[150] PEI,[151] to graft Ag onto MWNT, graphene nanosheet, graphene, or GO for improved antimicrobial performance. In all the above-mentioned investigations, an improved bacterial toxicity was observed for the carbon-based polymer nanocomposites as compared to their pristine counterparts. More importantly, even at lower loading concentrations of the antimicrobial material, the CBPN remained effective in preventing bacterial colonization.

12.3.3 Polymers as Matrix for Blends and Fiber Formation

The incorporation of antimicrobial agents in electrospun materials is useful for biotechnology and environmental applications, such as wound dressings, tissue-engineering scaffolds, drug delivery, and membrane water filtration. Since polymer-based fibers and blends are prone to bacterial colonization, similar to other biomaterials, grafting antimicrobial agents such as carbon-based nanomaterials have been vital in ensuring that microbial contaminations and infections are prevented.

In a handful of reports, polymers were also often combined with carbon nanomaterials to create antimicrobial polymer nanocomposite blends. Aslan and coworkers reported the fabrication of antimicrobial PLGA–SWNT films.[119] They tested the antimicrobial behavior of the PLGA–SWNT nanocomposite by varying its concentration and the length of SWNT. In their studies, bacterial inactivation was strongly influenced by SWNT length and concentration: *i.e.* short (*ca.* 300 nm) nanotubes at a concentration of 1/70 (w/w) yielded the highest degree of bacterial inactivation. In addition, they also noted that the highest rates of bacterial inactivation observed for the PLGA–SWNT were comparable to those observed for other antimicrobial materials. The authors proposed the antimicrobial properties of the PLGA–SWNT to be: 1) mechanical disruption of the cell wall, where nanotubes act to physically pierce or otherwise perturb the bacterial membrane; and 2) oxidative stress, in which the high reductive potential of nanoscale carbon acts to directly or indirectly (*e.g.* through formation of reactive oxygen species (ROS)) damage the cell constituents, such as proteins, lipids, nucleic acids.

Polymer blending was also utilized to incorporate SWNT and MWNT into biopolymers, such as agar.[152] Their investigation aimed to evaluate the effect of CNT–agar composites irradiated with near-infrared (NIR) light on *Streptococcus mutans*, as a potential photothermal antimicrobial nanotherapy. Their results showed that the CNT–agar (*i.e.* SWNT and MWNT) attain bactericidal activity after NIR light irradiation. Furthermore, their efficiency in bacterial killing is higher than that observed for graphite (GP)-agar and activated carbon (AC)-agar composites.

Carbon nanomaterials such as SWNT,[153] MWNT,[154] and graphene[155] were also reported to be incorporated into polysulfone, polyethersulfone, and chitosan-poly(vinyl alcohol) (PVA), respectively, to create mats and membranes. It is worth noting that in one of the reports, the electrospun chitosan-PVA containing graphene not only presented antimicrobial properties but also supported the view that the presence of the nanomaterial was beneficial to wound healing.[155] The authors suggested that this was due to a more favorable ROS transfer from the nanomaterial to prokaryotic cells than to eukaryotic cells upon contact with the nanocomposite. Since eukaryotic cells contain nuclear membranes, ROS originating from the nanomaterial will not penetrate the cells as easily and damage the DNA and genetic information. In contrast, a prokaryotic cell does not have a nuclear membrane, and thus ROS transfer can easily reach and damage the prokaryotic DNA. As a result, the presence of graphene in the polymer composite prevents bacterial proliferation and promotes wound healing.

12.4 Conclusion

In this chapter we have summarized the key approaches to synthesize carbon-based polymer nanocomposites in bulk solutions and thin films, and their antimicrobial properties. Methods for preparing bulk solutions can be done through solution mixing, melt blending, and *in situ* polymerization. Solution mixing, which involves shearing, stirring, or sonication is probably the simplest and yet is a very effective method for bulk material preparation. However, this method requires the use of solvents and can be especially costly for large-scale applications. An alternative method and more economical, which does not require any solvent, is melt blending. In this method, the polymer melts and the carbon-based nanomaterials are combined under high-shear conditions. Nevertheless, several reports have suggested that unlike the other bulk preparation techniques, melt mixing does not provide an equal degree of dispersion.

Thin films of these CBPNs, on the other hand, are generally prepared *via* casting, wet-phase inversion, LbL process, or electrodeposition. Casting is more commonly used especially for evaluating the properties of the bulk material on surfaces. Wet-phase inversion could be employed to produce membrane-type films. LbL gives the advantage of precise control in film thickness on the nanometer scale. Electrodeposition offers an opportunity to produce conducting polymers on surfaces while depositing the bulk material. Each of these techniques has their own advantages and limitations, but the selection of one over the other depends on the desired application.

Similar to the well-studied bacterial toxicity of CBNs, the incorporation of CBNs into polymers to produce CBPNs have also been reported to present antimicrobial properties. In some cases, the addition of polymers even produced greater bacterial inactivation due to better dispersion of the nanomaterials. In these studies, polymers not only served as dispersant agents of the CBN, but also facilitated the attachment of other antimicrobial materials and served as a matrix for producing antimicrobial membranes and fibers. Although no detailed mechanistic studies on antibacterial properties of these CBNs have been reported, several investigators have linked the CBPNs toxicity to cellular mechanical disruption upon contact of the CBN with the bacteria and oxidative stress. Carbon-based polymer nanocomposites have created a unique and new class of antibacterial agents. The antibacterial reports seem promising but it is clear that additional studies must be performed to further elucidate the exact mechanism of its toxicity so that its full potential application can be understood.

References

1. V. K. K. Upadhyayula, S. G. Deng, M. C. Mitchell and G. B. Smith, *Sci. Total Environ.*, 2009, **408**, 1–13.

2. E. R. Kenawy, S. D. Worley and R. Broughton, *Biomacromolecules*, 2007, **8**, 1359–1384.
3. I. Banerjee, R. C. Pangule and R. S. Kane, *Adv. Mater.*, 2011, **23**, 690–718.
4. M. S. Mauter and M. Elimelech, *Environ. Sci. Technol.*, 2008, **42**, 5843–5859.
5. A. Hucknall, S. Rangarajan and A. Chilkoti, *Adv. Mater.*, 2009, **21**, 2441–2446.
6. K. D. Park, Y. S. Kim, D. K. Han, Y. H. Kim, E. H. B. Lee, H. Suh and K. S. Choi, *Biomaterials*, 1998, **19**, 851–859.
7. J. D. Bryers, *Biotechnol. Bioeng.*, 2008, **100**, 1–18.
8. R. M. Donlan, *Emerg. Infect. Dis.*, 2001, **7**, 277–281.
9. R. M. Donlan, *Clin. Infect. Dis.*, 2001, **33**, 1387–1392.
10. A. Roosjen, H. C. van der Mei, H. J. Busscher and W. Norde, *Langmuir*, 2004, **20**, 10949–10955.
11. E. Ostuni, R. G. Chapman, M. N. Liang, G. Meluleni, G. Pier, D. E. Ingber and G. M. Whitesides, *Langmuir*, 2001, **17**, 6336–6343.
12. K. L. Prime and G. M. Whitesides, *Science*, 1991, **252**, 1164–1167.
13. J. P. Bearinger, S. Terrettaz, R. Michel, N. Tirelli, H. Vogel, M. Textor and J. A. Hubbell, *Nature Mater.*, 2003, **2**, 259–264.
14. J. H. Lee, J. Kopecek and J. D. Andrade, *J. Biomed. Mater. Res.*, 1989, **23**, 351–368.
15. K. L. Prime and G. M. Whitesides, *J. Am. Chem. Soc.*, 1993, **115**, 10714–10721.
16. J. F. Lutz, *Adv. Mater.*, 2011, **23**, 2237–2243.
17. N. Gilmartin and R. O'kennedy, *Enzyme Microb. Tech.*, 2012, **50**, 87–95.
18. P. Dallas, V. K. Sharma and R. Zboril, *Adv. Colloid Interface Sci.*, 2011, **166**, 119–135.
19. M. R. Yeaman and N. Y. Yount, *Pharmacol. Rev.*, 2003, **55**, 27–55.
20. D. J. Waxman and J. L. Strominger, *Annu. Rev. Biochem.*, 1983, **52**, 825–869.
21. D. J. Tipper and Stroming. Jl, *Proc. Natl. Acad. Sci. USA*, 1965, **54**, 1133.
22. B. Alcaide and P. Almendros, *Curr. Med. Chem.*, 2004, **11**, 1921–1949.
23. C. Marambio-Jones and E. M. V. Hoek, *J Nanopart Res*, 2010, **12**, 1531–1551.
24. Q. L. Feng, J. Wu, G. Q. Chen, F. Z. Cui, T. N. Kim and J. O. Kim, *J. Biomed. Mater. Res.*, 2000, **52**, 662–668.
25. S. Silver, L. T. Phung and G. Silver, *J. Ind. Microbiol. Biot.*, 2006, **33**, 627–634.
26. R. E. W. Hancock and R. Lehrer, *Trends Biotechnol.*, 1998, **16**, 82–88.
27. R. E. W. Hancock and H. G. Sahl, *Nature Biotechnol.*, 2006, **24**, 1551–1557.
28. T. F. C. Mah and G. A. O'Toole, *Trends Microbiol.*, 2001, **9**, 34–39.
29. S. B. Levy and B. Marshall, *Nature Med.*, 2004, **10**, S122–S129.
30. M. L. W. Knetsch and L. H. Koole, *Polymers*, 2011, **3**, 340–366.
31. P. L. Drake and K. J. Hazelwood, *Ann. Occup. Hyg.*, 2005, **49**, 575–585.
32. E. Ah, W. S. Lee, K. M. Kim and S. Y. Kim, *J. Dermatol.*, 2008, **35**, 759–760.
33. P. M. Ajayan, J. C. Charlier and A. G. Rinzler, *Proc. Natl. Acad. Sci. USA*, 1999, **96**, 14199–14200.

34. A. K. Geim and K. S. Novoselov, *Nature Mater.*, 2007, **6**, 183–191.

35. D. R. Dreyer, S. Park, C. W. Bielawski and R. S. Ruoff, *Chem. Soc. Rev.*, 2010, **39**, 228–240.

36. Y. W. Zhu, S. Murali, W. W. Cai, X. S. Li, J. W. Suk, J. R. Potts and R. S. Ruoff, *Adv. Mater.*, 2010, **22**, 5226–5226.

37. G. X. Zhao, J. X. Li, X. M. Ren, C. L. Chen and X. K. Wang, *Environ. Sci. Technol.*, 2011, **45**, 10454–10462.

38. K. Yang, L. Z. Zhu and B. S. Xing, *Environ. Sci. Technol.*, 2006, **40**, 1855–1861.

39. K. Yang and B. S. Xing, *Chem. Rev.*, 2010, **110**, 5989–6008.

40. W. R. Yang, K. R. Ratinac, S. P. Ringer, P. Thordarson, J. J. Gooding and F. Braet, *Angew. Chem. Int. Ed.*, 2010, **49**, 2114–2138.

41. J. Wang, *Electroanalysis*, 2005, **17**, 7–14.

42. P. Avouris, M. Freitag and V. Perebeinos, *Nature Photon.*, 2008, **2**, 341–350.

43. A. K. Geim, *Science*, 2009, **324**, 1530–1534.

44. P. Avouris, Z. H. Chen and V. Perebeinos, *Nature Nanotechnol.*, 2007, **2**, 605–615.

45. A. H. Castro Neto, F. Guinea, N. M. R. Peres, K. S. Novoselov and A. K. Geim, *Rev. Mod. Phys.*, 2009, **81**, 109–162.

46. K. P. Loh, Q. L. Bao, G. Eda and M. Chhowalla, *Nature Chem.*, 2010, **2**, 1015–1024.

47. Q. L. Li, S. Mahendra, D. Y. Lyon, L. Brunet, M. V. Liga, D. Li and P. J. J. Alvarez, *Water Res.*, 2008, **42**, 4591–4602.

48. S. Kang, M. S. Mauter and M. Elimelech, *Environ. Sci. Technol.*, 2009, **43**, 2648–2653.

49. W. B. Hu, C. Peng, W. J. Luo, M. Lv, X. M. Li, D. Li, Q. Huang and C. H. Fan, *ACS Nano*, 2010, **4**, 4317–4323.

50. O. Akhavan and E. Ghaderi, *ACS Nano*, 2010, **4**, 5731–5736.

51. M. Moniruzzaman and K. I. Winey, *Macromolecules*, 2006, **39**, 5194–5205.

52. T. Ramanathan, A. A. Abdala, S. Stankovich, D. A. Dikin, M. Herrera-Alonso, R. D. Piner, D. H. Adamson, H. C. Schniepp, X. Chen, R. S. Ruoff, S. T. Nguyen, I. A. Aksay, R. K. Prud'homme and L. C. Brinson, *Nature Nanotechnol.*, 2008, **3**, 327–331.

53. I. E. M. Carpio, C. M. Santos, X. Wei and D. F. Rodrigues, *Nanoscale*, 2012, **4**, 4746–4756.

54. F. Ahmed, C. M. Santos, R. A. Vergara, M. C. Tria, R. Advincula and D. F. Rodrigues, *Environ. Sci. Technol.*, 2012, **46**, 1804–1810.

55. H. Xia, Q. Wang and G. Qiu, *Chem. Mater.*, 2003, **15**, 3879–3886.

56. K. S. Suslick, *Science*, 1990, **247**, 1439–1445.

57. N. G. Sahoo, S. Rana, J. W. Cho, L. Li and S. H. Chan, *Prog. Polym. Sci.*, 2010, **35**, 837–867.

58. I. Zaman, H.-C. Kuan, Q. Meng, A. Michelmore, N. Kawashima, T. Pitt, L. Zhang, S. Gouda, L. Luong and J. Ma, *Adv. Funct. Mater.*, 2012, **22**, 2735–2743.

59. I. E. Mejias Carpio, C. M. Santos, X. Wei and D. F. Rodrigues, *Nanoscale*, 2012, **4**, 4746–4756.

60. M. Fang, K. Wang, H. Lu, Y. Yang and S. Nutt, *J. Mater. Chem.*, 2009, **19**, 7098–7105.

61. A. L. Higginbotham, J. R. Lomeda, A. B. Morgan and J. M. Tour, *ACS Appl. Mater. Interfaces*, 2009, **1**, 2256–2261.

62. R. Liu, S. Liang, X.-Z. Tang, D. Yan, X. Li and Z.-Z. Yu, *J. Mater. Chem.*, 2012, **22**, 14160–14167.

63. D. Chen, H. Zhu and T. Liu, *ACS Appl. Mater. Interfaces*, 2010, **2**, 3702–3708.

64. J. Liang, Y. Huang, L. Zhang, Y. Wang, Y. Ma, T. Guo and Y. Chen, *Adv. Funct. Mater.*, 2009, **19**, 2297–2302.

65. X. Zhao, Q. Zhang, D. Chen and P. Lu, *Macromolecules*, 2010, **43**, 2357–2363.

66. X. Yang, L. Li, S. Shang and X.-M. Tao, *Polymer*, 2010, **51**, 3431–3435.

67. Y. Xu, W. Hong, H. Bai, C. Li and G. Shi, *Carbon*, 2009, **47**, 3538–3543.

68. T. Ramanathan, A. A. Abdala, S. Stankovich, D. A. Dikin, M. Herrera Alonso, R. D. Piner, D. H. Adamson, H. C. Schniepp, X. Chen, R. S. Ruoff, S. T. Nguyen, I. A. Aksay, R. K. Prud'Homme and L. C. Brinson, *Nature Nano*, 2008, **3**, 327–331.

69. F. Ahmed, C. M. Santos, R. A. M. V. Vergara, M. C. R. Tria, R. Advincula and D. F. Rodrigues, *Environ. Sci. Technol.*, 2011, **46**, 1804–1810.

70. L. Chen, X.-J. Pang, M.-Z. Qu, Q.-t. Zhang, B. Wang, B.-L. Zhang and Z.-L. Yu, *Comps. Appl. Sci. Manuf.*, 2006, **37**, 1485–1489.

71. N. G. Sahoo, Y. C. Jung, H. J. Yoo and J. W. Cho, *Macromol. Chem. Phys.*, 2006, **207**, 1773–1780.

72. B. Vigolo, A. Pénicaud, C. Coulon, C. Sauder, R. Pailler, C. Journet, P. Bernier and P. Poulin, *Science*, 2000, **290**, 1331–1334.

73. N. Saran, K. Parikh, D.-S. Suh, E. MuÃoz, H. Kolla and S. K. Manohar, *J. Am. Chem. Soc.*, 2004, **126**, 4462–4463.

74. S. Barrau, P. Demont, E. Perez, A. Peigney, C. Laurent and C. Lacabanne, *Macromolecules*, 2003, **36**, 9678–9680.

75. X. Gong, J. Liu, S. Baskaran, R. D. Voise and J. S. Young, *Chem. Mater.*, 2000, **12**, 1049–1052.

76. E. Camponeschi, B. Florkowski, R. Vance, G. Garrett, H. Garmestani and R. Tannenbaum, *Langmuir*, 2006, **22**, 1858–1862.

77. W. Chen and X. Tao, *Macromol. Rapid Commun.*, 2005, **26**, 1763–1767.

78. D. Qian, E. C. Dickey, R. Andrews and T. Rantell, *Appl. Phys. Lett.*, 2000, **76**, 2868–2870.

79. M. Wong, M. Paramsothy, X. J. Xu, Y. Ren, S. Li and K. Liao, *Polymer*, 2003, **44**, 7757–7764.

80. A. K. Kota, B. H. Cipriano, M. K. Duesterberg, A. L. Gershon, D. Powell, S. R. Raghavan and H. A. Bruck, *Macromolecules*, 2007, **40**, 7400–7406.

81. J. Sandler, M. S. P. Shaffer, T. Prasse, W. Bauhofer, K. Schulte and A. H. Windle, *Polymer*, 1999, **40**, 5967–5971.

82. Y. S. Song and J. R. Youn, *Carbon*, 2005, **43**, 1378–1385.
83. M. S. P. Shaffer and A. H. Windle, *Adv. Mater.*, 1999, **11**, 937–941.
84. H. G. Chae, T. V. Sreekumar, T. Uchida and S. Kumar, *Polymer*, 2005, **46**, 10925–10935.
85. Y. Bin, M. Kitanaka, D. Zhu and M. Matsuo, *Macromolecules*, 2003, **36**, 6213–6219.
86. D. R. Paul and L. M. Robeson, *Polymer*, 2008, **49**, 3187–3204.
87. H. Kim, Y. Miura and C. W. Macosko, *Chem. Mater.*, 2010, **22**, 3441–3450.
88. H.-B. Zhang, W.-G. Zheng, Q. Yan, Y. Yang, J.-W. Wang, Z.-H. Lu, G.-Y. Ji and Z.-Z. Yu, *Polymer*, 2010, **51**, 1191–1196.
89. G. Ruess and F. Vogt, *Monatsh.*, 1948, **78**, 222–242.
90. H. C. Schniepp, J.-L. Li, M. J. McAllister, H. Sai, M. Herrera-Alonso, D. H. Adamson, R. K. Prud'homme, R. Car, D. A. Saville and I. A. Aksay, *J. Phys. Chem. B*, 2006, **110**, 8535–8539.
91. M. J. McAllister, J.-L. Li, D. H. Adamson, H. C. Schniepp, A. A. Abdala, J. Liu, M. Herrera-Alonso, D. L. Milius, R. Car, R. K. Prud'homme and I. A. Aksay, *Chem. Mater.*, 2007, **19**, 4396–4404.
92. W. D. Zhang, L. Shen, I. Y. Phang and T. Liu, *Macromolecules*, 2003, **37**, 256–259.
93. C. Xing, L. Zhao, J. You, W. Dong, X. Cao and Y. Li, *J. Phys. Chem. B*, 2012, **116**, 8312–8320.
94. S. Bocchini, A. Frache, G. Camino and M. l. Claes, *Eur. Polym. J.*, 2007, **43**, 3222–3235.
95. C.-S. Wu, *Carbon*, 2009, **47**, 3091–3098.
96. C.-S. Wu, *Polym. Int.*, 2011, **60**, 807–815.
97. M. A. L. Manchado, L. Valentini, J. Biagiotti and J. M. Kenny, *Carbon*, 2005, **43**, 1499–1505.
98. H. Zhang and Z. Zhang, *Eur. Polym. J.*, 2007, **43**, 3197–3207.
99. P. Pötschke, A. R. Bhattacharyya and A. Janke, *Polymer*, 2003, **44**, 8061–8069.
100. P. Pötschke, A. R. Bhattacharyya and A. Janke, *Eur. Polym. J.*, 2004, **40**, 137–148.
101. Z. Jin, K. P. Pramoda, G. Xu and S. H. Goh, *Chem. Phys. Lett.*, 2001, **337**, 43–47.
102. R. Haggenmueller, H. H. Gommans, A. G. Rinzler, J. E. Fischer and K. I. Winey, *Chem. Phys. Lett.*, 2000, **330**, 219–225.
103. J. Zeng, B. Saltysiak, W. S. Johnson, D. A. Schiraldi and S. Kumar, *Compos. Eng.*, 2004, **35**, 173–178.
104. Y. Zeng, Z. Ying, J. Du and H.-M. Cheng, *J. Phys. Chem. C*, 2007, **111**, 13945–13950.
105. E. J. Siochi, D. C. Working, C. Park, P. T. Lillehei, J. H. Rouse, C. C. Topping, A. R. Bhattacharyya and S. Kumar, *Compos. Eng.*, 2004, **35**, 439–446.
106. J. R. Potts, D. R. Dreyer, C. W. Bielawski and R. S. Ruoff, *Polymer*, 2011, **52**, 5–25.

107. P. Liu, K. Gong, P. Xiao and M. Xiao, *J. Mater. Chem.*, 2000, **10**, 933–935.
108. P. Xiao, M. Xiao, P. Liu and K. Gong, *Carbon*, 2000, **38**, 626–628.
109. H. Jin Yoo, Y. Chae Jung, N. Gopal Sahoo and J. Whan Cho, *J. Macromol. Sci., Part B*, 2006, **45**, 441–451.
110. N. Hu, L. Wei, Y. Wang, R. Gao, J. Chai, Z. Yang, E. S.-W. Kong and Y. Zhang, *J Nanosci. Nanotech.*, 2012, **12**, 173–178.
111. N. Hu, H. Zhou, G. Dang, X. Rao, C. Chen and W. Zhang, *Polym. Int.*, 2007, **56**, 655–659.
112. S. H. Lee, D. R. Dreyer, J. An, A. Velamakanni, R. D. Piner, S. Park, Y. Zhu, S. O. Kim, C. W. Bielawski and R. S. Ruoff, *Macromol. Rapid Commun.*, 2010, **31**, 281–288.
113. L. Ren, S. Huang, C. Zhang, R. Wang, W. Tjiu and T. Liu, *J. Nanopart. Res.*, 2012, **14:940**, 1–9.
114. G. L. Li, G. Liu, M. Li, D. Wan, K. G. Neoh and E. T. Kang, *J. Phys. Chem. C*, 2010, **114**, 12742–12748.
115. K. H. Jung, Y. Kim and Y. Lim, *Inter. J. Mod. Phys. B*, 2011, **25**, 4311–4314.
116. B. Zhang, Y. Chen, L. Xu, L. Zeng, Y. He, E.-T. Kang and J. Zhang, *J. Polym. Sci. Part A: Polym. Chem.*, 2011, **49**, 2043–2050.
117. Y. Yang, X. Song, L. Yuan, M. Li, J. Liu, R. Ji and H. Zhao, *J. Polym. Sci. Part A: Polym. Chem.*, 2012, **50**, 329–337.
118. G. Eda and M. Chhowalla, *Nano Lett.*, 2009, **9**, 814–818.
119. S. Aslan, C. Z. Loebick, S. Kang, M. Elimelech, L. D. Pfefferle and P. R. Van Tassel, *Nanoscale*, 2010, **2**, 1789–1794.
120. H. N. Lim, N. M. Huang and C. H. Loo, *J. Non-Cryst. Solids*, 2012, **358**, 525–530.
121. B. Yuan, T. Zhu, Z. Zhang, Z. Jiang and Y. Ma, *J. Mater. Chem.*, 2011, **21**, 3471–3476.
122. L. Brunet, D. Lyon, K. Zodrow, J. C. Rouch, B. Caussat, P. Serp, J. C. Remigy, M. Wiesner and P. Alvarez, *Environ. Eng. Sci.*, 2008, **25**, 565–576.
123. E.-S. Kim, G. Hwang, M. Gamal El-Din and Y. Liu, *J. Membr. Sci.*, 2012, **394–395**, 37–48.
124. A. A. Mamedov, N. A. Kotov, M. Prato, D. M. Guldi, J. P. Wicksted and A. Hirsch, *Nature Mater.*, 2002, **1**, 190–194.
125. J. H. Rouse and P. T. Lillehei, *Nano Lett.*, 2002, **3**, 59–62.
126. J. Shen, Y. Hu, C. Qin and M. Ye, *Langmuir*, 2008, **24**, 3993–3997.
127. N. A. Kotov, I. Dékány and J. H. Fendler, *Adv. Mater.*, 1996, **8**, 637–641.
128. T. Szabó, A. Szeri and I. Dékány, *Carbon*, 2005, **43**, 87–94.
129. A. Rani, K. A. Oh, H. Koo, H. j. Lee and M. Park, *Appl. Surf. Sci.*, 2011, **257**, 4982–4989.
130. S. Aslan, M. Deneufchatel, S. Hashmi, N. Li, L. D. Pfefferle, M. Elimelech, E. Pauthe and P. R. Van Tassel, *J. Colloid Interface Sci.*, 2012, **388**, 268–273.
131. K. D. Pangilinan, C. M. Santos, N. C. Estillore, D. F. Rodrigues and R. C. Advincula, *Macromol. Chem. Phys.*, 2013, **214**, 464–469.

132. A. Liu, C. Li, H. Bai and G. Shi, *J. Phys. Chem. C*, 2010, **114**, 22783–22789.
133. P. A. Mini, A. Balakrishnan, S. V. Nair and K. R. V. Subramanian, *Chem. Commun.*, 2011, **47**, 5753–5755.
134. S. Bhandari, M. Deepa, A. K. Srivastava, A. G. Joshi and R. Kant, *J. Phys. Chem. B*, 2009, **113**, 9416–9428.
135. G. Han, J. Yuan, G. Shi and F. Wei, *Thin Solid Films*, 2005, **474**, 64–69.
136. D.-W. Wang, F. Li, J. Zhao, W. Ren, Z.-G. Chen, J. Tan, Z.-S. Wu, I. Gentle, G. Q. Lu and H.-M. Cheng, *ACS Nano*, 2009, **3**, 1745–1752.
137. C. M. Santos, M. C. R. Tria, R. A. M. V. Vergara, K. M. Cui, R. Pernites and R. C. Advincula, *Macromol. Chem. Phys.*, 2011, **212**, 2371–2377.
138. K. M. Cui, M. C. Tria, R. Pernites, C. A. Binag and R. C. Advincula, *ACS Appl. Mater. Interfaces*, 2011, **3**, 2300–2308.
139. R. Pernites, A. Vergara, A. Yago, K. Cui and R. Advincula, *Chem. Commun.*, 2011, **47**, 9810–9812.
140. C. M. Santos, K. Milagros Cui, F. Ahmed, M. C. R. Tria, R. A. M. V. Vergara, A. C. de Leon, R. C. Advincula and D. F. Rodrigues, *Macromol. Mater. Eng.*, 2012, **297**, 807–813.
141. C. M. Santos, M. C. R. Tria, R. A. M. V. Vergara, F. Ahmed, R. C. Advincula and D. F. Rodrigues, *Chem. Commun.*, 2011, **47**, 8892–8894.
142. C. M. Santos, M. Joey, A. Farid, L. Alex, C. A. Rigoberto and F. R. Debora, *Nanotechnology*, 2012, **23**, 395101.
143. A. T. Sellinger, E. M. Leveugle, K. Gogick, L. V. Zhigilei and J. M. Fitz-Gerald, *J. Vac. Sci. Technol., A*, 2006, **24**, 1618–1622.
144. S. H. Domingues, R. V. Salvatierra, M. M. Oliveira and A. J. G. Zarbin, *Chem. Commun.*, 2011, **47**, 2592–2594.
145. A. Tiraferri, C. D. Vecitis and M. Elimelech, *ACS Appl. Mater. Interfaces*, 2011, **3**, 2869–2877.
146. J. Zhou and X. Qi, *Lett. Appl. Microbiol.*, 2010, **52**, 76–83.
147. X. B. Qi, G. Poernomo, K. A. Wang, Y. A. Chen, M. B. Chan-Park, R. Xu and M. W. Chang, *Nanoscale*, 2011, **3**, 1874–1880.
148. W. Yuan, G. H. Jiang, J. F. Che, X. B. Qi, R. Xu, M. W. Chang, Y. Chen, S. Y. Lim, J. Dai and M. B. Chan-Park, *J. Phys. Chem. C*, 2008, **112**, 18754–18759.
149. Z. X. Tai, H. B. Ma, B. Liu, X. B. Yan and Q. J. Xue, *Colloid Surf. B*, 2012, **89**, 147–151.
150. Z. Zhang, J. Zhang, B. Zhang and J. Tang, *Nanoscale*, 2013, **5**, 118–123.
151. X. Cai, M. S. Lin, S. Z. Tan, W. J. Mai, Y. M. Zhang, Z. W. Liang, Z. D. Lin and X. J. Zhang, *Carbon*, 2012, **50**, 3407–3415.
152. T. Akasaka, M. Matsuoka, T. Hashimoto, S. Abe, M. Uo and F. Watari, *Mater. Sci. Eng. B-Adv*, 2010, **173**, 187–190.
153. J. D. Schiffman and M. Elimelech, *ACS Appl. Mater. Interfaces*, 2011, **3**, 462–468.
154. V. Vatanpour, S. S. Madaeni, R. Moradian, S. Zinadini and B. Astinchap, *J. Membr. Sci.*, 2011, **375**, 284–294.
155. B. G. Lu, T. Li, H. T. Zhao, X. D. Li, C. T. Gao, S. X. Zhang and E. Q. Xie, *Nanoscale*, 2012, **4**, 2978–2982.

CHAPTER 13

Polymer/Copper-Based Materials for Antimicrobial Applications

HUMBERTO PALZA* AND KATHERINE DELGADO

Universidad de Chile, Beauchef 850, Santiago, Chile
*Email: hpalza@ing.uchile.cl

13.1 Introduction

There are several methods producing antimicrobial polymer materials. They can be prepared by the incorporation of the biocide agent into bulk polymers during processing of the material or by endowing a biocidal function to the polymer after processing, for example applying surface coatings having the agent.[1–3] A different approach is the polymerization of monomer-containing biocide groups or its copolymerization with another monomer producing a new family of polymers bearing the biocidal agents.[4] The grafting of antimicrobial agents into the polymers is yet another methodology to prepare bioactive materials.[4] Recently, the polymerization of the biocide polymer on the surface of commercial polymers by atom transfer radical polymerization has also been reported.[5,6] From all the above-mentioned methods, the direct addition of the biocide agent into the bulk polymers seems the most efficient and simplest, as this route can be easily implemented in the standard polymer processing units without any kind of postreactor chemical reactions on the polymer. In this context, polymer–metal composites prepared by melt blending are perceived as a useful way to produce biocidal polymers as there are no problems with the

RSC Polymer Chemistry Series No. 10
Polymeric Materials with Antimicrobial Activity: From Synthesis to Applications
Edited by Alexandra Muñoz-Bonilla, María L. Cerrada and Marta Fernández-García

degradations of the biocide metal agent under the standard processing conditions (~ 200 °C). Silver-based hybrid materials are one of the most studied polymeric composites because silver is a strong and versatile antimicrobial agent with low toxicity.[7–9] Moreover, silver presents high thermal stability, low volatility, and long-term activity allowing its use under standard processing conditions.[8,10] Nevertheless, copper has emerged during recent years as an excellent biocide material although its study in order to prepare biocidal polymeric materials has been less extensive. The goal of this chapter is to introduce the main concepts behind the biocide behavior of copper and to show examples of its use as filler or additive in polymeric matrices with a focus on nanoparticles.

13.2   Copper as a Biocide Material

Copper ions, either alone or in copper complexes, have been used for centuries to disinfect liquids, solids, and human tissues. For example, the ancient Aztecs used copper oxide and copper carbonate compounds for treating skin diseases whereas pieces of copper were put in water bottles in order to prevent dysentery in the Second World War. Today, copper is used as a water purifier, algaecide, fungicide, nematocide, molluscicide, and antibacterial and antifouling agent.[11,12] This knowledge is based on empirical evidences that have been rationalized during the last decades. For example, Keevil and coworkers studied a set of metallic surfaces showing that those containing copper are the most effective at reducing the survival of bacteria.[13] In pure copper surfaces, several strains of *Staphylococcus aureus* were killed in less than 90 min at 22 °C.[14] In contrast, viable organisms for all the strains were detected on stainless steel after 72 h at 22 °C. Moreover, when copper- and silver-based materials are compared it is concluded that the use of copper alloys as antimicrobial materials in indoor environments, such as hospitals, is favored.[15] These results come from the high efficacy displayed by the copper alloys, as compared with the silver-ion-containing materials, at temperature and humidity levels typical of indoor environments.

Copper can also be used as a bacterial inhibitor in various stages of food processing.[16] Noteworthy, metallic copper surfaces inhibited the growth of two of the more prevalent bacterial pathogens that cause foodborne diseases.[16] Similarly, the addition of copper to drinking glasses reduces biofilm formation of some micro-organisms, reducing the risk of oral infections.[17] Recognition of the bacteriostatic properties of copper has led to testing its capacity as a water purifier. Copper is one of the most toxic metals to heterotrophic bacteria in aquatic environments.[12] Similarly, water distribution systems made of copper have a greater potential for suppressing growth and for decreasing persistence of some micro-organisms in potable water than distribution systems constructed of plastic materials or galvanized steel.[12] For example, copper pipes displayed slower biofilm formation than the polyethylene (PE) ones during the first 200 days.[18] Moreover, the number of virus-like particles was lower in biofilms and outlet water from copper pipes. Another possible use of copper is related to allergies and asthma.[12] The inactivation of bacteriophages by copper was reported in 1961.[19] Several authors have shown the antiviral properties of copper derivatives, including hindering the spread of human immunodeficiency virus type 1 and waterbone visures.[12] The biocide characteristic of copper is valid even for macro-organisms as many algae are highly susceptible to copper sulfate. This bioactivity allows its use by water engineers to prevent the growth of algae in potable-water reservoirs.[12]

Examples currently found in the literature about the efficiency of different copper alloys against several bacteria are summarized in Table 13.1.

We would like to stress that copper is the only solid antimicrobial touch surface approved by the US Environmental Protection Agency (EPA) allowing the development of novel technologies with commercial success.[32]

The antimicrobial features of copper have been extended recently to nanoparticles. Several studies confirmed that copper nanoparticles exhibit a better antimicrobial efficacy than macro-micrometric copper particles due to its high surface/volume ratio and surface energy.[33,34] At the macroscale the production of cuprous oxide (Cu_2O) on the surface of copper by corrosion processes obstructs cupric ion release, having a negative influence on the antimicrobial effectivity.[34–36] However, the high surface energy of nanoparticles modifies the corrosion processes, reducing Cu_2O formation and improving the interaction between copper nanoparticles and water, thus producing ions easily and quickly.[33,37,38] Despite the improved corrosion performance at the nanoscale, some researchers suggest that a close interaction between copper nanoparticles and cell membranes is more important for the antimicrobial efficacy.[33] The antimicrobial effectiveness of metallic copper and copper oxides (Cu_2O and CuO) nanoparticles has been demonstrated against several pathogens involved in hospital infections.[33] A better antimicrobial effect against *Escherichia coli* and *Bacillus subtilis* has been reported for copper nanoparticles than for the silver ones, although the sizes of the particles were not the same.[39,40]

Copper nanoparticles either supported on various materials, such as silica, or embedded into mineral particles, such as sepiolite, have also shown antimicrobial behavior.[41,42] Deposition of copper nanoparticles on the surface of spherical SiO_2 nanoparticles was studied achieving a Cu–SiO_2 hybrid structure.[43] In this case, well dispersed copper nanoparticles with sizes less than 10 nm were observed on the silica surfaces showing excellent antibacterial ability.

13.3 Mechanisms behind Antimicrobial Copper

Regarding the mechanism of copper toxicity to micro-organisms, this action is due to several processes, such as displacement of essential metals from their native binding sites or through ligand interactions.[12] Toxicity also results from changes in the conformational structure of nucleic acids and proteins and from interference with oxidative phosphorylation and osmotic balance. Despite the several possible mechanisms, copper's initial site of action is thought to be at the plasma membrane as high concentrations lead to a rapid decline in membrane integrity, changing its permeability.[44] Independently of the mechanism, copper ions are the active agent able to either kill (cidal effect) or inhibit the growth (static effect) of micro-organisms by several processes, as summarized in Figure 13.1.

Table 13.1　Examples found in the literature showing the efficiency of different copper alloys against several bacteria.

Bacteria	Transmission Via	Copper Alloys[a]	References
Escherichia coli G⁻	food/water borne	Copper (99.9%) Silicon bronze (95%Cu) Red brass (93%Cu) Yellow brass (61%Cu) Ni-Al bronze (81%Cu) Al bronze (78%Cu) Ni silver (66%Cu) Brass (90%Cu,80%Cu,70%Cu) Bronze (95%Cu,97%Cu,90%Cu) Cu-Ni (90%Cu,80%Cu) Cu-Ni-Zn (65%Cu, 55%Cu)	14,20–25
Listeria monocytogenes G⁺	food borne	Copper (99.9%) Brass (90%Cu,80%Cu,70%Cu) Bronze (95%Cu,97%Cu,90%Cu) Cu-Ni (90%Cu,80%Cu) Cu-Ni-Zn (65%Cu, 55%Cu)	12,13
Bacillus subtilis G⁺	food borne	Copper (99.9%)	24
Salmonella enteric G⁻	food borne	Copper (99.9%) Brass (70%Cu)	26
Methicillin-resistant *Staphylococcus aureus* (MRSA) G⁺	hospital (nosocomial infection)	Copper (99.9%) Brass (90% Cu,80%Cu,76%Cu, 74%Cu,70%Cu) Bronze (95%Cu,97%Cu,90%Cu) Phosphor bronze (95%Cu) Cu-Ni (90%Cu,80%Cu) Cu-Ni-Zn (65%Cu, 55%Cu)	12,15,27–29
Mycobacterium tuberculosis G⁺	hospital	Copper (99.9%) Brass (70%Cu) Yellow brass (62%)	29
Ciprofloxacin-resistant *Staphylococcus* (CRS) G⁺	hospital	Copper (99.9%); Brass (76%Cu,74%Cu)	27
Enterobacter aerogenes G⁻	hospital	Brass (70%Cu) Phosphor bronze (95%Cu) Cu-Ni (89%Cu)	29
Vancomycin-resistant *Enterococcus faecalis* (VRE) G⁺	hospital	Brass (70%Cu, 60%Cu) Cu-Ni-Zn (65%Cu)	30,31
Klebsiella pneumoniae G⁻	hospital	Brass (70%Cu)	29
Pseudomonas aeruginosa G⁻	hospital	Brass (70%Cu)	29
Enterococcus faecium G⁺	hospital	Brass (70%Cu,65%Cu,60%Cu) Phosphor bronze (95%Cu) Cu-Ni (89%Cu)	16

G⁻: Gram-negative
G⁺: Gram-positive
a: Weight percent

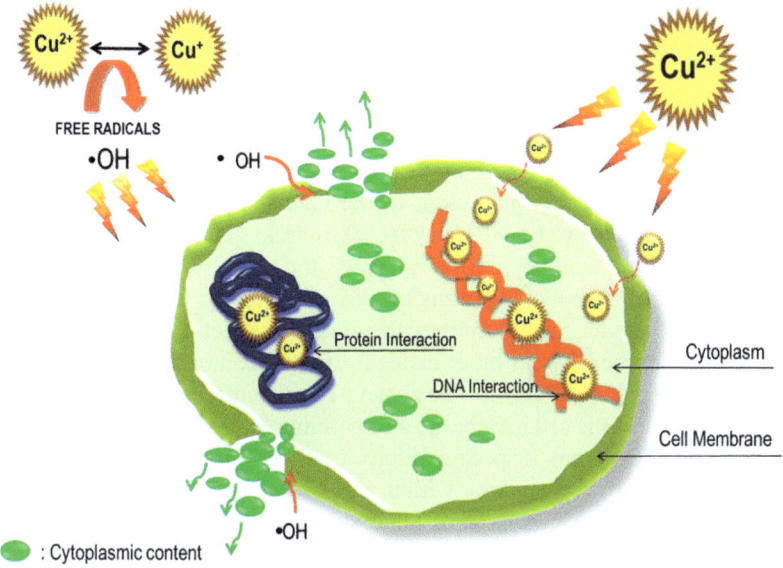

: Cytoplasmic content

Figure 13.1   Diagram showing the different mechanisms associated with the toxicity of copper against bacterial cells. Copper can form both free radicals disrupting the cell membrane and release ions able to interact with proteins and DNA molecules inside the cell.

The redox properties making some metals, including copper, essential elements in biological systems, may also contribute to their inherent toxicity. For example, redox cycling between Cu^{2+} and Cu^+ can catalyze the production of highly reactive hydroxyl radicals by Fenton-like reactions, which can subsequently damage lipids, proteins, DNA, and other biomolecules.[12,45] These reactions are based on copper reactivity with physiological hydrogen peroxide (H_2O_2) produced intracellularly through the oxidation of enzyme NADPH oxidase (see eqns (13.1) and (13.2)). Under these conditions, copper can begin redox cycling (see eqn (13.3)). These reactions are summarized as:[46]

$$NADPH + 2O_2 \rightarrow NADP^+ + H^+ + 2O_2^- \qquad (13.1)$$

$$2H^+ + 2O_2^- \rightarrow H_2O_2 + O_2 \qquad (13.2)$$

$$Cu^+ + H_2O_2 \rightarrow Cu^{2+} + OH^- + \bullet OH \qquad (13.3)$$

Copper can further catalyze another redox reaction when reacts with a lipo-peroxide (LOOH), producing peroxyradicals (LOO•) and reactivates alkoxy radicals (LO•), following eqns (13.4) and (13.5):

$$LOOH + Cu^{2+} \rightarrow LOO \bullet + Cu^+ + H^+ \qquad (13.4)$$

$$LOOH + Cu^+ \rightarrow LO \bullet + Cu^{2+} + OH^- \qquad (13.5)$$

These alkoxy radicals exacerbate the oxidative stress at the cellular membrane promoting lipid peroxidation.[45,47] The physical properties of a membrane are largely determined by its lipid composition. Therefore, these processes lead to loss of cell viability.

Cupric ions (Cu^{2+}) themselves also contribute to the copper toxicity since they are able to form organic complexes with sulfur-, nitrogen- or oxygen-containing functional groups present in the micro-organism. This may result in defects in the conformational structure of nucleic acids and proteins, besides interferring with oxidative phosphorylation and osmotic balance. It has been further suggested that Cu^{2+} cations caused selective lesions in the permeability barrier of the plasma membrane but they did not affect the permeability of the vacuolar membrane whereas other divalent cations did not display any effect. The biocidal effects of copper can also be achieved by the interaction of Cu^{2+} with SH^- moieties present in the cell membrane and within cells.[12] On the other hand, today it is clear that Cu^{2+} has a specific affinity for DNA and can bind and disorder helical structures by crosslinking within and between strands.[48] Moreover, Cu^{2+} reversibly denatures DNA in low ionic strength solutions competing with its hydrogen bonding. The binding of copper to DNA shows an unexpected high specificity when studied in the presence of other metallic ions. Furthermore, copper can alter proteins and inhibit their biological activities. For example, copper was the most potent metal inhibiting the protein tyrosine phosphatase, vaccinia H1-related (VHR), regulating cell growth and differentiation. Among the various metal ions examined for their inactivation effect on VHR, such as Fe^{3+}, Cu^{2+}, Zn^{2+}, and Cd^{2+}, copper ions were found to be the most potent inactivator.[49]

It has been recently reported[50] that the dielectric properties of the cellular components of *E. coli* were changed with the presence of Cu^{2+}. In particular, the permittivity increases with time and with copper concentration due to changes in the polarizability and conductivity in specific areas of *E. coli*. The mechanism of copper antifungal and antialgae activity has not been well studied. Copper ions could form electrostatic bonds with negatively charged regions on the cell wall of the micro-organisms.[12] These electrostatic bonds create stresses that lead to distorted cell wraps and increased permeability, while reducing the normal intake of life-sustaining nutrients. Once inside an algae cell, copper may attack the sulfur groups in amino acids contained in the proteins used for photosynthesis. As a result, photosynthesis is blocked and cell lysis and its death occur.[12]

As mentioned above, cupric ions (Cu^{2+}) are essentials for toxicity.[51] These ions come from a simple dissolution by water in the case of copper-based salts. Otherwise, these ions come from a more complex mechanism based on corrosion for metal or oxide copper materials. Copper releases Cu^{2+} in the presence of water and oxygen according to the following corrosion mechanisms:[35]

$$2Cu + H_2O \rightarrow Cu_2O + 2H^+ + 2e^- \qquad (13.6)$$

$$Cu_2O + 2H^+ \rightarrow 2Cu^{2+} + H_2O + 2e^- \qquad (13.7)$$

$$Cu_2O + H_2O \rightarrow 2Cu^{2+} + 2OH^- + 2e^- \qquad (13.8)$$

Reduction of oxygen dissolved in water also takes place (see eqn (13.9)):

$$2O_2 + 2H_2O + 4e^- \rightarrow 4(OH)^- \qquad (13.9)$$

13.4 Copper Toxicity to Humans

Copper is an essential micronutrient required to sustain life in humans and other organisms but it can also be toxic if present in excess. In this context, it is highlighted that these living systems have developed sophisticated mechanisms to maintain copper balance that respond to sub- or supraoptimal levels. Such mechanisms are related to the direct elimination of the excess from the human body. This equilibrium makes copper much less toxic to humans than to microorganisms. Actually, in healthy individuals consuming a balanced diet, copper deficiency or toxicity is rare, in part because of the ability of the body to regulate copper levels over moderate range of exposures.[52] Therefore, copper is considered a safe metallic compound with an extremely low probability of toxicity. Actually, a dietary requirement has been established in many jurisdictions and for adults usually ranges from 0.9 to 1.7 mg/day. Food, drinking water, and copper-containing diet supplements are the main sources of copper exposure, because intake of copper through inhalation or dermal routes is minimal for most persons. Moreover, it is unlikely to exceed the upper limits for copper, although drinking water from copper pipes and food can contribute significantly to daily copper intake.[52] Chronic ingestions of high amounts of copper result in copper accumulation in the liver although only limited data exist on adverse effects of chronic exposure to high levels of copper in healthy individuals.[52]

13.5 Polymer–Copper Nanocomposites with Antimicrobial Behavior

Considerable success in nanochemistry and nanotechnology has been achieved during recent years allowing the preparation of novel materials of great interest for catalyst, microelectronics, holography, packaging, car industry, aerospace applications, *etc*.[53,54] The impact of this technology is largely due to the synthesis of tailor-made nanoparticles and the specific properties that they possess, such as higher specific surface areas compared with conventional microparticles. As a result, the two novel areas of nanosize chemistry and functional materials sciences have appeared.[53] In polymer science, the impact of nanoscience has mainly come from the development and preparation of (nano)composites, which are hybrid materials containing nanometric inorganic fillers embedded in a polymeric matrix. These (nano)composites are opening up a new generation of macromolecular materials with low densities and multifunctional properties.[55,56] The main advantage that these (nano)composites present is the extremely low amount of filler needed to achieve the desired requirements that in many instances can be one or even two orders of magnitude lower than conventional microfillers.[55,56] In this way, novel polymeric materials have been obtained while maintaining the main properties of the matrix as for example efficient and low energy consumption during processing. Polymeric nanocomposites emerged with the recognition that exfoliated clays could yield significant mechanical advantages in polymeric systems at filler concentrations around 3 vol%.[55,57] Afterwards, it has been shown that the addition of a small number of nanoparticles modifies many of the polymer properties, such as: glass transition, crystallization kinetic, morphology, rheological behavior, electrical conductivity, and thermal stability. Today, organic-modified clays and carbon nanotubes have been largely tested in nanocomposites. However, despite the outstanding properties of metallic nanoparticles, such as electrical conductivity, catalytic activity, and antimicrobial behavior, they are less studied as fillers in polymeric composites. The impact of nanotechnology in polymer/metal antimicrobial materials in particular was shown by Damm *et al.* demonstrating that silver metallic nanoparticles embedded into polyamide matrices by melt-blending present a higher ion release rate than microparticles, due to their extremely high specific surface area.[7] Therefore, the use of polymeric composites based on metallic nanoparticles for biocidal applications should be highlighted, in particular hybrid materials based on copper nanoparticles (CNP). Based on the above, the addition of CNP into polymeric matrices can be a suitable route to prepare bioactive multifunctional polymer-based materials.

The preparation of polymer–CNP composites with antifungal and bacteriostatic properties has been reported using poly(vinylmethyl ketone) (PVMK), poly(vinyl chloride) (PVC), and poly(vinylidene fluoride) (PVDF) as polymeric matrices.[58] The biocide behavior of these composites against *Saccharomyces cerevisiae* yeast, *E. coli*, *S. aureus*, *Listeria monocytogenes*, and molds, depends on the specific properties of the polymeric matrix.[58] Moreover, a larger amount of copper is released by the PVMK and PVC nanocomposites compared with the PVDF one, showing the effect of the matrix on this property. The results of the biostatic activity correlate very well with the copper released directly into the yeast-free culture broth exposed to the nanocomposites for 4 h. It was also reported that the Cu^{2+} release rate rises with the increase of mass fraction of copper nanoparticles. The copper species released to the yeast-free culture broth are Cu^{2+} ions coming from the dissolution of the CuO shell present on the nanocomposites surface. It was stated as an hypothesis that the CuO water-insoluble layer formed can act as a natural barrier to the copper oxidation and dissolution. In particular, this surface layer produces a more controlled release of metal ions than other copper systems such as standard copper salts.

Polyethylene/CNP composites prepared by melt blending for intrauterine devices have been developed showing excellent bioactive properties.[37,38] All composites exhibit burst-release behavior of Cu^{2+} within the first month of incubation and after that they display near-zero-order releases for composites. The authors claim that the burst-release behavior is due to the fact that partial copper nanoparticles attached on surface of the composites can directly react with the solution. These surface particles release cupric ions that enter into the solution without diffusion limitations. Regarding the mechanisms, it was stated that the cupric ions generated by the corrosion of copper in composites can diffuse through the pore canals of polymer matrix into the solvent, in that case, simulated uterine solution. There are two types of pore canals: the immanent in polymer matrix and the acquired from erosion of copper nanoparticles. Only the reactants from the solvent can reach through the pore canals the copper nanoparticles in composites and initiate the corrosion reaction. The rate of corrosion in composites, therefore, depends on the porosity and tortuosity of the composites. X-ray diffraction shows that peaks associated with Cu_2O increase during the experiment, confirming the corrosion process in copper metal nanoparticles. The introduction of a porous structure can improve the cupric ions release rate of these nonporous Cu/PE composites.[59]

Polypropylene has also been mixed with copper metal nanoparticles by melt blending in order to produce antimicrobial plastic materials.[60,61] The results from composites with different amounts of CNP show that the time needed to reduce the *E. coli* bacteria to 50% dropped to half with only 1 v/v % of CNP, compared with the pure polymer. After 4 h, this composite killed more than 99.9% of the bacteria. The biocide kinetics can be controlled by the nanofiller content and composites with CNP concentrations higher than 10 v/v % eliminated 99% of the bacteria in less than 2 h. It is noteworthy that X-ray photoelectron spectroscopy (XPS) did not detect copper on the composite surface even when the sample is sputtered. This result confirms that the biocide behavior is attributed to copper in the bulk of the composite. Copper oxide nanoparticles (CuONP) were also embedded in polypropylene showing stronger antimicrobial behavior against *E. coli* than metal copper nanoparticles.[61] These results show that the antimicrobial property further depends on the type of copper particle. Moreover, composites with CuONP present a higher release rate than composites with metal nanoparticles in short times, explaining the antimicrobial tendency.[61]

A different route is the preparation of hybrid filler having CNP, such as bentonite-supported copper nanoparticles.[62] The sodium cations within the bentonite interlayers were exchanged by Cu^{2+} cations in an aqueous solution of copper sulfate. These exchanged copper cations were reduced by adding hydrazinium hydrate. Such aqueous bentonite/metal hybrid nanoparticle dispersions were blended with cationic poly(methyl methacrylate) (PMMA) latex to produce PMMA hybrid nanocomposites containing exfoliated polymer grafted organoclay together with bentonite-supported metal nanoparticles. The PMMA/bentonite/CNP hybrid nanocomposites exhibited high antimicrobial activity against the ubiquitous bacterium *S. aureus* although this activity was lower than for similar PMMA/bentonite/silver hybrid materials.

As discussed through this chapter, the antimicrobial properties of silver and copper are well established, especially when metals are applied in the form of nanoparticles. However, silver and copper nanoparticles or colloids can agglomerate, resulting in the decrease of their antimicrobial and antifungal properties, in particular in polymer matrices. The problem of nanoparticle stability can be solved by the development of silica nanospheres containing immobilized silver or copper nanoparticles as above explained.[41,42] These nanospheres were applied as effective antibacterial or antifungal additives for architectural paints and impregnates.[63] The most important characteristic of these additives lies in the stability of the metal nanoparticles as well as in the harmlessness to human beings. Silica nanospheres containing immobilized CNP were a more effective antifungal additive as compared to those containing immobilized silver nanoparticles. Moreover, it was observed that tested paints with silica nanospheres containing immobilized copper are especially effective against *Aspergillus niger* and *Chaetosphaeridium globosum*. Moreover, the growth of algae on samples containing 0.5 ppm of CNP was completely stopped, while control samples were intensively covered by algae. This route opens a new approach in the development of antimicrobial polymers with enhanced activity due to the stabilization of the metal nanoparticles. Another example is the use of this approach to produce antimicrobial textiles.[64] Silica sol using the sol-gel method was doped with two different amounts of Cu nanoparticles. Cotton fabric samples were impregnated by the prepared sols. The addition of 0.5 wt.% Cu into silica sol caused the silica nanoparticles to agglomerate in more grape-like clusters on cotton fabrics. However, the presence of higher amounts of Cu nanoparticles (2 wt.%) in silica sol resulted in the most slippery smooth surface on cotton fabrics. All fabricated surfaces containing Cu nanoparticles showed the perfect antibacterial activity against both of Gram-negative and Gram-positive bacteria.

13.6 General Mechanism for Polymer–Copper Composites

A general mechanism explaining the antimicrobial behavior of polymer/copper composites can be stated. The specific mechanism for the antimicrobial action of copper materials is known, as explained above. Copper materials release ions in the presence of water and oxygen, forming complexes with compounds present in the bacteria. These processes result in damage to the cell wall and proteins, both effects killing the bacteria. Therefore, any material able to release enough Cu^{2+} will be antimicrobial. In particular, similar to other copper-based materials, the release of copper ions from polymer–CNP composites is the main reason for the antimicrobial activity of these materials.[58,61] The biostatic activity in these polymeric composites correlates very well with the copper released.[58] It is noteworthy that an advantage of these systems is that the release rate of ions can be controlled by the amount of CNP present in the polymer, allowing the design of a new family of tailor-made biocide materials.[37,58] In this section, the idea is to understand the processes behind the experimental findings showing clearly that an antimicrobial material can be produced by adding copper nanoparticles to a polymer matrix.

Regarding the mechanisms from the polymeric point of view, we focus first on composites based on matrices having extremely low polarity (highly hydrophobic) and high crystallinity, such as polyethylene and polypropylene, and that are able to release copper ions. In this case, the release from copper particles present in the surface region of the composites seems to be the first approach to understand the experimental data. Moreover, these surface particles can account for the maximum in the ion release rate by the "burst" effect, as reported elsewhere.[37] However, X-ray photoelectron spectroscopy analysis shows that CNP are not present on the surface of these samples[60] or are present at concentrations lower than from the bulk of the material.[58] Based on the above, the only suitable mechanism for the release of copper ion is the corrosion of particles present in the bulk of the polymer matrix owing to the diffusion of water molecules interacting with the surface of these particles. Matrices such as polyethylene or polypropylene are highly nonpolar but water molecules can diffuse through it.[65] This diffusion is only through interconnected amorphous parts in polypropylene defining a "percolation network". Other authors talk about "pore channels", meaning the same as discussed here but in a more general way to include amorphous polymers. As the presence of nanoparticles further changes the morphology of the spherulites, in particular by increasing their nucleation density,[8,66] the diffusion of water will be higher in those nanocomposites because of the newly formed spherulite/spherulite amorphous interfaces. The CNP/polymer interface can further increase water diffusion through holes or micrometer-scale defects, allowing fast Knudsen diffusion. This mechanism is directly extended to more polar matrices. When poly(vinylmethyl ketone) was used as polymeric matrix, the minimum deep where the nanocomposite becomes hydrated, and eventually releases the soluble copper species, was estimated around 50 nm, which was about 1/10 of the total film thickness in that case.[58]

By showing that water diffuses through the polymer reaching the particles present in the bulk, the next mechanisms are simpler. When water with dissolved oxygen reaches the copper particles presented in the polymer bulk, the standard corrosion process occurs (see eqns (13.6)–(13.9)). Afterwards, copper ions, coming from corrosion or dissolution processes, can diffuse out through the polymer matrix and finally be released. This mechanism is confirmed for metal copper nanoparticles by comparing the X-ray diffraction analysis of the original composite with the diffraction of the same sample but immersed in water. In the last case, new diffraction peaks appear related to a Cu_2O layer formed on the surface of CNP.[37,38,61] This layer can be easily converted into Cu^{2+} because of its high reactivity.[37,67] A summary of the different mechanisms behind the release of copper ions from polymer/copper composites is displayed in Figure 13.2. Based on these results, we can state that the mechanisms are diffuse-controlled since the water diffusion through the polymer limited the whole process.

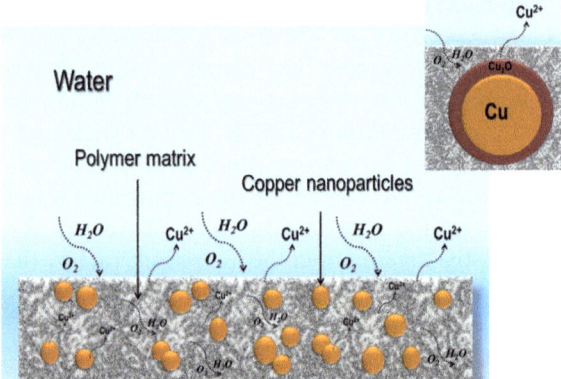

Figure 13.2   Diagram showing a summary of the general mechanism associated with the release of copper ions from copper embedded into a polymer matrix. Water having oxygen molecules can diffuse through the polymer matrix reaching a copper particle. Afterward, a standard copper corrosion process occurs, releasing copper ions that come out into the solution. These ion releases are responsible for the antibacterial behavior of polymer–copper composites. Upper-right diagram: zoom showing the specific processes on a single copper particle.

13.7   Other Antimicrobial Polymers Containing Copper

The incorporation of copper nanoparticles is not the only route taking advantage of the biocide properties of copper in polymers. For example, biocides polymeric hybrid materials based on copper has been prepared by the incorporation of particles on the polymer surface. Copper plasma immersion ion implantation (PIII) was utilized to produce an antibacterial surface on polyethylenes.[1] It was revealed that a relatively large amount of copper, about 11%, is implanted into the near surface region. At the same time, about 3% of copper is found to be also deposited on the surface by this methodology. All these samples presented antibacterial properties. With the same technique, but comparing different metals, the authors revealed that the Ag elemental depth profiles are similar to those of Cu but there is a larger amount of Ag at the surface compared with that of Cu possibly due to the different charge states in the plasma.[68]

A different approach allows the copper impregnation on the surface of cotton fibers, latex, and other polymeric materials.[11] These materials present broad-spectrum antimicrobial (antibacterial, antiviral, and antifungal) and antimite activities.[11] Moreover, this technology enabled the production of antiviral gloves and filters (which deactivate HIV-1 and other viruses), antibacterial self-sterilizing fabrics (which kill antibiotic-resistant bacteria), antifungal socks (which alleviate symptoms of athlete's foot), and antidust mite mattress covers (which reduce mite-related allergies). Copper oxide nanoparticles have been also deposited on the surface of cotton fabrics using ultrasound irradiation.[69] Optimization of the process turned out in a homogeneous distribution of CuO nanocrystals, 15 nm in size, on the fabric surface. It is noteworthy that these samples present strong antibacterial activities against *E. coli* and *S. aureus*. Copper alginate–cotton cellulose (CACC) composite fibers were prepared by immersing cotton fibers in aqueous solution of sodium alginate.[70,71] Afterwards, ionic crosslinking of alginate chains within the cotton cellulose fibers with Cu^{2+} ions was carried out to yield CACC composite fibers. In the last step, the CACC fibers were reduced with sodium borohydride to yield copper nanoparticles into the composite. CACC fibers possessed antibacterial action that depended upon the amount of Cu^{2+} loaded and concentration of alginate into cotton-cellulose fibers. The fibers showed higher Cu^{2+} release in physiological fluid than in distilled water. However, Cu-nanoparticles-loaded fibers by the standard process showed less biocidal action than CACC fibers. This can be related with the fact that CACC fibers just contain Cu^{2+} ions that are directly released from the fibers and act upon the bacterial cell, while in the case of copper-nanoparticle-loaded fibers the conversion of Cu into Cu^{2+} seem to be the governing factor to control their biocidal action.

The addition of additives based on zeolites exchanged with copper ions into polymeric matrices is another route allowing the production of antimicrobial materials.[72] In particular these modified zeolites were embedded into thermoplastic polyether-type polyurethanes resulting in a composite material with relevant antimicrobial activity against methicillin-resistant *S. aureus*, *P. aeruginosa* and *C. tropicalis*.

References

1. W. Zhang, Y. H. Zhang, J. H. Ji, J. Zhao, Q. Yan and P. K. Chu, *Polymer*, 2006, **47**, 7441.
2. D. S. Jones, J. Djokic and S. P. Gorman, *Biomaterials*, 2005, **26**, 2013.
3. Y. L. Yuan, F. Ai and X. P. Zang, *Colloid Surf. B*, 2004, **35**, 1.
4. E. R. Kenawy, S. D. Worley and R. Broughton, *Biomacromolecules*, 2007, **8**, 1359.
5. J. Huang, H. Murata, R. R. Koepsel, A. J. Russell and K. Matyjaszewski, *Biomacromolecules*, 2007, **8**, 1396.
6. S. B. Lee, R. R. Koepsel, S. W. Morley, K. Matyjaszewski, Y. Sun and A. J. Russell, *Biomacromolecules*, 2004, **5**, 877.
7. C. Damm, H. Münstedt and A. Rösch, *Mater. Chem. Phys.*, 2008, **108**, 61.
8. C. Radheshkumar and H. Münstedt, *React. Funct. Polym.*, 2006, **66**, 780.
9. C. Damm and H. Münstedt, *J. Mater. Sci.*, 2007, **42**, 6067.
10. H. M. C. de Azeredo, *Food Res. Int.*, 2009, **42**, 1240.
11. G. Borkow and J. Gabbay, *FASEB J.*, 2004, **18**, 1728.
12. G. Borkow and J. Gabbay, *Curr. Med. Chem.*, 2005, **12**, 2163.
13. S. A. Wilks, H. Michels and C. W. Keevil, *Int. J. Food Microb.*, 2005, **105**, 445.
14. J. O. Noyce, H. Michels and C. W. Keevil, *J. Hosp. Inf.*, 2006, **63**, 289.

15. H. T. Michels, J. O. Noyce and C. W. Keevil, *Lett. Appl. Microb.*, 2009, **49**, 191.
16. G. Faundez, M. Troncoso, P. Navarrete and G. Figueroa, *BMC Microb.*, 2004, **4**, 19.
17. A. M. Mulligan, M. Wilson and J. C. Knowles, *Biomaterials*, 2003, **24**, 1797.
18. M. J. Lehtola, I. T. Miettinen, M. M. Keinanen, T. K. Kekki, O. Laine, A. Hirvonen, T. Vartiainen and P. J. Martikainen, *Water Res.*, 2004, **38**, 3769.
19. N. Yamamoto, C. W. Hiato and W. Haller, *Biochem. Biophys. Acta*, 2001, **91**, 257.
20. H. T. Michels and D. G. Anderson, *Proceedings of the Sixth International Copper-Copper Conference*, 2007, **1**, 121.
21. S. L. Warnes and C. W. Keevil, *Appl. Environ. Microbiol.*, 2012, **14**, 1730.
22. A. Lewis, C. W. Keevil, *Antibacterial properties of alloys and its alloys in HVAC&R systems*, International Copper Association, New York, 2004.
23. A. J. Varkey, *Sci. Res. Essays*, 2010, **5**, 3834.
24. C. Espirito Santo, E. Wen Lam, C. G. Elowsky, D. Quantana, D. W. Domaille, C. J. Chang and G. Grass, *Appl. Environ. Microbiol.*, 2011, **77**, 794.
25. C. Espirito Santo, N. Taudte, D. H. Nies and G. Grass, *Appl. Environ. Microbiol.*, 2008, **74**, 977.
26. L. Zhu, J. Elguindi, C. Rensing and S. Ravishankar, *Food Microbiol.*, 2012, **30**, 303.
27. A. Mikolay, S. Huggett, L. Tikana, G. Grass, J. Braun and D. H. Nies, *Appl. Microbiol. Biotechnol.*, 2010, **87**, 1875.
28. A. L. Casey, D. Adams, T. J. Karpanen, P. A. Lambert, B. D. Cookson, P. Nightingale, L. Miruszenko, R. Shillam, P. Christian and T. S. J. Elliott, *J. Hosp. Inf.*, 2010, **74**, 72.
29. S. Mehtar, I. Wiid and S. D. Todorov, *J. Hosp. Inf.*, 2008, **68**, 45.
30. S. L. Warnes and C. W. Keevil, *Appl. Environ. Microb.*, 2011, **77**, 6049.
31. S. L. Warnes, S. M. Green, H. T. Michels and C. W. Keevil, *Appl. Environ. Microbiol.*, 2010, **76**, 5390.
32. www.antimicrobialcopper.org.
33. G. Ren, D. Hu, E. W. C. Cheng, M. A. Vargas-Reus, P. Reip and R. P. Allaker, *Int. J. Antimicrob. Ag.*, 2009, **33**, 587.
34. K. Midander, P. Cronholm, H. L. Karlsson, K. Elihn, L. Möller, C. Leygraf and I. O. Wallinder, *Small*, 2009, **5**, 389.
35. R. Francis, *The Corrosion of Copper and Its Alloys: A Practical Guide for Engineers* ed. Nace International, Houston, Texas, 2010, pp. 1–369.
36. R. Breur, *Fouling and Bioprotection of Metals*, ed. H. J. A. Breur, Netherlands, 2001, pp. 1–163.
37. S. Cai, X. Xia and C. Xie, *Biomaterials*, 2005, **26**, 2671.
38. T. Xu, H. Lei, S. Z. Cai, X. P. Xia and S. C. Xie, *Contraception*, 2004, **70**, 153.

39. K. Y. Yoon, J. H. Byeon, J. H. Park and J. Hwang, *Sci. Total Environ.*, 2007, **373**, 572.
40. J. P. Ruparelia, A. K. Chatterjee, S. P. Duttagupta and S. Mukherji, *Acta Biomater.*, 2008, **4**, 707.
41. O. Akhavan and E. Ghaderi, *Surf. Coat. Tech.*, 2010, **205**, 219.
42. E. A. Cubillo, C. Pecharroman, E. Aguilar, J. Santaren and J. S. Moya, *J. Mater. Sci.*, 2006, **41**, 5208.
43. Y. Kim, D. Lee, H. Cha, C. Kim, Y. Kang and Y. Kang, *J. Phys. Chem. B*, 2006, **110**, 24923.
44. Y. Ohsumi, K. Kitamoto and Y. J. Anraku, *J. Bacteriol.*, 1988, **170**, 2676.
45. S. V. Avery, N. G. Howlett and S. Radice, *Appl. Environ. Microbiol.*, 1996, **62**, 3960.
46. G. M. Teitzel, A. Geddie, S. K. De Long, M. J. Kirisits, M. Whitrlry and M. R. Parsek, *J. Bacteriol.*, 2006, **188**, 7242.
47. V. S. Lebedev, A. V. Veselovskii, E. Deinega and I. Fedorov, *Biofizika*, 2002, **47**, 295.
48. J. M. Rifkind, Y. A. Shin, J. M. Hiem and G. L. Eichorn, *Biopolymers*, 2001, **15**, 1879.
49. J. H. Kim, H. Cho, S. E. Ryu and M. U. Choi, *Arch. Biochem. Biophys.*, 2000, **382**, 72.
50. W. Bai, K. Zhao and K. Asami, *Colloid Surf. B*, 2007, **58**, 105.
51. A. O. Summers and S. Silver, *Annu. Rev. Microbiol.*, 1978, **32**, 637.
52. K. A. Cockell, J. Bertinato and M. R. L'Abbé, *Am. J. Clin. Nutr.*, 2008, **88**, 863S.
53. A. D. Pomogailo and V. N. Kestelman, *Metallopolymer Nanocomposites*, ed. Springer, Germany, 2005.
54. E. Thostenson, C. Li and T. Chou., *Comp. Sci. Tech.*, 2005, **65**, 491.
55. D. R. Paul and L. M. Robeson, *Polymer*, 2008, **49**, 3187.
56. M. Moniruzzaman and K. Winey, *Macromolecules*, 2006, **39**, 5194.
57. M. Kawasumi, N. Hasegawa, M. Kato, A. Usuki and A. Okada, *Macromolecules*, 1997, **30**, 633.
58. N. Cioffi, L. Torsi, N. Ditarantano, G. Tantalillo, L. Ghibelli, L. Sabbatini, T. Bleve-Zacheo, M. D'Alessio, P. G. Zambonin and E. Traversa., *Chem. Mater.*, 2005, **17**, 5255.
59. W. Zhang, X. Xia, C. Qi, C. Xie and S. Cai, *Acta Biomater.*, 2012, **8**, 897.
60. H. Palza, S. Gutiérrez, K. Delgado, O. Salazar, V. Fuenzalida, J. Avila, G. Figueroa and R. Quijada, *Macromol. Rapid Commun.*, 2010, **31**, 563.
61. K. Delgado, R. Quijada, R. Palma and H. Palza, *Lett. Appl. Microbiol.*, 2011, **53**, 50.
62. H. Weickmann, J. C. Tiller, R. Thomann and R. Mulhaupt, *Macromol. Mater. Eng.*, 2005, **290**, 875.
63. M. Zielecka, E. Bujnowska, B. Kepskaa, M. Wenda and M. Piotrowska, *Prog. Org. Coat.*, 2011, **72**, 193.
64. A. Berendjchi, R. Khajavi and M. E. Yazdanshenas, *Nanoscale Res. Lett.*, 2011, **6**, 594.
65. T. M. Ton-That and B. J. Jungnickel., *J. Appl. Polym. Sci.*, 1999, **74**, 3275.

66. H. Palza, J. Vera, M. Wilhelm and P. Zapata., *Macromol. Mater. Eng.*, 2011, **296**, 744.
67. X. Xia, C. Xie, S. Cai, Z. Yang and X. Yang, *Corros. Sci.*, 2006, **48**, 3924.
68. W. Zhang, Y. Luo, H. Wang, S. Pu and P. Chu, *Surf. Coat. Technol.*, 2009, **203**, 2550.
69. I. Perelshtein, G. Applerot, N. Perkas, E. Wehrschuetz-Sigl, A. Hasmann, G. Guebitz and A. Gedanken, *Surf. Coat. Technol.*, 2009, **204**, 54.
70. M. Grace, N. Chand and S. K. Bajpai, *J. Eng. Fib. Fabr.*, 2009, **4**, 24.
71. S. K. Bajpai, M. Bajpai and L. Sharma, *J. Appl. Polym. Sci.*, 2012, **126**, 318.
72. P. Kaali, M. M. Pérez-Madrigal, E. Strömberg, R. E. Aune, G. Czél and S. Karlsson, *Exp. Polym. Lett.*, 2011, **5**, 1028.

CHAPTER 14

Photocatalytic Oxide–Polymer (Nano)Composites for Antimicrobial Coatings and other Applications

ANNA KUBACKA, ANA IGLESIAS-JUEZ AND
MARCOS FERNÁNDEZ-GARCÍA*

Instituto de Catálisis y Petroleoquímica (ICP-CSIC), CSIC, C/Marie Curie 2,
28049-Madrid, Spain
*Email: mfg@icp.csic.es

14.1 Introduction

Organic/inorganic nanocomposite materials, where the organic major component is based on polymers, are a fast-growing area of research in order to meet new requirements and challenges in properties and performances of the resulting materials that allow a further development and progress of the present and future society. The excellent characteristics of titanium dioxide, TiO_2 (known also as titania) and zinc oxide, ZnO, make them ideal candidates as inorganic component in these promising oxide-based polymeric nanocomposites. Therefore, the primary aim of the present chapter consists in the description of model examples of these materials based in their advanced, light-triggered properties, with emphasis on those exhibiting biocidal or self-cleaning as well as self-degradation characteristics. As a prior stage, the striking photocatalytic properties of the oxides related to nanostructure are discussed.

RSC Polymer Chemistry Series No. 10
Polymeric Materials with Antimicrobial Activity: From Synthesis to Applications
Edited by Alexandra Muñoz-Bonilla, María L. Cerrada and Marta Fernández-García
© The Royal Society of Chemistry 2014
Published by the Royal Society of Chemistry, www.rsc.org

14.2 The Bare Oxides: Photocatalysis by Nanostructured Materials

An oxide can be excited by light energy higher than its bandgap inducing the formation of energy-rich electron–hole pairs. It is commonly understood that the term photocatalysis is referred to any chemical process catalyzed by a solid where the external energy source is an electromagnetic field with wave numbers in the UV-visible-infrared range.[1] In this section we outlined the effect of the nanostructure in the photocatalytic performance of the two more important photocatalytic oxides, TiO_2 and ZnO.

14.2.1 Titanium Oxide

The interpretation of the photocatalytic behavior of TiO_2 nanomaterials requires to take into account both intraparticle and interparticle effects after light excitation. These effects in turn are ultimately driven by two types of physicochemical variables: morphology and defect structure of the oxide. Intraparticle effects are rationalized in terms of primary particle size and shape effects on the structural and electronic characteristics of nanosize materials. In addition, interparticle effects, particularly secondary particle size and porosity, also appear to be important variables as detailed below. Size, shape, and porosity are, thus, key variables to control photoactivity and this occurs mainly through i) light absorption *vs.* dispersion; and ii) de-excitation pathways that dominate recombination.[1]

Concerning nanoanatase (the thermodynamically stable phase for nano-structured TiO_2), two limiting situations can be identified. These are a function of the best-known morphological variable, namely size, and concern the ranges 4–5 and 12–15 nm where differences in defect structure lead to distinctive electronic consequences. These, however, are concerned with localized gap states without the influence of band structure. Below 5 nm, as a result of cation vacancies and amorphization, a p-type behavior may be suggested. On the other hand, above that point n-type bulk-like behavior can be expected. In these particles, anion vacancies and, in specific cases, interstitial Ti are the most typical defects. Above 15 nm, the defect structure would be exclusively concerned with the surface. This contains *ca.* 5% of the atoms and essentially, it could be envisaged that bulk behavior would result. A summary of the effect of size and defect in the electronic structure of titania particles is summarized in Figure 14.1.

Shape is the other morphological variable with a known influence on the defect structure of titania. There appears to be an increase in the number of undercoordinated surface atoms from elongated to isotropic geometries. A beneficial effect on the photocatalytic activity, derived from presence of (001) surfaces, is generally reported[2,3] for particles with a size greater than 5 nm. This indicates that elongated shapes would be detrimental for photoactivity with respect to isotropically shaped ones or any other shape that maximizes the presence of (001) surfaces. The photocatalytic influence of this (001) surface

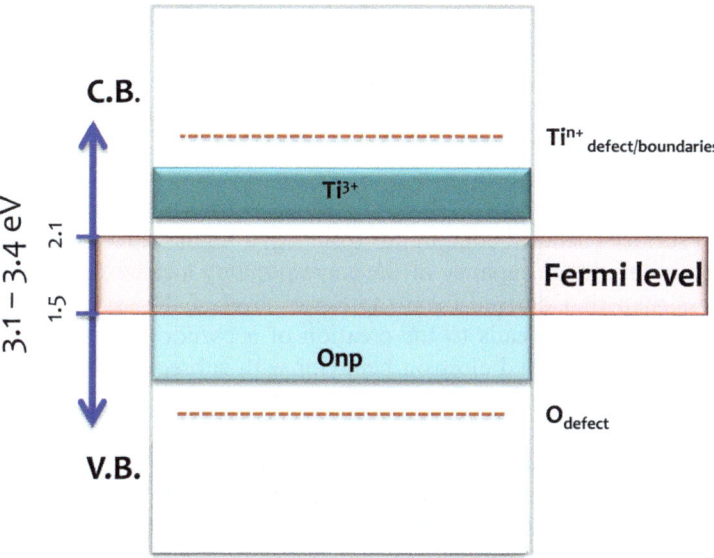

Figure 14.1 Schematic view of main electronic characteristics of nanometric TiO$_2$. See text for details.

may be a trade-off of two opposing facts: the larger the chemical reactivity of the more open surface and the higher charge recombination expected by a higher surface/near-surface density of defects. There is, however, the question of whether the presence of alien (mostly fluoride) species that stabilize such (001) surfaces may contribute to the high photocatalytic performance observed. Recent studies support both possibilities. Some suggest that the fluoride contribution may not be dominant,[4] while others tend to support the key catalytic role of fluoride.[5] Also, the influence of the defect nature and number in the electronic properties and photoactivity of (001) surfaces is under debate.[6]

Upon light excitation, nongeminate radiative losses, and hence charge carriers with potential for the correct chemistry, are usually found in the visible range above 2 eV. For example, within the high-energy region of the visible spectra, up to three processes are triggered by photons of *ca.* 2.0–2.2, 2.5, and 2.8–2.9 eV. Systems can involve one, two or the three processes depending on the method of preparation and other so-far unknown variables. Decay pathways that occur below an energy of about 2.0 eV are usually ascribed to Ti(III) centers.[7] Interpretation of the electronic structure behind these three de-excitation processes, which is relevant to this work, is not universal.[1,7] In spite of the lack of a unified interpretation, minimizing the number of defects, more specifically the anion bulk vacancies, appears essential for enhancing photocatalytic activity for both UV (ultraviolet) and visible-light excitations. Where the anion vacancies are concerned, their final role is linked to their structural/electronic energy-dispersion properties. A localized nature of their electronic states and ultimately their performance as recombination centers is the most likely case when they are located at bulk positions. In contrast, when anion bulk vacancies produce quasicontinuum states, the positive or negative role on charge recombination is likely related to their initial electronic occupation (or in others words to the number of defects and the corresponding electronic state energy position with respect to the Fermi edge).[1,7]

The electronic consequences of these defect structures after light excitation can be complex. In the case where the electron transfer from surface defect states to surface chemical compounds results in a higher reaction rate than the detrapping step, the occupancy of the corresponding localized states would be much less than that derived from the Fermi level of the system and Fermi–Dirac statistics.[8] This leads to the creation of a pseudo-Fermi edge that can control the defect-derived electron behavior with subsequent consequences in the dynamic behavior of the charge carriers.

With respect to nanosized semiconductors, there is another relevant aspect namely carrier multiplication. As shown by Schaller and Klimov, one photon with an excess of energy (over the bandgap) can produce not one but multiple excitons. Up to 4 excitons per photon are possible for high-energy photons. While the exact mechanism for this enhancement is still a subject of debate, high multiexciton yields in nanocrystals are likely due to factors such as the close proximity between interacting charges, reduced dielectric screening, and relaxation in translation-momentum conservation.[9]

Interparticle, or more generally speaking media, effects appear to be as important as intraparticle effects for the interpretation of light–matter interactions even in pure anatase nanomaterials. Due to the relatively poor control of particle size and morphology in crystalline oxide systems, materials with an average primary particle size in the 5–10 nm range can possess two hetero-size contact structures with particles either above or below 5 and 12 nm (Figure 14.2). The first possibility is rather important since this would lead to a p-n union between any particles of a size below or above 5 nm. Here, the

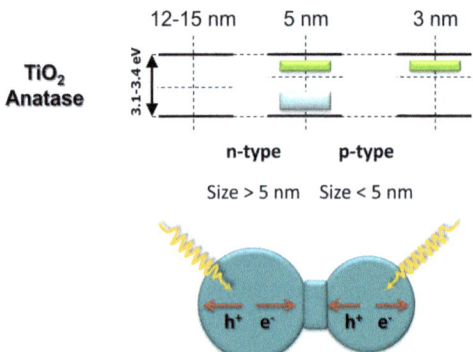

Figure 14.2   Top part: Size dependence of main electronic characteristics of nanometric TiO_2. Bottom part: change dynamics after light excitation as a result of a p-n junction present in nanometric TiO_2 as an inherent effect of polydispersity.

Schottky barrier acts as a charge accumulation center and allows long-lived charge carriers with the potential to undergo chemical processes (Figure 14.2). This type of complex interparticle contact that occurs around 5 nm, together with the fact that in nanomaterials nonradiative recombination is a multi-particle process that dominates the charge-carrier relaxation process support the view that morphology variables, such as size, shape, secondary particle size and porosity, play a key role in controlling light–matter interactions. The second critical state is for sizes around 12–15 nm. Here, the importance is in relation to the effect of surface states (as mentioned above). The heterocontact can potentially lead to a low-energy state for electrons located in the smallest particles. Such an electronic state may display its own "separate" electron dynamic behavior. Finally, we must consider the effect of grain frontier effects. Alimohammada and Fitchthorn[10] analyzed the attachment of well-defined nanoparticle surfaces. They observed oriented interfaces (*e.g.* attachment) driven by electrostatic forces between the (001) local planes and under-coordinated atoms along the edges of two (101) planes of other nanoparticles. Data regarding local order detail within such interfaces require further input. While poorly understood, specific local environments able to stabilize relatively unsaturated cation sites may be present in such grain interfaces and these sites may show specific photoactivity behavior.

The combination of intraparticle and interparticle effects discussed so far clearly emphasizes the fact that all the morphological variables concerning primary particle size and shape, particle-size distribution (or, in other words, polydispersity) as well as secondary particle size and porosity are important in any description of the photoactivity of TiO_2 materials.

14.2.2 Zinc Oxide

ZnO has been demonstrated to exhibit remarkable performance in electronics, optics and photonics. As in the case of TiO_2, ZnO is a wide bandgap (3.37 eV) semiconductor that makes this material appropriate for short-wavelength optoelectronic applications. In this sense, some studies have confirmed that ZnO exhibits a better efficiency than TiO_2 in photocatalytic degradation of some dyes in aqueous solution.[11,12] Unfortunately, the photocorrosion of ZnO that occurs with the UV-light irradiation, as well as the susceptibility of ZnO to facile dissolution at extreme pH values, have significantly decreased the pho-tocatalytic activity of ZnO in aqueous solution and blocked its application in photocatalysis.[13] In fact, ZnO is affected by dissolution under both acidic and highly alkaline conditions, and could even be dissolved in neutral solution under light illumination. Therefore, ZnO could become an excellent photo-catalyst if the photocorrosion and susceptibility can be improved or suppressed.

Regarding structural features, ZnO can be found in three crystalline forms: wurtzite, cubic zinc blende and rarely observed cubic rock salt. Among these structures, the wurtzite one is the most stable and presents photocatalytic ac-tivity. This structure can be described as a number of alternating planes composed of tetrahedrally coordinated O^{2-} and Zn^{2+} stacked alternately along the c-axis. Wurtzite zinc oxide has a hexagonal structure (space group *C6mc*) with lattice parameters $a = 0.3296$ and $c = 0.52065$ nm. The oppositely charged ions produce positively charged Zn-(0001) and negatively charged O-(000–1) surfaces, resulting in a normal dipole moment and spontaneous polarization along the c-axis as well as a divergence in surface energy. The presence of polar surfaces as well as the noncentral symmetry will have a strong influence in the crystal growth, inducing anisotropic morphologies, and, therefore, will affect the final photocatalytic behavior. Thus, different nanostructures can be formed by controlling the synthetic routes.

From these considerations, ZnO has, structurally, three types of fast growth directions: {2–1–10}, {01–10} and {0001}, leading to different morphologies widely reported in the literature. One of the key factors determining the crystal growth and the final morphology involves the relative surface energy of various crystalline facets. The surface energies (E) of the facets in ZnO crystals follow the sequence $E(0001) > E(10–1–1) > E(10–10) > E(10–11) > E(000–1)$, which is the same order as the crystal growth rates. Nonfaceted particles have higher surface energies than the faceted ones and, therefore, are not favored.[14] Macroscopically, the different crystal planes have specific kinetic parameters, which could be highlighted under specific growth conditions. Thus, a crystallite, after its initial nucleation step, will commonly develop into a three-dimensional structure with well-defined, low-index crystallographic faces during the crystallization process. In the case of ZnO, the structures would tend to maximize the areas of the {2–1–10}, {01–10} and {0001} facets because of their lower energy, leading to characteristic morphologies. Wurtzite ZnO usually tends to grow along the c-axis and maximizes the exposed areas of the {2–1–10} and {01–10} nonpolar facets.[15] Hence, controlling the polar (0001) facet of ZnO photocatalyst remains a challenge.[16]

Due to the particular structural disposition discussed above, the synthesis of ZnO nanostructures is currently attracting intense worldwide interest. Thus, numerous ZnO nanostructures are prepared, such as nanodots,[17] nanowires (NWs),[18] nanorods (NRs),[19] nanobelts,[20] nanorings,[21] nanotubes (NTs),[22] nanocages[23] and hierarchical patterns.[24] Among preparation methods, the wet chemical routes provide better control over size and morphology and, in addition, they would be considered more facile and promising for industrial-scale preparation. Moreover, recent studies have shown that the use of organic additives in the aqueous solution can effectively tune the shape of the product by selective adsorption and subsequent controlled removal of organic additives at interfaces.[25–27]

As already stated, ZnO particle morphologies are rather complex and diverse with respect to TiO_2. Different particle shapes as well as hierarchical super-structures can be achieved. Due to this special fact, structural features, as crystallinity (crystallite size and lattice strain) and morphology, arise as the most affecting parameters in detrimental to surface area.[28,29] Within this context, several studies reported the correlation of different morphologies on the final photocatalytic performance of ZnO.

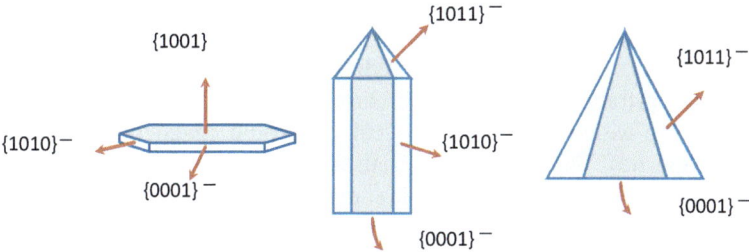

Figure 14.3 Typical morphological configurations observed for nanometric ZnO.

Morphology models of as-prepared ZnO flakes, columns, and pyramids are shown in Figure 14.3. If the different model morphologies for ZnO are considered, the dominating planes are (0001) and (000–1) (top and bottom surfaces, respectively) in the case of flakes, which correspond to the O-terminated and Zn-terminated facets. Secondly, ZnO with columnar shape, the dominating planes are nonpolar {10–10} plane family (*i.e.* six-sided surfaces). In pyramid-like ZnO, the mainly exposed planes are the O-terminated polar (000-1) plane (bottom surface) and the {10–11} plane family (six side surfaces).

Upon these structural considerations, Han *et al.* correlated the photocatalytic activity of ZnO prepared by different routes exhibiting the above model morphologies.[30] The catalytic results obtained were clearly in accordance with the exposed plane chemical reactivity. Thus, the most active facet was (0001) formed Zn^{2+} termination. The chemical adsorption of O_2 or OH^- is improved in this plane and, therefore, favors the combination with photogenerated electrons and holes. In the same way, Zhou *et al.* proposed the preparation of ZnO hexagonal micropyramids by using ionic liquids as solvents.[31] Such structure shows all their exposed facets consisting of polar $\pm(0001)$ and {10–11} planes.

In addition to the above-mentioned structural influence on the photocatalytic activity, the presence of native defects has also been discussed, leading to contradictory effects depending on its type and concentration.[32,33] An optimal surface oxygen vacancy concentration would result in an increased photoactivity.[34] Thus, Zheng *et al.* reported that the photoactivity of ZnO nanocrystals depends on the type and concentration of oxygen defects.[35] More generally speaking, the presence of donor defects (oxygen vacancies[36] as well as interstitial zinc[37]) has been proposed to have a strong influence on the photocatalytic activity. In this sense, some authors improved the photoactivity of ZnO by intentionally creating defects in the crystal lattice, for example by means of a rapid crystallization induced by microwave synthesis.[38] Finally, the presence of *p-n* contacts have been mentioned in bare ZnO as an effect of heterogeneity in size distribution for samples having particles with anion (fraction with the smallest size) and cation (large particle size) substitutional vacancies[39] but further experimental evidence is desirable.

14.3   Oxide–Polymer Nanocomposites

In recent years, organic–inorganic hybrid or nanocomposite materials that combine the attractive qualities of dissimilar components have received great attention for a wide range of mechanical, electronic, magnetic, biological, and optical properties. These novel materials are not merely physical mixtures but can be broadly defined as complex materials, having both organic and inorganic components intimately connected at nanometric length scale. The methodologies used for their preparation are mainly: sol-gel, *in situ* polymerization, solution and melt processes. They can be classified, as defined by Novak[40] and others,[41] as: type I for soluble, preformed organic polymers embedded in an inorganic network; type II for embedded, preformed organic polymer owing covalent bonds to the inorganic network; type III for mutually interpenetrating organic–inorganic networks; type IV for mutually interpenetrating networks with covalent bonds between the organic and inorganic phases; and type V for "nonshrinking" sol-gel composite materials. Obviously, there are materials falling between categories that would make their classification difficult.

Here, we focus in composite systems having light-advanced properties and particularly connected with oxide photocatalytic triggered properties with particular emphasis in the context of self-cleaning and self-degradation properties. As described in the previous section, light absorption promotes the existence of charge carriers, electrons and holes, at the semiconductor. Such charge carriers must leave the oxide component and reach the surface of the nanocomposite system in order to generate the above-mentioned properties. Such charge-carrier mobility processes are dominated by interface (*e.g.* oxide–polymer and polymer–media) effects, not fully understood at the moment. In spite of this, self-cleaning corresponds to an application field widely tested, with significant literature in the case of titania.[42] Self-cleaning properties are of importance in relation to the control and elimination of dangerous microorganisms. This point would be the main subject of the following sections. On the other hand, self-degradation of a polymer-based nanocomposite system is aimed to be attained after completion of the life cycle and using renewable energy sources such as sunlight. Self-degradation is also of importance in connection with the aging of the material suffered while exposing the system to UV or visible excitation during biocidal action. Contrarily to the self-cleaning case, the self-degradation properties of oxide–polymer nanocomposites have been explored to a rather modest degree.

14.3.1   Titanium Dioxide–Polymeric Nanocomposites

There is a significant interest in the development of antimicrobial materials for application in health and biomedical devices, food packaging, and personal hygiene industries.[43–57] Among several possibilities currently explored, titania can be noted as a potential candidate for polymer modification with a significant number of advantages, the most obvious resulting from the absence of releasing of dangerous materials to the media. This is a direct, positive input when compared with other current nanomaterials, like silver, with antimicrobial purposes. Moreover, the titania–polymer system seems to present an optimal response in the fight against all micro-organisms, irrespective of their nature, *e.g.* bacteria, virus and even algae or protozoa.[42]

Probably, the earliest work of photoactive TiO_2-containing polymer-based nanocomposites was carried out by Kim and coworkers, which aimed to reduce biofouling effects on polymer-based (polyamide) membrane systems.[58,59] The nanocomposite material showed efficient control of *E. coli* in a period of hours, thus establishing the potential of the system in micro-organism control and killing. The presence of TiO_2 on polymer-based membranes was recently reviewed, although it appears that, apart from the already mentioned contributions, there is an essential lack of information concerning disinfection applications of these materials.[60] As detailed in the review article, some disinfection studies considered a technology derived from the combination of TiO_2 slurries and polymer-based membranes but the use of nanocomposites was/is certainly scarce.

Only recently several groups tried to build up some general work aiming to testing the potential antimicrobial properties of polymer-based composite systems containing TiO_2 as a biocidal agent. In this line, Kubacka *et al.*[61] describe the preparation by a melting process of ethylene-vinyl alcohol copolymer, $EVOH/TiO_2$ nanocomposite films with different amounts of the inorganic TiO_2 component with anatase structure, primary particle size of ~9 nm and a BET surface area of 104 m^2/g. These nanoparticles are synthesized by a microemulsion method, and EVOH copolymers are selected because of their extensive commercial applications in food packaging, health and biomedical industrial areas. The use of compatibilizer or coupling agent(s) is not needed in these nanocomposites due to the amphiphilic nature of EVOH copolymer, which is able to include titania nanoparticles in the final material with a good adhesion at interfaces between the two components. These nanocomposites present extraordinary antimicrobial properties against a number of clinically and food-derived infection-relevant pathogens, such as Gram-positive and Gram-negative bacteria (*E. coli*, *E. caratovora*, *E. faecalis*, and *P. aeruginosa and fluorescens*), Gram-positive cocci (*S. aureus*), and yeasts (*Z. rouxii* and *P. jadini*). Figure 14.4 exemplifies the biokilling performance for a series of samples having different weight loadings of the inorganic component. The study showed that a key advantage of these polymer-based materials biocidal properties (with respect to oxide-alone ones or even of polymer-based systems incorporating biocidal agents like Ag) relies on the fact that the whole nanocomposite surface becomes biocidal and thus relaxes the requirement for direct biocidal agent to micro-organism contact. In such a way, these nanocomposites exhibit an astonishing biocidal behavior in comparison with that presented by well-known biocide agents such as Ag-based systems, AgBr particles coating with poly(vinylpyridine) or simple chemicals (glutaraldehyde, formaldehyde, H_2O_2, phenol, cupric ascorbate, or sodium hypochlorite).[51,54,62,63] The outstanding biocidal capability comes from a close

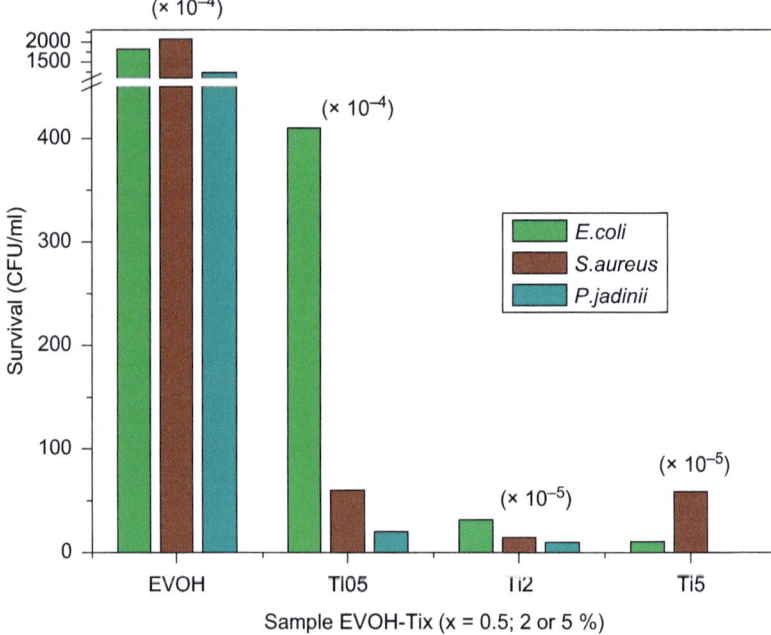

Figure 14.4 Photokilling performances of EVOH-TiO$_2$ composites with variable inorganic content.

contact at the nanometric scale between polymer and oxide nanoparticles, this contact being proved by the presence of new electronic states in the nanocomposite systems, not present in the individual components. The optimum handling of charge carriers through the organic/inorganic interface occurs for oxide loading between 2 and 5 wt.%. A compromise is found between the maximization of the components interarea and the aggregation of the inorganic component.[61,64]

Figure 14.5 displays the results of a SEM study illustrating how incorporation of TiO$_2$ nanoparticles influences the biokilling potential of the nanocomposites, not only affecting the cell viability but also the bacteria aggregation and biofilm formation. Biofilm formation control is a major issue in microbial control and killing, and the high effectiveness showed by these systems appears as a distinctive advantage of polymer–TiO$_2$ nanocomposites with respect to the bare oxide and/or traditional chemical agents. Note that the advantage against the pristine oxide comes directly from the fact that, as explained above, the composite system is a type of noncontact agent, relaxing some of the inherent limitations of the inorganic biocidal that require close contact with the micro-organism. In addition, these nanocomposites are easily photodegraded by exposure to sunlight (see below). Therefore, they could be considered as environmentally friendly polymeric nanomaterials with potential applicability in the productive sector.[61]

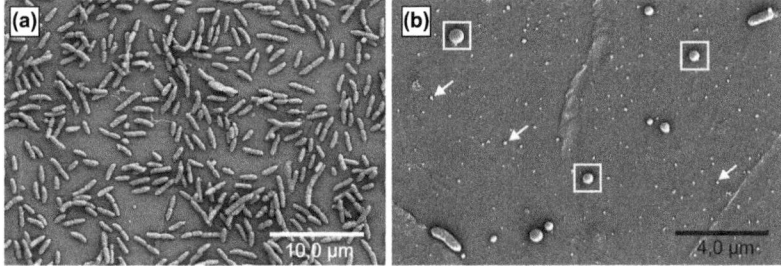

Figure 14.5   SEM images of the *P. aeruginosa* cells sited at the surface of the TiO$_2$-
EVOH nanocomposite with a 2 wt.% of nanoparticle content in the
absence (a) and presence of UV light (b).

The physicochemical characterization of these nanocomposites has additionally shown that the EVOH crystallinity does not practically vary by the titania nanoparticles incorporation, although the crystal size slightly rises with nanoparticle content at low compositions. The values of glass-transition and melting temperatures do not differ significantly with the increase of titania composition, indicating a slight effect of TiO$_2$ nanoparticles on the transitions related to the amorphous and crystalline phases within the polymeric component. However, a considerable microhardness improvement is observed as the TiO$_2$ amount rises in the nanocomposites.[65] Such features would indicate that the conventional applicability of the EVOH copolymer is not affected or even reasonably improved by the presence of the biocidal oxide.

In a recent work, the same group provided a detailed analysis of the effect of the oxide nanostructure in the self-cleaning and self-degradation properties of the TiO$_2$-EVOH material. Biocidal capability against *E. coli* and *S. aureus* was found to be maximized as a complex function of several morphological variables among which (oxide) secondary particle size and porosity were critical. Interconnected oxide particles, a parameter in turn correlated with primary particle size and shape as well as porosity, seem in fact necessary for charge-carrier species to reach the system surface. The study also analyzed the eco-disposal of the materials after light use with the help of sunlight as a renewable energy source. In such case, the key morphological parameter was oxide primary particle size. The analysis was able to show that the different morphological dependence of these two properties is intimately linked with the fact that different charge carriers are involved in them. While hole-related species are responsible of the self-cleaning properties, self-degradation properties require electron-related species.[66] The requirement of these two different charge species (electrons and holes) thus makes self-degradation and self-cleaning properties fully complementary as they make full use of all charge carriers produced by light excitation, limiting the recombination process between them.

In additional studies, Kubacka *et al.* have evaluated isotactic polypropylene, iPP, as a matrix and, once more, anatasa titania nanoparticles to verify their straightforward and cost-effective approach as well as the biocidal possibilities of these other nanocomposites[67,68] with also applications in the packaging sector. In this case, an interfacial agent (a low molecular weight polypropylene with maleic anhydride grafted, PP-*g*-MAH) is required to be added to improve the interfacial contact between the organic matrix and the nanoparticles due to the hydrophobic nature of iPP.[69] A similar approach was successfully followed by others with low-density polyethylene (LDPE).[70,71] The antimicrobial activity is substantially enhanced in the resulting nanocomposites by efficiently managing charge-carrier handling through the organic/inorganic interface, making biocidal the whole system. As occurred in EVOH/TiO$_2$ nanocomposites, the crystallinity degree of iPP or LDPE matrix remains practically unchanged by the presence of nanoparticles in the polymer.

The antimicrobial performance exhibited by these TiO$_2$–polymer nanocomposites surface can be even improved by extending the oxide absorption power into the visible region through an oxide-surface doping process. In this way, the light source for the disinfection process can be the ambient light of food stands at supermarkets or, after completion of lifetime of the material and considering its ecodisposal, such a feature can open the fruitful and economical use of sunlight. Consequently, as a previous step for producing advanced nanocomposite materials, TiO$_2$-doped nanoparticles have been prepared by photodeposition of metallic entities (*e.g.* silver) in a content of 1 wt.%[72] or by addition of a similar amount of certain oxides like Cu$_2$O/CuO or ZnO using impregnation procedures.[73,74] Afterwards, different amounts of these doped nanoparticles have been incorporated into an EVOH copolymer through, once more, a melting process without using any compatibilizer. These nanocomposites demonstrate very effective germicide activity toward Gram-negative, Gram-positive bacteria/cocci (*E. coli, P. putida, S. aureus*), and yeasts (*P. jadini*) using both UV- and visible-light sources, the latter up to a wavelength of 500 nm. Figure 14.6 compares the film performance of TiO$_2$-EVOH and doped-TiO$_2$-EVOH nanocomposites with a 2 wt.% of inorganic content for both the UV- and visible-light excitation cases. The plots display nanocomposites killing performances after 30 min contact with Gram-positive (*E. coli*) and Gram-negative (*S. aureus*) bacteria. As expected, the surface doping of the oxide is effective with respect to TiO$_2$ as inorganic phase of the nanocomposite upon all illumination conditions, but particularly while exposed to visible-light irradiation. Photoluminescence studies indicate that such a positive effect is based on the influence of the additives in charge recombination, allowing hole-related charge species to interact more efficiently with micro-organism targets. In addition, the materials show exceptional resistance to biofilm formation, which is (partially) responsible of micro-organisms' antibiotic and/or chemical resistance. In summary, the presence of minuscule amounts of metallic Ag or Cu$_2$O/CuO or ZnO oxides greatly boosts the antimicrobial power (in comparison to TiO$_2$-EVOH systems) through several optical effects under UV light and introduces the successful use of visible-light sources. Such optical effects concern visible-light plasmonic resonances for Ag metal and the optimization of the optical absorption capability in the UV/visible range for, respectively, ZnO and Cu$_2$O/CuO oxides but, common to all doping agents, is the already-mentioned adequate handling of charge

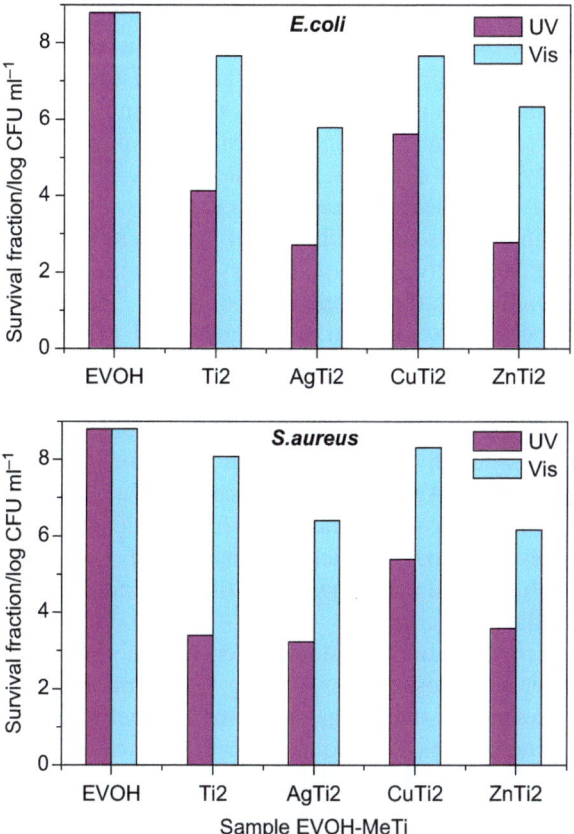

Figure 14.6   Photokilling performances of EVOH-TiO$_2$ composites with a modified TiO$_2$ component at 2 wt.% by surface addition of several inorganic phases (Ag, Cu or Zn). See text for details.

carriers generated upon light excitation through the corresponding inorganic/inorganic and inorganic/organic interfaces.

Other valuable works concern the use of an industrial TiO$_2$ material, called P25 from Evonik (previously Degussa), to produce a polyester-based composite film that was coated in the walls of a photocatalytic reactor to disinfect the air of a typical room (*ca.* 60 m^3), using a recirculation system that can be installed, for example, in air-conditioning apparatus. Using UV light, the system efficiently eliminated bacteria although it showed a limited performance against fungi.[75] Besides, Zhang *et al.*[76] reported the incorporation of titania to a polyurethane matrix using a surfactant (sodium dodecyl sulfate, SDS) as interfacial agent. The presence of the latter component also increases the hydrophilicity of the final nanocomposite material. The system was tested against bacteria (*E. coli, C. albicans*) and a fungus (*A. niger*) upon both UV and visible (typical indoor conditions from fluorescence illumination) excitations. *E. coli* seems to be efficiently eliminated, while some deficiencies were surprisingly observed in the photokilling of *C. albicans* and *A. niger*. Very recently, in 2010, two contributions described additional uses of these TiO₂-polymer composite materials. Tyllianakis and Sevastos[77] described the use of complex biocidal agents containing silver, titania and quaternary ammonium salts in order to obtain highly efficient nanocomposite films showing outstanding activity against *E. coli*, *S. epidermis*, *S. aureus*, and *C. paraprilosis* and potential use in biomedical applications as scaffolds. Excellent performance is observed both in the presence or absence of light excitation and is attributed to the combined effect of silver and quaternary salts, while under dark conditions and titania upon illumination. Similarly, Kong *et al.*[78] reported the preparation of titania nanocomposites exhibiting antimicrobial performances even under dark conditions, since the secondary amine-containing antifouling copolymer shell provides additional antimicrobial activity to the one provided by TiO₂ nanoparticles. These materials are obtained by photoinduced copolymerization of 2-(*tert*-butylamino)ethyl methacrylate and ethylene glycol dimethacrylate to form core–shell poly(*t*BAM-*co*-EGDMA)/commercial anatase needle-like TiO₂ nanoparticles. They show enhanced photocatalytic antibacterial properties compared to neat TiO₂ nanoparticles against both *E. coli* and *S. aureus* due to the synergic antibacterial performances of the biocidal polymer shell and light-activated biocidal TiO₂ core.

We can also mention that a system currently explored concerns the use of a natural polymer as chitosan incorporating the titania oxide as biocidal component. The presence of the inorganic component helps in mitigating chitosan biocidal deficiencies, typical of aqueous media at acidic pH and due to solubility deficiencies and limited availability of amino groups. Multilayer systems containing these two components[79] showed outstanding activity as well as that observed in more complex formulations using an additional layer containing Ag–AgBr under, respectively, UV- and visible-light excitation conditions for the inactivation of *E. coli*.[80] Also TiO₂-chitosan composites immobilized in cotton fibers were shown to obtain 2/3 log reductions for *E. coli* and *S. aureus* upon 12 h of visible-light interaction. A lower activity 97% was observed in the case of the fungus *A. niger*.[81] Some of the previous results on chitosan-based nanocomposite films containing additional (typically silver) antimicrobial agents were reviewed by Li *et al.*[45] Use of other biodegradable polymer matrices, such as polycaprolactone was recently reported for the effective elimination of *C. albicans* (ca. 95% in 2 h)[82] or *E. coli* (more than 5-log in 1 h).[83] In the last study, the composite system was able to show high biocidal capability upon both UV- and visible-light exposure, the latter being as important as titania is, as already mentioned for other titania–polymer systems, not a visible-light absorber. The use of visible-light wavelengths opens up as described earlier in this chapter, a series of advantages for the commercial use of the systems.

Finally, it should be noted that although the use of TiO₂ oxide as a UV blocker is well established, very few studies analyzed the degradation of polymer-based nanocomposites in the presence of titania by effect of an external UV or visible field. Apart from the single study previously mentioned,[66] we are only aware of studies concerning the photodegradation of PE, PP and polyvinyl chloride (PVC) materials.[84–86] In all cases and as mentioned, the formation of carbonyl moieties as an effect of electron interaction with the polymer in the presence of air appears as the initial step of the degradation mechanism irrespective of the light characteristics (UV, visible or sunlight). In Figure 14.7

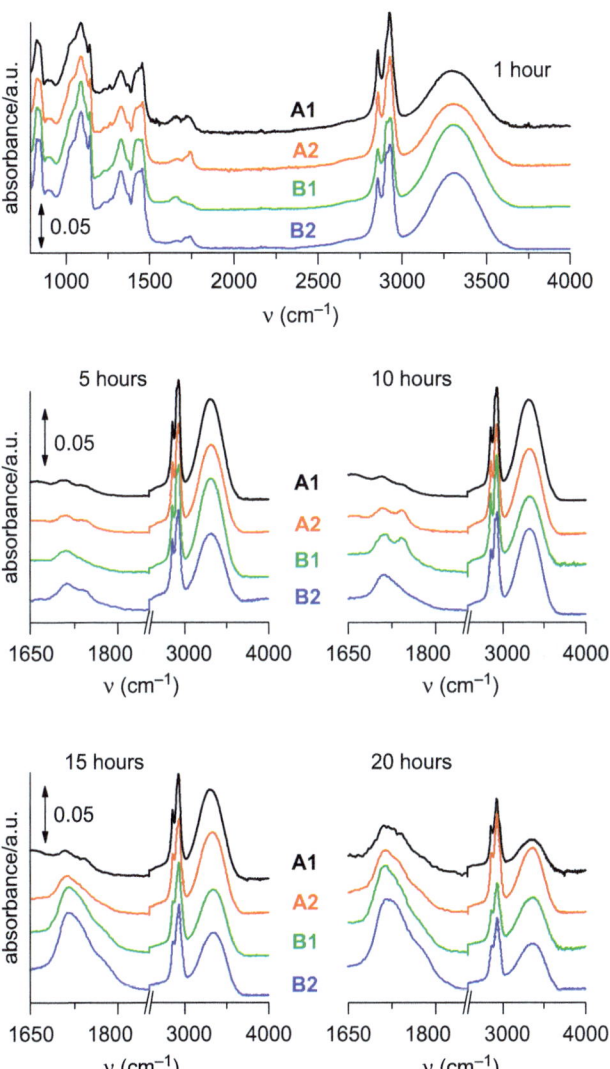

Figure 14.7 ATR-FTIR spectra of composite films containing 2 wt.% of TiO_2 subjected to sunlight aging for specific time treatments. Full spectra are presented at 1 h of treatment, while selected regions are presented for other times. Composite materials differ in the titania component. See text for details. Reproduced from ref. 66 with permission from The Royal Society of Chemistry.

are presented the infrared results concerning the degradation of an EVOH film subjected to an accelerated aging treatment, roughly corresponding to 1.5 months of sunlight exposure. Four different titanias were used in the study to a 2 wt.% loading in the composite. Irrespective of the differences between the inorganic phases, the most visible and common effects observed during the aging treatment are the modification of the -OH region (3000–3400 cm^{-1}) as well as the concomitant generation of the above-mentioned carbonyl groups (1700–1750 cm^{-1}).[61,66] Such degradation processes ended with formation of carbon dioxide and water as principal products but displaying an inherently slow kinetics with time dependence adequate for managing lifetime (for polymer-based industrial products) by proper control of the oxide loading at the nanocomposite.

14.3.2   Zinc Oxide–Polymeric Nanocomposites

Although the antibacterial activity of ZnO is well known,[1] only a few examples of polymeric nanocomposites zinc oxide nanoparticles have been reported. ZnO nanoparticles have been incorporated into thermoplastic polymers, polyamide 6 and low density PE.[87] In both cases and under dark or illuminated conditions, the nanocomposites showed great antimicrobial activity with low content of ZnO, 1% w/w, which is enhanced as the ZnO increases in the nanocomposite. However, the addition of ZnO nanoparticles does not produce any activity against *A. niger* fungus.

Other studies concern electrospun nanocomposites fibers of PU with ZnO over coat cotton substrates were developed to impart UV protection and antimicrobial function.[88] The authors concluded that the antimicrobial mechanism involves direct contact between the substrate and the microbes, acting as a barrier rather than by the diffusion of the cation agent. The inclusion of ZnO in cotton/polyester fabrics was also carried out, showing significant activity against *E. coli*,[89] *B. subtilis*[90] and to a lesser extent *A. niger*.[90] ZnO nanoparticles were assembled on the cotton fabrics by a layer-by-layer deposition process showed excellent antibacterial activity against *S. aureus*[91] bacteria. Particles were also deposited on fabrics by using ultrasound irradiation.[92] These materials presented antibacterial effectiveness against *E. coli* and *S. aureus*. Finally, dark activity against *S. aureus* by leaching of nanoparticles (leaching of Zn cations by themselves was discarded) was observed in ZnO surface-treated PVC[93–95] or against *E. coli* by ZnO coatings on chitosan bag matrices.[96] However, the composites with PVC were not effective against *A. flavus* and *P. citrinum*.[95] The biocidal properties derived from copresence of Ag and ZnO in LDPE polymer materials under dark conditions were also analyzed in relation to food (more concretely, fresh orange juice) preservation.[97]

Finally, we would like to mention the single contribution we are aware of studying the photo- or self-degradation of a ZnO-filled PP system.[98] This work presents results following the general trends already discussed in the previous section for systems containing TiO$_2$.

Acknowledgements

A. Kubacka and A. Iglesias-Juez acknowledge MICINN for financial support (Ramón y Cajal postdoctoral grants). Funding of MICINN is fully acknowledged by the authors (CTQ2010-14872/BQU, PLE2009-0037, PRI-PIBJP-2011-0914).

References

1. A. Kubacka, M. Fernández-García and G. Colón, *Chem. Rev.*, 2012, **112**, 1555–1614.
2. N. Sakai, K. Ebina, K. Takada and T. Sasaki, *J. Am. Chem. Soc.*, 2004, **126**, 5851–5858.
3. Y. Q. Dai, C. M. Cobley, J. Zeng, Y. M. Sun and Y. N. Xia, *Nano Lett.*, 2009, **9**, 2455–2459.
4. Z. Wang, K. Lv, G. Wang, K. Deng and D. Tang, *Appl. Catal. B.*, 2010, **100**, 378–385.
5. J. Pan, G. Liu, G. Qing, M. Lu and H.-M. Cheng, *Angew. Chem. Int. Ed.*, 2011, **50**, 2133–2137.
6. G. Liu, J. C. Yu, G. Q. Lu and H.-M. Cheng, *Chem. Commun.*, 2011, 6763–6783.
7. N. Serpone, *J. Phys. Chem. B*, 2006, **110**, 24287–24293.
8. J. Bisquert, *Phys. Chem. Chem. Phys.*, 2008, **10**, 3175–3194.
9. R. D. Schaller and V. I. Klimov, *Phys. Rev. Lett.*, 2004, **92**, 186601.
10. M. Alimohammadi and K. A. Fichthorn, *Nano Lett.*, 2009, **9**, 4198–4203.
11. S. Dindar and J. Icli, *J. Photochem. Photobiol. A: Chem.*, 2001, **140**, 263–268.
12. C. Lizama, J. Freer, J. Baeza and H. Mansilla, *Catal. Today*, 2002, **76**, 235–246.
13. N. Daneshvar, D. Salari and A. R. Khataee, *J. Photochem. Photobiol. A*, 2004, **162**, 317–322 .
14. S. Biswas, S. Kar and S. Chaudhuri, *Mater. Sci. Eng. B*, 2007, **142**, 69–77.
15. Z. L. Wang, *J. Phys. Condens. Matter*, 2004, **16**, R829–R858.
16. Y. X. Wang, X. Y. Li, G. Lu, X. Quan and G. H. Chen, *J. Phys. Chem. C*, 2008, **112**, 7332–7336.
17. W. L. Xu, M. J. Zheng, G. Q. Ding and W. Z. Shen, *Chem. Phys. Lett.*, 2005, **411**, 37–42.
18. M. H. Huang, Y. Wu, H. Feick, N. Tran, E. Weber and P. D. Yang, *Adv. Mater.*, 2001, **13**, 113–116.
19. L. Guo, Y. L. Ji, H. B. Xu, P. Simon and Z. Y. Wu, *J. Am. Chem. Soc.*, 2002, **124**, 14864–14865.
20. Z. W. Pan, Z. R. Dai and Z. L. Wang, *Science*, 2001, **291**, 1947–1949.
21. X. Y. Kong, Y. Ding, R. S. Yang and Z. L. Wang, *Science*, 2004, **303**, 1348–1351.
22. J. Zhang, L. D. Sun, C. S. Liao and C. H. Yan, *Chem. Commun.*, 2002, **3**, 262–263.
23. H. J. Fan, R. Scholz, F. M. Kolb, M. Zacharias and U. Gosele, *Solid State Commun.*, 2004, **130**, 517–521.

24. J. G. Wen, J. Y. Lao, D. Z. Wang, T. M. Kyaw, Y. L. Foo and Z. F. Ren, *Chem. Phys. Lett.*, 2003, **372**, 717.
25. G. Colón, M. C. Hidalgo, J. A. Navío, E. Pulido Melián, O. González Díaz and J. M. Doña Rodríguez, *Appl. Catal. B: Environ.*, 2008, **83**, 30–38.
26. H. Zhang, D. Yang, D. Li, X. Ma, S. Li and D. Que, *Cryst. Growth Des.*, 2005, **5**, 547–550.
27. Y. Feng, M. Zhang, M. Guo and X. Wang, *Cryst. Growth Des.*, 2010, **10**, 1500–1507.
28. D. Li and H. Haneda, *Chemosphere*, 2003, **51**, 129–137.
29. L. Xu, Y. L. Hu, C. Pelligra, C. H. Chen, L. Jin, H. Huang, S. Sithambaram, M. Aindow, R. Joesten and S. L. Suib, *Chem. Mater.*, 2009, **21**, 2875–2885.
30. X. G. Han, H. Z. He, Q. Kuang, X. Zhou, X. H. Zhang, T. Xu, Z. X. Xie and L. S. Zheng, *J. Phys. Chem. C*, 2009, **113**, 584–589.
31. X. Zhou, Z. X. Xie, Z. Y. Jiang, Q. Kuang, S. H. Zhang, T. Xu, R. B. Huang and L. S. Zheng, *Chem. Comm.*, 2005, 5572–5574.
32. J. Wang, P. Liu, X. Fu, Z. Li, W. Han and X. Wang, *Langmuir*, 2009, **25**, 1218–1223.
33. M. Y. Guo, L. M. Ching Ng, F. Liu, A. B. Djurišić, W. K. Chan, H. Su and K. S. Wong, *J. Phys. Chem. C*, 2011, **115**, 11095–11101.
34. H. H. Wang and C. S. Me, *Phys. E.*, 2008, **40**, 2724–2729.
35. Y. Zheng, C. Chen, Y. Zhang, X. Lin, Q. Zheng, K. Wei, J.M. Zhu and Y. Zhu, *Inorg. Chem.*, 2007, **46**, 6675–6682.
36. S. Baruah, S. S. Sinha, B. Ghosh, S. K. Pal, A. K. Raychaudhuri and J. Dutta, *J. Appl. Phys.*, 2009, **105**, 074308.
37. S. S. Warule, N. S. Chaudhari, B. B. Kale and M. A. More, *Cryst. Eng. Comm.*, 2009, **11**, 2776–2783.
38. S. Baruah, R. F. Rafique and J. Dutta, *Nano*, 2008, **3**, 399–407.
39. H. Zhang, *J. Mater. Chem.*, 2009, **19**, 5089–5121.
40. B. M. Novak, *Adv. Mater.*, 1999, **35**, 422–428.
41. C. Sanchez, B. Julian, Ph. Belleveille and M. Popall, *J. Mater. Chem.*, 2005, **15**, 3559–3592.
42. A. Muñoz-Bonilla and M. Fernández-García, *Prog. Polymer Sci.*, 2012, **37**, 281–339.
43. P. Appendini and J. H. Hotchkiss, *Innovative Food Sci. Emerg. Technol.*, 2002, **3**, 113–126.
44. F. Devlieghere, L. Vermeiren and J. Debevere, *Int. Dairy J.*, 2004, **14**, 273–285.
45. Q. Li, S. Mahendra, D. Y. Lyon, L. Brunet, M. V. Liga, D. Li and P. J. J. Álvarez, *Water Res.*, 2008, **42**, 4591–4602.
46. S. Noimark, C. W. Dunnill, M. Wilons and I. P. Parkin, *Chem. Soc. Rev.*, 2009, **138**, 3435–3445.
47. A. L. Brody, *Food Technol.*, 2003, **57**, 52–55.
48. H. Z. Tang, R. J. Doerksen and G. N. Tew, *Chem. Commun.*, 2005, **12**, 1537–1539.
49. M. Ignatova, S. Voccia, B. Gilbert, N. Markova, D. Cossement, R. Gouttebaron, R. Jérôme and C. Jérôme, *Langmuir*, 2006, **22**, 255–262.

50. R. Brayner, R. Ferrari-Iliou, N. Brivois, S. Djediat, M. F. Benedetti and F. Fievet, *Nano Lett.*, 2006, **6**, 866–870.
51. V. Sambhy, M. MacBride, B. R. Peterson and A. Sen, *J. Am. Chem. Soc.*, 2006, **128**, 9798–9808.
52. B. Dizman, M. O. Elasri and L. J. Mathias, *Biomacromolecules*, 2005, **6**, 514–520.
53. S. M. Iconomopoulou and G. A. Voyiatzis, *J. Controll. Release*, 2005, **103**, 451–464.
54. J. C. Sagripanti and A. Bonifacio, *J. AOAC Int.*, 2000, **83**, 1415–1422.
55. Y. Chen, S. D. Worley, J. Kim, T. Y. Wei, J. I. Santiago, J. F. Williams and G. Sun, *Ind. Eng. Chem. Res.*, 2003, **42**, 280–284.
56. M. E. Robbins, E. D. Hopper and M. H. Schoenfisch, *Langmuir*, 2004, **20**, 10296–12302.
57. A. M. Klibanov, *J. Mater. Chem.*, 2007, **17**, 2479–2482.
58. S. Y. Kwak and S. H. Kim, *Environ. Sci. Technol.*, 2001, **35**, 2388–2394.
59. S. H Kim, S. Y. Kwak, B. H. Sohn and T. H. Park, *J. Membr. Sci.*, 2003, **211**, 157–165.
60. J. Kim and B. der Bruggen, *Environ. Pollution*, 2010, **158**, 2335–2349.
61. A. Kubacka, C. Serrano, M. Ferrer, H. Lünsdorf, P. Bielecki, M. L. Cerrada, M. Fernández-García and M. Fernández-García, *Nano Lett.*, 2007, **7**, 2529–2534.
62. M. M. Cowan, K. Z. Abshire, S. L. Houk and S. M. Evans, *J. Ind. Microbiol. Biotechnol.*, 2003, **30**, 102–106.
63. P. G. Mazzola, A. M. S. Martins and T. C. V. Penna, *BMC Infect Dis.*, 2006, **6**, 131–136.
64. R. J. Jiménez Riobóo, A. De Andrés, A. Kubacka, M. Fernández-García, M. L. Cerrada, C. Serrano and M. Fernández-García, *Eur. Polym. J.*, 2010, **46**, 397–403.
65. M. L. Cerrada, C. Serrano, M. Sánchez-Chaves, M. Fernández-García, F. Fernández-Martín, A. de andres, R. J. Jiménez Riobóo, A. Kubacka, M. Ferrer and M. Fernández-García, *Adv. Funct. Mater.*, 2008, **18**, 1949–1960.
66. K. C. Christoforidis, A. Kubacka, M. Ferrer, M. L. Cerrada, M. Fernández-García and M. Fernández-García, *RSC Adv.*, 2013, **3**, 8541–8550.
67. A. Kubacka, M. L. Cerrada, C. Serrano, M. Fernández-García, M. Ferrer and M. Fernández-García, *J. Nanosci. Nanotechnol.*, 2008, **8**, 3241–3246.
68. A. Kubacka, M. Ferrer, M. L. Cerrada, C. Serrano, M. Sánchez-Chaves, M. Fernández-García, A. de Andres, R. J. Jiménez Riobóo, F. Fernández-Martín and M. Fernández-García, *Appl. Catal. B Environ.*, 2009, **89**, 441–447.
69. M. L. Cerrada, C. Serrano, M. Sánchez-Chaves, M. Fernández-García, A. De andres, R. J. Jiménez Riobóo, F. Fernández-Martín, A. Kubacka, M. Ferrer and M. Fernández-García, *Environ. Sci. Technol.*, 2009, **43**, 1630–1634.
70. P. A. Zapata, H. Palza, K. Delgado and F. M. Rabaglati, *Polym. Chem.*, 2013, **50**, 702–708.
71. H. Bodaghi, Y. Mostofi, A. Oromiehie, Z. Zamani, B. Ghanbarzadeh, C. Costa, A. Conte and M. A. Del Nobile, *LWT-Food Sci. Technol.*, 2013, **50**, 702–706.

72. A. Kubacka, M. L. Cerrada, C. Serrano, M. Fernández-García, M. Ferrer and M. Fernández-García, *J. Phys. Chem. C*, 2009, **113**, 9182–9190.

73. A. Kubacka, M. Ferrer, M. Fernández-García, C. Serrano, M.L. Cerrada and M. Fernández-García, *Appl. Catal. B Environ.*, 2011, **104**, 346–352.

74. A. Kubacka, M. Ferrer and M. Fernández-García, *Appl. Catal. B Environ.*, 2012, **121–122**, 230–238.

75. M. P. Paschoalino and W. F. Jardim, *Indoor Air*, 2008, **18**, 473–479.

76. X. Zhang, H. Su, Y. Zhao and T. Tan, *J. Photochem. Photobiol. A*, 2008, **199**, 123–129.

77. M. Tyllianakis and D. Sevastos, *J. Mater. Chem.*, 2010, **21**, 2201–2214.

78. H. Kong, J. Song and J. Shen, *Environ. Sci. Technol.*, 2010, **44**, 5672–5676.

79. W. Y. Yuan, J. Ji, J. H. Fu and J. C. Shen, *J. Biomed. Mater. Res. B*, 2008, **85**, 556–563.

80. W. Y. Yuan, J. Ji, Q. An, Y. F. Liu and J. C. Chen, *Nanotechnol.*, 2009, **20**, 245101.

81. T. Quian, H. Su and T. Tan, *J. Photochem. Photobiol. A*, 2011, **218**, 130–136.

82. M. Solimen, I. Talhdil, C. Breen and S. Akkam, *J. Hazard. Mater.*, 2011, **187**, 199–205.

83. A. Muñoz-Bonilla, M. L. Cerrada, M. Fernández-García, A. Kubacka, M. Ferrer and M. Fernández-García, *Int. J. Mol. Sci.*, 2013, **14**, 9249–9266.

84. X. Zhao, Z. Li, Y. Chen, L. Su and Y. Zhi, *J. Mol. Catal. A*, 2007, **268**, 101–106.

85. N. S. Allen, M. Edge, A. Ortega, G. Sandoval and R. B. McIntyre, *Polym. Deg. Stab.*, 2004, **85**, 927–946.

86. S. H. Kim, S.-Y. Kwak and T. Suzuki, *Polymer*, 2006, **47**, 3005–3016.

87. M. Ma, Y. Cheng, Z. Xu, P. Xu, H. Qu, Y. Fang, T. Xu and L. Wen, *Eur. J. Med. Chem.*, 2007, **42**, 93–98.

88. S. Lee, *J. Appl. Polym. Sci.*, 2009, **114**, 3652–3658.

89. I. Perelshtein, G. Applerot and A. Gedaken, *ACS Appl. Mater. Interf.*, 2009, **1**, 361–366.

90. M. H. Zohdy, H. A. Kareen, A. M. El-nagger and M. S. Hassan, *J. Appl. Polym. Sci.*, 2003, **89**, 2604–2610.

91. Ş. S. Uğur, M. Sarışık, A. H. Aktaş, M. C. Uçar and E. Erden, *Nanoscale Res. Lett.*, 2010, **5**, 1204–1210.

92. I. Perelshtein, G. Applerot, N. Perkas, E. Wehrschetz-Sigl, A. Hasmann, G. M. Guebitz and A. Gedanken, *ACS Appl. Mater. Interfaces*, 2009, **1**, 363–367.

93. J.T. Seil and T.J. Webster, *Acta Biomater.*, 2011, **7**, 2579–2584.

94. W. Li, X. Li, P. Zhang and Y. Xing, *Adv. Mater. Res.*, 2011, **152–153**, 489–492.

95. X. Li, Y. Xing, Y. Jiang, Y. Ding and W. Li, *Int. J. Food Sci. Technol.*, 2009, **44**, 2161–2168.

96. H. Li, J. C. Deng, H. R. Deng, Z. L. Liu and X. L. Li, *Chem. Eng. J.*, 2010, **160**, 378–382.

97. A. Emamifar, M. Kadivar, M. Shahedi and S. Soliemanian-Zad, *Food Control*, 2011, **22**, 408–413.

98. H. Zhao and R. K. Y. Li, *Polymer*, 2006, **47**, 3207–3217.

CHAPTER 15

Future Perspectives and Concluding Remarks

MARÍA L. CERRADA, ALEXANDRA MUÑOZ-BONILLA AND MARTA FERNÁNDEZ-GARCÍA*

Instituto de Ciencia y Tecnología de Polímeros (ICTP-CSIC), Juan de la Cierva 3, 28006 Madrid, Spain
*Email: martafg@ictp.csic.es

15.1 Brief Overview

At this point we have learnt about different methodologies to create antimicrobial polymeric materials and their feasible modes of action. All these approaches pointed out that there is not only one instrument of fighting against pathogen antimicrobials but, on the contrary, there are numerous methods to be applied depending on the microbes and the final applications of the antimicrobial systems. Natural and synthetic products that exhibit inherent antimicrobial properties have been addressed. Moreover, mimic host-defense peptides development is another way to imitate the nature against mainly bacteria, although nylon-3 polymers with selective antifungal activity have recently appeared,[1] announcing a wider potential of these materials since fungi are more difficult to eliminate because these pathogens are eukaryotes. Moreover, the incorporation of antimicrobial organic and/or inorganic materials to generate activities in the final products is one of the most exploited options.[2–4] Despite these approaches, we would like to mention in this chapter other strategies nowadays used to combat the adverse consequences that the presence of these pathogens generates in our surroundings.

RSC Polymer Chemistry Series No. 10
Polymeric Materials with Antimicrobial Activity: From Synthesis to Applications
Edited by Alexandra Muñoz-Bonilla, María L. Cerrada and Marta Fernández-García
© The Royal Society of Chemistry 2014
Published by the Royal Society of Chemistry, www.rsc.org

15.2 Other Antimicrobial Polymeric Systems

As has been indicated throughout this book, there are numerous natural and synthetic agents based on quaternized antimicrobials. Among all of them, there are two groups consisting in biodegradable and cationic quaternary ammonium-containing polymers, synthetic polycarbonates[5] and natural chitosans,[6] reported as effective antimicrobial materials. Biodegradable resins are another important group described in the literature. Natural resin acids from pine and conifer trees (also termed rosins) are abundant as well as low cost and are based on diterpene resin acids, such as abietic, levopimaric, and pimaric acids. They have a characteristic bulky hydrophenanthrene structure with the empirical formula $C_{20}H_{30}O_2$ (Scheme 15.1).[7–12] This hydrophenanthrene moiety, consisting of fused cycloaliphatic and aromatic structures, provides resin acids with substantial hydrophobicity, a property that has facilitated their use in marine antifouling coating materials for decades and in biocides, though lacking rational design and investigation of selectivity and biocompatibility.[12] Wang *et al.*[12] have recently prepared various resin-acid-derived monomers and polymers, which showed excellent hydrophobicity, biodegradability and biocompatibility. The synthesis of quaternary ammonium-containing resin-acid-derived antimicrobial compounds and polymers is shown in Scheme 15.2. This synthetic protocol starts with a highly efficient Diels–Alder reaction between levopimaric acid and maleic anhydride to produce maleopimaric acid, followed by an amidation reaction with *N,N*-dimethylaminoethylamine to yield compound **1**. Quaternary ammonium-containing resin acid **2** is then prepared by a quaternization reaction with ethyl bromide. An esterification between compound **2** and propargyl alcohol leads to the formation of quaternary ammonium-containing resin propargyl ester **3**. Posterior click reaction of compound **3** and azide substituted polycaprolactone (PCL) renders compound **4**.

The authors concluded that these antimicrobial materials possessed excellent antimicrobial activity and high selectivity against *Staphylococcus aureus* and *Escherichia coli* bacteria over mammalian cells, *i.e.* mouse red blood cells (Figure 15.1).

Later, this group prepared antibacterial effective systems with quaternary ammonium containing poly(*N,N*-dimethylaminoethyl methacrylate) and

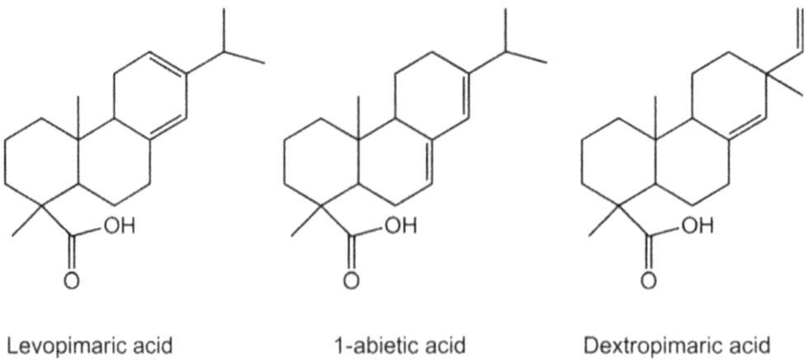

Levopimaric acid   1-abietic acid   Dextropimaric acid

Scheme 15.1   Resin-acid structures that differ only in the position of the conjugated double bond.

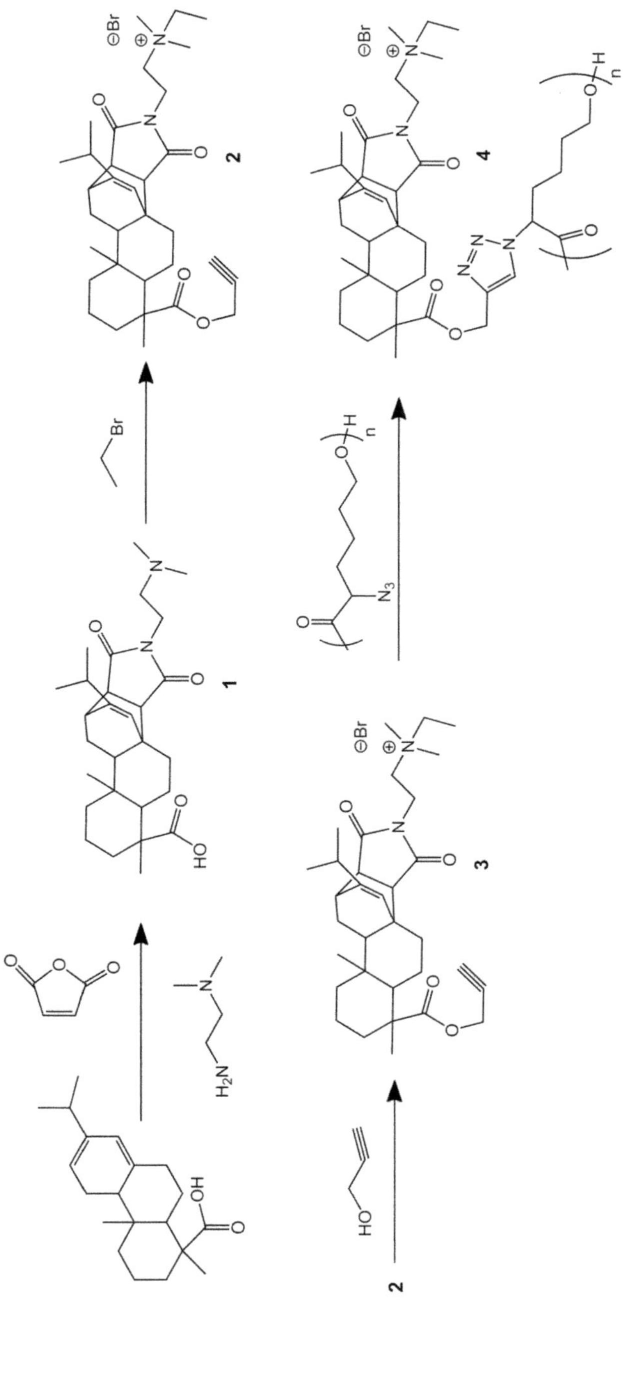

Scheme 15.2 Synthesis of quaternary ammonium-containing resin-acid-derived antimicrobial compounds and polymers.[12]

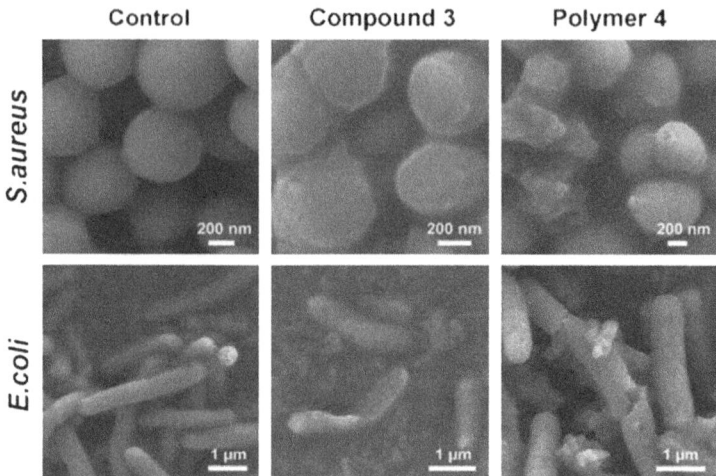

Figure 15.1   FE-SEM morphology images of *S. aureus* and *E. coli* cells exposed to
3 and 4.[12]
Reproduced by permission of The Royal Society of Chemistry.

natural rosin as the pendant group (PDMAEMA-*g*-rosin). Their approach
consisted in a hydrophilic cationic group sandwiched between the polymer
backbone and bulky hydrophobic hydrophenanthrene side groups. The results
indicated that conformation of hydrophobic group (particularly steric
hindrance) played a role in dictating antibacterial efficacy against *E. coli* and
S. aureus bacteria. The comparison between two amphipathic structures having
cationic charges at different locations with respect to the rosin moiety and
polymer backbone can be easily observed in Scheme 15.3.[13]

    Another family of antimicrobial cationic polymers is the polyethylenimines
(PEIs). They are formed as either linear, branched or hyperbranched molecules.
Like other cationic polymers, PEIs present considerable antibacterial
activity.[14,15] They have also been used successfully to transfer genes into living
cells because they have a high cationic charge density, hence, they have strong
DNA-binding ability.[16–18] In addition, it has been proved that PEIs are strong
inhibitors of human papillomavirus (HPV) and cytomegalovirus (HCMV)
infections without being cytotoxics by blocking the primary binding of HPV
and HCMV virions to target cells.[19] This polymeric family has been recently
tested against herpes simplex virus type 2 (HSV-2),[20] the constant infection of
genital herpes, which is a common sexually transmitted infection and is usually
a recurrent infection. It was found that vaginal administration of 1% PEI
combined with liposomes and with a decrease in the number of primary amine
residues, resulting from ethylene carbonate treatment, strongly enhanced the
therapeutic antiviral efficiency against HSV-2 in a mouse model.

    Antimicrobial hydrogels are also proposed for combating drug-resistant
infections, since gels exhibit many polymer characteristics without becoming
dissolved, remaining in place under physiological conditions while maintaining

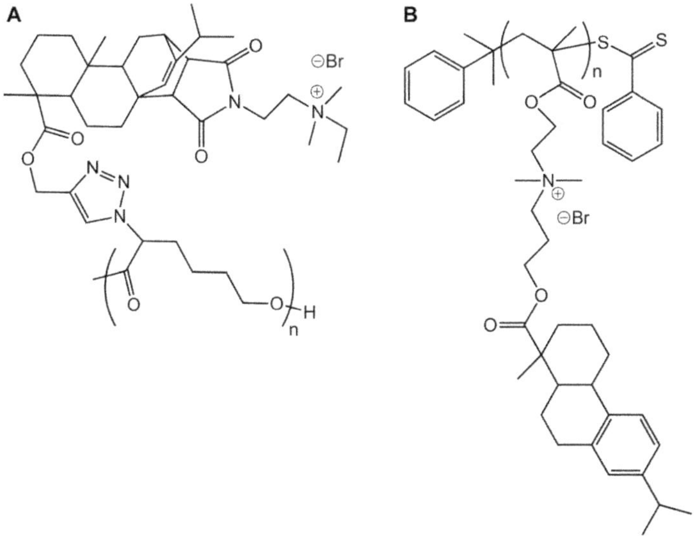

Scheme 15.3 (A) Polymer with cationic charges located at the periphery[12] and (B) polymer with cationic charges embedded inside.[13]

antimicrobial activity. Therefore, development of synthetic, cost-effective, and biodegradable antimicrobial hydrogels would be highly useful. These agents should also be moldable, suitable for processing, thus allowing *in situ* applications. Furthermore, the antimicrobial hydrogel must remain stable and active for the duration of its purpose. Very recently, Li *et al.*[21] have reported stimulus-responsive antimicrobial gels formed from stereocomplexation[22] of different combinations between biodegradable poly(L-lactide)-*b*-poly(ethylene glycol)-*b*-poly(L-lactide) (PLLA-PEG-PLLA) and a charged biodegradable poly(D-lactide)-*b*-cationic poly(carbonate)-*b*-poly(D-lactide) (PDLA-CPC-PDLA) triblock copolymer. These materials exhibited shear-thinning properties because of the nature of their noncovalent associations. In particular, these were tested against methicillin resistant *S. aureus* (MRSA, Gram-positive), vancomycin resistant *enterococci* (VRE, Gram-positive), *Pseudomonas aeruginosa* (Gram-negative), *Acinetobacter baumannii* (Gram-negative, resistant to most antibiotics), *Klebsiella pneumoniae* (Gram-negative, resistant to carbapenem), and *Cryptococcus neoformans*. The gels completely inhibited growth and showed a nearly perfect killing efficiency on all microbial cells tested.

On the other hand, nitric oxide (NO) is an endogenous diatomic free radical involved in several physiological processes, including vasodilation, immune response, neurotransmission, cytoprotection and wound healing.[23,24] NO is produced in large quantities by human macrophages during infection and other inflammatory conditions;[25] therefore, it exhibits antimicrobial properties against a broad spectrum of bacteria. It is reported that NO is able to kill bacterial cells by direct or indirect oxidation mainly through formation of peroxynitrite ($^-$OONO), the product resulting from the reaction of NO with superoxide anion $(O_2{}^{\bullet-})$.[23] This peroxynitrite is able to nitrosate cysteine sulfhydryl groups, nitrate tyrosine and tryptophan residues of proteins, and oxidize methionine residues to methionine sulfoxide, which negatively impact the integrity of the bacteria membrane and cell function, and then results in a broad-spectrum antibacterial activity.[26] However, its ability to mediate these reactions is strongly influenced by the presence of physiological concentrations of CO_2.[27] In addition, a significantly low level of NO can prevent biofilm formation and disperse established biofilm through a signaling process.[28] Although the antimicrobial and antibiofilm properties of NO-releasing polymeric coatings have been evaluated, most *in vitro* studies have been performed under "static" conditions.[29] These are not normally correlated to the natural life cycle of bacteria.[30] Schoenfisch's group has used a parallel-plate flow cell that creates shear forces to mimic natural environment encountered by microbes, and studied the decrease in microbial attachment on NO release polymeric coatings.[29,31] Compared with this and other previous reported systems, a flow-through CDCP (Centers for Disease Control and Prevention) biofilm reactor, capable of creating both renewable nutrient sources and shear forces, is regarded as a standard tool to study biofilm formation under natural conditions for longer time periods with repeatable results.[30,32]

Diethylenetriamine was modified to a diazeniumdiolate NO donor form (DETA/NO). In general, diazeniumdiolate NO donors are readily synthesized and release NO in a predictable mode. Hetrick and Schoenfisch[29] covalently attached diazeniumdiolates to the backbone of sol–gel derived (*i.e.* xerogel) polymers to limit the leaching of amine decomposition products. Exposure to high pressures of NO ($\sim$ 5 atm) facilitated the synthesis of diazeniumdiolate NO-donors at secondary amines throughout the xerogel (Scheme 15.4A). In the presence of a proton source, such as water or buffer, diazeniumdiolates spontaneously decompose to yield two equivalents of NO and the parent amine precursor (Scheme 15.4B).[33]

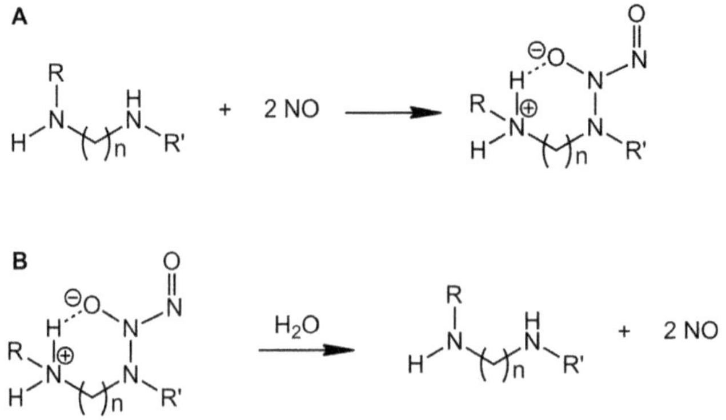

Scheme 15.4   (A) Reaction of NO with amines to produce diazeniumdiolate NO donors and (B) subsequent diazeniumdiolate decomposition and NO release in the presence of water.

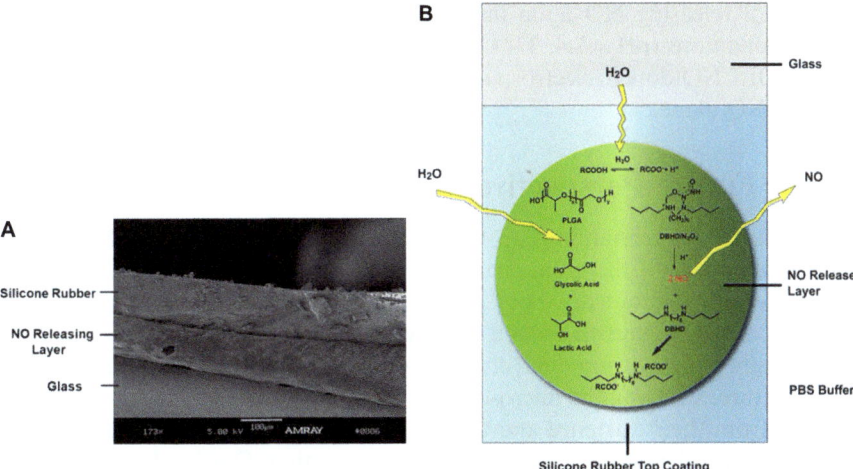

Figure 15.2 (A) SEM image of a cross section of DBHD/N$_2$O$_2$-PLGA-SR-based
NO-releasing coating. The film has a two-layer configuration with a base
layer of DBHD/N$_2$O$_2$ mixed in PLGA and silicone rubber as the top
coating; (B) schematic illustration of DBHD/N$_2$O$_2$-PLGA-SR-based
NO-releasing coating.[32]

Nitric-oxide-releasing films with a bilayer configuration have been recently
fabricated by doping dibutyhexyldiamine diazeniumdiolate (DBHD/N$_2$O$_2$) in a
poly(lactic-*co*-glycolic acid) (PLGA) layer and this base layer further
encapsulated with a silicone rubber top coating (Figure 15.2).[32] By incorporating pH-sensitive dyes within the films, pH changes in the PLGA layer are
visualized and correlated with the NO release profiles (flux *vs.* time). It is
demonstrated that PLGA acts as both a promoter and controller of NO release
from the coating by providing protons through its intrinsic acid residues (both
end groups and monomeric acid impurities) and hydrolysis products (lactic acid
and glycolic acid). Control of the pH changes within the PLGA layer can be
achieved by adjusting the ratio of DBHD/N$_2$O$_2$ and utilizing PLGAs with
different hydrolysis rates. The different NO-releasing coatings when incubated
in a CDC flow bioreactor for a one-week period at RT or 37 °C, exhibit
considerable antibiofilm properties against Gram-positive *S. aureus* and
Gram-negative *E. coli*. In particular, compared with the silicone rubber surface
alone, an NO-releasing film with a base layer of 30 wt.% DBHD/N$_2$O$_2$ mixed
with poly(lactic acid) exhibits an ∼98.4% reduction in biofilm biomass of
S. aureus and ∼99.9% reduction for *E. coli* at 37 °C. The authors claimed that
these diazeniumdiolate-doped PLGA-based NO-releasing coatings can be
useful for a variety of indwelling biomedical devices (*e.g.*, catheters).[32]

Furthermore, polymer solutions of Tecophilic® (a thermoplastic polyurethane) Tecoflex® (an aliphatic polyether-based thermoplastic polyurethane)
and poly(vinyl chloride) containing disodium 1-[2-(carboxylato)pyrrolidin-1-
yl]diazen-1-ium-1,2-diolate (PROLI/NO) (a low molecular weight NO donor)
were electrospun to generate fibers ranging from 100–3000 nm in diameter

capable of releasing NO upon immersion in aqueous solutions under physio-logical conditions (pH = 7.4, 37 °C). In these conditions, the NO release half-life for PROLI/NO-doped electrospun fibers is 2–200 times longer than that of PROLI/NO alone.[34]

15.3 Future Perspectives and Concluding Remarks

The complete understanding on how Nature works is going to be crucial in combating many of the great challenges that society is facing. As we have al-ready seen throughout this book, there are many approaches to fight against harmful micro-organisms, which behave in different ways as consequence of their nature and environmental circumstances. Therefore, society needs to learn from Nature, studying the micro-organism growing mechanisms, the outcomes derived from their existence and the manner to eliminate or prevent their col-onization in order to be able to live in a safer setting. Our world is constantly changing and, hence, all the organisms that live on it simultaneously evolve. This is something that people can constantly hear from the news. Today, on April 7th 2013, we have been informed from Gregory Hartl, a WHO media officer, that influenza A (H7N9) virus has been confirmed in seven cases of human infection in China although there is no evidence of ongoing human-to-human transmission, at this time. This makes us consider the design of new powerful antimicrobial polymeric materials or improvement of the already tested systems. Structural parameters, in terms of hydrophobicity/hydro-philicity balance, molecular weight, counterion and quaternization compound, are essential and require to be adjusted to achieve optimal antimicrobial ac-tivity and selectivity. At the same time, their obtainment through more ecoef-ficient and safer processes, minimizing their waste and toxicity is rather critical and mandatory in their whole design. As a result, the scientific and technolo-gical communities are using click chemistry, green chemistry, sustainable chemistry and other methodologies as tools to develop these new materials. In addition, the cytotoxicity also plays a decisive role, depending on the ultimate applicability, to implement these materials in the market, taking into account that *in vivo* results[35] are indispensable to be achieved In this sense, the anti-microbial studies exhibit another problem, which concerns the lack of man-datory rules to perform both *in vitro* and *in vivo* tests. It is rather difficult to compare different measurements, conditions, strains and so on. Although some protocols established by the Clinical and Laboratory Standards Institute (CLSI) and the European Committee on Antimicrobial Susceptibility Testing (EUCAST) have been attained by the scientific community, there is still con-troversy. Asín *et al.*[36] studied the difference on the susceptibility breakpoints based on pharmacokinetic/pharmacodynamic models and Monte Carlo simu-lation with those defined by the CLSI and the EUCAST for antibiotics used for the treatment of infections caused by Gram-positive bacteria. The authors proposed the promotion of prospective trials that include pharmacokinetic analysis, microbiological susceptibility and clinical data, which will help to establish a correlation between breakpoints and clinical outcome. This could also be another strategy to diminish antibiotic resistance, by optimization of the dosing regimen of available antimicrobial polymeric materials. In addition, the conditions under which these antimicrobial systems are measured are important to address.

Anionic polymers are also considered as candidate microbicides not only by their efficacy against human immunodeficiency virus type 1 (HIV-1) infection but also against a broad spectrum of sexually transmitted infection pathogens.[37,38] The most mentioned polyanionic candidate microbicides include cellulose sulfate, carrageenan, naphthalene sulfonate, cellulose acetate phthalate and poly(styrene sulfonate) or poly(styrene sulfonic acid).[37-40] However, there is still controversy about the adequacy of these systems in treatments *in vivo*, since clinical trials demonstrated that these products were ineffective and may have increased the risk of HIV-1 infection,[41-44] under some circumstances. Patel *et al.* revealed an enhancement of virus replication, in particular, cellulose sulfate system inhibited viral binding if the virus is introduced in phosphate-buffered saline (PBS) or hydroxyethyl piperazine ethane sulfonic acid (HEPES) media, while the antibinding activity is lost if the virus is introduced in seminal plasma.[41] Tan *et al.* also demonstrated that tested polyanion-based microbicides were able to accelerate the formation of seminal fluid-derived amyloid fibrils by electrostatic interaction, resulting an enhancement of HIV-1 infection, which may be one of the causes of their failure in clinical trials.[42] Qiu *et al.* established, however, the *in vivo* efficacy of poly(4-styrenesulfonic acid-*co*-maleic acid) gel formulations against HIV-1 and HSV-2, genital herpes, infections in mice.[45] Consequently, the enhancement effect is influenced by the assay conditions and is not an intrinsic property of these polyanions.[43]

Another disadvantage is the long lasting experiments required to obtain the antimicrobial material effectiveness. Wang's group[46,47] described a simple, rapid, sensitive, accurate, cost-effective system using conjugated polymers (CP), which have excellent light-harvesting ability to enhance the sensitive of biological detections, and an efficient electron and/or energy transfer for evaluating antimicrobial susceptibility and screening antibiotics in a throughput approach. This protocol is generated from the photodynamic antimicrobial chemotherapy (PACT) technique,[48] which is mainly used to fight against chronic and fatal infections of fungi. Those are more resistant to antifungal agents due to their structural cell membrane made of glycoproteins and polysaccharide polymers, which provide a barrier for drug penetration. Lambrechts *et al.* demonstrated that under dark conditions cationic porphyrin 5-phenyl-10,15,20-Tris(*N*-methyl-4-pyridyl)porphyrin chloride (TriP[4]) binds to the cell envelope of *Candida albicans*, and none or very little TriP[4] enters the cell. Upon illumination, the cell membrane is damaged and ultimately becomes permeable for TriP[4]. After lethal membrane damage, a massive invasion of TriP[4] into the cell occurs. The photosensitizer, light, and molecular oxygen generate reactive oxygen species (ROS) that are toxic and kill cells. The toxicity of these systems, however, limits their applications.

Wang's group have created electrostatic complex of water-soluble anionic polythiophene (PTP) and a cationic porphyrin (TPPN) (Figure 15.3) as an

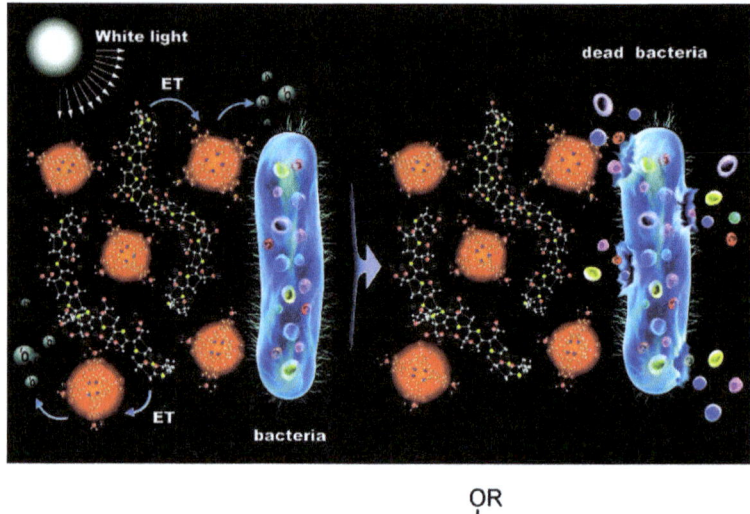

Figure 15.3 Schematic antibacterial mechanism of water soluble complex and chemical structures of anionic polythiophene (PTP) and a cationic porphyrin (TPPN). ET = energy transfer.[49]
Adapted with permission from *J. Am. Chem. Soc.*, **131** (36), 13117–13124. Copyright (2009) from the American Chemical Society.

antibacterial agent for efficient activity against *E. coli* and *Bacillus subtilis* with white light at a fluence rate of 27 J cm^{-2} (70% of *E. coli* reduction within 5 min).[49] Polythiophene-porphyrin CP (Scheme 15.5) also presents efficient antifungal activity against *Aspergillus niger*. Nevertheless, this activity requires photoexcitation, and subsequent energy transfer from polythiophene to the porphyrin sites.[47]

Other CPs based on a poly(phenylene ethynylene) repeat unit structure and tetraakylammonium side groups (Scheme 15.6) have also been described by the group of Whitten.[50–53] These cationic systems can bind to the surface of bacteria and work as singlet oxygen photosensitizers to kill the bacteria under dark and white-light irradiation.

Later, Wang and coworkers[54] prepared water-soluble conjugated polymer containing fluorene and 4-difluoro-4-bora-3a,4a-diaza-*s*-indacene (BODIPY)

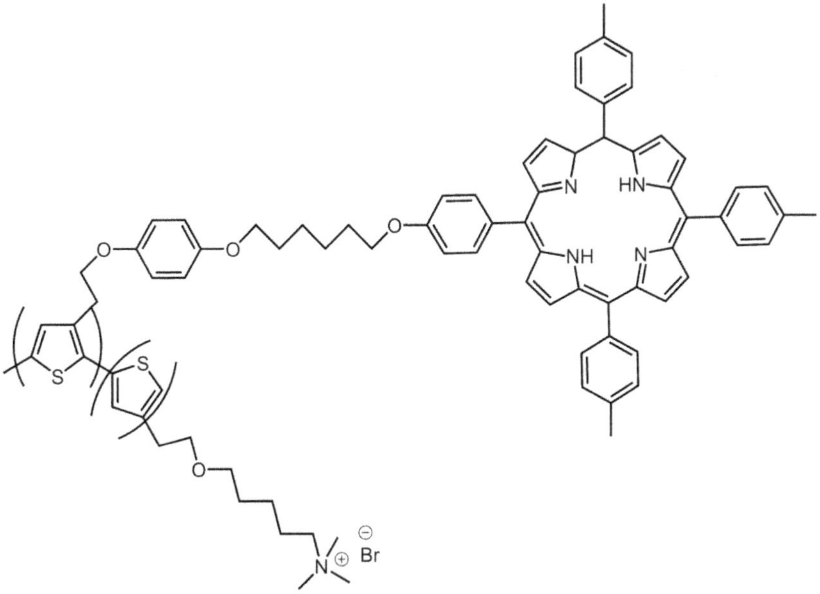

Scheme 15.5 Water-soluble CP with a molecular design centered on the polythiophene-porphyrin (PTPP) dyad.

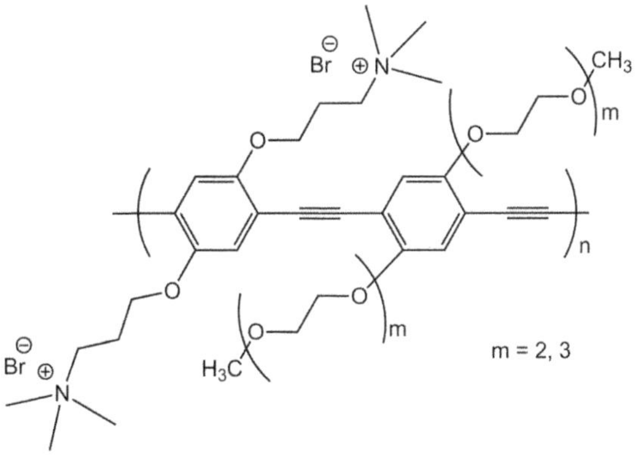

Scheme 15.6 Structure of cationic conjugated polyelectrolytes based on poly(phenylene ethynylene).

in the main chain, and tetramethyl ammonium pendant groups (PBF) that exhibits red emission (Scheme 15.7). The polymer forms nanoparticles in the presence of negatively charged disodium salt 3,3'-dithiodipropionic acid (SDPA) with a size of about 100 nm. Upon photoexcitation with white light, PBF nanoparticles can sensitize the oxygen molecule to readily produce ROS for rapidly killing neighboring *E. coli* bacteria cells.

The same group[55] has developed a system in which the photosensitizer is activated by chemical molecules instead of a light source through a

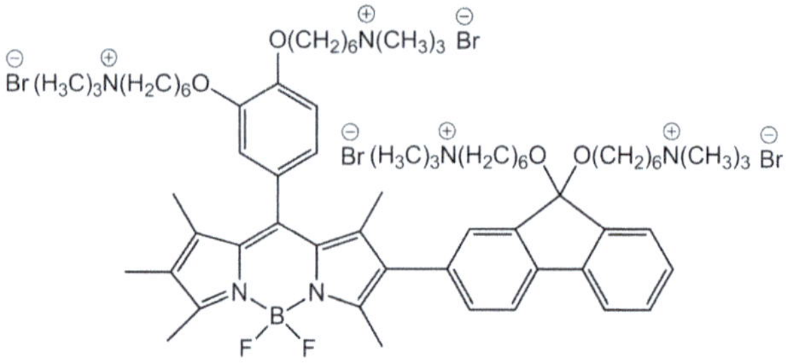

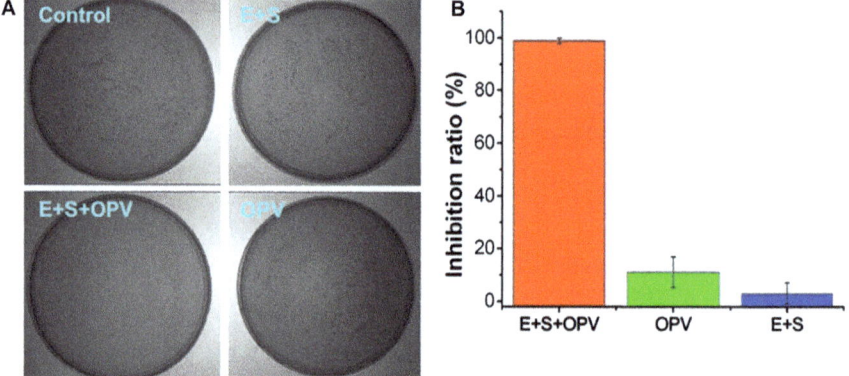

Scheme 15.7   Structure of water-soluble polymer PBF.

Figure 15.4   (A) Plate photographs for *C. albicans* on yeast extract peptone dextrose (YTD) agar plate treated with OPV in the absence and presence of the luminol luminescence system. (B) Biocidal activity of OPV toward *C. albicans* in the absence and presence of the luminol luminescence system (E + S).[55]
Reprinted with permission from *J. Am. Chem. Soc.*, **134** (34), 13184–13187. Copyright (2012) from the American Chemical Society.

bioluminescence resonance energy transfer (BRET) process. In this system, luminol, hydrogen peroxide, and horseradish peroxidase were used as bioluminescent molecules and a cationic oligo(*p*-phenylene vinylene) (OPV) was used as the photosensitizer. The excited OPV by BRET from luminol sensitizes oxygen molecules in the surroundings to produce ROS that kill *C. albicans* pathogenic fungi (Figure 15.4). The advantage of BRET systems is that it can work *in vivo* even in the deeper tissue, which overcomes the drawback of the deep-tissue penetration with light irradiation.

On the other hand, the minimum inhibitory concentration (MIC) is widely utilized as a standard laboratory method to quantify the bacterial susceptibility to antibiotics. MIC does not provide, however, a true estimation of the concentration of antimicrobial required to treat a biofilm. The assays to measure the minimum biofilm eradication concentration (MBEC), which is the concentration of an antimicrobial agent required to kill a bacterial biofilm, provide a means to investigate biocides activity against microbial biofilms.[56,57] As can also be realized through this book, the colonization of surfaces by micro-organisms is known to adversely affect the function of a variety of specific interfaces, such as those found in petroleum pipelines, paper mills, heat exchangers and aquatic flow systems, textiles, contact lenses, and medical implants. This process, colonization, could permit the bacteria to invade mucosal cells, alter calcium flux in epithelial cells, and release toxins. These bacteria are thought to produce proteases and other exoproducts that interfere with cytokine signalling pathways and other host factors used to mount a resistance response against the bacterial invader. Surface attachment is certainly an essential phase in biofilm formation and precipitates chemical signalling pathways within and between bacterial cells.[58] Substrate stiffness has been suggested to affect the density of surface colonization.[59,60] It was pointed out that the adhesion of viable colony-forming, *Staphylococcus epidermidis* and *E. coli* bacteria, correlates positively with increasing elastic modulus (E) of weak polyelectrolyte multilayered substrata over the range 1 MPa $< E <$ 100 MPa.[59] Moreover, photocrosslinkable or not polyelectrolyte films showed that *E. coli* and *Lactococcus lactis* are highly sensitive to the film stiffness. While the *L. lactis* displayed a slow growth on both films, independently of their rigidity, *E. coli* exhibited a more rapid development on noncrosslinked and, then, softer films compared with the stiffer ones (Figure 15.5).

To eliminate or substantially reduce the extent of microbial attachment and biofilm formation on these surfaces, extensive efforts have been focused on the fabrication of new surfaces, or on the improvement of existing antimicrobial surfaces by, *e.g.* applying surface coatings, or modifying surface architecture/topology.[61] Those surfaces can repel micro-organism cells, preventing their attachment, or inactivating/killing cells. For that, the study of surface topography, especially at the nanoscale, is a crucial task to understand the antimicrobial behavior.[62] Just very recently, an example of naturally existing surface with a physical structure that exhibits effective bactericidal properties has been reported.[61,63,64] Cicada (*Psaltoda claripennis*) wings, consisting of an array of nanoscale pillars, with approximately hexagonal spacing, are extremely effective at killing *P. aeruginosa* cells; the wing surface was able to kill individual cells in *ca.* 3 min. This bactericidal ability is mostly due to a physicomechanical effect, which is retained when the surface chemistry is substantially altered through microwave irradiation.

Combination therapy is one strategy for stemming the emergence of resistant species.[65] Gram-negative species are generally more susceptible to the individual antimicrobial agents than the Gram-positive bacteria, while Gram-positive bacteria are more susceptible to combination therapy. The synergistic activity between nitric oxide (NO) released from diazeniumdiolate modified proline (PROLI/NO) and silver(I) sulfadiazine (AgSD) was evaluated against *E. coli,* vancomycin-susceptible *Enterococcus faecalis, Proteus mirabilis, P. aeruginosa,* methicillin-susceptible and methicillin-resistant *S. aureus* and *S. epidermidis.*[66] The authors demonstrated that the combination of AgSD and PROLI/NO is synergistic across a wide range of Gram-positive bacteria, including antibiotic-resistant "super bugs". Furthermore, by varying the interval

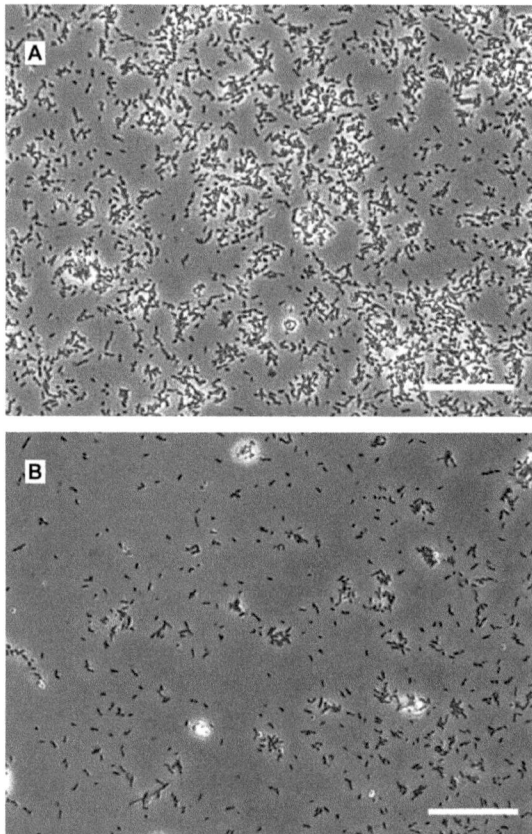

Figure 15.5   Microscopy images showing *E. coli* after 4 h of growth on noncrosslinked
(A) and crosslinked films (B). The scale bars of 10 μm.[60]
Reprinted with permission from *Biomacromolecules*, **14**(2), 520–528.
Copyright (2013) from the American Chemical Society.

of addition of AgSD and PROLI/NO, they established that the duration between dosing of individual agents is important in eliciting maximum synergistic
activity. When the concentration of AgSD or PROLI/NO is subbactericidal,
the delay in the incorporation of the second agent should not exceed some
threshold time limit, presumably as the bacteria are capable of repairing the
membrane damage or dividing to form healthy cells, even under stress. Combining quaternary ammonium-functionalized silica nanoparticles with nitric
oxide (NO) release capabilities turned out in an increase in bactericidal efficacy
against Gram-positive *S. aureus*; no change was observed, however, in activity
against Gram-negative *P. aeruginosa*.[67] Moreover, silver nanoparticles, which
are very commonly used as antimicrobial agents, are stabilized and reduced to
produce colloid silver nanoparticles (Ag-pep) using a cell-penetrating peptide
GGGRRRRRRYGRKKRRQRR (G3R6TAT).[68] These nanocomposites
demonstrated a distinctly enhanced biocidal effect toward *B. subtilis* and *E. coli*

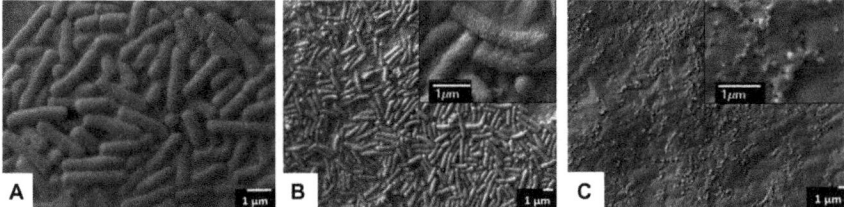

Figure 15.6 SEM results of *E. coli* before (A) and after treatment with Ag stabilized with sodium dodecyl sulfate (B) and Ag-pep (C).[69]
Reproduced by permission of The Royal Society of Chemistry.

bacteria and pathogenic *C. albicans* yeast compared with silver nanoparticles stabilized with sodium dodecyl sulfate (Figure 15.6).[69]

The effect of combined treatment of antimicrobial coatings and γ-irradiation on reduction of food pathogens, such as *Listeria monocytogenes*, *E. coli*, and *Salmonella Typhimurium* has also been studied by Takala *et al.*[70] They used a methylcellulose-based coating containing various mixtures of antimicrobial agents and γ-irradiation with various doses (0–3.3 kGy). The irradiation produced an increase of micro-organism sensitivity, which depends on the mixture of coating used. The sensitivity of *L. monocytogenes*, Gram-positive, is similar for all the coatings, while it is relatively higher for the Gram-negative *E. coli* and *S. Typhimurium*.

The integration of the advantages of conventional polymer membrane technology and hierarchically structured TiO_2/ZnO nanomaterials (TiO_2 nanofiber acts as trunk and ZnO nanorods as branches) has been also applied in the water treatment[71] to produce clean water at a constant high flux with no membrane-fouling problem and in an energy-saving manner.

As has been revealed throughout the whole book, there is not a unique flank of attack to battle the hostile microbial effects and there are still different approaches that have to be improved/created/combined to solve the existing problems in addition to those that will be developed in either a near or medium-term future.

A last thought or consideration that we would like people to be aware of is the link between the microbial infection and cognition.[72,73] Dementia is a public health problem because of its frequency and medicosocial care cost. Alzheimer's disease (AD) is the leading type of dementia. There is not currently any curative treatment for AD, and the identified risk factors are generally irreversible (*e.g.*, advanced age, female sex, and presence of the apolipoprotein E allele $\varepsilon4$ (APOE$\varepsilon4$)). A connection between persisting chronic infections (*e.g.*, HSV, *Chlamydia pneumoniae*) and AD has been explored. These infectious agents may elicit neuroinflammation and worsen AD lesions. *Helicobacter pylori*, a spiral-shaped Gram-negative bacterium, may participate in AD pathophysiology by causing lifelong gastric inflammation. *H. pylori* infection acquired in childhood can induce a cascade of events leading to gastroduodenal diseases (ulcers, cancers) and extradigestive pathologies, such as idiopathic thrombocytopenic purpura and may be involved in cardiovascular disease.[74] In addition, infections of the brain – by bacteria, viruses, and parasites – can be associated with seizures. These seizures are, sometimes, the result of the infection; but in other cases, the evidence suggests that seizures are a result of the brain reaction to the infection. In the case of neurocysticercosis (which is caused by accidental ingestion of eggs of *Taenia solium, i.e.* pork tapeworm), the infection is usually due to contamination of food by people with taeniasis. In developing countries, neurocysticercosis is the most common parasitic disease of the nervous system and it is the main cause of acquired epilepsy; the data demonstrate that a substance in the granuloma that forms around a dying parasite is a major contributor to the seizures. The mechanisms underlying the connection between seizures and infection are not known, but there is clear proof for a contribution by inflammatory mediators.[75] Furthermore, it is pointed out in a recent review that there is a large amount of evidence that suggests a linkage between specific infectious pathogens to AD.[76] Therefore, scientists as a group of different experts should combine our strengths and look for answers to avoid or eradicate the problems that society has nowadays and will have in the coming future.

Acknowledgements

MINECO is fully acknowledged for financial support (MAT2010-17016 and MAT2010-19883). A. Muñoz-Bonilla also acknowledges MINECO for her Juan de la Cierva postdoctoral contract.

References

1. R. Liu, X. Chen, Z. Hayouka, S. Chakraborty, S. P. Falk, B. Weisblum, K. S. Masters and S. H. Gellman, *J. Am. Chem. Soc.*, 2013, **135**, 5270–5273.
2. Y. Zhang, M. W. Lee, S. An, S. Sinha-Ray, S. Khansari, B. Joshi, S. Hong, J.-H. Hong, J.-J. Kim, B. Pourdeyhimi, S. S. Yoon and A. L. Yarin, *Catal. Commun.*, 2013, **34**, 35–40.
3. G. Zhang, H. Morikawa, Y. Chen and M. Miura, *Mater. Lett.*, 2013, **97**, 184–186.
4. I. Sawada, R. Fachrul, T. Ito, Y. Ohmukai, T. Maruyama and H. Matsuyama, *J. Membr. Sci.*, 2012, **387–388**, 1–6.
5. F. Nederberg, Y. Zhang, J. P. K. Tan, K. Xu, H. Wang, C. Yang, S. Gao, X. D. Guo, K. Fukushima, L. Li, J. L. Hedrick and Y.-Y. Yang, *Nature Chem.*, 2011, **3**, 409–414.
6. P. Li, Y. F. Poon, W. Li, H.-Y. Zhu, S. H. Yeap, Y. Cao, X. Qi, C. Zhou, M. Lamrani, R. W. Beuerman, E.-T. Kang, Y. Mu, C. M. Li, M. W. Chang, S. S. Jan Leong and M. B. Chan-Park, *Nature Mater.*, 2011, **10**, 149–156.
7. S. Maiti, S. S. Ray and A. K. Kundu, *Prog. Polym. Sci.*, 1989, **14**, 297–338.
8. X. Liu, W. Xin and J. Zhang, *Green Chem.*, 2009, **11**, 1018–1025.

9. P. A. Wilbon, Y. Zheng, K. Yao and C. Tang, *Macromolecules*, 2010, **43**, 8747–8754.

10. J. Wang, K. Yao, A. L. Korich, S. Li, S. Ma, H. J. Ploehn, P. M. Iovine, C. Wang, F. Chu and C. Tang, *J. Polym. Sci., Part A: Polym. Chem.*, 2011, **49**, 3728–3738.

11. K. Yao, J. Wang, W. Zhang, J. S. Lee, C. Wang, F. Chu, X. He and C. Tang, *Biomacromolecules*, 2011, **12**, 2171–2177.

12. J. Wang, Y. P. Chen, K. Yao, P. A. Wilbon, W. Zhang, L. Ren, J. Zhou, M. Nagarkatti, C. Wang, F. Chu, X. He, A. W. Decho and C. Tang, *Chem. Commun.*, 2012, **48**, 916–918.

13. Y. Chen, P. A. Wilbon, Y. P. Chen, J. Zhou, M. Nagarkatti, C. Wang, F. Chu, A. W. Decho and C. Tang, *RSC Adv.*, 2012, **2**, 10275–10282.

14. I. Yudovin-Farber, J. Golenser, N. Beyth, E. I. Weiss and A. J. Domb, *J. Nanomater.*, 2010, **2010**, 826343.

15. D. K. Shvero, M. P. Davidi, E. I. Weiss, N. Srerer and N. Beyth, *J. Biomed. Mater. Res.B*, 2010, **94B**, 367–371.

16. W. T. Godbey, K. K. Wu and A. G. Mikos, *J. Controll. Release*, 1999, **60**, 149–160.

17. K. Utsuno and H. Uludağ, *Biophys. J.*, 2010, **99**, 201–207.

18. J. Shi, B. Chou, J. L. Choi, A. L. Ta and S. H. Pun, *Mol. Pharm.*, 2013, **10**, 2145–2156.

19. G. A. Spoden, K. Besold, S. Krauter, B. Plachter, N. Hanik, A. F. M. Kilbinger, C. Lambert and L. Florin, *Antimicrob. Agents Chemother.*, 2012, **56**, 75–82.

20. Y. Maitani, K. Ishigaki, Y. Nakazawa, D. Aragane, T. Akimoto, M. Iwamizu, T. Kai and K. Hayashi, *J. Controll. Release*, 2013, **166**, 139–146.

21. Y. Li, K. Fukushima, D. J. Coady, A. C. Engler, S. Liu, Y. Huang, J. S. Cho, Y. Guo, L. S. Miller, J. P. K. Tan, P. L. R. Ee, W. Fan, Y. Y. Yang and J. L. Hedrick, *Angew. Chem. Int. Ed.*, 2013, **52**, 674–678.

22. S. Brochu, R. E. Prud'homme, I. Barakat and R. Jerome, *Macromolecules*, 1995, **28**, 5230–5239.

23. F. C. Fang, *J. Clin. Invest.*, 1997, **99**, 2818–2825.

24. D. A. Wink and J. B. Mitchell, *Free Radical Biol. Med.*, 1998, **25**, 434–456.

25. F. C. Fang and A. Vazquez-Torres, *Am. J. Physiol. Lung Cell. Mol. Physiol.*, 2002, **282**, L941–L943.

26. F. C. Fang, *Nature Rev. Microbiol.*, 2004, **2**, 820–832.

27. E. R. Stadtman, *Curr. Med. Chem.*, 2004, **11**, 1105–1112.

28. N. Barraud, D. J. Hassett, S.-H. Hwang, S. A. Rice, S. Kjelleberg and J. S. Webb, *J. Bacteriol.*, 2006, **188**, 7344–7353.

29. E. M. Hetrick and M. H. Schoenfisch, *Biomaterials*, 2007, **28**, 1948–1956.

30. D. L. Williams and R. D. Bloebaum, *Microsc. Microanal.*, 2010, **16**, 143–152.

31. B. J. Privett, S. T. Nutz and M. H. Schoenfisch, *Biofouling*, 2010, **26**, 973–983.

32. W. Cai, J. Wu, C. Xi and M. E. Meyerhoff, *Biomaterials*, 2012, **33**, 7933–7944.

33. S. M. Marxer, A. R. Rothrock, B. J. Nablo, M. E. Robbins and M. H. Schoenfisch, *Chem. Mater.*, 2003, **15**, 4193–4199.
34. P. N. Coneski, J. A. Nash and M. H. Schoenfisch, *ACS Appl. Mater. Interfaces*, 2011, **3**, 426–432.
35. A. C. Engler, N. Wiradharma, Z. Y. Ong, D. J. Coady, J. L. Hedrick and Y.-Y. Yang, *Nano Today*, 2012, **7**, 201–222.
36. E. Asín, A. Isla, A. Canut and A. Rodríguez Gascón, *Int. J. Antimicrob. Agents*, 2012, **40**, 313–322.
37. N. D. Christensen, C. A. Reed, T. D. Culp, P. L. Hermonat, M. K. Howett, R. A. Anderson and L. J. D. Zaneveld, *Antimicrob. Agents Chemother.*, 2001, **45**, 3427–3432.
38. J. A. Simoes, D. M. Citron, A. Aroutcheva, R. A. Anderson, C. J. Chany, D. P. Waller, S. Faro and L. J. D. Zaneveld, *Antimicrob. Agents Chemother.*, 2002, **46**, 2692–2695.
39. B. C. Herold, N. Bourne, D. Marcellino, R. Kirkpatrick, D. M. Strauss, L. J. D. Zaneveld, D. P. Waller, R. A. Anderson, C. J. Chany, B. J. Barham, L. R. Stanberry and M. D. Cooper, *J. Infect. Dis.*, 2000, **181**, 770–773.
40. S. Garg, K. Vermani, A. Garg, R. A. Anderson, W. B. Rencher and L. J. D. Zaneveld, *Pharm Res*, 2005, **22**, 584–595.
41. S. Patel, E. Hazrati, N. Cheshenko, B. Galen, H. Yang, E. Guzman, R. Wang, B. C. Herold and M. J. Keller, *J. Infect. Dis.*, 2007, **196**, 1394–1402.
42. S. Tan, L. Lu, L. Li, J. Liu, Y. Oksov, H. Lu, S. Jiang and S. Liu, *PloS one*, 2013, **8**, e59777.
43. S. Sonza, A. Johnson, D. Tyssen, T. Spelman, G. R. Lewis, J. R. Paull and G. Tachedjian, *Antimicrob. Agents Chemother.*, 2009, **53**, 3565–3568.
44. V. Pirrone, S. Passic, B. Wigdahl and F. C. Krebs, *Virol. J*, 2012, **9**, 33.
45. M. Qiu, Y. Chen, S. Song, H. Song, Y. Chu, Z. Yuan, L. Cheng, D. Zheng, Z. Chen and Z. Wu, *Antiviral Res.*, 2012, **96**, 138–147.
46. C. Zhu, Q. Yang, L. Liu and S. Wang, *Angew. Chem. Int. Ed.*, 2011, **50**, 9607–9610.
47. C. Xing, G. Yang, L. Liu, Q. Yang, F. Lv and S. Wang, *Small*, 2012, **8**, 525–529.
48. S. A. G. Lambrechts, M. C. G. Aalders and J. Van Marle, *Antimicrob. Agents Chemother.*, 2005, **49**, 2026–2034.
49. C. Xing, Q. Xu, H. Tang, L. Liu and S. Wang, *J. Am. Chem. Soc.*, 2009, **131**, 13117–13124.
50. L. Lu, F. H. Rininsland, S. K. Wittenburg, K. E. Achyuthan, D. W. McBranch and D. G. Whitten, *Langmuir*, 2005, **21**, 10154–10159.
51. S. Chemburu, T. S. Corbitt, L. K. Ista, E. Ji, J. Fulghum, G. P. Lopez, K. Ogawa, K. S. Schanze and D. G. Whitten, *Langmuir*, 2008, **24**, 11053–11062.
52. T. S. Corbitt, J. R. Sommer, S. Chemburu, K. Ogawa, L. K. Ista, G. P. Lopez, D. G. Whitten and K. S. Schanze, *ACS Appl. Mater. Interfaces*, 2008, **1**, 48–52.
53. T. S. Corbitt, L. Ding, E. Ji, L. K. Ista, K. Ogawa, G. P. Lopez, K. S. Schanze and D. G. Whitten, *Photochem. Photobiol. Sci.*, 2009, **8**, 998–1005.

54. H. Chong, C. Nie, C. Zhu, Q. Yang, L. Liu, F. Lv and S. Wang, *Langmuir*, 2011, **28**, 2091–2098.
55. H. Yuan, H. Chong, B. Wang, C. Zhu, L. Liu, Q. Yang, F. Lv and S. Wang, *J. Am. Chem. Soc.*, 2012, **134**, 13184–13187.
56. H. Ceri, M. Olson, D. Morck, D. Storey, R. Read, A. Buret and B. Olson, *Methods Enzymol.*, 2001, **337**, 377–385.
57. L. P. Girard, H. Ceri, A. P. Gibb, M. Olson and F. Sepandj, *Periton. Dialysis Int.*, 2010, **30**, 652–656.
58. M. E. Davey and G. A. O'toole, *Microbiol. Mol. Biol. Rev.*, 2000, **64**, 847–867.
59. J. A. Lichter, M. T. Thompson, M. Delgadillo, T. Nishikawa, M. F. Rubner and K. J. Van Vliet, *Biomacromolecules*, 2008, **9**, 1571–1578.
60. N. Saha, C. Monge, V. Dulong, C. Picart and K. Glinel, *Biomacromolecules*, 2013, **14**, 520–528.
61. J. Hasan, R. J. Crawford and E. P. Ivanova, *Trends Biotechnol.*, 2013, **31**, 295–304.
62. A. I. Hochbaum and J. Aizenberg, *Nano Lett.*, 2010, **10**, 3717–3721.
63. E. P. Ivanova, J. Hasan, H. K. Webb, V. K. Truong, G. S. Watson, J. A. Watson, V. A. Baulin, S. Pogodin, J. Y. Wang, M. J. Tobin, C. Löbbe and R. J. Crawford, *Small*, 2012, **8**, 2489–2494.
64. S. Pogodin, J. Hasan, VladimirA. Baulin, HaydenK. Webb, ViK. Truong, The H. Phong Nguyen, V. Boshkovikj, Christopher J. Fluke, Gregory S. Watson, Jolanta A. Watson, Russell J. Crawford and Elena P. Ivanova, *Biophys. J.*, 2013, **104**, 835–840.
65. G. Cottarel and J. Wierzbowski, *Trends Biotechnol.*, 2007, **25**, 547–555.
66. B. J. Privett, S. M. Deupree, C. J. Backlund, K. S. Rao, C. B. Johnson, P. N. Coneski and M. H. Schoenfisch, *Mol. Pharm.*, 2010, **7**, 2289–2296.
67. A. W. Carpenter, B. V. Worley, D. L. Slomberg and M. H. Schoenfisch, *Biomacromolecules*, 2012, **13**, 3334–3342.
68. L. Liu, K. Xu, H. Wang, J. T. K., W. Fan, S. S. Venkatraman, L. Li and Y.-Y. Yang, *Nature Nano.*, 2009, **4**, 457–463.
69. L. Liu, J. Yang, J. Xie, Z. Luo, J. Jiang, Y. Y. Yang and S. Liu, *Nanoscale*, 2013, **5**, 3834–3840.
70. P. N. Takala, S. Salmieri, K. D. Vu and M. Lacroix, *Radiat. Phys. Chem.*, 2011, **80**, 1414–1418.
71. H. Bai, Z. Liu and D. D. Sun, *Appl. Catal., B*, 2012, **111–112**, 571–577.
72. N. Dunn, M. Mullee, V. H. Perry and C. Holmes, *Alzheimer Dis. Assoc. Disord.*, 2005, **19**, 91–94.
73. S. Walter and E. B. Sandra, *Nature Rev. Neurol.*, 2013, **9**, 301–302.
74. C. Roubaud Baudron, L. Letenneur, A. Langlais, A. Buissonnière, F. Mégraud, J.-F. Dartigues, N. Salles and Q. S. for the Personnes Agées, *J. Am. Geriatr. Soc.*, 2013, **61**, 74–78.
75. J. L. Stringer, in *Encyclopedia of Basic Epilepsy Research*, Editor-in-Chief:A. S.Philip, Academic Press, Oxford, 2009, pp. 584–588.
76. F. Mawanda and R. Wallace, *Epidemiol. Rev.*, 2013, **35**, 161–180.

Subject Index